NANOWIRES AND NANOBELTS

Materials, Properties and Devices

Volume 1: Metal and Semiconductor Nanowires

NANOWIRES AND NANOBELTS

Materials, Properties and Devices

Volume 1: Metal and Semiconductor Nanowires

edited by

Zhong Lin Wang
Georgia Institute of Technology

Library of Congress Cataloging-in-Publication Data

Nanowires and nanobelts: materials properties and devices – Metal and Semiconductor nanowires
Edited by Zhong Lin Wang.

ISBN 10 1-4020 7443 3 (Hard Cover – 2 vol set)
ISBN 13 9781402074431

ISBN 10 0-387-28705-1 (Soft Cover)
ISBN 13 9780387287058

ISBN 0-38728745-0 (E-book)

Printed on acid-free paper

First softcover printing, 2006

Printed in the United States of America.

9 8 7 6 5 4 3 2 1 SPIN 11545026
springeronline.com

Contents

Preface

Nanowires, nanobelts, nanoribbons, nanorods ..., are a new class of quasi-one-dimensional materials that have been attracting a great research interest in the last few years. These non-carbon based materials have been demonstrated to exhibit superior electrical, optical, mechanical and thermal properties, and can be used as fundamental building blocks for nano-scale science and technology, ranging from chemical and biological sensors, field effect transistors to logic circuits. Nanocircuits built using semiconductor nanowires demonstrated were declared a "breakthrough in science" by *Science* magazine in 2001. *Nature* magazine recently published a report claiming that "Nanowires, nanorods, nanowhiskers, it does not matter what you call them, they are the hottest property in nanotechnology" (*Nature* **419** (2002) 553). There is no doubt that nanowire based quasi-one-dimensional materials will be the new focal point of research in the next decades.

Volume 1: Metal and Semiconductor Nanowires covers a wide range of materials systems, from noble metals (such as Au, Ag, Cu), single element semiconductors (such as Si and Ge), compound semiconductors (such as InP, CdS and GaAs as well as heterostructures), nitrides (such as GaN and Si_3N_4) to carbides (such as SiC). The objective of this volume is to cover the synthesis, properties and device applications of nanowires based on metal and semiconductor materials. The volume starts with a review on novel electronic and optical nanodevices, nanosensors and logic circuits that have been built using individual nanowires as building blocks. Then, the theoretical background for electrical properties and mechanical properties of nanowires is given. The molecular nanowires, their quantized conductance, and metallic nanowires synthesized by chemical technique will be introduced next. Finally, the volume covers the synthesis and properties of semiconductor and nitrides nanowires.

Volume 2: Nanowires and Nanobelts of Functional Materials covers a wide range of materials systems, from functional oxides (such as ZnO, SnO_2 and In_2O_3), structural ceramics (such as MgO, SiO_2 and Al_2O_3), composite materials (such as Si-Ge, SiC–SiO_2), to polymers. This volume focuses on the synthesis, properties and applications of nanowires and nanobelts based on functional materials. Novel devices and applications made from functional oxide nanowires and nanobelts will be presented first, showing their unique properties and applications. The majority of the text will be devoted to the synthesis and properties of nanowires and nanobelts of functional oxides. Finally, sulphide nanowires, composite nanowires and polymer nanowires will be covered.

The materials covered in both volumes are very diverse and rich in properties. Most of the nanowire and nanobelt structures are structurally controlled with respect to growth directions and side surfaces, resulting in controlled and tunable electrical and optical properties, which offer huge advantages for applications in nanotechnology.

The chapters were written by leading scientists worldwide whose groups have been the pioneers in the field and have done substantial work in their specific disciplines. Both volumes review the most up-to-date progress in nanowires and nanobelts. The books are intended as research books for advanced students and researchers with background in physics, chemistry, electrical engineering, mechanical engineering, chemical engineering, biology and bioengineering.

Z. L. WANG

List of Contributors

Amma, Achim
Department of Chemistry
The Pennsylvania State University
152 Davey Laboratory
University Park, PA 16802
USA

Björk, Mikael T.
Lund University,
Solid State Physics/the Nanometer Consortium
Box 118, S-221 00 Lund
Sweden

Boussaad, S.
Department of Electrical Engineering &
The Center for Solid State Electronics Research
Arizona State University
Tempe, AZ 85287
USA

Chen, C.-C.
Department of Chemistry,
National Taiwan Normal University,
and Institute of Atomic and Molecular Sciences Academia Sinica, Taipei
Taiwan
e-mail: t42005@cc.ntnu.edu.tw

Chen, K. H.
Institute of Atomic and Molecular Sciences, Academia Sinica,
and Center for Condensed Matter Sciences
National Taiwan University, Taipei
Taiwan
e-mail: chenkh@po.iams.sinica.edu.tw

Chen, L. C.
Center for Condensed Matter Sciences
National Taiwan University, Taipei
Taiwan
e-mail: chenlc@ccms.ntu.edu.tw

Cui, Yi
Department of Chemistry and Chemical Biology
Harvard University
Cambridge, MA 02138 USA

Duan, Xiangfeng
Department of Chemistry and Chemical Biology
Harvard University, Cambridge
MA 02138
USA

Guo, Guolin
State Key Laboratory for Structural Chemistry of Unstable and Stable Species
College of Chemistry and Molecular Engineering
Peking University
Beijing 100871
P.R. China

Guo, Hong
Center for the Physics of Materials and Department of Physics McGill University

Montreal, PQ
Canada H3A 2T8
e-mail: guo@physics.mcgill.ca

He, H. X.
Department of Electrical
Engineering & The Center for
Solid State Electronics Research
Arizona State University
Tempe, AZ 85287
USA

Huang, Yu
Department of Chemistry and
Chemical Biology
Harvard University
Cambridge, MA 02138
USA

Lee, S. T.
Center for Super-Diamond and
Advanced Films (COSDAF) &
Department of Physics and
Materials Science
City University of Hong Kong
Hong Kong SAR
China
e-mail: APANNALE@cityu.edu.hk

Liang, Wuwei
The George W. Woodruff School of
Mechanical Engineering
Georgia Institute of Technology
Atlanta, GA 30332-0405
USA

Lieber, Charles M.
Department of Chemistry and
Chemical Biology, and Division of
Engineering and Applied Sciences
Harvard University
Cambridge
MA 02138
USA
e-mail: cml@cmliris.harvard.edu

Lifshitz, Y.
Center for Super-Diamond and
Advanced Films (COSDAF) &
Department of Physics and
Materials Science
City University of Hong Kong
Hong Kong SAR
China

Mehrez, Hatem
Department of Physics
Harvard University
Cambridge, MA 02138 and
Center for the Physics of Materials and
Department of Physics
McGill University, Montreal
PQ, Canada H3A 2T8

Mallouk, Thomas E.
Department of Chemistry
The Pennsylvania State University
152 Davey Laboratory
University Park, PA 16802, USA
e-mail: tom@chem.psu.edu

Meng, Guowen
Institute of Solid State Physics
Chinese Academy of Sciences
P.O. Box 1129, Hefei 230031
P.R. China
e-mail: gwmeng@mail.issp.ac.cn

Ohlsson, B. Jonas
Lund University,
Solid State Physics/
the Nanometer Consortium
Box 118
S-221 00 Lund, Sweden
also at QuMat Technologies AB
Malmö, Sweden

Rodrigues, Varlei
Laboratório Nacional de Luz Síncrotron
Caixa Postal 6192—CEP 13084-971
Campinas, São Paulo, Brazil
e-mail: varlei@lnls.br

Sun, Yugang
Department of Chemistry
University of Washington
Seattle
WA 98195-1700
USA

Samuelson, Lars
Lund University
Solid State Physics/
the Nanometer Consortium
Box 118, S-221 00 Lund
Sweden
e-mail: lars.samuelson@ftf.lth.se

Tao, Nongjian
Department of Electrical
Engineering & The Center for Solid
State Electronics Research
Arizona State University
Tempe, AZ 85287
USA
e-mail: nongjian.tao@asu.edu

Tomar, Vikas
The George W. Woodruff School of
Mechanical Engineering
Georgia Institute of Technology
Atlanta, GA 30332-0405
USA

Ugarte, Daniel
Laboratório Nacional de Luz Síncrotron
Caixa Postal 6192—CEP 13084-971
Campinas, São Paulo, Brazil
e-mail: ugarte@lnls.br

Xia, Younan
Department of Chemistry
University of Washington
Seattle, WA 98195-1700
USA
e-mail: xia@chem.washington.edu

Xu, B. Q.
Department of Electrical Engineering &
The Center for Solid State Electronics
Research Arizona State University
Tempe, AZ 85287 USA

Xu, Dongsheng
State Key Laboratory for
Structural Chemistry of Unstable and
Stable Species
College of Chemistry and
Molecular Engineering
Peking University
Beijing 100871, P.R. China
e-mail: dsxu@chem.pku.edu.cn

Xu, Hongqi
Lund University, Solid State Physics/
the Nanometer Consortium
Box 118, S-221 00 Lund, Sweden

Yu, Dapeng
School of Physics, Electron Microscopy
Laboratory, and State Key Laboratory
for Mesoscopic Physics
Peking University
Beijing 100871, P.R. China
e-mail: yudp@pku.edu.cn

Zhang, Lide
Institute of Solid State Physics
Chinese Academy of Sciences
P.O. Box 1129, Hefei 230031
P.R. China
e-mail: ldzhang@mail.issp.ac.cn

Zhang, R. Q.
Center for Super-Diamond and
Advanced Films (COSDAF) &
Department of Physics and
Materials Science
City University of Hong Kong
Hong Kong SAR, China

Zhou, Min
The George W. Woodruff School of
Mechanical Engineering
Georgia Institute of Technology
Atlanta, GA 30332-0405, USA
e-mail: min.zhou@me.gatech.edu

Part I

Nanodevices and Nanocircuits Based on Nanowires

Chapter 1

Nanowires as Building Blocks for Nanoscale Science and Technology

Yi Cui,[1] Xiangfeng Duan,[1] Yu Huang,[1] and Charles M. Lieber[1,2]

[1]*Department of Chemistry and Chemical Biology;*
[2]*Division of Engineering and Applied Sciences, Harvard University, Cambridge, MA 02138, USA*

1. Introduction

1.1. Bottom-up paradigm for nanoscale science and technology

The field of nanotechnology represents an exciting and rapidly expanding research area that crosses the borders between the physical, life and engineering sciences [1, 2]. Much of the excitement in this area of research has arisen from recognition that new phenomena and unprecedented integration density are possible with nanometer scale structures. Correspondingly, these ideas have driven scientists to develop methods for making nanostructures. In general, there are two philosophically distinct approaches for creating small objects, which can be characterized as top-down and bottom-up. In the top-down approach, small features are patterned in bulk materials by a combination of lithography, etching and deposition to form functional devices. The top-down approach has been exceedingly successful in many venues with microelectronics being perhaps the best example today. While developments continue to push the resolution limits of the top-down approach, these improvements in resolution are associated with a near exponential increase in cost associated with each new level manufacturing facility. This economic limitation and other scientific issues with the top-down approach have motivated efforts worldwide to search for new strategies to meet the demand for nanoscale structures today and in the future [3–5].

The bottom-up approach, in which functional electronic structures are assembled from chemically synthesized, well-defined nanoscale building blocks, much like the way nature uses proteins and other macromolecules to construct complex biological systems, represents a powerful alternative approach to conventional top-down methods [4, 6, 7]. The bottom-up approach has the potential to go far beyond of the limits

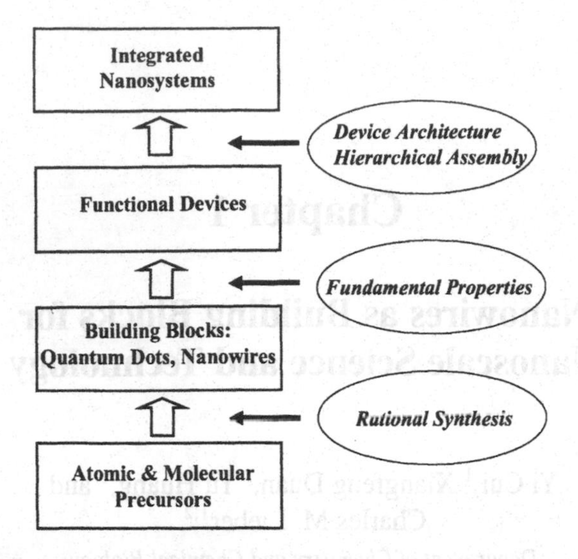

Fig. 1. Schematic outlining key challenges (open rectangular) and specific research areas (ellipses) required to enable the bottom-up approach to functional nanosystems.

of top-down technology by defining key nanometer scale metrics through synthesis and subsequent assembly—not by lithography.

To enable this bottom-up approach for nanotechnology requires a focus on three key areas, which are at the heart of devices and integration (Fig. 1). First, the bottom-up approach necessitates nanoscale building blocks with precisely controlled and tunable chemical composition, structure, size, and morphology since these characteristics determine their corresponding physical properties. To meet this goal requires developing methods that enable rational design and predictable synthesis of building blocks. Second, it is critical to develop and explore the limits of functional devices based on these building blocks. Nanoscale structures may behave in ways similar to current electronic and optoelectronic devices, although it is also expected that new and potentially revolutionary concepts will emerge from these building blocks, for example, due to quantum properties. Third and central to the bottom-up concept, will be the development of architectures that enable high-density integration with predictable function, and the development of hierarchical assembly methods that can organize building blocks into these architectures.

Addressing and overcoming the hurdles in these three major areas of the bottom-up approach could revolutionize a wide range of technologies of today. Moreover, it is very likely that the bottom-up approach will enable entirely new device concepts and functional systems, and thereby create technologies that we have not yet imagined. For example, it is possible to combine seamlessly chemically distinct nanoscale building blocks, which could not be integrated together in top-down processing, and thus obtain unique function and/or combinations of function in an integrated nanosystem. Small and highly perfect building blocks may also lead to quantum electronic devices that enable quantum computing in architectures that have common features to digital systems.

1.2. Nanowire building blocks

Individual molecules [8–12] and quantum dots [13–15], which can be classified as zero dimension (0D) structures, are attractive building blocks for bottom-up assembly of nanoscale electronics. These 0D structures have been intensively pursued over the past decade since they represent the smallest building blocks with corresponding high potential for integration. However, the use of individual molecules or quantum dots in integrated systems, has been limited by challenges in establishing reliable electrical contact to individual molecules or quantum dots. It has thus been difficult to elucidate and understand the intrinsic properties of individual devices, and moreover, to develop and demonstrate realistic schemes for scalable interconnection and integration of 0D devices into functional architectures.

One-dimensional (1D) nanostructures have also been the focus of extensive studies worldwide due to their unique physical properties and potential to revolutionize broad areas of nanotechnology [4]. First, 1D nanostructures represent the smallest dimension structure that can efficiently transport electrical carriers, and thus are ideally suited to the critical and ubiquitous task of moving charges in integrated nanoscale systems. Second, 1D nanostructures can also exhibit device function, and thus can be exploited as both the wiring and device elements in architectures for functional nanosystems [4, 16]. In this regard, two classes of materials, carbon nanotubes [16–27] and semiconductor nanowires [28–35], have shown particular promise.

Single-walled carbon nanotubes (NTs) can exhibit either metallic or semiconducting behavior depending on diameter and helicity [17]. The unique electronic properties of NTs open up the possibility of creating a number of different devices that could have potential in nanoelectronics [16, 18–20]. For example, single-walled NTs have been used to fabricate room-temperature field effect transistors (FETs) [21, 22], diodes [23, 24] and recently, logic circuits [25, 26]. However, the inability to control whether NT building blocks are semiconducting or metallic makes specific device fabrication largely a random event. Hence, moving beyond proof-of-concept single device elements to the integrated arrays required for nanosystems poses a serious issue for NT based approaches. A creative solution to the problem of coexisting metallic and semiconducting NTs involves selective destruction of metallic tubes [27], although such an approach requires extensive top-down lithography and subsequent processing to implement and may not be practical for highly integrated nanoelectronics systems.

Semiconductor nanowires (NWs) [4, 28, 29] represent another important and broad class of nanometer scale wire structure. In contrast to NTs, however, semiconductor NWs can be rationally and predictably synthesized in single crystal form with all key parameters controlled during growth: chemical composition, diameter, length, and doping [30–32]. Semiconductor NWs thus represent one of best-defined and controlled classes of nanoscale building blocks, which correspondingly have enabled a wide-range of devices and integration strategies to be pursued in a rational manner. For example, semiconductor NWs have been assembled into nanometer scale FETs [32, 33], p–n diodes [33, 34], light emitting diodes (LEDs) [33], bipolar junction transistors [34], complementary inverters [34], complex logic gates and even computational circuits that have been used to carry out basic digital calculations [35]. In contrast to NTs, NW devices can be assembled in a rational and predictable manner because the size, interfacial properties, and electronic properties of the NWs can be precisely controlled during synthesis, and moreover, reliable methods exist for their parallel assembly [36]. In addition, it should be recognized that it is possible to combine distinct NW building blocks in ways not possible in conventional electronics and to leverage the knowledge base that exists for the chemical modification of inorganic surfaces [37, 38] to produce semiconductor NW devices that achieve new function and correspondingly could lead to unexpected device concepts.

1.3. Overview of the chapter

In this chapter, we will describe a broad range of studies drawn primarily from the authors' laboratory addressing the key issues of semiconductor NWs, including synthesis, fundamental properties, hierarchical assembly and device properties. First, we will discuss rational design of synthetic strategies and the synthesis of semiconductor NWs with controlled chemical, structural, and electronic properties via a vapor-liquid-solid (VLS) growth mechanism. Second, we will apply VLS method to the synthesis of axial and radial NW heterostructures. Third, we will review electrical and optical properties of individual NWs and NW heterostructures, and moreover, will examine nanoscale devices formed using individual NWs as building blocks. Fourth, we will demonstrate how the unique properties of NWs can be applied in ultrasensitive chemical and biological sensing. Fifth, we present approaches for the hierarchical assembly of NWs into well-defined arrays with controlled orientation and spatial location. Lastly, we address the critical issue of integrated nanosystems with crossed NW structures and the assembly of integrated nanoscale devices. We conclude with a brief summary and perspective on future opportunities.

2. Rational Synthesis of Single Component Nanowires

Rational design and synthesis of nanoscale materials is critical to work directed towards understanding fundamental properties, creating nanostructured materials, and developing nanotechnologies. To explore the diverse and exciting opportunities in 1D systems requires materials for which the chemical composition, diameter, length, electronic, and optical properties can be controlled and systematically varied. To meet these requirements we have focused our efforts on developing a general and predictive approach for the synthesis of 1D structures, much as molecular beam epitaxy has served as an all-purpose method for the growth of two-dimensional (2D) structures. Specifically, we set as a goal an ability to design and synthesize rationally NWs with predictable control over the key structural, chemical, and physical properties, since such control enables studies designed to understand the intrinsic behavior of 1D structures and to explore them as building blocks for nanotechnologies.

2.1. Symmetry breaking: The key concept for 1D growth

In general, the growth of 1D materials requires that two dimensions are restricted to the nanometer regime, while the third dimension extends to macroscopic dimensions. This requirement is considerably more difficult than the corresponding constraints needed for successful growth of 0D and 2D structures [39, 40]. For example, many important semiconductor materials adopt a cubic, zinc blende structure, and thus when growth is stopped at an early stage, the resulting nanoscale structures are 3D polyhedron or nanocrystals and not 1D NWs. To achieve 1D growth in systems where atomic bonding is relatively isotropic requires that the symmetry must be broken during growth rather than simply arresting growth at an early stage.

Over the past decade, considerable effort has been placed on the bulk synthesis of NWs, and various strategies have been developed to break the growth symmetry either "physically" or "chemically". A common theme in many of these studies has been the use of linear templates, including the edges of surface steps [41], nanofibers [42, 43], and porous membranes [44], to direct chemical reactions and material growth in 1D. This strategy is conceptually simple and has been used to prepare a wide range of NW materials. Despite the simplicity, template mediated growth is also limited in that the resulting NWs are usually polycrystalline, which could limit their potential for both fundamental studies and many applications. Another general strategy that has received increasing focus over that past several years involves exploiting a 'catalyst' to confine growth in 1D. Depending on the phases involved in the reaction, this approach is typically defined as vapor-liquid-solid (VLS) [45, 46], solution-liquid-solid (SLS) [47, 48] or vapor-solid (VS) [49, 50] growth.

2.2. Catalytic growth: Concepts and synthetic design

Catalytic growth, where the catalyst is used to direct 1D growth of single crystal materials via a VLS mechanism (Fig. 2a), is a powerful concept for NW synthesis [28, 29]. Here the catalyst is envisioned as a nanocluster or nanodroplet that defines the diameter of and serves as the site that directs preferentially the addition of reactant to the end of a growing NW much like a living polymerization catalyst directs the addition of monomers to a growing polymer chain.

This synthetic concept is especially important since it readily provides the intellectual underpinning needed for the specification of the catalyst and growth conditions required for predictable NW growth. First, equilibrium phase diagrams (Fig. 2b) are used to determine catalyst materials that form a liquid alloy with the NW material of interest. The phase diagram is then used to choose a specific composition and growth temperature such that there is a coexistence of a liquid alloy and solid NW phases. The liquid catalyst alloy cluster serves as the preferential site for absorption of reactant since the sticking coefficient is much higher on liquid versus solid surfaces and, when supersaturated, the nucleation site for crystallization. Preferential 1D growth occurs in the presence of reactant as long as the catalyst nanodroplet remains in the liquid state. Within this framework it is straightforward to synthesize NWs with different diameters and composition, if appropriate nanometer scale diameter catalyst clusters are available. Several methods that exploit this general approach are described below with an emphasis on laser-assisted catalytic growth and metal-catalyzed chemical vapor deposition (Fig. 3).

2.3. Laser-assisted catalytic growth

A straightforward and general approach for producing the nanometer scale clusters required to nucleate and direct the growth of NWs is laser ablation and

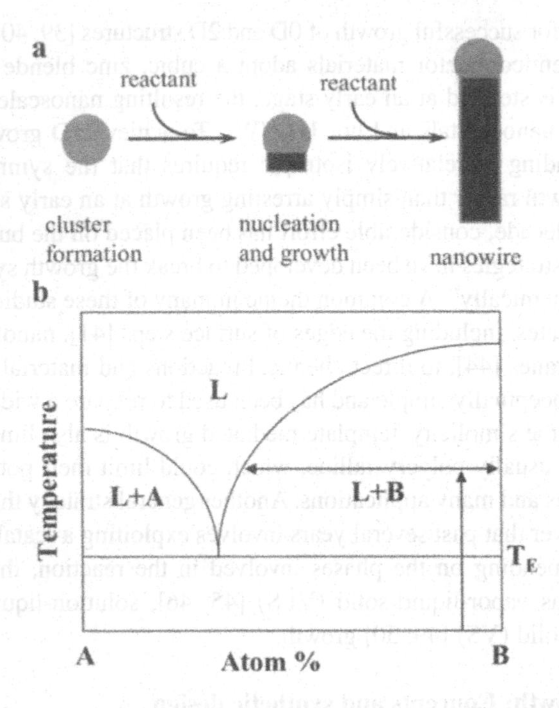

Fig. 2. Catalytic growth of NWs. (a) Schematics illustrating the underlying concept for catalytic growth of NWs. Liquid catalytic clusters act as the energetically favored site for localized chemical reaction, absorption of vapor phase reactant and crystallization of crystalline NWs. (b) Binary A-B phase diagram used as guide for choosing a catalyst for NW growth. The vertical arrow represents a specific composition of catalyst (A) to NW (B) with end point corresponding to the growth temperature. The horizontal arrow defines the composition of the catalyst liquid (L) catalyst-NW (A-B) nanodroplet and shows that pure solid NW (B) is the only solid phase at this temperature.

condensation [51]. In this context, NW growth can be readily achieved by laser ablation of a composite target containing the catalyst and NW material in a heated flow tube (Fig. 3a). The background pressure within the flow reactor is used to control condensation of the ablated material and the cluster size, while the temperature is varied to maintain the catalyst cluster/reactant alloy in the liquid state as defined by the corresponding phase diagram. When the laser-generated clusters become supersaturated with the desired NW material, a nucleation event occurs producing a (NW) solid-liquid (NW-catalyst alloy) interface. To minimize the interfacial free energy subsequent solid growth/crystallization occurs at this initial interface, which thus imposes the highly anisotropic growth constraint required for one-dimensional nanoscale wires. We have termed this method laser-assisted catalytic growth (LCG) [30]. This approach has proven to be very general for the synthesis of semiconductor NW materials, including group IV elemental [52], group IV alloys, group III-V, II-VI, and IV-VI compound semiconductor NW materials [30, 53–55]. Below we demonstrate first illustrate the basic principles underlying this approach with the rational and predictable growth of silicon NWs (SiNWs).

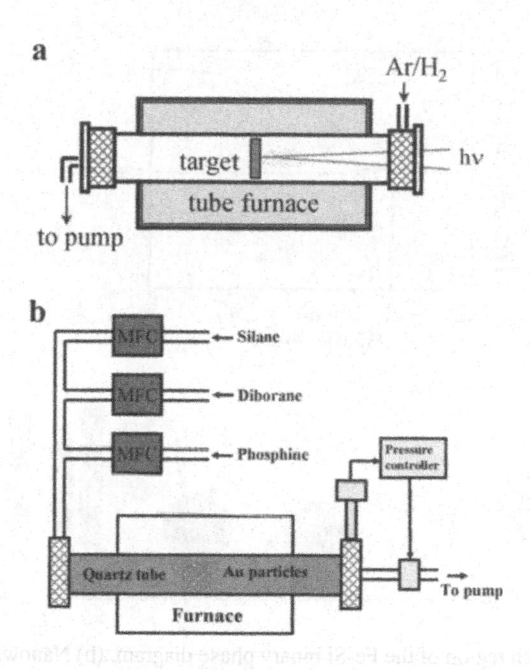

Fig. 3. Schematics for (a) a laser-based NW growth apparatus and (b) a metal-catalyzed CVD NW growth system.

The critical catalyst for SiNW growth, the Si:catalyst composition, and the growth temperature can be determined by examining the Si-rich regions of binary metal-Si phase diagrams [52]. For example, the Fe-Si phase diagram (Fig. 4a) shows that there is a broad area above 1200 °C in the Si-rich region where $FeSi_x$ *(l)* and Si *(s)* coexist. Within the framework of our approach, SiNW synthesis is achieved by laser ablation of a $Si_{0.9}Fe_{0.1}$ target at temperatures ≥1200 °C (Fig. 4b). Laser ablation of the $Si_{1-x}Fe_x$ target produces a vapor of Si and Fe that rapidly condenses into Si-rich liquid Fe-Si nanoclusters, and when the nanoclusters become supersaturated in Si, the coexisting pure Si phase precipitates and crystallizes as nanowires [52]. Ultimately, the growth terminates when the gas flow carries the nanowires out of the hot zone of the furnace.

Transmission electron microscopy (TEM) studies of the product obtained after laser ablation confirmed these predictions and showed that primarily wire-like structures with remarkably uniform diameters on the order of 10 nm with lengths >1 μm were produced by this approach (Fig. 5a) [52]. The TEM images recorded on individual nanowires further show that the nanowires consist of very uniform diameter crystalline cores, which are typically 6–20 nm in diameter, surrounded by an amorphous coating (Fig. 5b, c). By recording electron diffraction patterns perpendicular to the wire axes and lattice resolved images, it is also possible to determine that the nanowires grow preferentially along the [111] direction at ≥1200 °C. We elucidated the composition of the core and amorphous sheath using energy-dispersive X-ray (EDX) analysis of individual nanowires before and after etching with HF to remove the amorphous coating. These experiments demonstrate that the crystalline core is

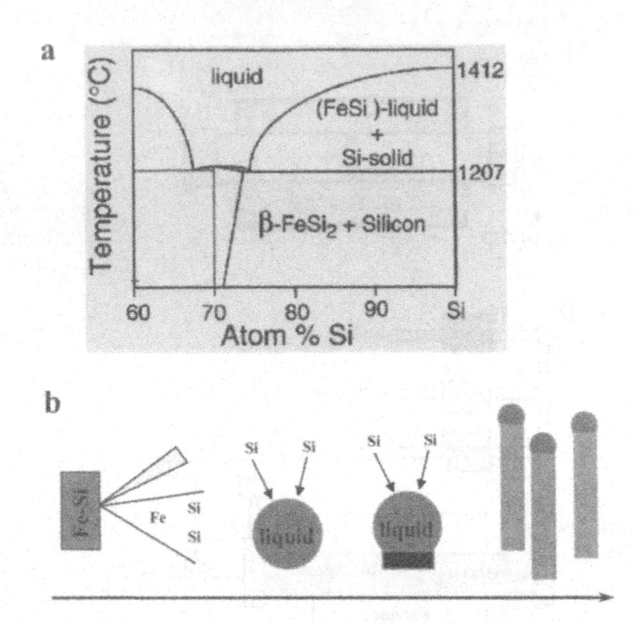

Fig. 4. (a) Silicon-rich region of the Fe-Si binary phase diagram. (b) Nanowire growth process. Laser ablation of the Fe-Si target creates a dense, hot vapor that condenses into nanoclusters as the Fe and Si species cool through collisions with the buffer gas. The furnace temperature is controlled to maintain the Fe-Si nanocluster in a liquid state. Nanowire growth begins after the liquid becomes supersaturated in Si, and continues as long as the Fe-Si nanoclusters remain in a liquid state and Si reactant is available. Growth terminates when the nanowire passes out of the hot zone of the reactor. Adapted from [29].

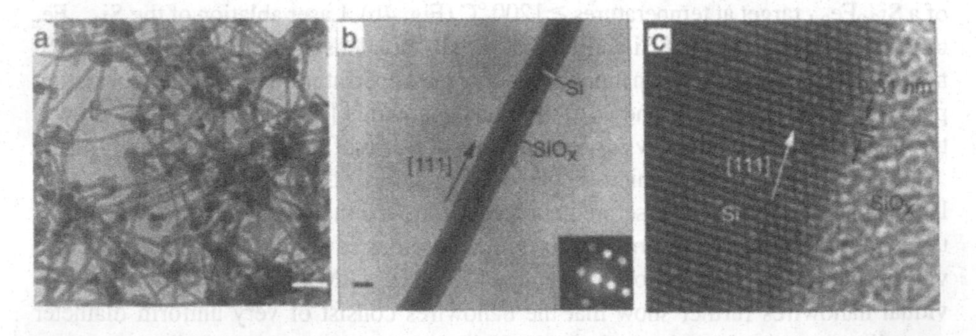

Fig. 5. (a) TEM image of the nanowires produced following ablation of a $Si_{0.9}Fe_{0.1}$ target. The white scale bar corresponds to 100 nm. (b) Diffraction contrast TEM image of a silicon nanowire crystalline material (the silicon core) appears darker than amorphous material (SiO_2 sheath) in this imaging mode. The scale bar corresponds to 10 nm. Inset: Electron diffraction recorded along the [211] zone axis. (c) High-resolution TEM image of the crystalline silicon core and amorphous SiO_2 sheath. The (111) planes (spacing 0.31 nm) are oriented perpendicular to the growth direction (white arrow). Adapted from [52].

pure Si without detectable Fe and that the amorphous coating is SiO_2. Taken together, these data have shown that the nanowires produced using our method consist of single-crystal silicon cores that are surrounded by an amorphous and insulating silicon oxide sheath.

The successful synthesis of SiNWs by LCG demonstrates that we can rationally design the growth of NWs by exploiting binary diagrams. In addition, the LCG approach can be extended to the synthesis of compound semiconductor NWs. For example, pseudobinary and more complex phase diagrams can be used to define clearly catalyst, composition, and growth conditions required for successful compound semiconductor NW growth. We have also pointed out [30] that catalysts for LCG can be chosen in the absence of detailed phase diagram data by identifying metals in which the NW component elements are soluble in the liquid phase but do not form solid compounds more stable than the desired NW phase; that is, the ideal metal catalyst should be "physically active" but "chemically stable". This basic guiding principle suggests that the noble metals such as Au should represent good starting points for many semiconductor materials. Indeed, Fig. 6 summarizes the broad range of NW materials, with different structure types and chemical properties, synthesized as high quality single crystals using Au as the catalyst. These results demonstrate unambiguously the generality of our approach.

Gold nanoclusters are unable to catalyze GaN NW growth due to the low solubility of nitrogen [54]. These results are consistent with the basic requirement for a catalyst: physically active and chemically stable. In this context, Au is not physically active. A good catalyst should form a miscible liquid phase with GaN but not lead to a more stable solid phase under the NW growth conditions. This guiding principle suggests that Fe and Ni, which dissolve both gallium and nitrogen, and do not form a more stable compounds with these components than GaN will be good catalysts for GaN NW growth.

Significantly, we found that LCG using either a GaN/Fe or GaN/Ni targets produces a high yield of nanometer diameter wire-like structures [54]. Fig. 7a shows a diffraction contrast TEM image of a GaN NW synthesized using an iron catalyst. The uniform contrast of the NW suggests that it has a single crystal structure. Electron

IV-IV Group	III-V Group Binary	III-V Group Ternary	II-VI Group Binary	IV-VI Group Binary
Si Ge $Si_{(1-x)}Ge_x$	GaP GaAs InP InAs	$Ga(As_{(1-x)}P_x)$ $In(As_{(1-x)}P_x)$ $(Ga_{(1-x)}In_x)P$ $(Ga_{(1-x)}In_x)As$ $(Ga_{(1-x)}In_x)(As_{(1-x)}P_x)$	ZnS ZnSe CdS CdSe	PbSe PbTe

Fig. 6. Table summarizing NW materials grown by the LCG approach. All of the NWs were synthesized using gold as the catalyst except for the case of GaN (see text).

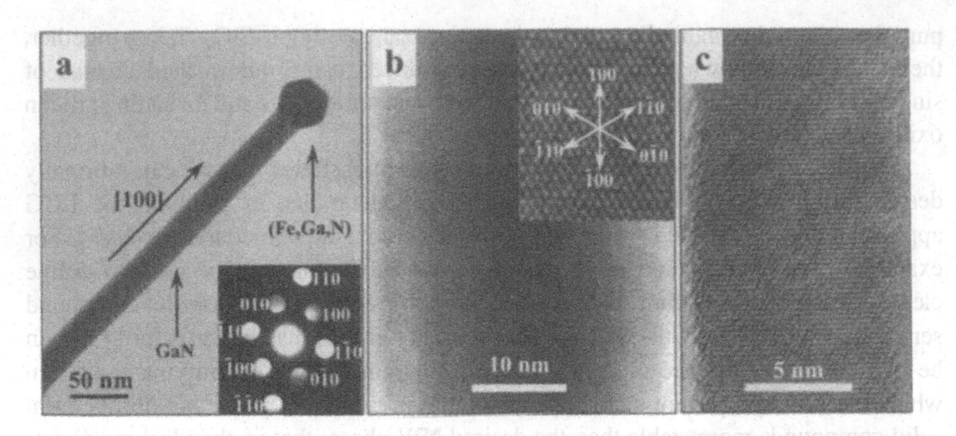

Fig. 7. LCG of GaN NWs. (a) Diffraction contrast TEM image of a GaN NW that terminates in a faceted nanoparticle of higher (darker) contrast. (inset) CBED pattern recorded along <001> zone axis. (b, c) Lattice resolved TEM images of GaN NW with a diameters of ca. 30 and 10 nm, respectively. The image was taken along <001> zone axis. Adapted from [54].

diffraction confirmed that the NWs are single crystals with a wurtzite structure and a [100] growth direction. Images of the NW ends also exhibited nanoclusters consisting primarily of iron, consistent with a VLS growth mechanism. Lastly, high-resolution TEM images (Fig. 7b, c) showed confirmed the single crystal structure and growth direction, and moreover, demonstrated that the surfaces of the GaN NWs terminate sharply with only 1-2 atomic layers of amorphous material. This latter observation has important implications for electronic nanodevice fabrication as described below.

2.4. Metal-catalyzed chemical vapor deposition

In the LCG method, laser ablation is used to simultaneously generate nanoscale metal catalyst clusters and semiconductor reactant that produce NWs via a VLS growth mechanism. A major advantage of this approach is its flexibility and generality since laser ablation can be used to produce nanoclusters of virtually any material, thereby enabling the growth of a very wide range of NWs. To fully exploit the potential of these NW materials also requires control of physical dimensions (i.e., diameter and length) and electronic properties (e.g., doping).

An alternative to the LCG implementation of catalytic NW growth is nanocluster catalyzed chemical vapor deposition (CVD), in which the reactants and dopants are well-defined gas sources (Fig. 3b) [29]. The catalytic CVD method enables the NW size, composition and doping level to be controlled in a very precise manner [58].

A clear illustration of the potential of this method is the size-controlled growth of Si NWs. In VLS mechanism, the diameter of the NWs is controlled by three factors [46]: (1) The size of nanoclusters catalysts define the size of the eutectic droplet for nucleation; (2) The growth temperature determines the solubility of Si reactant in the nanocluster and thus further the size of eutectic droplets; and (3) The vapor pressure of the reactant can affect the diameter when it exceeds critical value for homogeneous deposition of reactant onto NW surfaces. Recognizing these key contributions to the diameter during growth, we have exploited distinct monodisperse Au nanoclusters as catalysts for diameter-controlled SiNW growth (Fig. 8a) [58]. Field-emission scanning electron microscopy (FE-SEM) images of the Si NWs grown from Au nanoclusters dispersed on SiO_2 planar surfaces showed a comparable NW density to the starting nanocluster density (Fig. 8b). Qualitatively, these images also show that the Si NWs grown from Au nanoclusters are nearly monodisperse with diameters

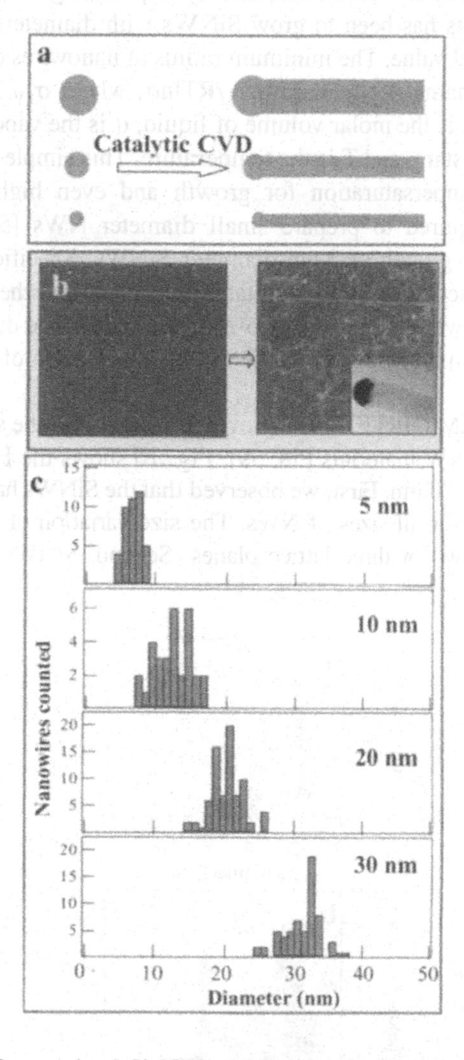

Fig. 8. Catalytic CVD growth of Si NWs. (a) Schematic of size-controlled NW synthesis using monodisperse Au nanocluster catalysts. (b) Atomic force microscope (AFM) image of Au nanoclusters deposited on a silicon substrate, and the Si NWs grown by CVD. (c) Distributions of NW diameters obtained from 5, 10, 20, and 30 nm diameter nanoclusters. Adapted from [58].

determined by the nanoclusters. This latter point was confirmed from TEM images (inset, Fig. 8b), which show directly that the Au particles at the NW ends are similar to the NW diameters. Histograms summarizing extensive TEM studies of the NW diameters obtained from different diameter nanoclusters also show that Si NWs grown from 5 (4.9 ± 1.0), 10 (9.7 ± 1.5), 20 (19.8 ± 2.0), and 30 nm (30.0 ± 3.0 nm) Au nanoclusters have average NW diameters of 6.4 ± 1.2, 12.3 ± 2.5, 20 ± 2.3, and 31.1 ± 2.7 nm, respectively. Significantly, the dispersion of the Si NW diameters mirrors that of Au catalysts, suggesting that the NW dispersity is limited only by the dispersity of Au nanocluster catalysts.

The Bohr radius of Si is ~5 nm, and thus an important goal for observing quantum confinement effects has been to grow SiNWs with diameters comparable to or smaller than this critical value. The minimum radius of nanowires can be predicted by equilibrium thermodynamics: $r_{min} = 2\sigma_{LV}V_L/RT\ln\sigma$, where σ_{LV} is the liquid-vapor surface free energy, V_L is the molar volume of liquid, σ is the vapor phase supersaturation, R is the gas constant, and T is the temperature. This simple model suggest that a high vapor phase supersaturation for growth and even higher supersaturation for nucleation are required to prepare small diameter NWs [59]. We have used this approach with the growth of 3 nm diameter SiNWs. Specifically, a high silane partial pressure was used initially to facilitate nucleation, and then the pressure was reduced during the growth phase to avoid overcoating [59]. The diameter distribution of SiNWs 3.3 ± 1.5 nm corresponds well with the distribution of Au clusters, 3.0 ± 1.8 nm (Fig. 9).

High resolution TEM (HRTEM) has provided insight into the structures exhibited by the CVD-grown SiNW materials [58, 59]. Fig. 10 shows the HRTEMs of SiNWs with diameter from 2 to 20 nm. First, we observed that the SiNWs have a uniform diameter along their length for all sizes of NWs. The size variation of single crystal cores is in all cases within two or three lattice planes. Second, SiNWs have single-crystal

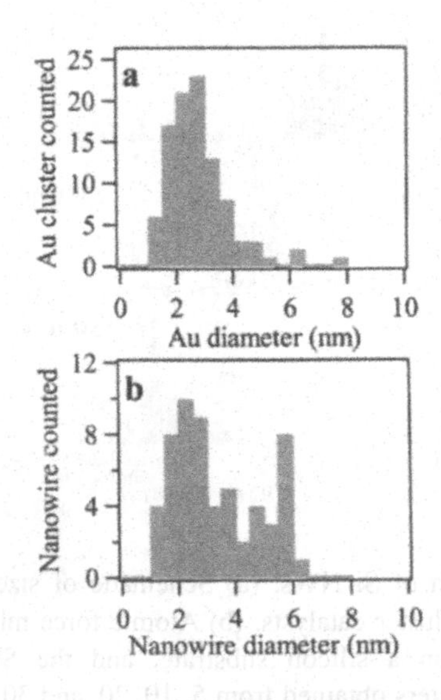

Fig. 9. Distribution of 3 nm Au cluster (a) and SiNW diameters (b).

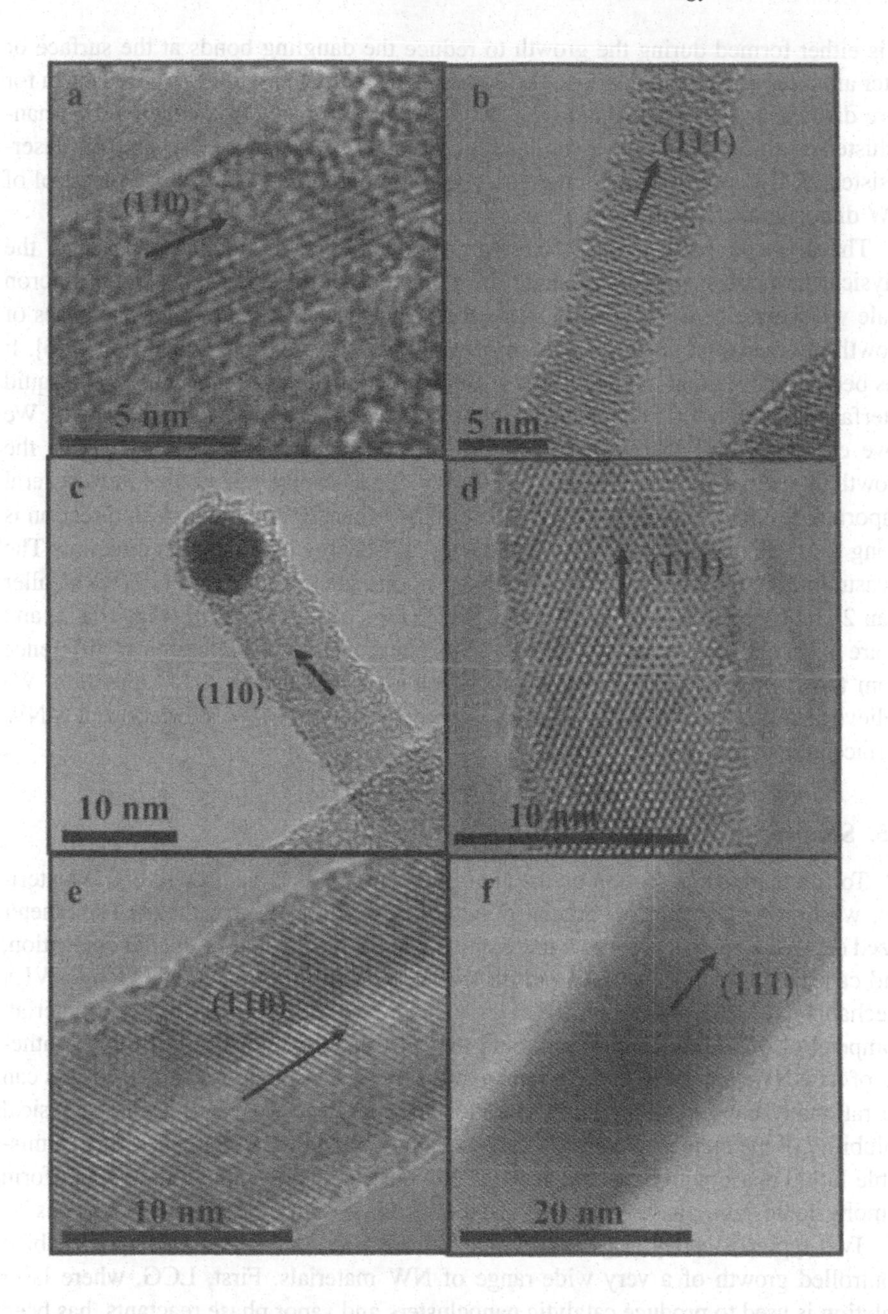

Fig. 10. Lattice resolved TEM images of SiNW grown via CVD method with different diameters. The growth direction is marked with a arrow.

cores sheathed with 1–3 nm thick SiO_x. For example, the SiNW with 2 nm core in Fig. 10a has ~3 nm oxide, which is the same as in a typical 20 nm diameter SiNW (Fig. 10f). The fact that different diameter nanowires all have amorphous layer with similar thickness indicates that amorphous layer is not intrinsic to the growth.

It is either formed during the growth to reduce the dangling bonds at the surface or after air exposure upon oxidation. Third, the SiNWs remain highly crystalline even for core diameters as small as 2 nm (Fig. 10a). Overall, the SiNWs synthesized by nanocluster catalyzed CVD have similar high crystallinity as those prepared by laser-assisted catalyzed growth, although the CVD approach enables far greater control of NW diameter and length.

The crystallographic growth direction of NWs is an important issue since the physical properties may differ along different growth directions. In silicon micron scale whiskers, the growth direction is usually along <111>, although rare cases of growth in other directions are also observed under nonisothermal conditions [46]. It has been proposed that this particular growth direction arises because the solid–liquid interface is a single <111> plane, which is kinetically most stable during growth. We have exploited HRTEM and the electron diffraction techniques to characterize the growth directions of different diameter SiNWs. Significantly, our studies show several important results [59]: (1) for NW diameter larger than 20 nm, the growth direction is along [111]. Fig. 10f shows a 20 nm diameter NW having [111] growth direction. The measurements on more than 20 NWs all show consistent results. (2) For NWs smaller than 20 nm, growth directions of either [111] (Fig. 10b, d) or [110] (Fig. 10a, c, and e) are observed. The growth direction of SiNWs growth by CVD has clear difference from those grown using LCG method, which is dominated by [111] direction. We believe that these preferences reflect competing catalyst/SiNW interface and SiNW surface energetics.

2.5. Summary

To summarize this section on the controlled and predictable growth of 1D materials, we first review the two crucial points of catalytic NW growth: (1) Nanometer sized catalytic clusters serve as a preferential site for reactant addition and nucleation, and can thereby define the size and direct the growth of crystalline NWs by a VLS mechanism; (2) Equilibrium phase diagrams can be used to predict catalyst material, composition and growth conditions, and thus enable rational and predictable synthesis of new NW materials. In the absence of detailed phase diagram data, catalysts can be rationally chosen through consideration of the chemical reactivity and physical solubility of the elements. An ideal catalyst should be physically active—form a miscible liquid phase with elements of the NW—but chemically stable; that is, not form a more stable solid phase than the desired NW phase under the growth conditions.

Two types of catalytic growth methods have been developed and have enabled controlled growth of a very wide-range of NW materials. First, LCG, where laser ablation is used to produce catalytic nanoclusters and vapor phase reactants, has been shown to be a general technique for rapidly exploring the growth of new NW materials. Since laser ablation can be used to produce nanoclusters of virtually any material, LCG has perhaps the greatest flexibility and generality for NW synthesis. Second, metal-catalyzed CVD, in which gas reactant sources are combined with well-defined nanocluster catalysts, provides exquisite control over NW growth since all of the key parameters, including catalytic nanocluster size, reactant ratios, and temperature, can be independently and precisely controlled.

3. Synthesis of Nanowire Heterostructures and Superlattices

Modulated nanostructures in which the composition and/or doping are varied on the nanometer scale represent important targets of synthesis since they could enable new and unique function and potential for integration in functional nanosystems. An initial indication of the significance of this idea was the growth of carbon nanotube-SiNW heterojunctions, which created nanometer scale metal-semiconductor junctions showing current rectification [60]. In this section, we will explore the synthesis of both longitudinal and radial NW heterostructures.

3.1. Axial nanowire heterostructures

Our approach to axial NW heterostructure growth (Fig. 11) [61] exploits metal-catalyzed nanowire synthesis. We introduce vapor-phase semiconductor reactants required for nanowire growth by either laser ablation of solid targets or vapor-phase molecular species. To create a single junction within the nanowire, the addition of the first reactant is stopped during growth, and then a second reactant is introduced for the remainder of the synthesis (Fig. 11b); repeated modulation of the reactants during growth produces nanowire superlattices (Fig. 11c). In principle, this approach can be successfully implemented if a nanocluster catalyst suitable for growth of the different superlattice components under similar conditions is found; our previous studies suggested that Au nanoclusters meet this requirement for a wide range of group III–V and group IV materials.

Gallium arsenide (GaAs)/gallium phosphide (GaP) superlattices have been grown by laser-assisted catalytic growth using GaAs and GaP targets [61]. Fig. 12 shows TEM images of the products of this synthesis. It is relatively straightforward to focus on the junction area as the NW lengths can be controlled directly by growth times. High-resolution TEM images of a typical GaAs/GaP junction region (Fig. 12a) exhibit a crystalline NW core without obvious defects, and show that the NW axes lies along the <111> direction, in agreement with previous studies of single-component systems. Two-dimensional Fourier transforms calculated from high-resolution images containing the junction region (Fig. 12a, inset) show pairs of reciprocal lattice peaks along the different lattice directions, while such transforms calculated from the regions above and below the junction (not shown) exhibit only single reciprocal lattice peaks. Analysis of these peak data yield lattice constants, indexed to the zinc

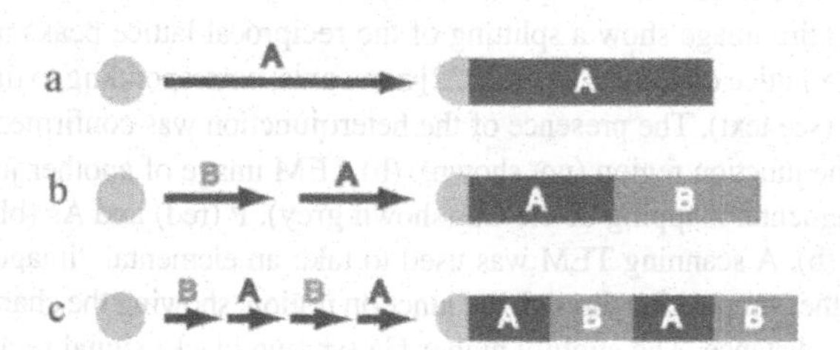

Fig. 11. Schematics illustrating the synthesis of axial heterostructures: (a) single component NW. (b) heterostructure. (c) superlattice. Adapted from [61].

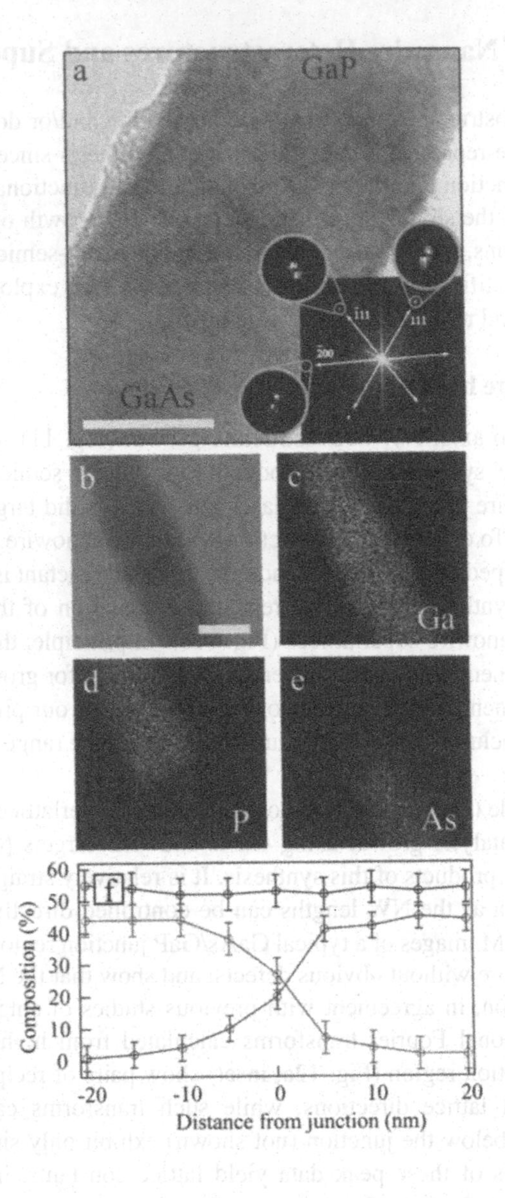

Fig. 12. GaAs/GaP nanowire junctions. (a) High-resolution TEM of a GaAs/GaP junction grown from a 20-nm gold nanocluster catalyst. Scale bar, 10 nm. Inset, two-dimensional Fourier transforms of the entire image show a splitting of the reciprocal lattice peaks along the <111>, <$\bar{1}$11> and <$\bar{2}$00> lattice directions in the [0$\bar{2}$2] zone axis, corresponding to the lattice constants for GaAs and GaP (see text). The presence of the heterojunction was confirmed by EDS analysis above and below the junction region (not shown). (b) TEM image of another junction. Scale bar, 20 nm. (c, d, e) Elemental mapping of the Ga (shown grey), P (red) and As (blue) content of the junction shown in (b). A scanning TEM was used to take an elemental "image" of the junction. (f) Line profiles of the composition through the junction region, showing the change in composition as a function of the distance. The slightly higher Ga (shown black) signal relative to the P (red) and As (blue) signals may be due to uncertainties in the detector calibration or the presence of an amorphous gallium oxide layer around the crystalline nanowire core. Adapted from [61].

blende structures of GaP and GaAs, of 0.5474 ± 0.0073 nm and 0.5668 ± 0.0085 nm, and are in good agreement with the values for both GaP (0.5451 nm) and GaAs (0.5653 nm), respectively.

The TEM structural data suggest that the GaP/GaAs junctions could be abrupt, and thus we have carried out local elemental mapping of the heterojunction by energy dispersive X-ray spectroscopy (EDS) to address composition variation across the junction (Fig. 12b–e) [61]. These elemental maps show that Ga is uniformly distributed along the length of the NW, while P (Fig. 12d) and As (Fig. 12e) appear localized in the GaP and GaAs portions of the NW heterostructure, respectively. Quantitative analysis of the P and As composition variation (Fig. 12f) shows, however, that the junction is not atomically abrupt, but rather makes the transition between GaP and GaAs phases over a length scale of 15–20 nm. This length scale is reasonable considering that the ~20 nm diameter Au catalyst must re-alloy with GaP after initial GaAs growth. The observed composition variation has several potentially important implications. First, composition variation at the interface can relieve strain, and may enable the defect-free junctions and superlattices that we observe in this system—which has a relatively large lattice mismatch. We note, however, that simple estimates of the length between dislocations suggest that defect-free, atomically abrupt interfaces may be possible in wires of diameter less than 20 nm. Second, there are photonic and electronic applications where abrupt interfaces are important. The observed composition variation could be substantially reduced in smaller-diameter NWs; that is, a 5 nm diameter NW superlattice should have variations of <5 nm across the junction interfaces. Alternatively, it should be possible to use different nanocluster catalysts or variations in the growth temperature when reactants are switched to obtain sharper interfaces.

We find that this approach can produce compositionally modulated NW superlattices in which the number of periods and repeat spacing can be readily varied during growth [61]. TEM images of a six-period structure corresponding to a $(GaP/GaAs)_3$ superlattice (Fig. 13a) show that the 20 nm diameter NW are highly uniform over their ~3 μm lengths. Spatially resolved EDS measurements of composition (Fig. 13b) further demonstrate that the P and As regions are distinct from one another, and that there is minimal cross-contamination. Moreover, these data show that each GaP and GaAs NW segment has a length of about 500 nm, and are thus consistent with the equal growth times used for each segment, but also show that growth rates remain relatively constant during the entire NW synthesis.

This methodology for growth of superlattice structures can be generalized in many materials systems. Actually, we fabricated p–n junctions within individual SiNWs by Au-nanocluster-catalysed chemical vapor deposition and dopant modulation [61]. Other systems were also demonstrated by other groups: Si–Ge [62], InAs-InP [63]. These superlattice structures greatly increase the versatility and power of NW building blocks for nanoscale electronic and photonic applications, such as nanobarcodes, injection lasers, and engineered one-dimensional waveguides.

3.2. Radial nanowire homo- and heterostructures

The growth of crystalline overlayers on nanostructure surfaces is important for controlling surface properties and for enabling new function. We have recently

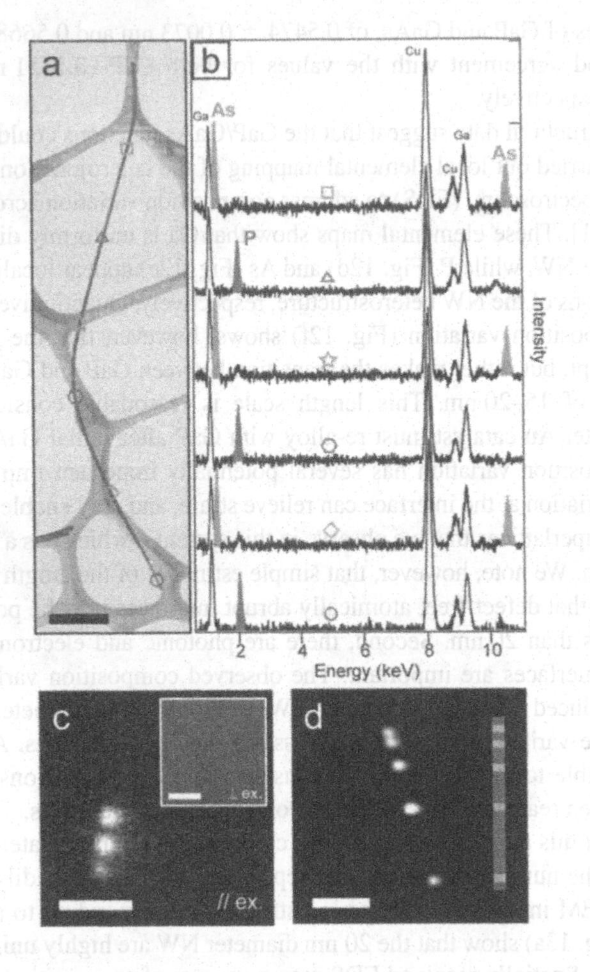

Fig. 13. Nanowire superlattice structures. (a) TEM image of a ~20-nm-diameter $(GaP/GaAs)_3$ nanowire superlattice. Scale bar, 300 nm. (b) Elemental profile of the superlattice along the nanowire length measured by EDS analysis. The symbols in (a) show the location of each EDS spectrum along the nanowire and the colour of the symbol indicates GaP (red) and GaAs (blue) regions. The P Kα peak (2.015 kV) is shown in red and the As Kα (10.543 kV) and Lα (1.282 kV) peaks in blue for clarity. The spectra show clearly a distinct, periodic modulation of the nanowire composition along its entire length, with three uniform periods of GaP, separated by three uniform periods of GaAs. (c) Photoluminescence image (// excitation) of a nanowire from the same sample as shown in (a) and (b). The three bright regions correspond to the three GaAs (direct bandgap) regions, while the dark segments are from the GaP (indirect bandgap) regions. Inset, no photoluminescence is observed above background for perpendicular (⊥) excitation due to the dielectric contrast between the nanowire and its surroundings. Scale bar, 5 μm. (d) Photoluminescence image of a 40-nm-diameter GaP(5)/GaAs(5)/GaP(5)/GaAs(5)/GaP(10)/GaAs(5)/GaP(20)/GaAs(5)/GaP(40)/GaAs(5)/GaP(5) superlattice; the numbers in parentheses correspond to the growth times in seconds for each layer. Inset, diagram showing the relative lengths of GaAs (blue) and GaP (red) layers. Scale bar, 5 μm. (e), Photoluminescence image of a 21-layer superlattice, $(GaP/GaAs)_{10}GaP$, showing a group of four equally spaced spots on the left, two in the middle with larger gaps, and another set of four with equal spacing on the end. The superlattice is ~25 μm in length. Adapted from [61].

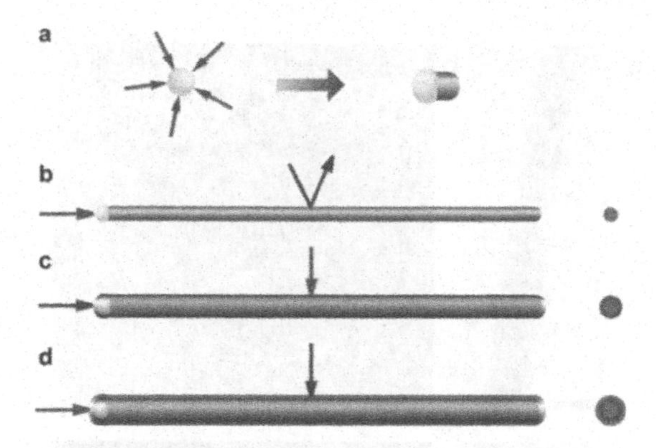

Fig. 14. Synthesis of core-shell nanowires by chemical vapour deposition. (a) Gaseous reactants (red) catalytically decompose on the surface of a gold nanocluster leading to nucleation and directed nanowire growth. (b) One-dimensional growth is maintained as reactant decomposition on the gold catalyst is strongly preferred. (c) Synthetic conditions are altered to induce homogeneous reactant decomposition on the nanowire surface, leading to a thin, uniform shell (blue). (d) Multiple shells are grown by repeated modulation of reactants. Adapted from [64].

demonstrated this concept with the synthesis of silicon and germanium core-shell and multi-shell NW homo and heterostructures using the CVD method applicable to a variety of nanoscale materials [64]. Our approach to the synthesis of core-shell NW structures is based upon control of radial versus axial growth (Fig. 14). Axial growth is achieved when reactant activation and addition occurs at the catalyst site and not on the NW surface (Fig. 14b). Correspondingly, it is possible to drive conformal shell growth by altering conditions to favor homogeneous vapor phase deposition on the NW surface (Fig. 14c). Subsequent introduction of different reactants and/or dopants produces multiple shell structures of nearly arbitrary composition, although epitaxial growth of these shells requires consideration of lattice structures. This approach to core-shall NW heterostructures is elaborated below for the technologically important silicon (Si) and germanium (Ge) systems.

Homoepitaxial Si-Si core-shell NWs were grown by chemical vapor deposition (CVD) using silane as the silicon reactant (Fig. 15). Intrinsic silicon (i-Si) NW cores were prepared by gold nanocluster directed axial growth, which yields single crystal structures with diameters controlled by the nanocluster diameter, and then boron-doped (p-type) silicon (p-Si) shells were grown by homogeneous CVD, where the shell thickness was directly proportional to the growth time. Radial shell growth can be "turned-on" by the addition of diborane, which serves both to lower the decomposition temperature of silane and acts as a p-type dopant. Transmission electron microscopy (TEM) images of the i-Si/p-Si product obtained from constant temperature growth shows a uniform core-shall structure consisting of a crystalline Si core and amorphous Si shell (Fig. 15a), where the core diameter, 19 nm, is consistent with the 20 nm nanocluster used in the initial axial growth step. TEM images show reproducible crystalline faceting at the core-shell interface (Fig. 15b). This faceting

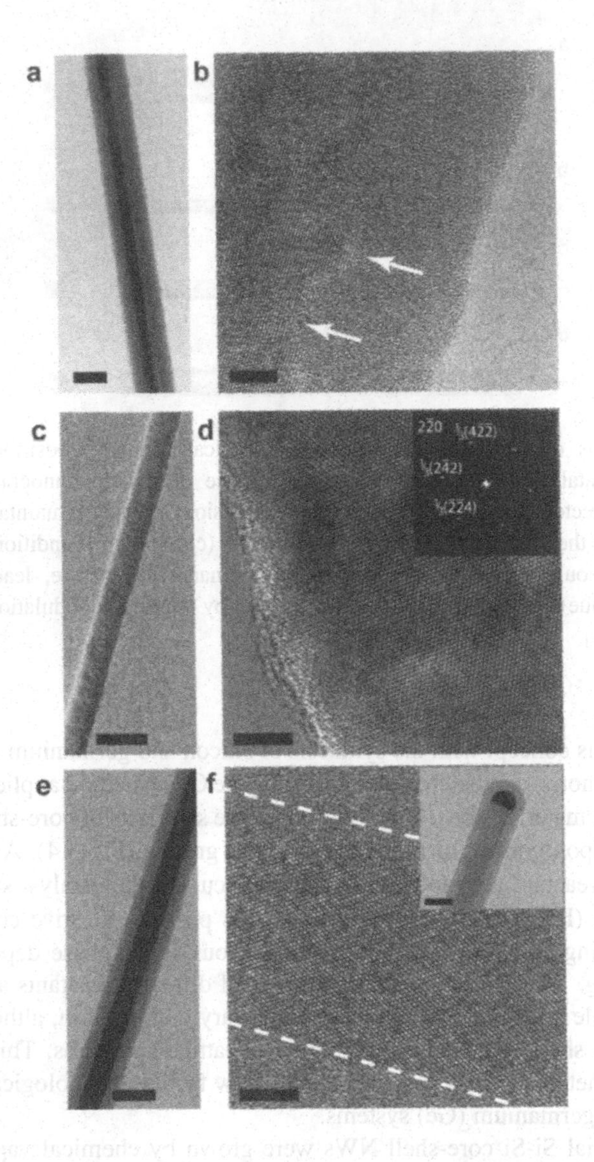

Fig. 15. Si–Si homoepitaxial core–shell nanowires. (a, b) Diffraction contrast and high-resolution TEM images, respectively, of an unannealed intrinsic silicon core and p-type silicon shell nanowire. Crystal facets in the high-resolution TEM image designated by arrows indicate initially epitaxial shell growth at low temperature. Scale bars are 50 nm and 5 nm, respectively. (c, d) TEM images (analogous to (a and b)) of an i-Si/p-Si core–shell nanowire annealed at 600 °C for 30 min after core–shell growth at 450 °C. Inset, two-dimensional Fourier transforms of the image depicting the [111] zone axis of the single crystal nanowire. (e, f) TEM images of an i-Si/SiO_x/p-Si nanowire. The oxide layer is too thin (<1 nm) to discern in the high-resolution image, but the sharp interface (dashed line) between the crystalline core and amorphous overcoat clearly differs from the faceting seen in (b) and illustrates the disruption of epitaxy. Inset, TEM image of p-Si coating the nanowire and the Au nanocluster tip. Scale bar is 50 nm. Adapted from [64].

suggests that the NW surfaces are sufficiently clean following axial growth to nucleate epitaxial shell growth.

To understand and control Si on Si homoepitaxy in core-shell N structures we carried out several distinct experiments. First, i-Si/p-Si core-shell NWs prepared as above were annealed *in situ* at 600 °C. TEM images recorded on the annealed samples exhibited no diffraction contrast between the core and shell (Fig. 15c), and lattice resolved images and electron diffraction data further show that the shell crystallizes to yield a single crystal structure (Fig. 15d). Second, the importance of the initial nucleation for achieving epitaxy in the shell was probed using *in situ* oxidation of the silicon core, which produces a thin amorphous silicon oxide layer at the surface of the core, prior to silicon shell growth. Significantly, TEM images of i-Si/SiO_x/p-Si core-shell-shell structures show a smooth and abrupt interface between the crystalline core and amorphous shell (Fig. 15e, f). The low roughness of the interface is comparable to that observed in NWs after only axial growth, and contrasts sharply with the faceted interface of the low-temperature homoepitaxy (Fig. 15b). These results show that the thin oxide layer completely disrupts homoepitaxy; further annealing and TEM studies show that oxidation inhibits crystallization of the shell under annealing conditions that lead to complete crystallization in samples without the oxide layer.

Radial heteroepitaxy of Si on Ge was pursued to explore the potential of our approach to core-shell structures in materials of rapidly increasing scientific and technological importance. For example, the energy band offsets in Si–Ge heterostructures produce internal fields that drive charge carrier redistribution, which can enable high mobility devices. Single crystal Ge NWs were defined via gold nanocluster directed axial growth, and then boron-doped p-Si shells were grown by homogeneous CVD (see Methods). Bright field TEM images (Fig. 16a) reveal a core-shell structure that is consistent with Ge-core (dark) and Si-shell (light) structure, and we confirmed this assignment by elemental mapping (Fig. 16b, c), which shows a localized Ge core and Si shell. High-resolution TEM images of i-Ge/p-Si core-shell NWs in which the p-Si shell was deposited at low temperature without annealing show a crystalline Ge core and predominantly amorphous Si shell (Fig. 16d). Analysis of cross-sectional elemental mapping data (Fig. 16e) shows that the Ge core is ~26 nm, the Si shell is ~15 nm, and the Ge–Si interface width is believed to be <1 nm on the basis of the electron beam width and modeling.

Significantly, we find that the amorphous Si shell can be completely crystallized following *in-situ* thermal annealing at 600 °C [64]. Lattice-resolved TEM images of Ge–Si core-shell structures following this thermal treatment exhibit a uniform crystalline Si shell (Fig. 16f), and suggest that thin regions of epitaxially grown Si are present in the unannealed wires. Higher silicon deposition temperatures might make the annealing step unnecessary by improving the surface mobility of adsorbed silicon. Elemental mapping (Fig. 16g) confirms that the contrast in high-resolution TEM images is consistent with an abrupt (<1 nm) Si–Ge interface. Although the present measurements cannot rule out some interface mixing, previous studies of planar growth under similar conditions found no evidence for Si–Ge interdiffusion. Notably, preliminary electrical transport studies of the i-Ge/p-Si core-shell NWs provide good evidence for internal field driven carrier redistribution and we believe these systems will be exciting to investigate in the future.

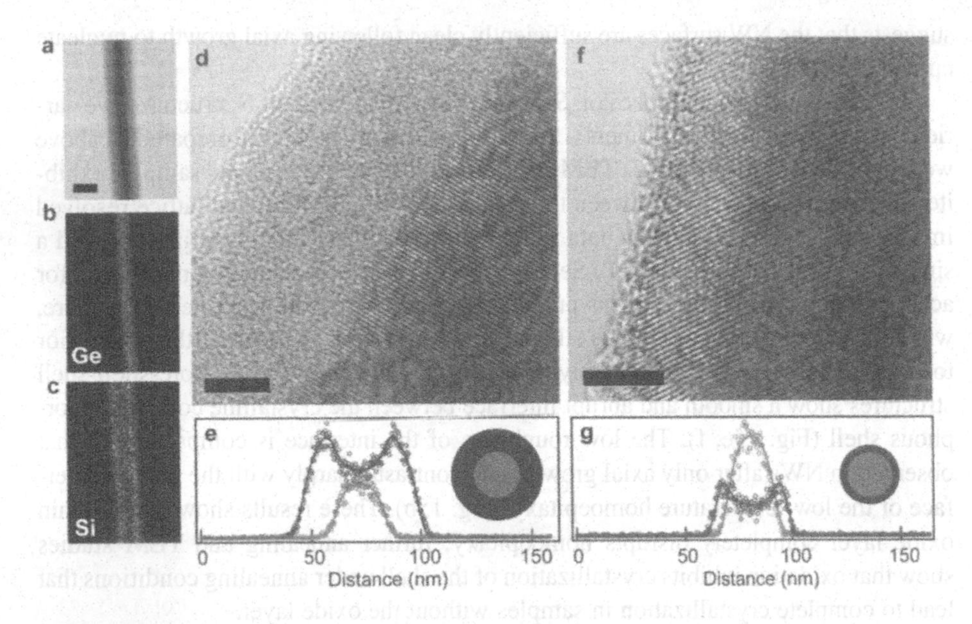

Fig. 16. Ge–Si core–shell nanowires. (a) Bright-field image of an unannealed Ge–Si core-shell nanowire with an amorphous p-Si shell. Scale bar is 50 nm. (b, c) Scanning TEM elemental maps of Ge (red) and Si (blue) concentrations, respectively, in the nanowire of (a). (d) High-resolution TEM image of a representative nanowire from the same synthesis as the wire in (a–c). Scale bar is 5 nm. (e) Elemental mapping cross-section showing the Ge (red circles) and Si (blue circles) concentrations. The solid lines show the theoretical cross-section for a 26-nm-diameter core, 15-nm-thick shell and <1 nm interface according to the model described in the Methods section. (f) High-resolution TEM image of annealed Ge–Si core–shell nanowire exhibiting a crystalline p-Si shell. Scale bar is 5 nm. (g) Elemental mapping cross-section gives a 5-nm shell thickness with a sharp interface consistent with the TEM image, suggesting that the Ge and Si do not interdiffuse substantially during the annealing process. Adapted from [64].

Synthetic control of Si–Ge core-multi-shell NW structures could be used to explore a variety of fundamental phenomena and new device concepts, although achieving this goal will require the ability to prepare an essentially arbitrary sequence of Si, Ge, and alloy overlayers on both Si and Ge nanowire cores. To this end, we also studied Ge deposition on Si NW cores [64]. TEM images and composition mapping (Fig. 17a) show the Si–Ge core-shell structure with a sharp (<1 nm) interface, and demonstrate that the Ge shell is fully crystallized for the low-temperature growth (Fig. 17b), presumably due to the higher surface mobility of Ge adatoms. In addition, diffraction data are consistent with coherently strained epitaxial overgrowth (inset, Fig. 17b); that is, a single diffraction peak is observed along the axial direction, which is indicative of compressively strained Ge and tensily strained Si. Two peaks, which can be indexed to the Ge (5.657 Å) and Si (5.431 Å) lattice constants, are also observed in the radial direction and indicate relaxation normal to the interface. Preliminary transport studies of these structures provide evidence for hole injection from the p-Si core to i-Ge shell as expected from valence band offsets. Finally, we

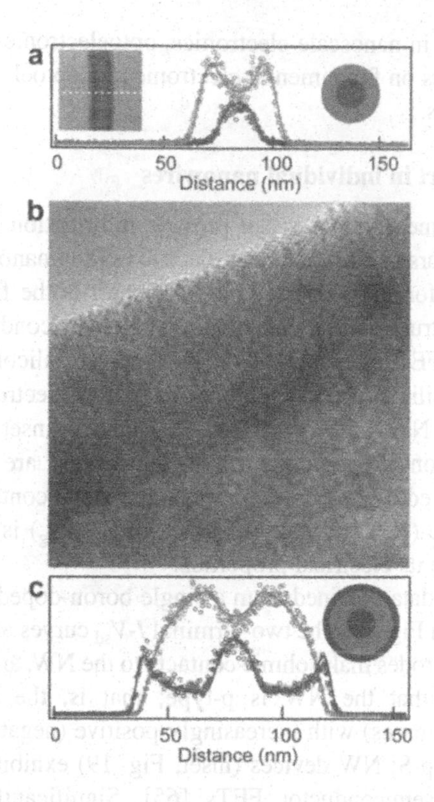

Fig. 17. Si–Ge and Si–Ge–Si core-shell nanowires. (a) Elemental mapping cross-section indicating a 21-nm-diameter Si core (blue circles), 10-nm Ge shell (red circles) and <1 nm interface. Inset, TEM image of the corresponding Si–Ge core-shell nanowire. The white dashed line indicates mapping cross-section. (b) High-resolution TEM image of a representative crystalline nanowire core and shell from the same synthesis as the wire in (a). Scale bar is 5 nm. Inset, two-dimensional Fourier transform of the real-space image showing the [111] zone axis. The split lattice reflections perpendicular to the interface can be indexed to the Ge and Si lattice constants (5.657 Å and 5.431 Å, respectively). (c) Cross-sectional elemental mapping of a double-shell structure with an intrinsic silicon core (diameter, 20 nm), intrinsic germanium inner shell (thickness, 30 nm), and p-type silicon outer shell (4 nm); silicon is blue circles and germanium is red circles. Adapted from [64].

have also explored the growth of more complex multi-shell structures; composition mapping of a Si–Ge–Si core-double-shell structure (Fig. 17c) demonstrates the clear potential for extending our approach in this direction.

4. Fundamental Properties of Single Nanowire Structures

The availability of a wide range of NW materials with controlled chemical composition, physical size and electronic properties opens up many exciting opportunities ranging from fundamental studies of the role of dimensionality on physical properties to potential applications in nanoscale electronics, optoelectronics and other areas. In the section, we will focus on fundamental electronic and optoelectronic properties of individual NW structures.

4.1. Electrical transport in individual nanowires

Electrical transport measurements can provide information about the electronic structure and the behaviors of carriers under electric field in nanoscale materials. One powerful configuration for studying electrical transport is the field effect transistor (FET). The basic FET structure fabricated from single semiconducting NWs is illustrated in Fig. 18. The FET is supported on an oxidized silicon substrate with the underlying conducting silicon used as a global back gate electrode to vary the electrostatic potential of the NW. In a typical NW-FET device (inset, Fig. 18), two metal contacts, which correspond to source and drain electrodes, are defined by electron beam lithography followed by evaporation of suitable metal contacts. Current (I) versus source-drain voltage (V_{sd}) and I versus gate voltage (V_g) is then recorded for a NW-FET to characterize its electrical properties.

Typical I versus V_{sd} data obtained from a single boron-doped Si NW-FET [32] at different V_g are shown in Fig. 19. The two-terminal I-V_{sd} curves are linear, which indicates that the metal electrodes make ohmic contacts to the NW, and moreover, the gate response demonstrates that the NW is p-type; that is, the conductance of the p-Si NW decreases (increases) with increasingly positive (negative) V_g. The transfer characteristics, I-V_g, of p-Si NW devices (inset, Fig. 19) exhibit behavior typical of p-channel metal-oxide-semiconductor FETs [65]. Significantly, the conductance modulation of the p-Si NW-FET exceeds 10^3, where the V_g required for switching (-10 to 10 V) could be reduced significantly by reducing the thick (600 nm) oxide dielectric layer in these back-gated devices (see below).

Gate-dependent measurements have also been used to estimate the hole concentration in p-channel NW-FETs. The total NW charge can be expressed as $Q = C \cdot V_{th}$, where C is the nanowire capacitance and V_{th} the threshold gate voltage required to deplete completely the NW. The capacitance is given by $C \cong 2\pi\varepsilon\varepsilon_0 L/\ln(2h/r)$, where ε is the effective gate oxide dielectric constant, h is the thickness of the SiO_2 layer on the substrate, L is the NW length and r is the NW radius. The hole density,

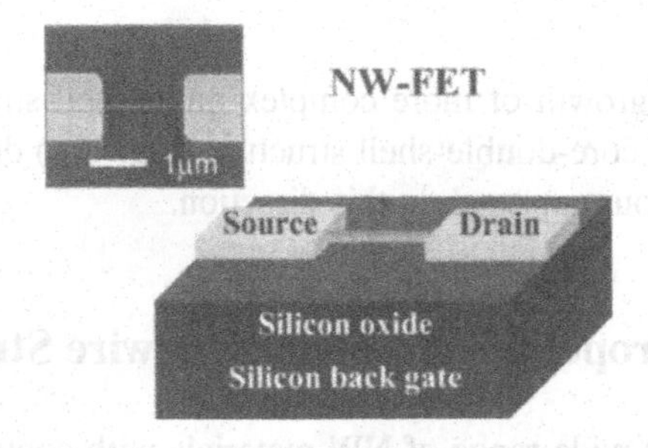

Fig. 18. Schematic of a NW-FET. (inset) SEM image of a NW-FET; two metal electrodes, which correspond to source and drain, are visible at the left and right sides of the image.

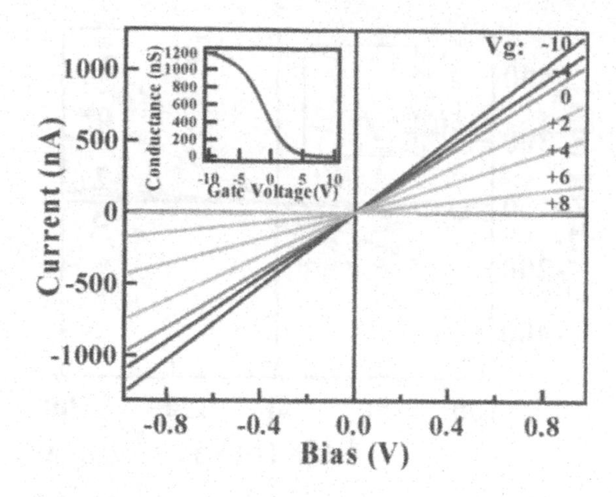

Fig. 19. Current versus voltage for a p-type Si NW-FET. The numbers inside the plot indicate the corresponding gate voltages (V_g). The inset shows current versus V_g for V_{sd} of 1 V.

$n_h = Q/(e \cdot \pi r^2 L)$, is estimated to be $\sim 10^{18}/cm^3$ for the device shown in Fig. 19. In addition, it is possible to estimate the carrier mobility of the NW FETs from the transconductance $dI/dV_g = \mu(C/L^2)V_{sd}$, where μ is the carrier mobility. Plots of dI/dV_g versus V_{sd} are linear for Si NWs, as expected for this model, yield hole mobilities in the range of 50–300 $cm^2/V \cdot s$. Significantly, the p-Si NW-FET mobilities are comparable to or larger than the best p-Si planar devices, 40–100 $cm^2/V \cdot s$, at comparable hole densities ($p \sim 10^{18}$–$10^{19}/cm^3$) [66].

It is also possible to assemble n-channel NW-FETs in a similar way from n-type NWs. For example, gate dependent I-V_{sd} data recorded from an InP NW-FET exhibits increased conductance for positive V_g and decreased conductance for negative V_g (Fig. 20), as expected for an n-channel device. The n-InP NW-FET transfer characteristics (I-V_g) show that the current increases rapidly from below 1 nA at $V_g = -2$ V to above 400 nA at $V_g = +2$ V (inset, Fig. 20), and tends to saturate at higher voltages, which can be attributed to contact resistance and other factors. Nevertheless, the conductance changes up to three orders of magnitude for only a few volt change in the gate voltage in these unoptimized devices.

The electron concentration and mobility in the n-channel NW-FETs have been estimated as described above for p-channel devices. For the n-InP NW-FET shown in Fig. 20, the electron mobility is 2200 $cm^2/V \cdot s$ for an electron concentration of $\sim 10^{18}/cm^3$. Studies of a number of different devices yields mobility values from 400 to 3000 $cm^2/V \cdot s$, which is comparable to or larger than bulk InP, 1000–2000 $cm^2/V \cdot s$, at similar carrier concentrations [66].

We have also carried out similar measurements on many types of NWs, such as GaN [67], CdSe [68], ZnO [68]. Their mobility values are all comparable to or larger than the corresponding bulk value. This suggests that when carriers transport through NWs, they do not meet significant boundary scattering and these single-component NW structures are very effective for carrier transport.

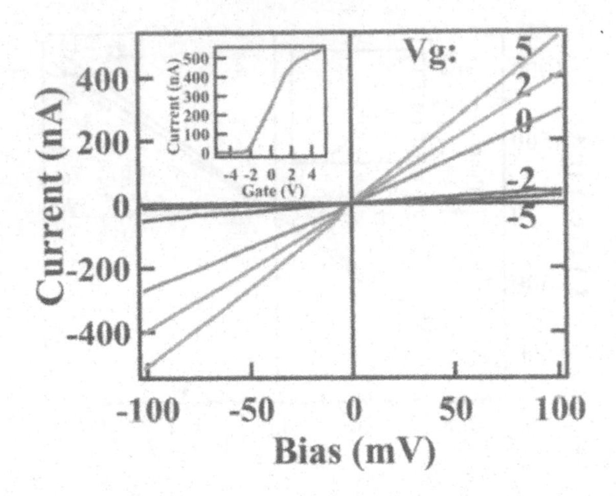

Fig. 20. Current versus voltage for an n-type InP NW-FET. The numbers inside the plot indicate the corresponding gate voltages (V_g). The inset shows current versus V_g for V_{sd} of 0.1 V.

4.2. High mobility nanowires

We have also investigated how chemical passivation of the SiO_x shell surrounding single crystal SiNW cores affects transport behavior (Fig. 21a) [69], since the Si/SiO_x interface and SiO_x surface defects could compensate the applied voltage, and trap and scatter carriers. Surface modification was carried out by reaction with 4-nitrophenyl octadecanoate. This specific reagent was chosen since it will lead to a stable Si-O-C ester linkage. Conductance (I/V_{sd}) versus backgate voltage (V_g) measurements were carried out before and after modification, to assess clearly the effect of surface chemistry on characteristics of a specific SiNW FET. In a typical device (Fig. 21b), the conductance responds weakly to V_g before modification. In contrast, the conductance is extremely sensitive to V_g after modification and can be shut off at ~2.5 *V* with on/off ratio over four orders of magnitude. The transconductance after modification is an order of magnitude larger than before modification. Using a cylinder on an infinite plate model, we obtain a hole mobility of 1000 $cm^2/V \cdot s$. This mobility is substantially larger than obtained in conventional Si devices. To assess reproducibility of this surface chemistry and corresponding dramatic improvements in device behavior, we carried out experiments on a number of distinct SiNW devices. The summary of these results (Fig. 21c, d) shows that transconductance and mobility increase an order of magnitude after modification. Significantly, the highest and the average hole mobility values of 1350 and 560 $cm^2/V \cdot s$ in p-SiNWs are about an order of magnitude larger than the bulk Si value, ~40 $cm^2/V \cdot s$, at the same doping concentration. In addition, we should note that value reported here are the lower-bound values since the contact and surface modification are far from optimum.

To obtain additional information about the origin of these passivation effects, we have carried out two additional experiments [69]. First, we investigated the effect of chain length on characteristics of SiNW FETs. Immediately after modification, 18-carbon and 6-carbon chains gave similar transport results although the lifetime of

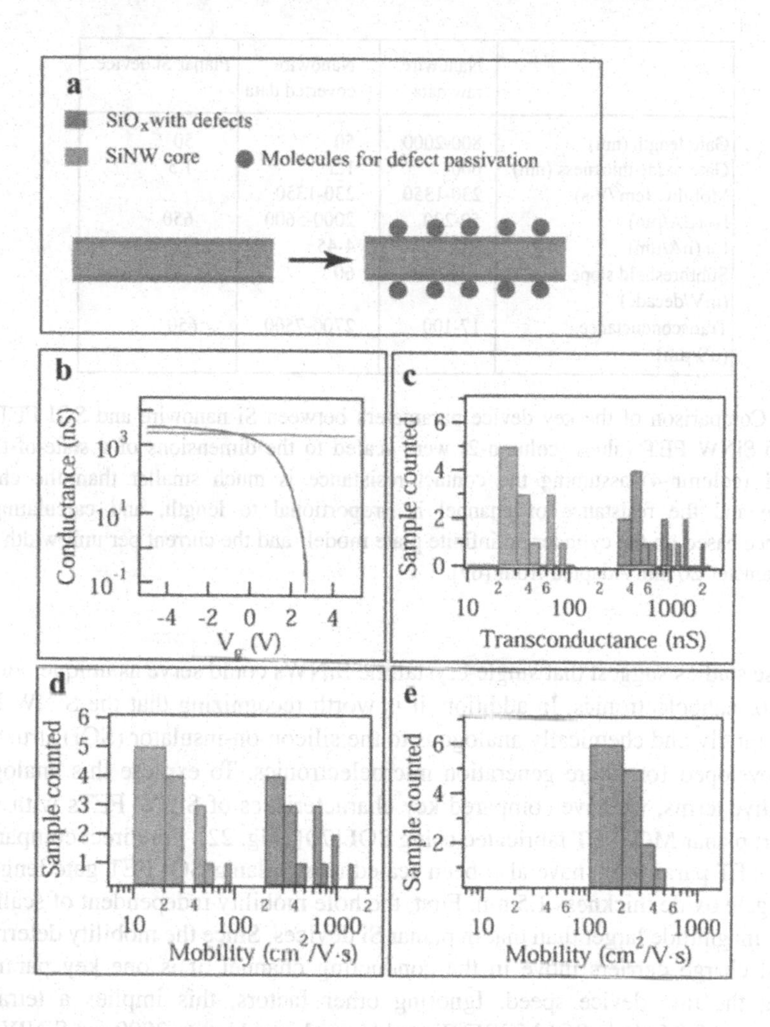

Fig. 21. (a) Schematic illustrating surface defect passivation. (b) Conductance versus V_g measured on the same SiNW before and after 4-nitrophenyl octadecanoate modification. Histograms of SiNW transconductance (c) and mobility (d) before and after 4-nitrophenyl octadecanoate modification. (d) Histogram of SiNW mobility values after modification with tetraethylammonium bromide. In plots (b) to (e), the green (red) color designates before (after) surface modification. Adapted from [69].

the observed improvement was >1 week versus 1 day, respectively. These results suggest that accessibility of surface to water hydrolysis may be important. Second, to follow up this observation, we have modified SiNW surface with tetraethylammonium bromide solution. Significantly, tetraethylammonium-modified NWs also show the significant increase of transconductance and mobility as shown in Fig. 21e. We believe that in both cases, the passivation reduces compensation of applied V_g through covalent or ionic reaction with surface SiO^- groups, although additional experiments will be needed to define the mechanism.

	Nanowire raw data	Nanowire coverted data	Planar Si device
Gate length (nm)	800-2000	50	50
Gate oxide thickness (nm)	600	1.5	1.5
Mobility (cm^2/V·s)	230-1350	230-1350	
I_{on} (μA/μm)	50-200	2000-5600	650
I_{off} (nA/μm)	2-50	4-45	9
Subthreshold slope (mV/decade)	174-609	60	70
Transconductance (μS/μm)	17-100	2700-7500	650

Fig. 22. Comparison of the key device parameters between Si nanowire and SOI FET. The measured SiNW FET values (column-2) were scaled to the dimensions of a state-of-the-art MOSFET (column-4) assuming the contact resistance is much smaller than the channel resistance and the resistance of channel is proportional to length, and calculating the capacitance based on the cylinder on infinite plate model[1] and the current per unit width using a NW diameter 20 nm. Adapted from [69].

These studies suggest that single-crystalline SiNWs could serve as unique building blocks for nanoelectronics. In addition, it is worth recognizing that the SiNW FETs are structurally and chemically analogous to the silicon-on-insulator (SOI) structures, being developed for future generation microelectronics. To explore this analogy in quantitative terms, we have compared key characteristics of SiNW FETs with state-of-the-art planar MOSFET fabricated using SOI [70] (Fig. 22). For direct comparison, the NW FET parameters have also been scaled using planar SOI FET gate length 50 nm and gate oxide thickness 1.5 nm. First, the hole mobility independent of scaling is an order magnitude larger than that in planar Si devices. Since the mobility determines how fast charge carriers move in the conducting channel, it is one key parameter affecting the raw device speed. Ignoring other factors, this implies a tetrahertz frequency for 0.85 V_{sd} in SOI MOSFET could be achieved in a ~2000 nm SiNW FET. Second, the on-state current (I_{on}) in SiNW FET is larger than planar Si FETs, which can also facilitate the high-speed operation. Third, the average subthreshold slope is ~5 times smaller and the average transconductance is ~30 times larger in SiNW FET, indicating the higher gain in SiNW FETs, which is important in logic circuits. The SiNW FET devices have larger leakage current but the issue could be addressed by using pn-diodes at the source and drain contacts as in conventional MOSFETs. This comparison suggests that efforts to make smaller SiNW FETs will be important in the future.

4.3. Nanowire-molecule structures

To address the issue of creating new functional devices, we have been exploring the combination of NWs and molecular building blocks and herein fabricate bistable nanoscale switches assembled using semiconductor NWs and redox active molecules as building blocks [71]. The overall configuration and operating principle of our NW devices are illustrated in Fig. 23. In a NW-FET functionalized with redox active

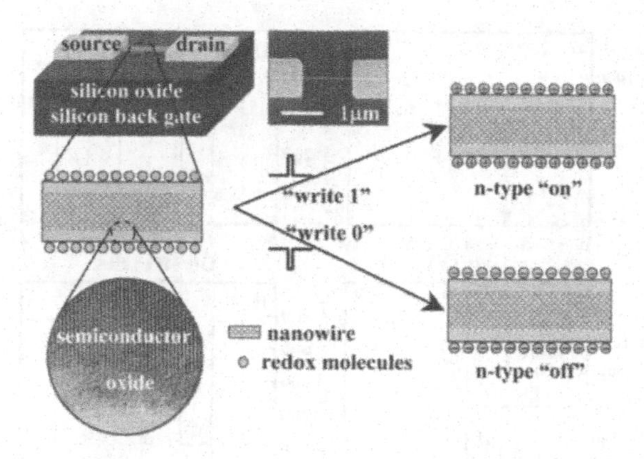

Fig. 23. NW-based nonvolatile devices. The devices consist of a semiconductor NW configured as a FET with the oxide surface functionalized with redox active molecules. The top-middle inset shows an SEM image of a device, and the lower circular inset shows a TEM image of an InP NW highlighting the crystalline core and surface oxide. Positive or negative charges are injected into and stored in the redox molecules with an applied gate or bias voltage pulse. In an n-type NW, positive charges create an *on* or logic "1" state, while negative charges produce an *off* or logic "0" state. Adapted from [71].

molecules, an applied gate voltage (V_g) or source-drain voltage (V_{sd}) pulse injects net positive or negative charges on the molecular layer. The oxide layer on the NW surface, which can be controlled synthetically, serves as a barrier to reduce charge leakage between the molecules and NW, and thus maintain the charge state of the redox molecules. The charged redox molecules gate the NW-FET to a logic *on* state with higher channel conductance or logic *off* state with lower channel conductance. For example, positive charges, like a positive gate, will lead to accumulation of electrons and an *on* state in n-channel NW-FETs, and depletion of holes and an *off* state in p-channel NW-FETs.

The NWs used in this study include p-type Si, n-type InP, and n-type GaN, with diameters ranging from 10 to 30 nm. Several different redox active molecules were examined including ferrocene, zinc tetrabenzoporphine, and cobalt phthalocyanine (CoPc). Below we focus our discussion on n-channel InP NW-FETs functionalized with CoPc redox molecules, although similar results consistent with the model presented in Fig. 23 were obtained for the different NWs and redox molecules. The NW-FETs were functionalized by spin coating chlorobenzene solutions of CoPc, which produces a uniform layer of CoPc on NW surfaces with layer thicknesses greater than one monolayer. Typical conductance, $G=I/V_{sd}$ (I is current), versus V_g data for an n-InP NW-FET before and after modification with a CoPc layer are shown in Fig. 2a. Before addition of the CoPc layer, the response is characteristic of an n-type FET; that is, little or no hysteresis is observed in G-V_g for positive/negative variations in V_g. Significantly, a large hysteresis in G-V_g is reproducibly observed after the NW surface is modified with CoPc (red curve, Fig. 24a). As V_g is increased from negative to positive, the channel conductance increases at a more negative value than for the unmodified NW, and as V_g is then cycled to negative values the channel

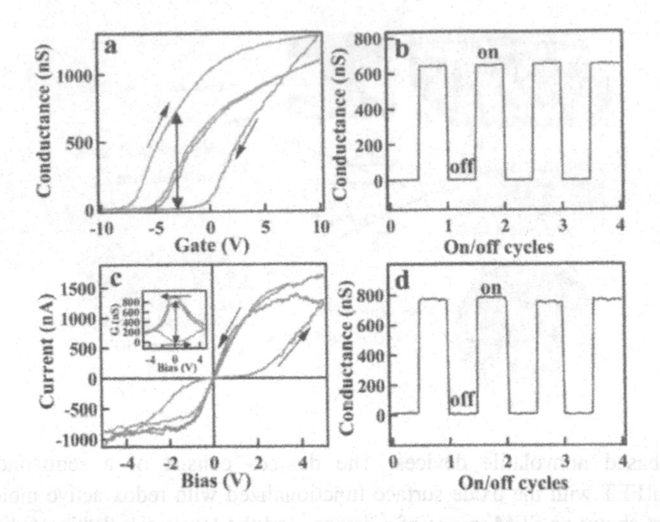

Fig. 24. (a) G versus V_g for an n-InP NW-FET before (green) and after (red) surface modification with CoPc recorded with V_{sd}=0.1 V. Single-headed arrows show gate sweep directions, and the double-headed arrow indicates a position of large *on/off* hysteresis. (b) Reversible *on* and *off* switching of the device in (a) using +10 and −10 V, respectively, 1 s gate pulses; *on/off* states were measured with V_{sd}=0.1 V and V_g=−3 V. (c) I versus V_{sd} for an n-InP NW-FET before (green) and after (red) surface modification with CoPc; (inset) G versus V_{sd}. Single-headed arrows show bias sweep directions, and the double-headed arrow (inset) indicates the position of maximum hysteresis. (d) *On* and *off* switching of the device in (c) using +5 and −5 V, respectively, 1 s bias pulses; *on/off* states were measured with V_{sd}=0.1 V. Adapted from [71].

conductance decreases at a more positive value than in the unmodified NW. The hysteresis defines the two states of a bistable system and has been exploited to configure a three-terminal switch or memory device. The low conductance *off* state and high conductance *on* state of the NW-FET switch are continuously monitored by measuring G at a fixed V_{sd} and V_g, while the NW-FET is switched between *on* and *off* states by a V_g pulse (e.g., ±10 V). A typical CoPc-modified NW FET operated in this way (Fig. 24b) shows reversible switching over many cycles between the *on* and *off* states, with the conductance change of nearly 10^4 (i.e., from 600 to <0.1 nS) being maintained for at least 100 cycles. Similar results have been obtained on more than 80 devices.

The three-terminal devices demonstrate that the molecule-gated NW-FETs can function as bistable *on/off* switches, although the operation of these devices with the global back gate (substrate) precludes the assembly of integrated and individually addressable device arrays. To overcome this important limitation, we have investigated the possibility of switching devices in a two-terminal configuration, where a V_{sd} pulse is used to inject positive or negative charges on the redox active molecules. Current (or G) versus V_{sd} measurements exhibit a large hysteresis in CoPc-modified NW-FETs (Fig. 24c), while no hysteresis is observed in the unmodified NW-FETs. Similar to our studies of three-terminal devices, we find that it is possible to reversibly switch between *on/off* states using 2–5 V pulses (Fig. 24d). Typically, we find that devices can be switched at least 60 times before any degradation in *on/off* ratio is observed; the retention time within the *on* and *off* states is at least 20 min.

4.4. Photonic properties of single nanowires

In the case of optical properties, most previous studies of one-dimensional nanostructures have focused primarily on lithographically and epitaxially defined quantum wires, such as "T-shaped" and "V-groove" quantum wires. Quantum wires differ substantially from NWs since these wire-like nanostructures are embedded in a substrate of similar dielectric constant with a noncylindrical confining potential. In NWs, the potentials for electrons and holes are radially symmetric, and the surrounding medium (air or vacuum) has a considerably lower dielectric constant, which can introduce new and interesting physical properties [72].

To explore the potentially unique optical properties of NW building blocks, we have studied the size-dependent PL properties of InP NWs [73]. PL images and spectra have been recorded on individual, isolated InP nanowires at both room temperature (RT) and 7 K. A typical PL image of a single InP NW recorded at room temperature (RT) is shown in Fig. 25a. The recorded emission intensity is quite uniform ($\pm 5\%$ variation) over the entire nanowire length within the ~ 1 μm spatial resolution of our experiments. In addition, luminescence spectra obtained at different positions along the nanowire axis (Fig. 25b) show nearly identical emission energies (less than 1% variation) and line shapes. Uniform PL is also observed in measurements recorded at low temperature (7 K). These results, which show uniform PL and little variation for a given diameter, indicate that the InP NWs are of good quality and can provide insight into size-dependent properties.

The InP NWs should exhibit size-dependent PL for diameters less than ca. the bulk exciton diameter, 19 nm, due to quantum confinement. To address this fundamental issue directly, the PL spectra of nanowires with diameters of 50, 20, 15, and 10 nm were collected. Representative spectra recorded at RT (Fig. 26a) and 7 K (Fig. 26b) exhibit a systematic shift to higher energy as the nanowire diameters are

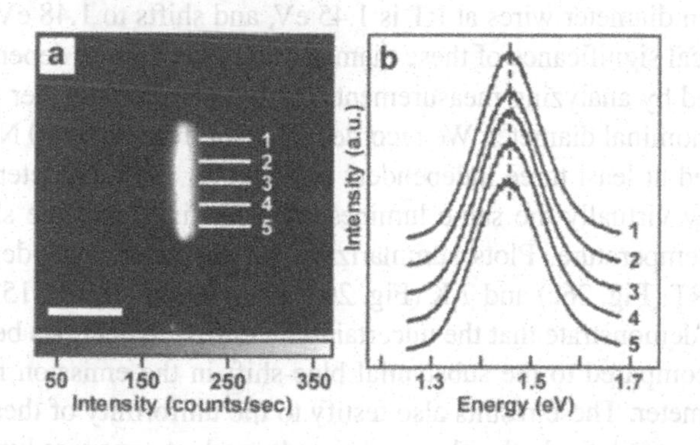

Fig. 25. (a) Typical PL image recorded on an individual InP NW. Scale bar is 5 μm. (b) PL spectra taken at the positions indicated in (a). The spectra were shifted vertically for clarity. The PL image and spectra shown here were taken at RT from an ca. 15 nm diameter NW. Adapted from [73].

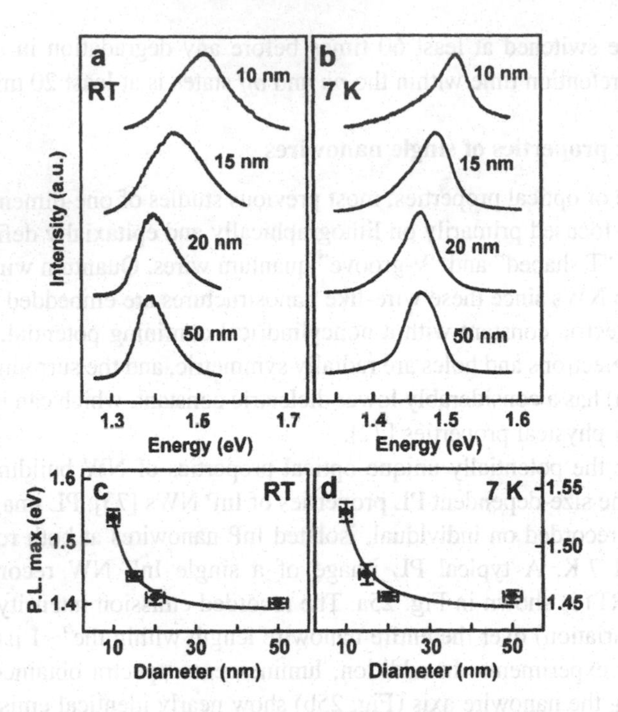

Fig. 26. (a) PL spectra taken at RT from single InP NWs with nominal diameters of 50, 20, 15, and 10 nm. (b) Single NW PL spectra at 7 K. (c) Emission energy maxima at RT versus NW diameter. (d) Emission energy maximum at 7 K. PL spectra in (a) and (b) were normalized to the peak maxima and shifted upward for clarity. The experimental data in (c) and (d) were fit using the EMM (solid line). Adapted from [73].

reduced below 20 nm, in agreement with the concept of quantum confinement. In addition, these experiments show that all of the diameter-dependent spectra recorded at 7 K shift to higher energy, consistent with the shift of the bulk band gap from 1.35 to 1.42 eV as the temperature is reduced from RT to 7 K. For example, the emission energy of 15 nm diameter wires at RT is 1.45 eV, and shifts to 1.48 eV at 7 K.

The statistical significance of these diameter and temperature dependent data have been determined by analyzing measurements carried out on a number of independent NWs for each nominal diameter. We recorded PL data from 20 to 50 NWs for a given sample and used at least three independent samples for each diameter. Significantly, these data show virtually the same luminescence maxima and line shapes for each diameter and temperature. Plots summarizing the diameter dependent PL maxima determined at RT (Fig. 26c) and 7 K (Fig. 26d) from nominally 10, 15, 20 and 50 nm diameter NWs demonstrate that the uncertainty in emission energies between wires is small (~1%) compared to the substantial blue-shift in the emission maximum with decreasing diameter. These results also testify to the uniformity of these NWs.

To explore quantitatively the observed size-dependent nanowire luminescence, we compared these data to an effective mass model (EMM) developed for a cylindrical potential for electrons and holes [73]. We find that our data are well fit by the EMM at RT (Fig. 26c) and 7 K (Fig. 26d) using the reduced effective mass, m^*, as the primary fitting parameter. The good fit of the experimental data suggests that this model captures the essential physics of the system. The reduced effective mass at RT determined from the fit, 0.052 m_0 (m_0, the free electron mass), is in reasonable agreement with the literature value of 0.065 m_0 (calculated using m_e=0.078 and the geometric mean of the anisotropic hole masses, m_h=0.40) for bulk InP [74]. Interestingly, the smaller effective mass determined from our data can be attributed to the crystalline orientation of the nanowires; that is, the <111> nanowire growth axis corresponds to the heavy hole direction in InP. The smaller observed effective mass is thus consistent with confinement perpendicular to the growth direction, where the hole mass is reduced. The value of the reduced mass determined from the 7 K data, 0.082 m_0, is larger than the RT value but consistent with the observation that the effective carrier masses in InP increase with decreasing temperature within the accuracy of the known data.

An important feature of free-standing NW structures is the large variation in the dielectric constant between the NW and surrounding (air) medium. The large dielectric variation and quantum confinement discussed above lead to striking phenomena in these NW materials. For example, photoluminescence studies of individual InP NWs show a giant polarization anisotropy with a polarization ratio $((I_{//}-I_{\perp})/(I_{//}+I_{\perp}))$ of 0.91 ± 0.07 (Fig. 27) [75]. This large anisotropy implies that photoluminescence can be essentially turned from on (Fig. 27a) to off (Fig. 27b) as the excitation polarization is rotated from parallel to perpendicular. The large and unprecedented polarization anisotropy can be accounted for quantitatively by considering the dielectric contrast

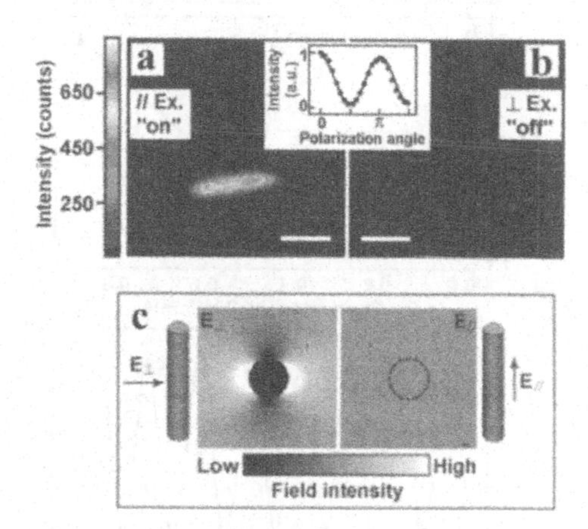

Fig. 27. (a) PL image of a single 20-nm InP nanowire with the exciting laser polarized along the wire axis. Scale bar, 3 μm. (b) PL image of the same nanowire as in (a) under perpendicular excitation. Inset, variation of overall photoluminescence intensity as a function of excitation polarization angle with respect to the nanowire axis. The PL images were recorded at room temperature with integration times of 2 s. (c) Dielectric contrast model of polarization anisotropy. The nanowire is treated as an infinite dielectric cylinder in a vacuum while the laser polarizations are considered as electrostatic fields oriented as depicted. Field intensities ($|E|^2$) calculated from Maxwell's equations clearly show that the field is strongly attenuated inside the nanowire for the perpendicular polarization, $E_{\perp}$, whereas the field inside the nanowire is unaffected for the parallel polarization, $E_{//}$. Adapted from [75]. See color plate 1.

between a NW and its air or vacuum surroundings; that is, the perpendicular electric field amplitude is attenuated according to $E_i = (2\varepsilon_0/(\varepsilon+\varepsilon_0))E_e$, where E_i is the electric field inside the cylinder, E_e is the excitation field, and $\varepsilon(\varepsilon_0)$ is the dielectric constant of the cylinder (vacuum) (Fig. 27c) [75]. This model yields a theoretical polarization ratio, $\rho = 0.96$, in good agreement with the experimental results for InP NWs.

In addition to generating photons via carrier recombination in nanoscale NW devices, absorbed photons can produce electrical carriers, which when measured, serve as the basis for Lilliputian photodetectors. The striking PL polarization anisotropy makes these NWs ideally-suited for polarization sensitive photonic devices such as photodetectors and optically gated switches. For example, a polarization-sensitive photodetector can be easily fabricated by making metallic contacts to both ends of individual NWs (Fig. 28a). The conductance (G) of individual NWs was found to increase by 2–3 orders of magnitude with increasing laser intensity (Fig. 28b), and these changes were reproducible and reversible, which indicate that the increases in G are due to direct carrier collection at the NW-metal contacts.

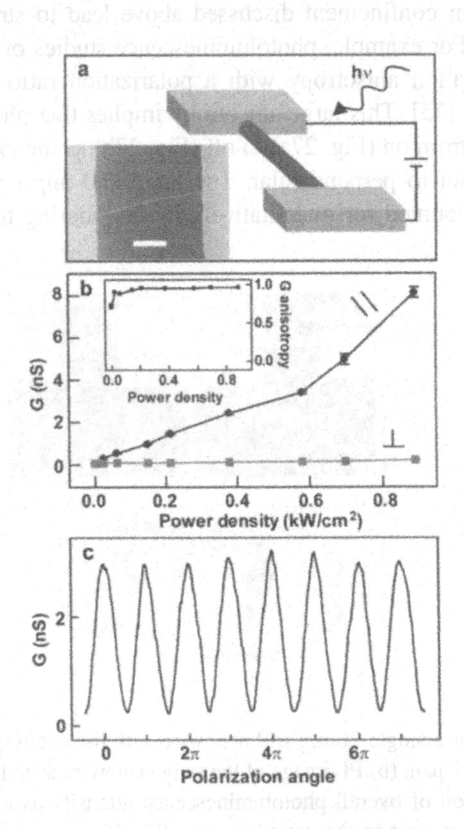

Fig. 28. Polarized photodetection using individual InP NWs. (a) Schematic depicting the use of a NW as a photodetector. Inset, SEM image of a 20-nm-diameter NW and contact electrodes for photoconductivity (PC) measurements. (b) Conductance, G, versus excitation power density for excitation light polarized parallel and perpendicular to the NW axis. Inset, PC anisotropy versus excitation power. (c) Conductance versus polarization angle as the polarization was manually rotated while measuring the PC. Adapted from [75].

A striking polarization anisotropy was also observed in the photoconductivity (Fig. 28c) [75]. The photoconductivity anisotropy, $\sigma=(G_{//}-G_{\perp})/(G_{\perp}+G_{\perp})$, values were similar to the PL anisotropy and the theoretical predictions described above. The large anisotropy shows that the conductance of devices is essentially switched on and off while the polarization of the incident light is changed. The overall sensitivity of the photodetector can be gauged by the responsivity, which is the ratio of photocurrent versus absorbed optical power. For the device shown in Fig. 28, the responsivity is about 3000 amperes/watt, which is an impressive number considering this device was not optimized [74].

There are several unique features of these NW based photodetectors that could enable applications. First, the sensitivity and polarization anisotropy is nearly independent of excitation wavelength for energies larger than band gap. In addition, it is easily possible to make a device that simultaneously measures intensity and polarization by using two crossed NWs and measuring their photoconductivity independently. Second, the active device element in the NW-based photodetector is substantially smaller than other polarization sensitive quantum-well based detectors [76, 77], which are not smaller than 50 by 50 micrometers and usually only sensitive to a very specific wavelength. Lastly, the extremely small size of these devices may open the possibility of creating ultra high-speed detectors.

4.5. Axial heterostructures: Electronic and photonic properties

We fabricated p–n junctions within individual SiNWs by Au-nanocluster-catalyzed CVD and dopant modulation. These nanowire p–n junctions were characterized at the single nanowire level by a variety of electrical measurements (Fig. 29) [61] because EDS is insufficiently sensitive to characterize dopant profiles. Current (I) versus voltage (V_{SD}) measurements showed rectifying behavior consistent with the presence of an intra-NW p–n junction (Fig. 29a). To establish that the current rectification was due to intra-NW p–n junctions, we characterized the local NW potential and gate response by electrostatic force microscopy (EFM) and scanned gate microscopy (SGM), respectively, An EFM image of a typical p–n junction in reverse bias showed that the entire voltage drop occurs at the p–n junction itself (Fig. 29b); EFM measurements showed no potential drop at the contact regions under forward or reverse bias (not shown), ruling out the contact/nanowire interface as the source of rectification in the I–V_{SD} behavior. In addition, SGM images recorded with the NW device in forward bias and the scanned tip–gate positive (Fig. 29c) show enhanced conduction to the right of the junction, indicating an n-type region, and reduced conduction to the left of the junction, indicating depletion of a p-type region. The abrupt change in majority carrier type coincides with the location of the intra-NW junction determined by EFM, and thus further confirms that the diode behavior results from well controlled dopant modulation.

The ability to synthesize modulation doped NW superlattices opens up new opportunities ranging from ultra-sensitive biological and chemical detectors to bipolar transistors and highly integrated logic gates for nanoelectronics. Moreover, the direct growth of modulation-doped nanowires eliminates the lithographic steps used to create doped nanotube p–n junctions, and thus facilitates the bottom-up assembly of

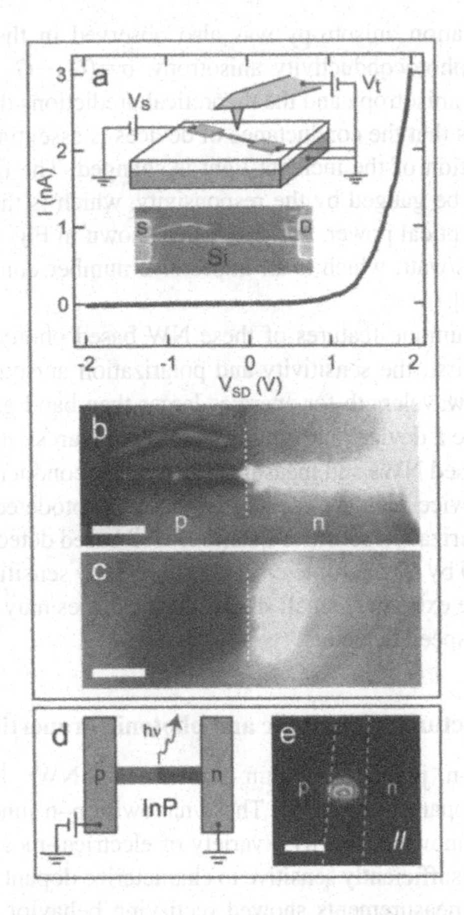

Fig. 29. Modulation-doped nanowires. (a) Insets: a diagram illustrating single-nanowire electrical characterization by transport and probe microscopy, and a scanning electron micrograph of the silicon nanowire device with source (S) and drain (D) electrodes indicated. Main panel, I versus V_{SD} for the silicon nanowire p–n junction. (b) EFM phase image of the nanowire diode under reverse bias with tip bias (V_t) at +3 V and the drain (right) at +2 V. The signal is proportional to the square of the tip–surface potential difference, and shows an abrupt drop in the middle of the wire at the junction. (c) SGM image showing the source–drain current as the tip (at +10 V) is scanned across the device. With the drain biased at −2 V (V_{SD}=+2 V), bright (dark) regions correspond to an increase (decrease) in the positive quantity I_{SD}. Vertical dashed white lines indicate the junction in (b) and (c). Scale bars, 500 nm. (d) Schematic of a InP nanowire LED. (e) Polarized emission from the LED along the nanowire axis. Dashed white lines indicate the edges of the electrodes in (d) and were determined from a white light image. No electroluminescence was detected with perpendicular polarization. Scale bar, 3 μm. Adapted from [61].

complex functional structures when combined with recent advances in the directed *en mass* organization of NW structures.

The use of single InP NW pn junctions as nanoscale LEDs has also been investigated (Fig. 29d) [61]. The single InP NW pn junction was fabricated using dopant diffusion from p- and n-type metal contact leads. I-V_{sd} measurements of InP NW pn

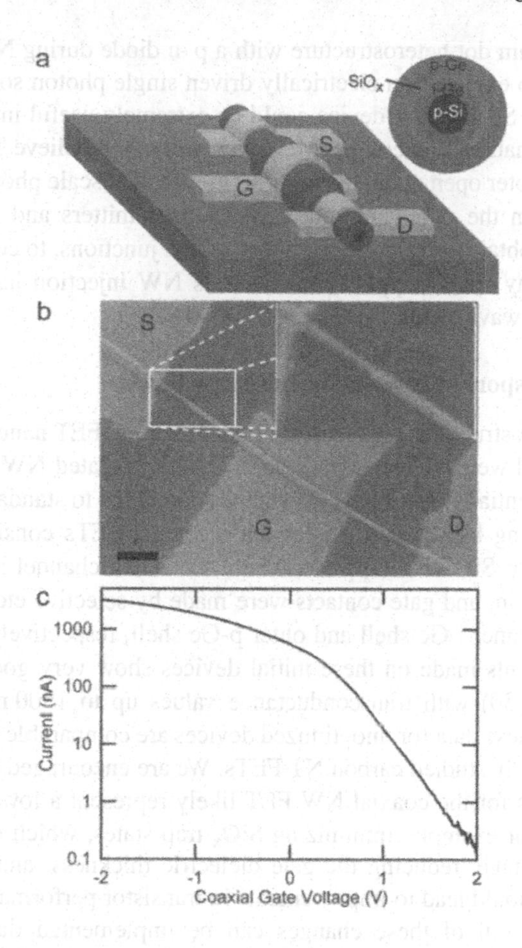

Fig. 30. Coaxially-gated nanowire transistors. (a) Device schematic showing transistor structure. The inset shows the cross-section of the AS-grown nanowire, starting with a p-doped Si core (blue, 10 nm) with subsequent layers of i-Ge (red, 10 nm), SiO_x (green, 4 nm), and p-Ge (5 nm). The source (S) and drain (D) electrodes are contacted to the inner i-Ge core, while the gate electrode (G) is in contact with the outer p-Ge shell and electrically isolated from the core by the SiO_x layer. (b) Scanning electron micrograph (SEM) of a coaxial transistor. Source and drain electrodes were deposited after etching the Ge (30% H_2O_2, 20 s) and SiO_x layers (buffered HF, 10 s) to expose the core layers. The etching of these outer layers is shown clearly in the inset and is indicated by the arrow. The gate electrodes were defined in a second step without any etching before contact deposition. Scale bar is 500 nm. (c) Gate response of the coaxial transistor at $V_{SD}=1$ V, showing a maximum transconductance of 1500 nA V^{-1}. Charge transfer from the p-Si core to the i-Ge shell produces a highly conductive and gateable channel. Adapted from [64].

structures exhibit rectification similar to that described above for the silicon NW pn junctions. In forward bias, individual InP NW devices exhibit light emission from p–n junctions that is both highly polarized and blue-shifted due to the one-dimensional structure and radial quantum confinement, respectively (Fig. 29e). The efficiency of these intra-NW LEDs is ca. 0.1%, although it can be increased substantially.

By defining a quantum dot heterostructure with a p–n diode during NW synthesis, it should be possible to engineer an electrically driven single photon source with well-defined polarization. Such a NW device could be extremely useful in quantum cryptography and information processing. More generally, we believe that the results described in this chapter open up many opportunities in nanoscale photonics and electronics, ranging from the relatively simple nanoscale emitters and complementary logic, which can be obtained from single nanowire p–n junctions, to complex periodic superlattices that may enable applications such as NW injection lasers and "engineered" 1D electron waveguides.

4.6. Electrical transport in core-shell heterostructures

Radial NW heterostructures enable a number of exciting FET nanostructures to be explored. To this end we have recently explored coaxially-gated NW FETs (Fig. 30) since these offer potentially significant advantages compared to standard planar gates. The nanowire building blocks used to fabricate coaxial FETs consisted of a core-multishell structure: p-Si/i-Ge/SiO_x/p-Ge, where the active channel is the i-Ge shell [64]. The source, drain, and gate contacts were made by selective etching and metal deposition onto the inner i-Ge shell and outer p-Ge shell, respectively. Significantly, transport measurements made on these initial devices show very good performance characteristics (Fig. 30) with transconductance values up to 1500 nA/V for a 1 V source-drain bias. These data for unoptimized devices are comparable to recent results reported for intensively studied carbon NT FETs. We are encouraged by this comparison since the values for the coaxial NW FET likely represent a lower limit to what may be achieved. For example, minimizing SiO_x trap states, which can compensate the applied gate voltage, reducing the gate dielectric thickness, and/or substituting high-K dielectrics should lead to improvements in transistor performance. We believe it is significant that all of these changes can be implemented during the initial synthesis stage, and thus integration of this advanced device structure is essentially no more difficult than a single component semiconducting NW or NT.

5. Nanowire Chemical and Biological Sensors

5.1. Underlying principle

Developments in nanotechnology will not only lead to improved electronics and make quantum jumps in electronics miniaturization, but could also enable significant advances in medicine and healthcare, which benefit mankind as a whole. For example, nanoscale devices could open up fundamentally new ways to probe single molecule properties, and allow *in-vivo* monitoring of diseases, such as cancer, which could lead to detection at the earliest and most treatable stage.

To this end, we have exploited our NW-based FETs for highly sensitive and selective detection of chemical and biological species [78]. Planar semiconductors have been used as the basis for many types of chemical and biological sensors in which detection can be monitored electrically and/or optically. For example, planar FETs can be configured as sensors by modifying the gate oxide with molecular receptors or a

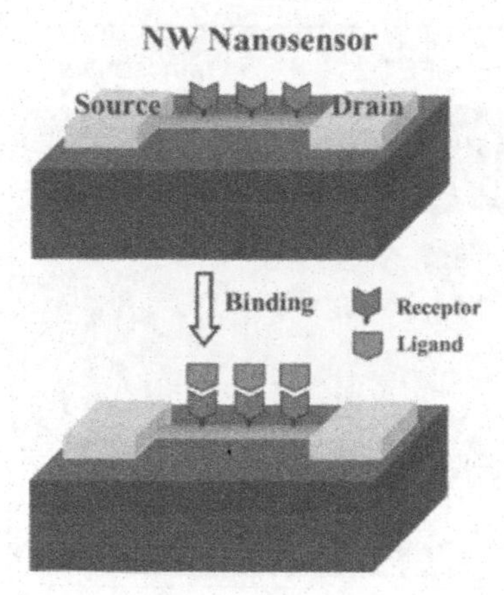

Fig. 31. Underlying concepts for a nanosensor based on NW-FETs. A solid state NW-FET is transformed into a sensor by modifying the NW surface with molecular receptors. Binding of a charged species (ligand) then results in a depletion or accumulation of carriers within the transistor structures, and changes the conductance of the NW-FET.

selective membrane for the analyte of interest [79, 80]. Binding of a charged species then results in a depletion or accumulation of carriers within the transistor structures. An attractive feature of such chemically sensitive FETs is that binding can be monitored by a direct change in conductance or related electrical property, although the sensitivity and potential for integration are limited. Our nanoscale sensors are based on a similar framework using nanowire-based FETs (Fig. 31), but have some obvious advantages compared to planar semiconductor sensors. First, the large surface to volume ratio can dramatically increase the sensitivity. Binding of chemical or biological species to the surface of NWs can lead to depletion or accumulation of carriers in the "bulk" of nanometer diameter structures versus only the surface region in a planar device, and increase the sensitivity to the single-molecule level. Second, high density arrays of sensors could be readily prepared owing to the extremely small size of the NWs and the rational assembly techniques discussed in earlier sections of this review. Lastly, the small size of these nanosensors could enable minimally invasive *in-vivo* sensors that could be readily coupled to other device functionality, such as telemetry and drug delivery systems. We have used Si-NW based FETs to demonstrate the principle and potential of such a nanoscale sensor approach.

5.2. Chemical sensors

The underlying concept for the NW nanosensor can be illustrated in general for the case of pH sensor (Fig. 32) [78]. Here a Si-NW solid state FET whose conductance is modulated by applied gate, is transformed into a nanosensor by modifying the silicon oxide surface with 3-aminopropyltriethoxysilane (APTES). The APTES provides a surface that can undergo protonation and deprotonation, where changes in surface charge (protonation) chemically gate the Si NW (Fig. 32a). The response of APTES-modified Si NWs to changes in solution pH (Fig. 32b) demonstrate that the

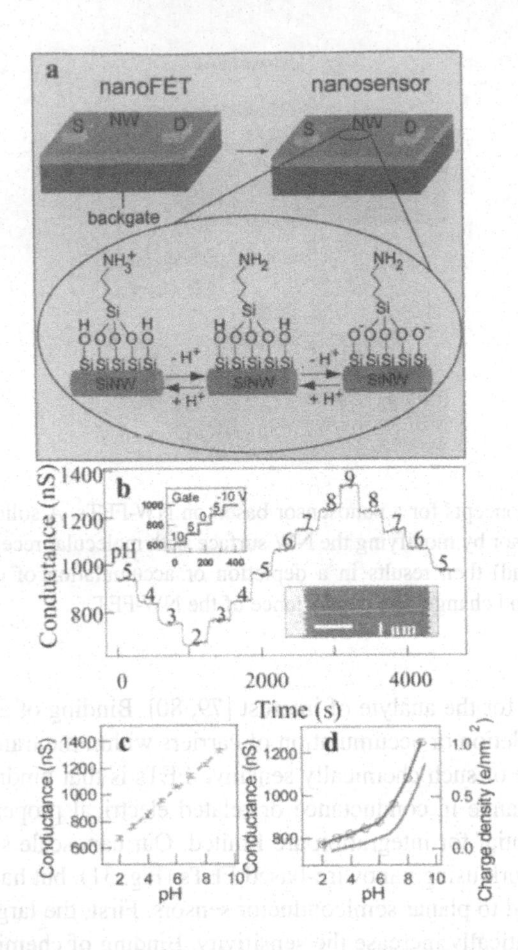

Fig. 32. NW pH detection. (a) Schematic illustrating the conversion of a NW-FET into a NW pH sensor. (b) Real-time reversible detection of the conductance changes of a modified Si NW as pH is changed from 2 to 9; the pH values are indicated on the conductance plot. (inset, top) Plot of the time-dependent conductance of a Si NW FET as a function of the back-gate voltage. (inset, bottom) SEM image of a typical device. (c) Plot of the conductance versus pH; the dashed line is linear fit through this data. (d) The conductance of an unmodified Si NW versus pH. The dashed curve is a plot of the surface charge density for silica as a function of pH. Adapted from [78].

NW conductance increases stepwise with discrete changes in pH from 2 to 9 and that the conductance is constant for a given pH. The changes in conductance are also reversible for increasing and/or decreasing pH. A typical plot of the conductance versus pH (Fig. 32c) shows that the pH dependence is linear over the pH 2 to 9 range and thus suggests that modified SiNWs could function as nanoscale pH sensors.

These results have been explained by considering the mixed surface functionality of the modified Si NWs. Covalently linking APTES to SiNW oxide surface results in a surface terminating in both $-NH_2$ and $-SiOH$ groups (Fig. 32a), which have different dissociation constants, p*K*a [38, 81]. At low pH, the $-NH_2$ group is protonated

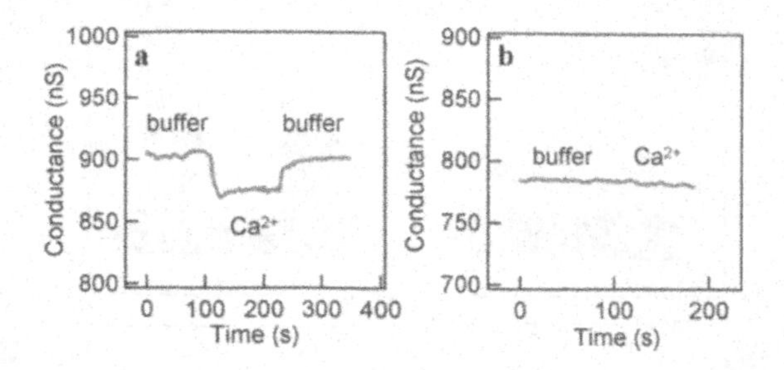

Fig. 33. Real-time detection of Ca^{2+} ions. (a) Plot of conductance versus time for a calmodulin- terminated Si NW, where "Ca^{2+} χορρεσπονδσ το τηε αδδιτιον οφ 25 μM Ca^{2+} solution. (b) Conductance versus time for an unmodified Si NW. Adapted from [78].

to $-NH_3^+$ and acts as a positive gate, which depletes hole carriers in the p-type SiNW and decreases the conductance. At high pH, –SiOH is deprotonated to $-SiO^-$, which correspondingly causes an increase in conductance. The observed linear response can be attributed to an approximately linear change in the total surface charge density (versus pH) because of the combined acid and base behavior of both surface groups. This point was supported with pH dependent measurements on unmodified (only –SiOH functionality) SiNWs, which showed a nonlinear pH dependence (Fig. 32d) similar to previous measurements of the pH-dependent surface charge density of silica [82].

The specific chemical detection in NW sensors can be realized by modifying NW with specific receptors or ligands [78]. For example, they can be used to sense calcium ions (Ca^{2+}), which are important for activating biological processes such as muscle contraction, protein secretion, cell death, and development [83]. To this end, a Ca^{2+} sensor was fabricated by immobilizing the calcium binding protein calmodulin onto Si NW surfaces. Conductance measurements made with such modified SiNW devices (Fig. 33a) showed a drop in the conductance upon addition of a 25 μM Ca^{2+} solution and a subsequent increase when a Ca^{2+}-free buffer was subsequently flowed through the device. The observed conductance decrease is consistent with expected chemical gating by positive Ca^{2+}, and the estimated dissociation constant, 10^{-5}–10^{-6} M, is in good agreement with the reported K_d for calmodulin. The control experiment on unmodified NW (Fig. 33b) show no conductance change when flowing in Ca^{2+} solution, demonstrating the detection is specific.

5.3. Biological sensors

The above results clearly demonstrate that NW-based FETs can be configured as chemical sensors. Since many biological molecules, such as proteins and DNA, are charged under physiological conditions, it is also possible to detect the binding of these macromolecules to NW surfaces. We have demonstrated detection of biologically relevant species through the detection of proteins and antibodies [78]. These critical examples are reviewed briefly below.

Initial investigations of protein detection by SiNW nanosensors were carried out for the well-characterized ligand-receptor binding of biotin-streptavidin (Fig. 34a)

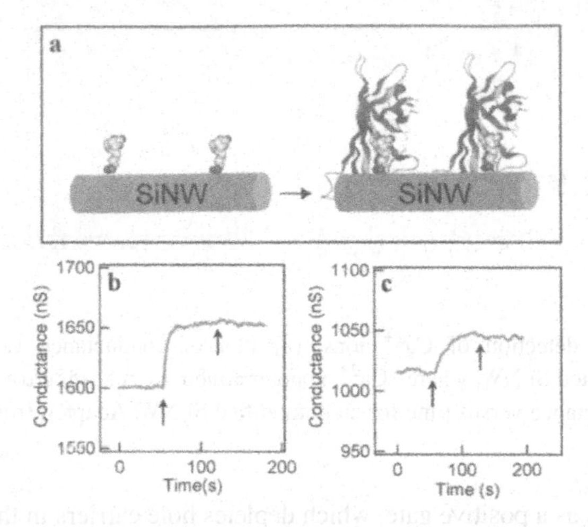

Fig. 34. Detection of protein binding. (a) Schematic illustrating a biotin-modified Si NW (left) and subsequent binding of streptavidin to the Si NW surface (right). The Si NW and streptavidin are drawn approximately to scale. (b) Plot of conductance versus time for a biotin-modified Si NW, where region first arrow corresponds to the addition of 250 nM streptavidin, and the second arrow corresponds to the addition of pure buffer solution. (c) Conductance versus time for a biotin-modified Si NW, where the two arrows have the same designation. The streptavidin solution had a concentration of 25 pM. Adapted from [78].

using biotin-modified NW surfaces [78]. Measurements showed that the conductance of biotin-modified SiNWs increased rapidly to a constant value upon addition of streptavidin solutions and that this conductance value was unchanged after the addition of pure buffer solution (Fig. 34b). The increase in conductance upon addition of streptavidin is consistent with binding of a negatively charged species to the p-type SiNW surface and the fact that streptavidin ($pI \sim 5$–6) [84] is negatively charged at the pH of these measurements. The absence of a conductance decrease with addition of pure buffer also agrees well with the small dissociation constant ($K_d \sim 10^{-15}$ M) and correspondingly small dissociation rate of biotin-streptavidin [84].

Significantly, the biotin-modified Si NWs nanosensors were also shown to exhibit very high sensitivity. Experiments indicate that it is possible to detect streptavidin binding down to a concentration at least as low as 10 pM (Fig. 34c). This detection level is substantially lower than the nanomolar range demonstrated recently by stochastic sensing of single molecules [85]. We note that a time dependent increase in the conductance can be resolved immediately after streptavidin addition at very low concentrations, which reflects contributions from the forward binding rate and/or diffusion, although future studies will be required to resolve these contributions.

The above studies demonstrate that our NW nanosensors are capable of highly sensitive and selective real-time detection of proteins, although the essentially irreversible biotin-streptavidin binding interaction precludes real-time monitoring of varying protein concentration. To explore this possibility, we studied the reversible binding of monoclonal antibiotin (m-antibiotin) to biotin [78]. Time-dependent conductance

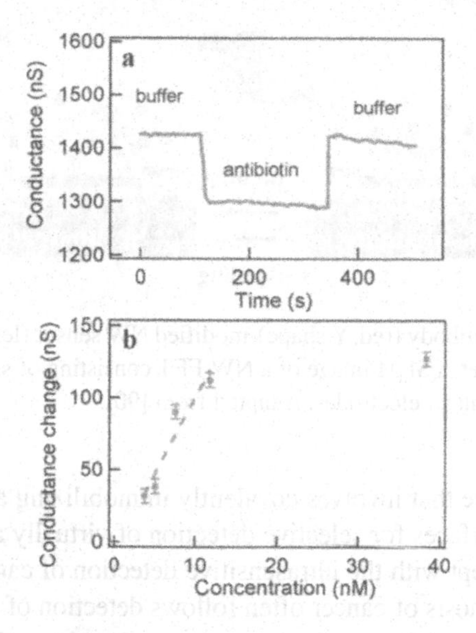

Fig. 35. Real-time detection of reversible protein binding. (a) Plot of conductance versus time for a biotin-modified SiNW for detection of ~3 μM m-antibiotin. (b) Plot of the conductance change of a biotin modified Si NW versus m-antibiotin concentration; the dashed line is a linear fit to the four low concentration data points. Adapted from [78].

measurements made on biotin-modified Si NWs (Fig. 35a) exhibit a well-defined decrease after addition of m-antibiotin solution followed by an increase in the conductance to about the original value upon addition of pure buffer solution. The decrease in conductance upon m-antibiotin addition is consistent with the binding m-antibiotin, which was positively charged at the pH 7 of these experiments [86]. The time scale for the increase in conductance (seconds), which we associate with m-antibiotin dissociation, was also consistent with the reported dissociation rate constant [87].

This reversible, real-time detection of m-antibiotin was further extended to monitor the Si NW sensor conductance as a function of m-antibiotin concentration in these studies [78]. These measurements showed a linear change in the conductance as a function of m-antibiotin concentration below ~10 nM and saturation at higher values (Fig. 35b), and thus demonstrate several important facts. First, from the linear regime, we estimate that the dissociation constant, K_d, is on the order of 10^{-9} M, which is in good agreement with the value, $\sim 10^{-9}$ M, determined previously [86]. Second, we can monitor protein concentration in real time, which could have important implications in basic research, for example, in monitoring protein expression, as well as in medical diagnostics. Lastly, preliminary experiments also suggest that the linear response dynamic range can be extended by applying a back gate to the NW, where gating can either increase or decrease the intrinsic binding constant.

5.4. Detection of cancer marker proteins

The capability of nanowire sensors for ultrasensitive detection opens up exciting opportunities in disease diagnostics and *in vivo* detection. We have recently developed

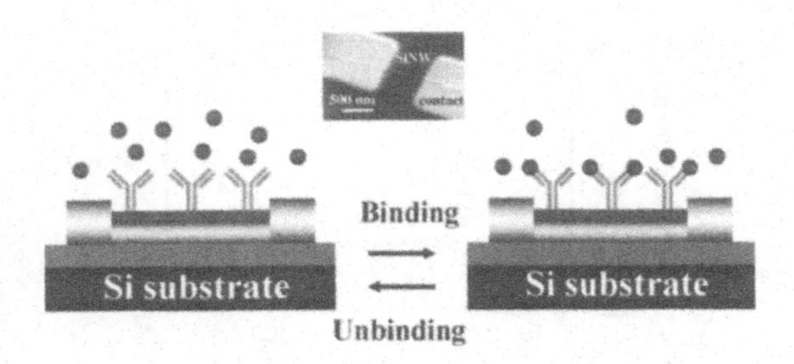

Fig. 36. Diagram of an antibody (red, Y shape)-modified NW sensor (left) for detecting antigen (blue) binding (right). Inset, a SEM image of a NW FET consisting of single SiNW covered by two (source and drain) contact electrodes. Adapted from [90].

a general sensor scheme that involves covalently immobilizing antibodies onto semiconductor nanowire surfaces for selective detection of virtually any protein (Fig. 36). We illustrate this concept with the ultrasensitive detection of cancer marker proteins.

Currently, the diagnosis of cancer often follows detection of a mass that is resolvable by anatomic imaging, frequently many years after the earliest stages of cancer development [88]. It is based on gross changes in size and shape of tissues and cells, rather than fundamental changes in the molecular processes that underlie the development of disease. Specificity and effectiveness of diagnosis could be dramatically improved if they were directed towards critical changes in certain molecular processes that result in the cancer phenotype. For instance, prostate cancer is the most common cancer in men and the second leading cause of cancer death in United States. Prostate specific antigen (PSA) level in serum has proved to be an extremely useful marker for early detection of prostate cancer and in monitoring patients for disease progression and the effects of treatments. It is highly desirable to have sensors to detect serum free PSA (fPSA) and PSA-complex (PSA-ACT) level at ng/ml range for clinical screening and very sensitive range (pg/ml) for detecting the prostate cancer at its earliest stage and monitoring the effect of treatment. The current PSA detection techniques are enzyme-linked immunosorbent assay (ELISA) or radioimmunoassay (RIA) [88, 89], both of which require cumbersome fluorescence or radioactive labeling. Herein, we demonstrate the label-free, fast, ultrasensitive multiplexing detection of cancer markers with NW sensors.

The underlying strategy for a generic protein sensor scheme (Fig. 36) is to immobilize their antibodies onto the nanowire surface and use them for specific antigen (protein) detection [90]. Since virtually antibody to any protein can be raised with current biotechnology, NW can be modified to detect almost any protein.

As a first example of cancer marker detection, we have looked at fPSA sensing using monoclonal anti-fPSA antibody (PSA-AB1) modified nanowires [90]. Measurements of nanowire conductance as a function of time (Fig. 37a) demonstrate that the conductance increases rapidly to a nearly constant value upon flowing in 500 pg/ml fPSA (region 1) and returns to the initial low level after replacing fPSA with buffer solution. The increase in conductance is consistent with the binding of a negatively charged species to the p-type SiNW surface and the fact that fPSA (p*I* 6.1) [91] is negatively charged at the pH of our measurements. The decrease of conductance

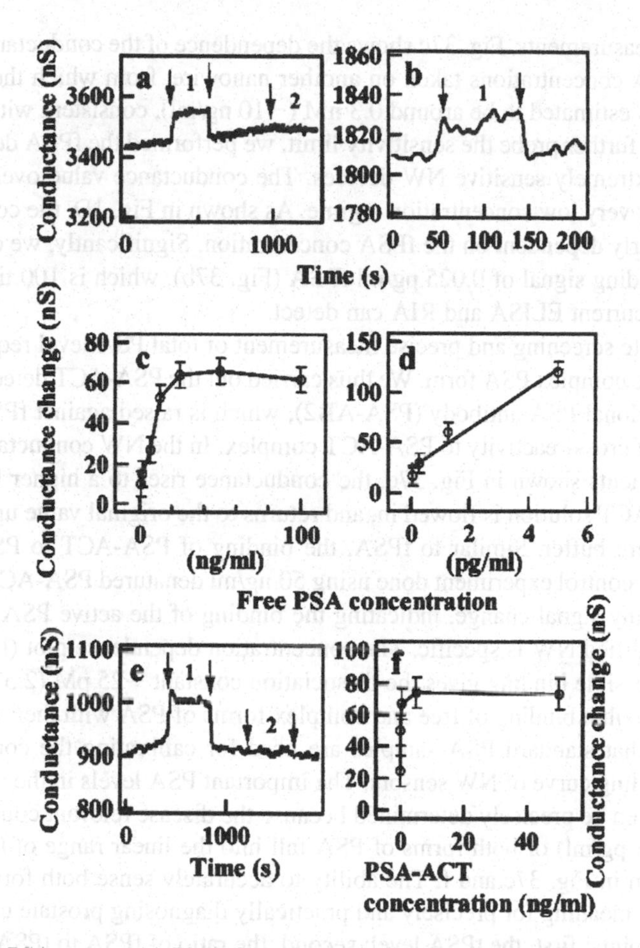

Fig. 37. NW PSA sensor. (a) Plot of conductance versus time for fPSA antibody-modified SiNW. Region 1 corresponds to 500 pg/ml fPSA contacting with NW, region 2 corresponds to 50 ng/ml BSA. (b) Conductance versus time curve showing the signal of the lowest concentration detectable where region 1 is 0.025 pg/ml fPSA, which corresponds to the lowest concentration data point in (d). (c) Concentration-dependence curve for fPSA antibody-modified SiNW. Each data point has a standard deviation (SD) of 6 nS. (d) Probing the detection limit of NW fPSA sensor below 10 pg/ml (SD=10 nS). The line is the linear fit with R^2 value 0.97. (e) Conductance versus time for PSA-ACT antibody-modified SiNW. Region 1 corresponds to 5 pg/ml PSA-ACT, region 2 corresponds to 50 ng/ml denatured PSA-ACT. (f) Concentration-dependence curve for PSA-ACT antibody-modified SiNW (SD=10 nS). For (a), (b), and (e), unlabeled regions are buffer and arrows indicate the points when solutions were changed. Adapted from [90].

with addition of pure buffer suggests that the binding between fPSA and PSA-AB1 is reversible. To ensure that the binding signal is due to specific binding, we carried out a control experiment: flowing 50 ng/ml bovine serum albumin (BSA) (region 2) did not cause any conductance change. This demonstrates that the binding signal is due to specific binding of fPSA onto its PSA-AB1 instead of unspecific adsorption. The reversible binding between them allows us to do the real-time concentration dependence measurements. Fig. 37c shows the dependence of the conductance change values on fPSA concentrations taken on another nanowire, from which the dissociation constant is estimated to be around 0.3 nM (~10 ng/ml), consistent with the literature value. To further probe the sensitivity limit, we performed the fPSA detection on some of our extremely sensitive NW devices. The conductance value over time was recorded under very low concentration regime. As shown in Fig. 1D, the conductance change is linearly dependent on the fPSA concentration. Significantly, we can clearly resolve the binding signal of 0.025 pg/ml fPSA (Fig. 37b), which is 100 times lower than what the current ELISA and RIA can detect.

The accurate screening and precise measurement of total PSA level requires us to be able to sense complex PSA form. We thus carried out the PSA-ACT detection using another monoclonal PSA antibody (PSA-AB2), which is raised against fPSA but has some degree of cross-reactivity to PSA-ACT complex. In the NW conductance versus time measurements shown in Fig. 37e, the conductance rises to a higher level when 5 pg/ml PSA-ACT solution is flowed in, and returns to the original value upon changing back to pure buffer. Similar to fPSA, the binding of PSA-ACT to PSA-AB2 is reversible. The control experiment done using 50 ng/ml denatured PSA-ACT complex did not show any signal change, indicating the binding of the active PSA-ACT onto PSA-AB2-modified NW is specific. The concentration dependence plot (Fig. 37f) of PSA-ACT reversible binding gives the dissociation constant ~25 pM (2.5 ng/ml).

In the reversible binding of free and complex forms of PSA with their antibodies, it is possible that standard PSA samples are used for calibrating the concentration dependent binding curve of NW sensors. The important PSA levels in the real analyte samples can then be precisely determined because the disease relevant concentrations (~4 ng/ml and pg/ml) of both forms of PSA fall into the linear range of the binding curve as shown in Fig. 37c and f. The ability to accurately sense both forms of PSA has significant meaning for precisely and practically diagnosing prostate cancer from several standpoints: first, the tPSA level; second, the ratio of fPSA to tPSA; third, the real time evolution of PSA level and the ratio of fPSA to tPSA.

The above results demonstrate that our individual NW sensors modified with antibody are capable of highly sensitive and selective real-time detection of any protein. To take this a step further forward, we need to integrate the nanowires into array-based high throughput sensing, which is important for multiple serum tumor marker screening, currently cost prohibitive. The schematics of this important sensing concept using NW sensors is illustrated in Fig. 38a. The nanowires are assembled into regular arrays and then selectively functionalized with different specific antibodies targeting different proteins. This strategy can enable us to identify each component in complex mixtures (such as human blood and cell lysate) simultaneously. To prove this concept, we investigated the multiplexing detection of fPSA and PSA-ACT. The two NWs on the wafer were independently modified with PSA-AB1 and -AB2 using two separate microfluidic channels, followed by washing thoroughly with buffer. In the subsequent detection, the flow solution in both microfluidic channels was kept identical at any given time. Measurements of conductance versus time with changes of the solution were recorded as shown in Fig. 38b. When 50 ng/ml fPSA was flowed in (region 1), the conductances of both NWs went up and were stable at a higher value and decreased back to the original low value after changing back to pure buffer solution.

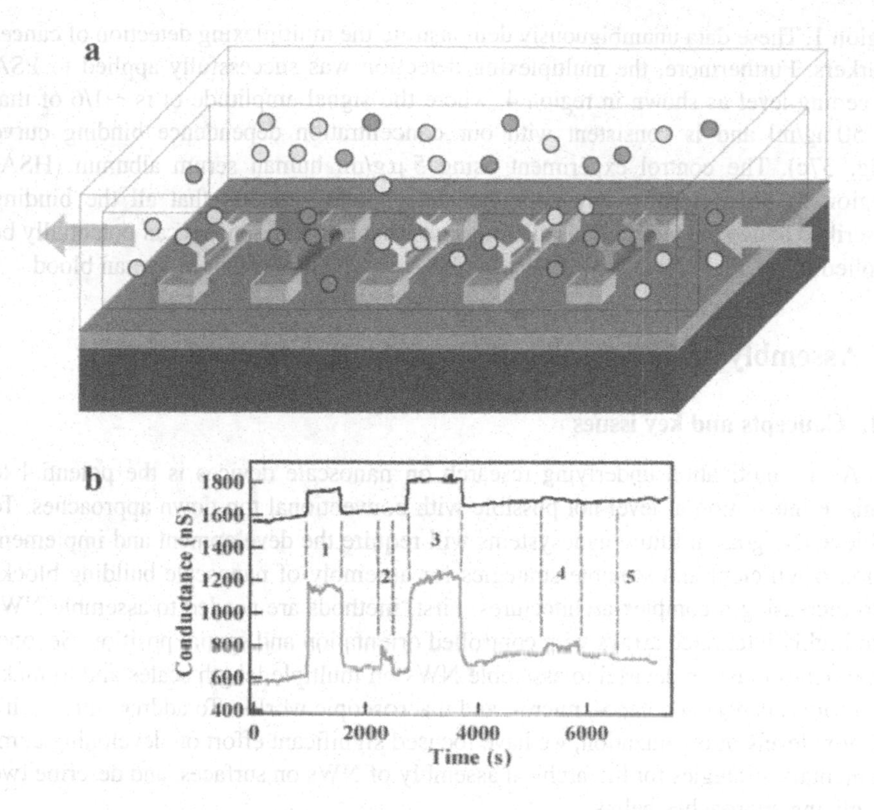

Fig. 38. NW multiplexing sensor. (a) Schematic illustrating the concept of multiplexing. Multiple SiNWs can be differentially modified and the analyte solution can be introduced to all the NWs simultaneously with a common microfluidic channel. The transparent part stands for a microfluidic channel and the blue arrow points toward the flow direction. (b) Plot of conductance versus time for simultaneous real-time detection of fPSA and PSA-ACT. The fPSA antibody-modified SiNW (black) used an antibody specific for fPSA only whereas the PSA-ACT antibody-modified SiNW (red) used an antibody raised against fPSA, but has cross-reactivity with PSA-ACT. Region, 1, 2, 3, 4, and 5 correspond to 50 ng/ml fPSA, 50 ng/ml PSA-ACT, 50 ng/ml fPSA+50 ng/ml PSA-ACT, 4 ng/ml fPSA, and 5 μg/ml HAS, respectively. Unlabeled regions correspond to buffer. Arrows indicate the points when solutions were changed. Adapted from [90]. See color plate 2.

This indicates that both AB1 and AB2 have reversible binding affinity to fPSA consistent with our previous results (Fig. 37). Significantly, adding 50 ng/ml PSA-ACT complex (region 2) and subsequent buffer flow results in the conductance of AB2-modified NW (red curve) increasing to a stable level and subsequently decreasing to the initial low value while that of AB1-modified NW remained stable at the same value. This agrees well with that AB2 has cross reactivity with PSA-ACT but AB1 does not. The signal of PSA-ACT binding onto AB2-nanowire is smaller than that of fPSA at the same concentration, which is due to the partial cross-reactivity of PSA-AB2 with PSA-ACT. In region 3, flowing in the mixture of 50 ng/ml fPSA and 50 ng/ml PSA-ACT produces the conductance increase on both NWs similar as in region 1. These data unambiguously demonstrate the multiplexing detection of cancer markers. Furthermore, the multiplexing detection was successfully applied to PSA screening level as shown in region 4, where the signal amplitude of is ~1/6 of that of 50 ng/ml and is consistent with our concentration dependence binding curve (Fig. 37c). The control experiment using 5 μg/ml human serum albumin (HSA) (region 5) did not show any noticible change and supports that all the binding described above is specific. This also suggests that our NW sensors can potentially be applied to the real blood sample because HSA is a major protein in human blood.

6. Assembly and Hierarchical Organization of Nanowires

6.1. Concepts and key issues

A key motivation underlying research on nanoscale devices is the potential to achieve integration at level not possible with conventional top-down approaches. To achieve this goal in future nanosystems will require the development and implementation of efficient and scalable strategies for assembly of nanoscale building blocks into increasingly complex architectures. First, methods are needed to assemble NWs into highly integrated arrays with controlled orientation and spatial position. Second, approaches must be devised to assemble NWs on multiple length scales and to make interconnects between nano-, micro-, and macroscopic worlds. To address these critical next levels of organization, we have focused significant effort on developing complementary strategies for hierarchical assembly of NWs on surfaces, and describe two promising approaches below.

6.2. Electrical field directed assembly

Applied electric fields (E-fields) can be used effectively to attract and align NWs due to their highly anisotropic structures and large polarizabilities (Fig. 39) [33]. This underlying idea of E-field directed assembly can be readily seen in images of NW solutions aligned between parallel electrodes (Fig. 39b), which demonstrate that virtually all of the NWs aligned in parallel along the E-field direction. E-field directed assembly can also be used to position individual NWs at specific positions with controlled directionality. For example, E-field assembly of NWs between an array of electrodes (Fig. 39c) clearly shows that individual NWs can be positioned to bridge pairs of diametrically-opposed electrodes and form a parallel array. In addition, by changing the E-field direction with sequential NW solutions, the alignment can be carried out in a layer-by-layer fashion to produce crossed NW junctions (Fig. 39d). These results demonstrate clearly that E-field directed assembly can be used to align and position individual NWs into parallel and crossed arrays, which correspond to two basic geometries for integration, and thus provide one robust approach for rational and parallel assembly of nanoscale device arrays.

6.3. Fluid flow directed assembly

E-field directed assembly, which represents the first approach described for assembly of 1D nanostructures, also has limitations, including (i) the need for

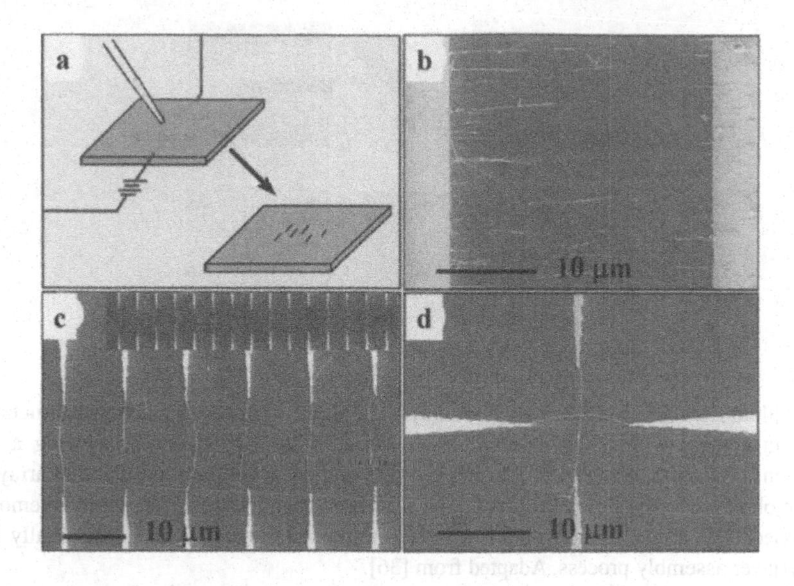

Fig. 39. E-field directed assembly of NWs. (a) Schematic view of E-field alignment. (b) Parallel array of NWs aligned between two parallel electrodes. (c) Spatially positioned parallel array of NWs obtained following E-field assembly. The top inset shows 15 pairs of parallel electrodes with individual NWs bridging each diametrically opposed electrode pair. (d) Crossed NW junction obtained using layer-by-layer alignment with the E-field applied in orthogonal directions in the two assembly steps. Adapted from [33].

substantial conventional lithography to pattern microelectrode arrays used to produce aligning fields, and (ii) the deleterious effect of fringing electric fields at the submicron length scales. To achieve a greater flexibility in rational, parallel assembly of ID nanostructures into nanosystems, we have developed a powerful new approach called fluidic flow directed assembly [36]. In this method, NWs (or NTs) can be easily aligned by passing a suspension of NWs through microfluidic channel structures, for example, formed between a poly (dimethylsiloxane)(PDMS) mold and a flat substrate (Fig. 40). Parallel and crossed NW arrays can be readily created using single (Fig. 40a) and sequential crossed (Fig. 40b) flows, respectively, for the assembly process.

Images of NWs assembled on substrate surfaces (Fig. 41a) within microfluidic flows demonstrate that virtually all of the NWs are aligned along the flow direction. This alignment readily extends over hundreds of micrometers (Fig. 41b). Indeed, alignment of the NWs has been found to extend up to millimeter length scales, and is limited only by the size of the fluidic channels used. The alignment of NWs within the channel flow has been explained within the framework of shear flow [92, 93]. Specifically, the channel flow near the substrate surface resembles a shear flow and aligns the NWs in the flow direction before they are immobilized on the substrate. This idea readily provides the intellectual underpinning needed for controlling the degree of alignment and average separation of the NWs.

The fluidic flow assembly approach can be used to organize NWs into more complex crossed structures, which are critical for building dense nanodevice arrays, using

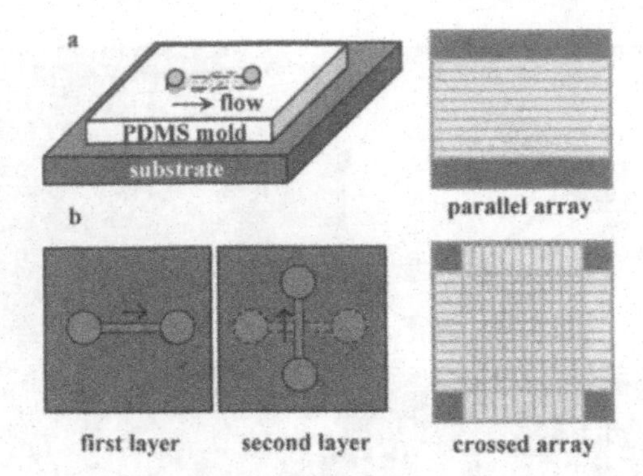

Fig. 40. Fluid flow directed assembly NWs. (a) A channel is formed when a trench structure is brought in contact with a flat substrate. NW assembly is carried out by flowing a NW suspension through the channel at a controlled rate and for a set duration. Parallel arrays of NWs are observed in the flow direction on the substrate when the trench structure is removed. (b) Crossed NW arrays can be obtained by changing the flow direction sequentially in a layer-by-layer assembly process. Adapted from [36].

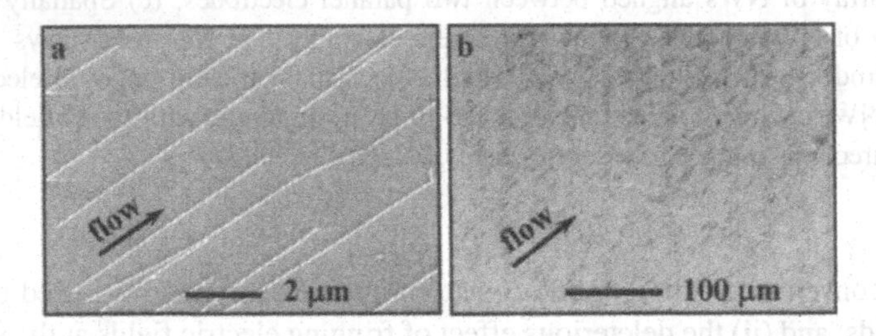

Fig. 41. Parallel assembly of NW arrays. (a) SEM image of a parallel array of InP NWs aligned by flow. (b) Optical microscope image of a parallel array of InP NWs aligned over very large area. Adapted from [36].

a layer-by-layer deposition process (Fig. 40b). The formation of crossed and more complex structures requires that the nanostructure-substrate interaction is sufficiently strong that sequential flow steps do not affect preceding ones: we find that this condition is readily achieved. For example, alternating the flow in orthogonal directions in a two-step assembly process yields crossbar structures in high yield (Fig. 42a, b). These data demonstrate that crossbars extending over 100s of microns on a substrate with only 100s of nanometers separation between individual cross points are obtained through a very straightforward, parallel low cost and fast process.

Fluidic flow directed assembly of multiple crossed NW arrays offers significant advantages over previous efforts. First, it is intrinsically very parallel and scalable with the alignment readily extending over very large length scales. Second, this approach is general for virtually any elongated nanostructure including carbon nanotubes [94] and DNA molecules [95]. Third, it allows for the directed assembly of geometrically complex structures by simply controlling the angles between flow

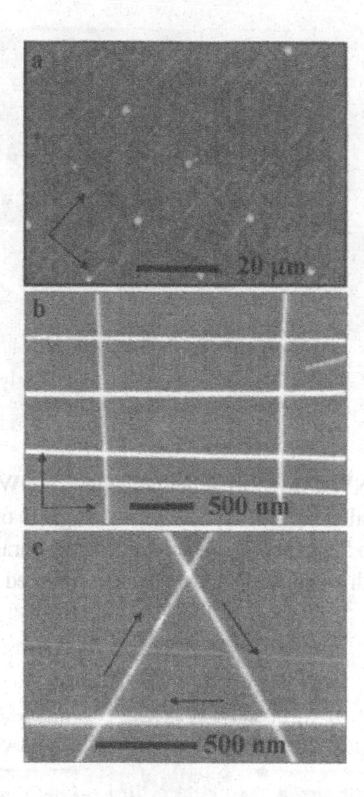

Fig. 42. Layer-by-layer assembly of crossed NW arrays. (a, b) Typical optical microscope and SEM images of crossed arrays of InP NWs obtained in a two-step assembly process with orthogonal flow directions for the sequential steps. Arrows indicate the two flow directions. (c) An equilateral triangle of GaP NWs obtained in three-step assembly process, with flow directions highlighted by arrows. Adapted from [36].

directions in sequential assembly steps. For example, equilateral triangles (Fig. 42c) were easily assembled in a three-layer deposition sequence using 60° angles between the three flow directions. The method of flow alignment thus provides a flexible way to meet the requirements of many device configurations in the future. An important feature of this layer-by-layer assembly scheme is that each NW layer can be independent of the preceding one(s), and thus a variety of homo- and hetero-junction configurations can be obtained at each crossed point by simply changing the composition of the NW suspension used for each flow step. For example, it should be possible to assemble directly and subsequently address individual nanoscale devices using our approach with n-type and p-type NWs, in which the NWs act as both the wiring and active device elements (see below).

The above results demonstrate clearly the power of the fluidic assembly approach, although to enable systems organization with greatest control requires in many cases that the spatial position also be defined. To realize this additional constraint on the assembly process we have explored complementary chemical interactions between chemically patterned substrates and NWs (Fig. 43a). Substrates for alignment are first patterned with two different functional groups, with one of the functional groups

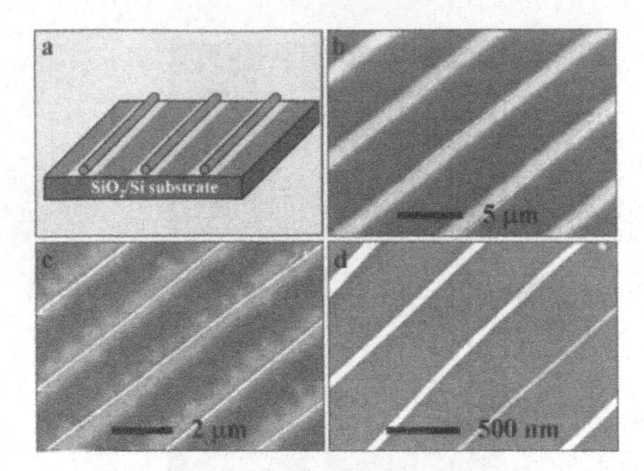

Fig. 43. Assembly of periodic NW arrays. (a) Schematic view of NW assembly onto a chemically patterned substrate. (b, c) Parallel arrays of GaP NWs aligned on poly(methylmethacrylate) (PMMA) patterned surface with 5 μm and 2 μm separation. (d) Parallel arrays of GaP NWs with 500 nm separation obtained with a patterned SAM surface. Adapted from [36].

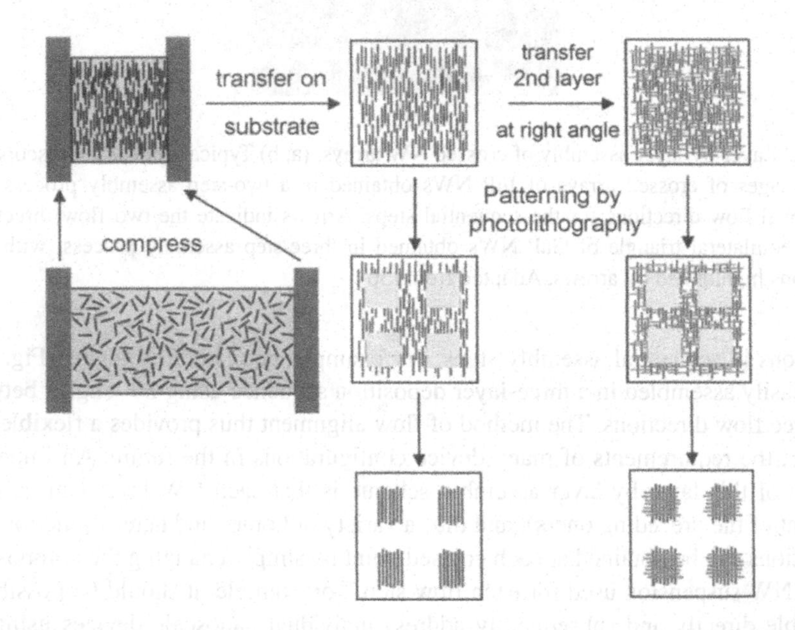

Fig. 44. The schematic showing the LB assembly process. Adapted from [96].

designed to have a strong attractive interaction with the NW surface, and then following flow alignment, regular, parallel NW arrays with lateral periods the same as those of the surface patterns are produced (Fig. 43b–d). These data demonstrate that the NWs are preferentially assembled at positions defined by the chemical pattern, and moreover, show that the periodic patterns can organize the NWs into regular superstructures. In addition, periodic crossed NW arrays can also be envisioned using a substrate with a crossed pattern of chemical functionality.

6.4. Large scale integration: Langmuir-Blodgett techniques

Microfluidic method enables the alignment of NWs up to ca. the mm scale. Assembly on larger scales and with precise spacing down to nanometer level requires alternative strategies. The Langmuir-Blodgett (LB) technique, in which an ordered monolayer is formed over a large area, represents an alternative method that may achieve this goal. Here we explore the assembly of nanowires using the LB technique [96].

The underlying concept of our approach is illustrated in Fig. 44. In general, we induce alignment of NWs by uniaxial compression of the monolayer. Fig. 45 shows aligned NWs transferred onto a modified silicon (Si/SiO_2) substrates. The LB aligned

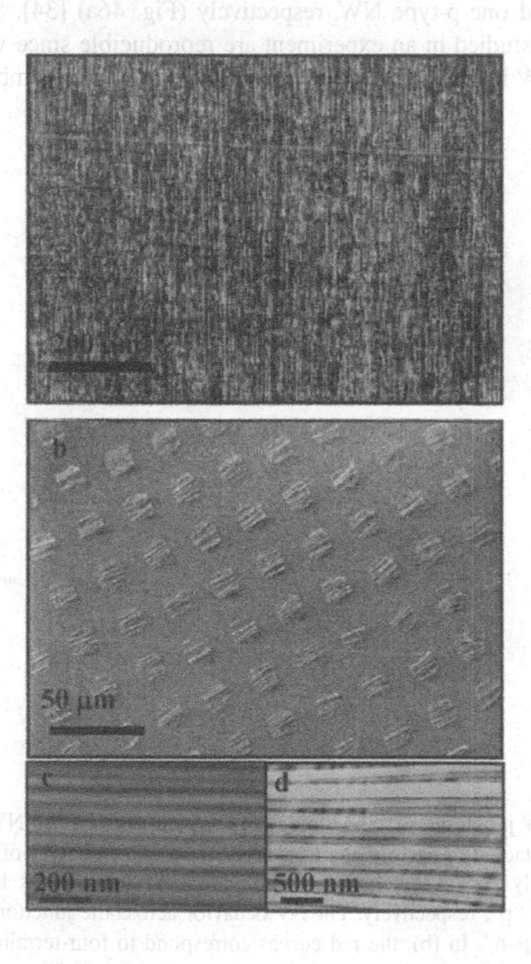

Fig. 45. The assembled NW arrays using LB method. Parallel arrays on (a) large area and (b) patterned substrate. The nanometer scale pitch can be obtain in (c) parallel and (d) crossed arrays. Adapted from [96].

NWs cover ~2×10 cm^2 area, although the approach can be easily applied on much larger surfaces. Moreover, hierarchical patterning of the parallel and crossed arrays can be readily implemented using this approach (Fig. 45b), and the NW separation can be controlled down to the nanometer level (Fig. 45c, d).

7. Crossed and Integrated Nanowire Devices

7.1. Crossed nanowire p–n junctions

The availability of well-defined n- and p-type NW building blocks opens up the possibility of creating complex functional devices by forming junctions between two or more wires. To explore this exciting opportunity, we have studied the transport behavior of n–n, p–p, and p–n junctions formed by crossing two n-type, two p-type, and one n-type and one p-type NW, respectively (Fig. 46a) [34]. Significantly, the types of junctions studied in an experiment are reproducible since we can select the specific type of NW used at each of the two stages of device assembly.

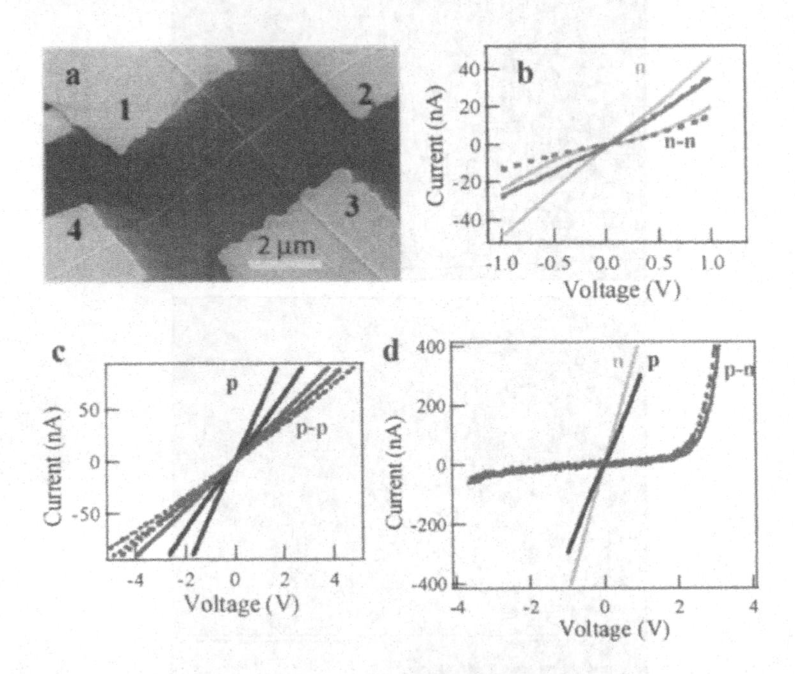

Fig. 46. Crossed NW junctions. (a) SEM image of a typical crossed InP NW device with four metal electrodes contacted to each of the four arms. (b–d) *I-V* behavior of n–n, p–p and p–n junctions, respectively. The *I-V* behavior of individual n- and p-NWs in the junctions is indicated by "n" and "p", respectively. The *I-V* behavior across the junctions is designated by "n–n", "p–p"' and "p–n". In (b), the red curves correspond to four-terminal *I-V* through the p-n junction; the current values are multiplied by 10. The solid lines represent transport behavior across one pair of adjacent arms, and the dashed lines represent that of the other three pairs of adjacent arms. Adapted from [34].

First, *I-V* data recorded on the individual NWs in n–n and p–p crossed junctions show linear or nearly linear *I-V* behavior (Fig. 46b, c), indicating that the metal electrodes used in the experiments make ohmic or nearly ohmic contact to the NWs. This point is important since it shows that the NW-metal contacts will not make nonlinear contributions to the *I-V* measurements across the nanoscale junctions. In general, transport measurements recorded across the n–n and p–p junctions show linear or nearly linear behavior. These results indicate that interface oxide between individual NWs does not produce a significant tunneling barrier since a tunneling barrier would lead to highly non-linear *I-V* behavior. In addition, the *I-V* curves recorded through each pairs of adjacent arms show similar current levels, which are smaller than that of the individual NWs themselves, demonstrating that the junction dominates the transport behavior. Taken as a whole, these data show that individual NWs can make good electrical contact with each other, despite the small contact area (10^{-12}–10^{-10} cm^2) and simple method of junction fabrication.

Initial studies designed to probe the utility of this new approach for creating functional devices were focused on p–n junctions from crossed p- and n-type NWs. These junctions can be made reproducibly by sequential deposition of dilute solutions of n- and p-type NWs with intermediate drying. Typical *I-V* behavior of a crossed Si NW p–n junction is shown in Fig. 46d. The linear *I-V* of the individual n- and p-type NWs components indicates ohmic contact between the NWs and metal electrodes, while transport across the p–n junction shows clear current rectification; that is, little current flows in reverse bias, while there is a sharp current onset in forward bias. Significantly, this behavior is similar to conventional semiconductor p–n junctions. In a standard p–n junction, rectification arises from the potential barrier formed at the interface between p- and n-type materials. In the case of our crossed nanowire p–n junctions, this picture is probably modified due to the presence of some interface oxide, although a thin oxide will not change substantially the overall *I-V* response.

The assignment of the observed rectification to the p–n junction formed at the crossing point between p- and n-type InP NWs was further supported by several other pieces of evidence. First, the linear or nearly *I-V* behavior of individual p- and n-type NWs shows that ohmic contacts were been made between the NWs and metal electrodes, and thus exclude the possibility that rectification arises from metal-semiconductor Schottky diodes [97]. Second, the *I-V* behavior of the junction determined through each pair of adjacent electrodes exhibit similar rectification and current level, which is also much smaller than the current through individual NWs, demonstrating that the junction dominates the *I-V* behavior. Third, four terminal measurements (Fig. 46d) in which current is passed through two adjacent electrodes while the junction voltage drop is measured across the two remaining electrodes exhibit similar I-V and rectification with only a slightly smaller voltage drop (0.1–0.2 V) compared to two terminal measurements at the same current level. Fourth, measurements made on over forty independent Si crossed p–n junctions showed similar rectification in the I-V data. Lastly, the formation of crossed NW p–n junctions is not by any means restricted to Si NWs, and is general to the wide-range of materials described in section 2. For example, p–n junctions have been assembled from p-InP/n-InP [33], p-Si/n-InP [98], and p-Si/n-GaN [35] NWs, and transport measurements have demonstrated that all of these crossed p–n junctions show consistent current rectification behavior.

7.2. Bipolar junction transistors

Since p–n junctions represent a basic element in many functional electronics devices, including amplifiers and switches, we have explored the possibility of assembling such devices at nanometer scale using the well-defined p- and n-type NW materials. As an example, integrated bipolar transistors [34], which are active devices capable of current gain, have been assembled from three distinct types of Si NWs in the form of two crossed junctions (Fig. 47a). A conventional bipolar transistor requires three distinct material types. For example, in n^+-p–n structure, a highly doped

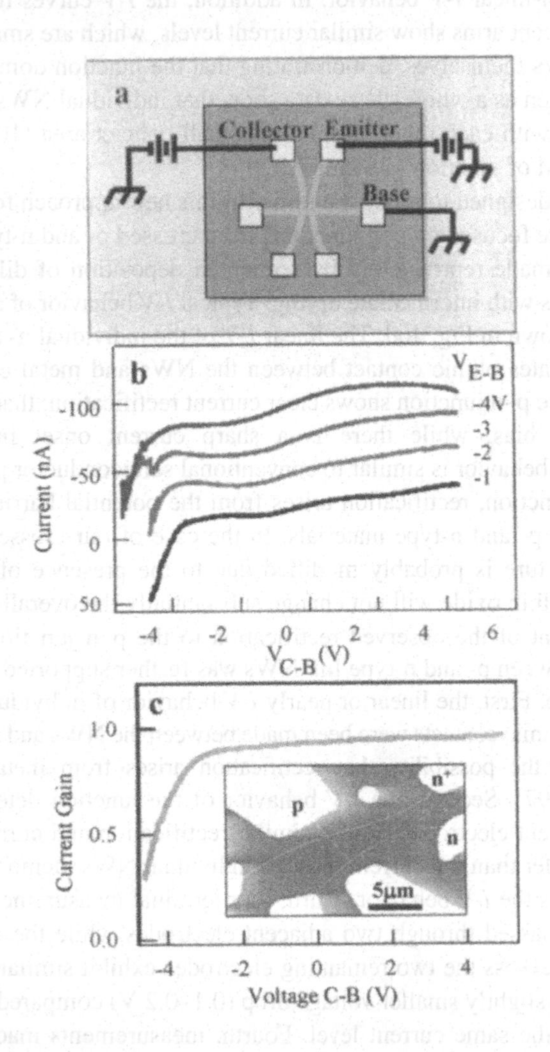

Fig. 47. Bipolar junction transistors. (a) Schematic illustrating the common base configuration of an n^+-p-n bipolar transistor built from crossed Si NWs. (b) Collector current versus collector-base voltage recorded on an n^+-p-n transistor with emitter and collector Si NWs 15 µm apart. The numbers inside the plot indicate the corresponding emitter-base voltage. (c) The common base current gain versus collector-base voltage. Inset: typical SEM image of Si-NW bipolar transistors. Adapted from [34].

n^+ layer is used as an emitter (E), a p-type layer for the base (B), and an n-type layer for the collector (C) [97]. Significantly, this n^+-p–n basic structure can be easily assembled with Si NWs since NWs with controlled doping type and doping concentration are available (inset, Fig. 47c) [32].

To characterize the electrical behavior of assembled NW bipolar transistors, the individual Si NW building blocks and p–n^+ and p–n junctions were first tested, and found to exhibit ohmic or nearly ohmic metal contacts and rectification, respectively. The bipolar transistor characteristics were then assessed from measurements of the collector current as a function of C-B voltage (Fig. 47b), while the n^+ Si NW emitter was biased at different values. In general, the collector current is relatively constant (versus C-B voltage) in the region from 0 to 6 V, which corresponds to the collector in reverse bias with only a very small leakage current, and this current value increases as the emitter forward bias/injected current is increased. The large collector current in reverse bias demonstrates these simple Si-NW-based bipolar transistors exhibit behavior similar to that found in standard planar devices, and moreover, can exhibit very good current gain.

The common base current gain, which is defined as the ratio of the collector current to emitter current (Fig. 47c), and the common emitter current gain, which is defined as the ratio of the collector current to base current, were found to be 0.94 and 16, respectively. The relatively large current gain observed in these simple devices suggests several important points. First, the efficiency of electron injection from emitter to base must be quite high, and can be attributed to the controlled NW doping that yields the desired n^+-p E–B junction. Second, large current gains have been achieved in devices with large (e.g., 15 μm) base widths. This fact suggests that the mobility of injected electrons can be quite high in the Si NWs, and is consistent with the direct mobility studies described above. These observations also indicate clear directions for improving the Si NW bipolar transistors. For example, it will be interesting to study the current gain as a function of base width, because it is easily possible to assemble structures with separations of the n^+ and n NWs on the order of 100 nm or less.

7.3. Crossed nanowire FETs

A major motivation underlying research on nanoscale devices is to achieve integration at densities higher than possible with current technologies. The NW-FETs discussed above represent nanoscale analogs to conventional MOSFETs, and have been very useful for testing basic device behavior (e.g., doping type and carrier mobility). However, the basic device structure of the NW-FETs and similar nanotube FETs [21, 22] requires lithography to define metallic source-drain electrodes, and use either a global back gate (i.e., the doped silicon substrate) or lithography to define a more local gate. These design and fabrication features pose serious problems for integration. First, lithographically defined metal electrodes (i.e., source, drain, and gate) will limit integration to a level similar to that of conventional silicon technology. Moreover, the use of global back gate electrodes eliminates the possibility of independently addressing individual devices, and thus is incompatible with integration in most architectures.

Direct assembly of highly integrated functional electronic circuits based on NWs requires the development of new device concepts that are amenable to scalable integration. To this end we recently developed a novel crossed NW based FET [35].

A crossed NW FET (cNW-FET) is assembled from two NWs where one or both have an oxide coating that serves as the gate dielectric (Fig. 48a, b). This approach is quite flexible since nano-FETs can be readily assembled with p- or n-type active channel NWs and the gate NW can also be p- or n-type independent of the channel. For example, using n-type GaN crossed NW as the gate for a p-type Si-NW, a p-channel cNW-FET is formed with both a nanoscale channel and a nanoscale gate. Typical I-V_{sd} data recorded for different NW V_g's resemble the characteristics of a conventional depletion mode p-channel FET device (Fig. 48c). Notably, the conductance of the Si NW responds very sensitively to the voltage applied to n-NW gate, and can be changed by more than five orders of magnitude with a 1–2 volt variation in the NW gate (Fig. 48d). In contrast, the conductance changed less than a factor of 10 for this same device when similar gate voltages were applied to the global back gate (Fig. 48d). The improved sensitivity can be attributed to the intrinsically thin gate dielectric between the two NWs. In addition, there is no leakage current from the n-NW gate when the cNW-FET is operated at low source drain bias in the deletion mode since the crossed p–n junction is always reverse biased, and in this regard the device is similar to junction field effect transistors (JFET) [97].

The cNW-FET represents an important new transistor concept for nanoelectronics. With this concept, three critical nanometer scale device metrics are naturally

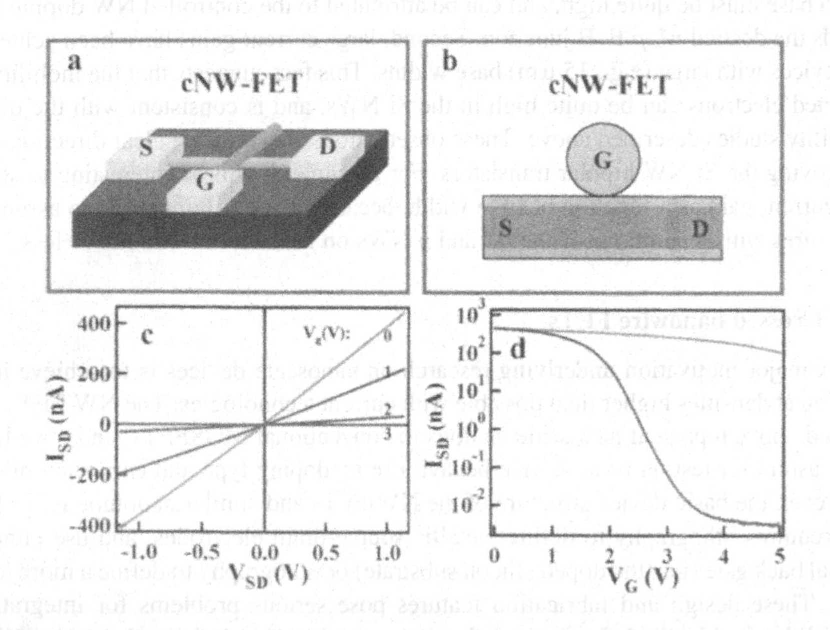

Fig. 48. Crossed NW FET (cNW-FET). (a) Schematics illustrating the cNW-FET concept. A nanoFET with both nanoscale conducting channel and nanoscale gate is obtained with one NW used as the gate for the other NW in a crossed configuration. (b) Schematic showing the critical device dimensions of the cNW-FET. Three intrinsic nanometer scale metrics are naturally defined by the structure (see text). (c) Gate-dependent I-V_{sd} characteristics of a crossed NW-FET. The NW gate voltage for each I-V curve is indicated (0, 1, 2, and 3 V). (d) The curves showing I versus V_g for n-NW and global back (light gray) gates for V_{sd} of 1 V. Adapted from [35].

defined in assembled circuits without lithography: (1) a nanoscale channel width determined by the diameter of the active NW; (2) a nanoscale channel length defined by the crossed NW gate diameter; and (3) a nanoscale gate dielectric thickness determined by the NW surface oxide. Significantly, these distinct nanometer scale metrics are determined and can be controlled with near atomic precision during NW synthesis and subsequent assembly, and should enable higher gain, higher speed, and lower power dissipation devices than possible by conventional approaches. Moreover, the cNW-FET concept can be readily integrated in a parallel manner without lithography, thus enabling one to envision a straightforward way to nanometer scale integrated electronics of the future. Examples of more complex devices—logic gates—assembled from cNW-FET elements are discussed below.

7.4. Crossed p–n junction LEDs

In direct band gap semiconductors like InP, the p–n junction also forms the basis for the critical optoelectronic devices, including LEDs and lasers. To assess whether our crossed NW devices might behave similarly, we have studied the photoluminescence (PL) and electroluminescence (EL) from crossed NW p–n junctions. Significantly, EL can be readily observed from these nanoscale junctions in forward bias [33]. A 3D plot of the EL intensity taken from a typical NW p–n junction at forward bias (Fig. 49a) shows the emitted light comes from a point-like source, and moreover, comparison of EL and PL images recorded on the same sample show that the position of the EL maximum corresponds to the crossing point in the PL image. These data thus demonstrate that the emitted light indeed comes from the crossed NW p–n junction.

Characterization of the EL intensity as a function of forward bias of the junction shows that significant light can be detected with our system at a voltage as low as 1.7 V

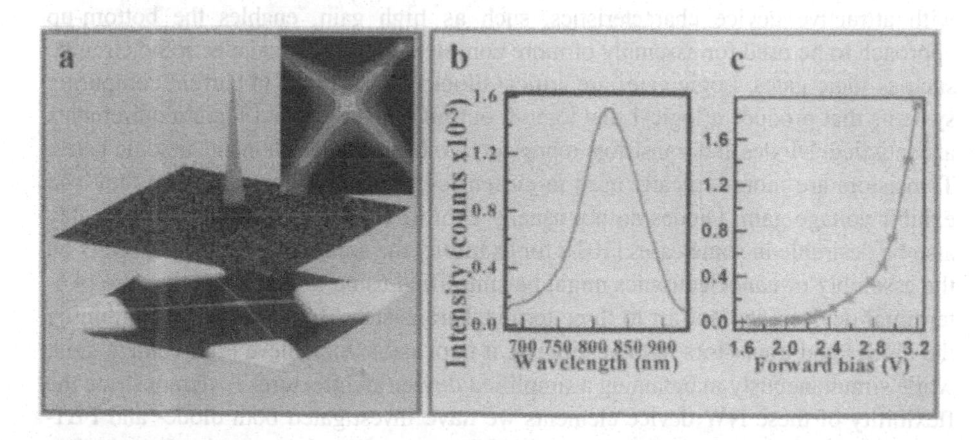

Fig. 49. Crossed NW LED. (a) (top) Three-dimensional (3D) plot of light intensity of the electroluminescence from a crossed NW LED. Light is only observed around the crossing region. (bottom) 3D atomic force microscope image of a crossed NW LED. (b) Spectrum of the emission shows a peak at ca. 820 nm. (c) Plot of integrated intensity versus forward bias voltage. See color plate 3.

(Fig. 49c). Further increases in forward bias beyond this "turn-on" voltage produce rapid, superlinear increase in the EL intensity. In addition, EL spectra recorded from the cNW-LEDs exhibit blue-shifts relative to the bulk band gap of InP (925 nm) (Fig. 49b). These blue-shifts are due in part to quantum confinement of excitons, although other factors may also contribute. Significantly, PL studies of individual InP NWs demonstrated that the PL peak can be systematically blue-shifted as the NW diameter is decreased, and thus these results provide a means for controlling the color of the LEDs in a well-defined way. Indeed, EL results recorded from p–n junctions assembled from smaller (and larger) diameter NWs show larger (smaller) blue-shifts. The ability to tune color with size in these nanoLEDs might be especially useful in the future nanophotonic applications.

In addition, the color of light emitted by cNW-LEDs can be varied by using chemically distinct semiconductors that have different band gaps. Considering the broad range of III–V and II–VI NW materials prepared to date, it should be possible to create nanoLEDs with emission from the UV to IR regions of the electromagnetic spectrum. Significantly, recent studies have demonstrated that p-Si/n-GaN cNW-LEDs have clear band edge emission in the UV at 370 nm [99]. In addition, single NW nanoLEDs have been demonstrated with GaN, CdS, CdSe [99] and InP [61] NWs by either creating an intrawire p–n junction or a Schottky diode at one of the metal contact electrodes. These studies and recent optical pumping measurements [101] suggest that if nonradiative processes can be strongly suppressed, single NW injection lasers should be possible. Finally, the cNW-LED structure can be assembled into a point-addressable array format [98], which could have a number of applications including optical storage and communications.

7.5. Integrated nanowire logic devices

The controlled high-yield assembly of crossed NW p–n diodes and cNW-FETs with attractive device characteristics, such as high gain, enables the bottom-up approach to be used for assembly of more complex and functional electronic circuits, such as logic gates. Logic gates are critical blocks of hardware in current computing systems that produce a logic-1 and logic-0 output when the input logic requirements are satisfied. Diodes and transistors represent two basic device elements in logic gates. Transistors are more typically used in current computing systems because they can exhibit voltage gain. Diodes do not usually exhibit voltage gain, although they may also be desirable in some cases [101]; for example, the architecture and constraints on the assembly of nanoelectronics might be simplified using diodes since they are two-terminal devices, in contrast to three-terminal transistors. In addition, by combining the diodes and transistors in logic circuits, it is possible to achieve high voltage gain, while simultaneously maintaining a simplified device architecture. To demonstrate the flexibility of these NW device elements we have investigated both diode- and FET-based logic. We have created AND, OR, NOR gate and used them to implement simple basic computation [35]. Here we will give NOR gate and half adder as examples.

FET-based logic NOR gates have been studied using assembled 1(p-Si) $\times$ 3 (n-GaN) cNW-FET arrays (Fig. 50a). Typically, NOR gates were configured with 2.5 V applied to one cNW-FET to create a constant resistance ~100 Mohms, and with the p-SiNW channel biased at 5 V. The two remaining n-GaN NW inputs were used as gates for two cNW-FETs in series. In this way, the output depends on the resistance ratio of the two cNW-FETs and the constant resistor. The V_o–V_i relation (Fig. 50b) shows constant low V_o when the other input is high, and a nonlinear response with large change in V_o when the other input is set low. The logic-0 is observed when either one or both of the inputs is high (Fig. 50c). In this case, one or both of the transistors are off and have resistances much higher than the constant resistor, and thus most of the voltage drops across the transistors. A logic-1 state can only be achieved when both of the transistors are on; that is, both inputs low. Analysis of the V_o–V_i data demonstrates that these 2-input NOR gates routinely exhibit gains in excess of five, which is substantially larger than the gain reported for complementary inverters based on Si-NWs [34] and carbon NTs [25]. High gain is a critical characteristic of gates since it enables interconnection of arrays of logic gates without signal restoration at each stage. The experimental truth table for this NW device (Fig. 50d) summarizes the V_o–V_i response and demonstrates that the device behaves as a logic NOR gate. Lastly, these multiple input logic NOR gates can be configured as simple NOT gates (inverters) by eliminating one of the inputs.

Logic OR, AND, and NOR (NOT) gates form a complete set of key logic elements and enable the organization of virtually any logic circuit. To demonstrate the potential

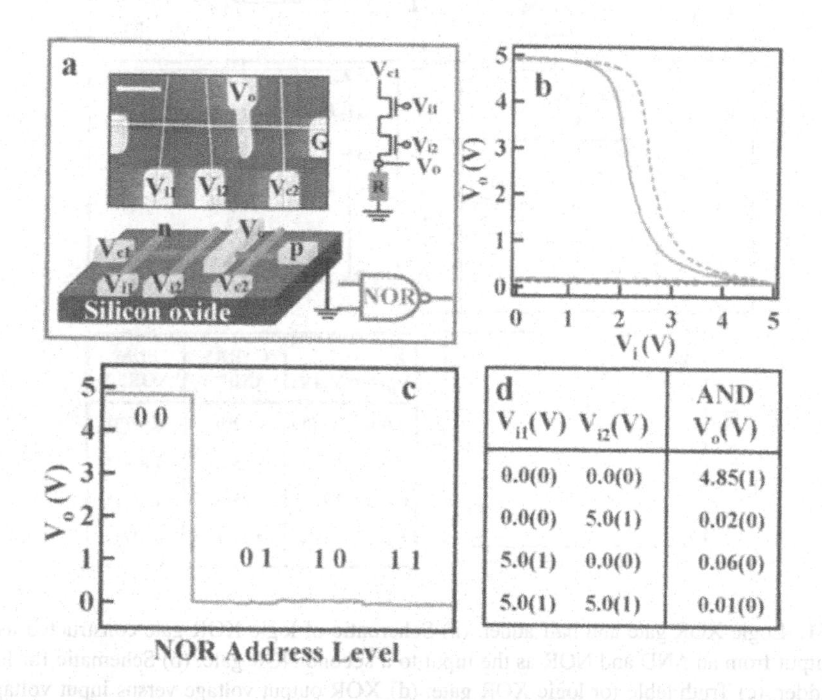

V_{i1}(V)	V_{i2}(V)	AND V_o(V)
0.0(0)	0.0(0)	4.85(1)
0.0(0)	5.0(1)	0.02(0)
5.0(1)	0.0(0)	0.06(0)
5.0(1)	5.0(1)	0.01(0)

Fig. 50. Logic NOR gate. (a) Schematic of logic NOR gate constructed from a 1 × 3 crossed NW junction array. The insets show an example SEM image (bar is 1 μm) and symbolic electronic circuit. (b) V_o–V_i relation; the slope of the data shows that device voltage gain is larger than 5. (c) The output voltage versus the four possible logic address level inputs. (d) The measured truth table for the NOR gate. Adapted from [35].

for integration of such logic gates, we have interconnected multiple AND and NOR gates to implement basic computation in the form of an XOR gate (Fig. 51a), which corresponds to the binary logic function SUM, and a half adder (Fig. 51b), which corresponds to the addition of two binary bits. The XOR gate is configured by using the output from AND and NOR gates as the input to a second NOR gate, while the logic half adder uses an additional logic AND gate as the CARRY. The truth table for the XOR structure (Fig. 51c) confirms this logic operation. Significantly, the measured V_o–V_i transport data for the XOR device (Fig. 51d, e) show (1) that the output is logic-0 or low when the inputs are both low or high, and logic-1 or high when one input is low and the other is high, and (2) that the response is highly nonlinear. The linear response region corresponds to a voltage gain of in excess of ten and is typical of the devices measured to date. We note that this large gain is achieved in an XOR configured from a low gain diode AND gate, and is due to the high gain of the cNW-FET

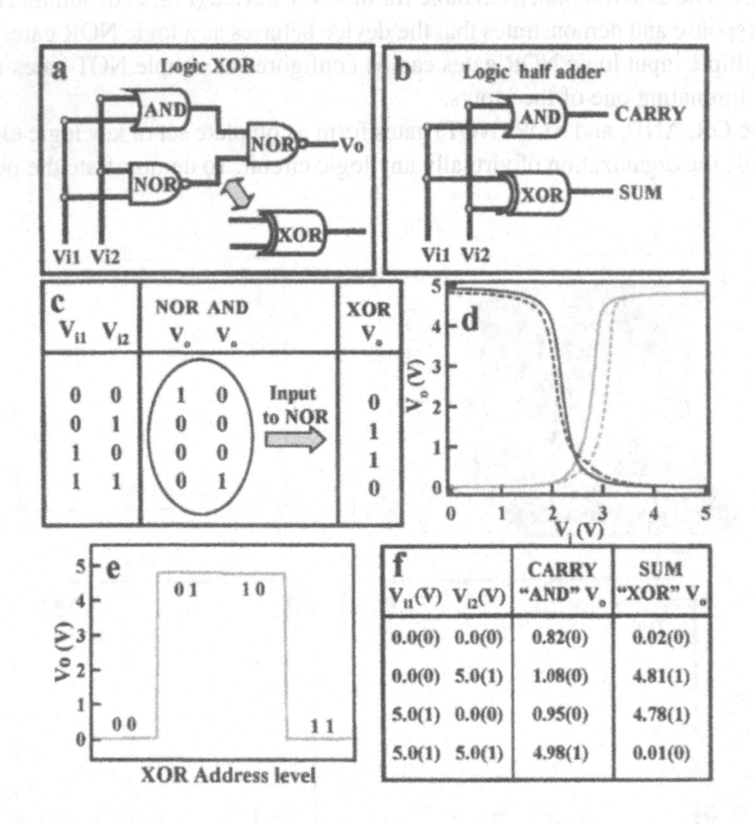

Fig. 51. Logic XOR gate and half adder. (a) Schematic of logic XOR gate constructed using the output from an AND and NOR as the input to a second NOR gate. (b) Schematic for logic half adder. (c) Truth table for logic XOR gate. (d) XOR output voltage versus input voltages. The slope of the $V_o - V_i$ data shows that the gain exceeds 10. The XOR gate was achieved by connecting the output electrodes of an AND and NOR gate to two inputs of another NOR gate. (e) The output voltage versus the four possible logic address level inputs for the XOR gate. (f) Experimental truth table for the logic half adder. The logic half adder was obtained by using the XOR gate as the SUM, and an AND gate as the CARRY. Adapted from [35].

NOR gate. These data are summarized in the experimental truth table (Fig. 51f), which demonstrates that the response is that of the binary logic SUM operation. Significantly, with another AND gate as CARRY, a logic HALF ADDER is formed and can be used to carry out calculations in a manner similar to conventional electronics.

7.6. Other directions

The crossed NW architecture is a powerful one when combined with highly-controlled synthesis of NWs and NW heterostructures. For example, a simple cross array can also function as memory [102], decoder [102], or multicolor LED arrays [103] depending on the NW materials used. When the NW building blocks are specifically-designed core-shell multilayer heterostructures, the NW crosses are enabled with bistability and the arrays have nonvolatile memory or programmable logic function. In contrast, controlling the threshold voltage with surface chemistry in p- and n-type NW crossed arrays allows the realization of an address decoder. Combining different bandgap NW materials in crossed pn junction arrays produce multiple color emission.

8. Summary and Outlook

We have illustrated and addressed the key issues in bottom-up assembly of functional nanosystems using semiconductor nanowires. The synthesis of semiconductor NWs and nanowire heterostructures with controlled chemical, structural, and electronic properties could be designed via a vapor-liquid-solid (VLS) growth mechanism. The electrical and optical property measurements all demonstrate NWs have high quality for carrier transport. Many types of nanoscale functional devices using individual NWs and NW heterostructures as building blocks are thus realized. The unique properties of NWs are also applied in ultrasensitive chemical and biological sensing. Moreover, we have developed approaches for the hierarchical assembly of NWs into well-defined arrays with controlled orientation and spatial location. Lastly, we have demonstrated that the crossed NW structures achieved by the assembly methods exhibit a broad range of electronic and photonic function that can be readily scaled into highly integrated nanoscale structures. These results provide strong justification for the bottom-up paradigm, and suggest a very bright future for both fundamental science at the nanoscale and applications in nanoelectronics, nanophotonics, nanosensors and other areas as we move to future.

Acknowledgements

We thank M. Bockrath, M. S. Gudiksen, J.-L. Huang, W. Liang, L. Lincoln, H. Park, J. Wang, and Q. Wei for helpful discussion and their contributions on the work described in this chapter. C.M.L. acknowledges generous support of this work by the Air Force Office of Scientific Research, Defense Advanced Projects Research Agency National Cancer Institute, National Science Foundation and Office of Naval Research.

References

1. G. Timp, *Nanotechnology*, Springer-Verlag, New York (1999).
2. For introduction see: *Sci. Am.*, September (2001).
3. J. D. Meindl, Q. Chen and J. A. Davis, *Science* **293** (2001) 2044.
4. C. M. Lieber, *Sci. Am.* September (2001) 58.
5. S. Polyviou and M. Levas, *The future of Moore's law*, http://www.iis.ee.ic.ac.uk/~frank/surp98/report/sp24/.
6. R. Dagani, *Chem. Eng. News* (October 16, 2000) 27.
7. J. R. Heath, P. J. Kuekes, G. S. Snider and R. S. Williams, *Science* **280**, 1716 (1998).
8. M. A. Reed and J. A. Tour, *Sci. Am.* June (2000) 86.
9. C. Joachim, J. K. Gimzewski and A. Aviram, *Nature* **408** (2000) 541.
10. C. P. Collier, E. W. Wong, M. Belohradsky, F. M. Raymo, J. F. Stoddart, P. J. Kuekes, R. S. Williams and J. R. Heath, *Science* **285** (1999) 391.
11. C. P. Collier, G. Mattersteig, E. W. Wong, Y. Luo, K Beverly, J. Sampaio, F. M. Raymo. J. F. Stoddart and J. R. Heath, *Science* **289** (2000) 1172.
12. M. A. Reed, J. Chen, A. M. Rawlett, D. W. Price and J. M. Tour, *Appl. Phys. Lett.* **286** (2001) 1550.
13. A. P. Alivisatos, *Science* **271** (1996) 933.
14. David L. Klein, Richard Roth, Andrew K. L. Lim, A. Paul Alivisatos and Paul L. McEuen, *Nature* **389** (1997) 699.
15. M. H. Devoret and R. J. Schoelkopf, *Nature* **406** (2000) 1039.
16. P. G. Collins and P. Avouris, *Sci. Am.*, December (2000) 62.
17. T. W. Odom, J.-L. Huang, P. Kim and C. M. Lieber, *Nature* **391** (1998) 62.
18. T. Odom, J. Huang, P. Kim and C. M. Lieber, *J. Phys. Chem. B* **104** (2000) 2794–2809.
19. C. Dekker, *Phys. Today* **52** (5) (1999) 22.
20. H. Dai, J. Kong, C. Zhou, N. Franklin, T. Tmobler, A. Cassell, S. Fan and M. Chapline, *J. Phys. Chem. B* **103** (1999) 11246.
21. S. J. Tans, R. M. Verschueren and C. Dekker, *Nature* **393** (1998) 49.
22. R. Martel, T. Schmidt, H. R. Shea, T. Hertel and P. Avouris, *Appl. Phys. Lett.* **73** (1998) 2447.
23. Z. Yao, H. W. C. Postma, L. Balents and C. Dekker, *Nature* **402** (1999) 273.
24. M. S. Fuhrer, J. Nygrad, L. Shih, M. Forero, Y. G. Yoon, M. S. C. Mazzoni, H. J. Choi, J. Ihm, S. G. Louie, A. Zettl and P. L. McEuen, *Science* **288** (2000) 494.
25. V. Derycke, R. Martel, J. Appenzeller and P. Avouris, *Nano Letters* **1** (2001) 453.
26. A. Bachtold, P. Hadley, T. Nakanishi and C. Dekker, *Science* **294** (2001) 1317.
27. P. G. Collins, M. S. Arnold and P. Avouris, *Science* **292** (2001) 706.
28. C. M. Lieber, *Solid State Commun.* **107** (1998) 106.
29. J. Hu, T. W. Odom and C. M. Lieber, *Acc. Chem. Res.* **32** (1999) 435.
30. X. Duan and C. M. Lieber, *Adv. Mater.* **12** (2001) 298.
31. M. S. Gudiksen, J. Wang and C. M. Lieber, *J. Phys. Chem. B* **105** (2001) 4062–4064.
32. Y. Cui, X. Duan, J. Hu and C. M. Lieber, *J. Phys. Chem. B* **104** (2001) 5213.
33. X. Duan, Y Huang, Y. Cui, J. Wang and C. M. Lieber, *Nature* **409** (2001) 66.
34. Y. Cui and C. M. Lieber, *Science* **291** (2001) 851.
35. Y. Huang, X. Duan, Y. Cui, L. Lauhon, K. Kim and C. M. Lieber, *Science* **294** (2001) 1313.
36. Y. Huang, X. Duan, Q. Wei and C. M. Lieber, *Science* **291** (2001) 630.
37. F. Seker, K. Meeker, T. F. Kuech and A. B. Ellis, *Chem. Rev.* **100** (2000) 2505.
38. R. K. Iler, *The Chemistry of Silica*, Wiley, New York (1979).
39. C. B. Murray, D. J. Norris and M. G. Bawendi, *J. Am. Chem. Soc.* **115** (1993) 8706–8715.
40. L. Esaki, *Science and Technology for Mesoscopic Structures*, Edited by S. Namba, C. Hamaguchi and T. Audio, Springer Verlag, Tokyo (1992).

41. F. J. Himpsel, T. Jung, A. Kirakosian, J. L. Lin, D. Y. Petrovykh, H. Rausher and J. Viernow, *MRS Bulletin* **24**(8) (1999) 20.
42. H. Dai, E. W. Wong, Y. Z. Lu, S. Fan and C. M. Lieber, *Nature* **375** (1995) 769.
43. W. Han, S. Fan, W. Li and Y. Hu, *Science* **277** (1997) 1287.
44. C. R. Martin, *Science* **266** (1994) 1961.
45. R. S. Wagner and W. C. Ellis, *Appl. Phys. Lett.* **4** (1964) 89.
46. R. S. Wagner, in *Whisker Technology*, Edited by A. P. Levitt, Wiley, New York (1970).
47. T. J. Trentler, K. M. Hickman, S. C. Goel, A. M. Viano, P. C. Gibbons and W. E. Buhro, *Science* **270** (1995) 1791.
48. T. J. Trentler, S. C. Goel, K. M. Hickman, A. M. Viano, M. Y. Chiang, A. M. Beatty, P. C. Gibbons and W. E. Buhro, *J. Am. Chem. Soc.* **119** (1997) 2172.
49. P. Yang and C. M. Lieber, *Science* **273** (1996) 1836.
50. P. Yang and C. M. Lieber, *J. Mater. Res.* **12** (1997) 2981.
51. M. S. El-Shall and A. S. Edelstein, in *Nanomaterials: Synthesis, Properties and Applications*, Edited by A. S. Sedelstein and R. C. Cammarata, Institute of Physics, Philadephia (1996).
52. A. M. Morales and C. M. Lieber, *Science* **279** (1998) 208.
53. X. Duan, J. Wang and C. M. Lieber, *Appl. Phys. Lett.* **76** (2000) 1116.
54. X. Duan and C. M. Lieber, *J. Am. Chem. Soc.* **122** (2000) 188.
55. Q. Wei and C. M. Lieber, *Mat. Res. Soc. Symp. Proc.* **581** (2000) 219–223.
56. M. B. Panish, *J. Electrochem. Soc.* **114** (1969) 517.
57. M. S. Gudiksen and C. M. Lieber, *J. Am. Chem. Soc.* **122** (2000) 8801.
58. Y. Cui, L. J. Lauhon, M. S. Gudiksen, J. Wang and C. M. Lieber, *Appl. Phys. Lett.* **78** (2001) 2214.
59. Y. Cui and C. M. Lieber et al. (unpublished results).
60. J. Hu, M. Ouyang, P. Yang and C. M. Lieber, *Nature* **399** (1999) 48–51.
61. M. S. Gudiksen, L. J. Lauhon, J. Wang, D. Smith and C. M. Lieber, *Nature* **415** (2002) 617–620.
62. Y. Wu, R. Fan and P. Yang, *Nano Lett.* **2** (2002) 83.
63. M. T. Bjork, B. J. Ohlosson, T. Sass, A. I. Persson, C. Thelander, M. H. Magnusson, K. Deppert, L. R. Wallenberg and L. Samuelson, *Nano Lett.* **2** (2002) 87.
64. L. Lauhon, M. S. Gudiksen, D. Wang and C. M. Lieber, *Nature* **420** (2002) 57.
65. S. M. Sze, *Semiconductor Devices, Physics and Technology*, Wiley, New York (1985) pp. 208–210.
66. O. Madelung, in *LANDOLT-BORNSTEIN New Series: Vol III/22a, Semiconductors: Intrinsic properties of Group IV Elements and III-V and II-VI and I-VII Compounds*, Edited by O. Madelung, Springer-Berlin, Heidelberg (1987).
67. Y. Huang, X. Duan, Y. Cui and C. M. Lieber, *Nano Lett.* **2** (2002) 101.
68. C. M. Lieber et al. (unpublished results).
69. Y. Cui, Z. Zhong, D. Wang, W. U. Wang and C. M. Lieber (submitted).
70. R. Chau, *Proc. IEDM* **2001**, 621.
71. X. Duan, Y. Huang and C. M. Lieber, *Nano Lett.* **2** (2002) 487.
72. G. Goldoni, F. Rossi and E. Molinari, *Phys. Rev. Lett.* **80** (1998) 4995.
73. M. S. Gudiksen, J. Wang and C. M. Lieber, *J. Phys. Chem. B* **106** (2002) 4036.
74. H. Fu and A. Zunger, *Phys. Rev. B* **55** (1997) 1642.
75. J. Wang, M. S. Gudiksen, X. Duan, Y. Cui and C. M. Lieber, *Science* **293** (2001) 1455.
76. S. Ura, H. Sunagawa, T. Suhara and H. Nishihara, *J. Lightwave Tech.* **6** (1988) 1028.
77. C. J. Chen, K. K. Choi, L. Rokhinson, W. H. Chang and D. C. Tsui, *Appl. Phys. Lett.* **74** (1999) 862.
78. Y.Cui, Q. Wei, H. Park and C. M. Lieber, *Science* **293** (2001) 1289.

79. P. Bergveld, *IEEE Trans. Biomed. Eng.* **BME-19** (1972) 342.
80. G. F. Blackburn, in *Biosensors: Fundamentals and Applications*, Edited by A. P. F. Turner, I. Karube and G. S. Wilson, Oxford University Press, Oxford (1987) pp. 481–530.
81. D. V. Vezenov, A. Noy, L. F. Rozsnyai and C. M. Lieber, *J. Am. Chem. Soc.* **119** (1997) 2006.
82. G. H. Bolt, *J. Phys. Chem.* **61** (1957) 1166.
83. C. B. Klee and T. C. Vanaman, *Adv. Protein Chem.* **35** (1982) 213.
84. M. Wilchek and E. A. Bayer, *Methods Enzymol.* **184** (1990) 49.
85. L. Movileanu, S. Howorka, O. Braha and H. Bayley, *Nature Biotechnol.* **18** (2000) 109.
86. H. Bagci, F. Kohen, U. Kuscuoglu, E. A. Bayer and M. Wilchek, *FEBS Lett.* **322** (1993) 47.
87. R. C. Blake II, A. R. Pavlov and D. A. Blake, *Anal. Biochem.* **272** (1999) 123.
88. M. D. Abeloff, J. O. Armitage, A. S. Licbter and J. E. Niederbuber, *Clinical Oncology*, Churchill Livingstone, New York (2000).
89. A. M. Ward, J. W. F. Catto and F. C. Hamdy, *Ann. Clin. Biochem.* **38** (2001) 633.
90. Y. Cui, W. U. Wang, L. Huynh and C. M. Lieber (submitted).
91. J. T. Wu, B. W. Lyons, G. H. Liu and L. L. Wu, *J. Clin. Lab. Anal.* **12** (1998) 6.
92. C. A. Stover, D. L. Koch and C. Cohen, *J. Fluid Mech.* **238** (1992) 277.
93. D. L. Koch and E. S. G. Shaqfeh, *Phys. Fluids A* **2** (1990) 2093.
94. Y. Huang, X. Duan and C. M. Lieber (unpublished results).
95. J. Hahm, Y. Huang and C. M. Lieber (unpublished results).
96. D. Whang, Y. Wu and C. M. Lieber (unpublished results).
97. S. M. Sze, *Physics of Semiconductor Devices*, Wiley, New York (1981).
98. Y. Huang, X. Duan and C. M. Lieber (unpublished results).
99. Y. Huang, X. Duan and C. M. Lieber (unpublished results).
100. M. Huang, S. Mao, H. Feick, H. Yan, Y. Wu, H. Kind, E. Weber, R. Russo and P. Yang, *Science* **292** (2001) 1897.
101. P. Horowitz and W. Hill, *The Art of Electronics*, Cambridge University Press, Cambridge (1989).
102. D. Wang, Z. Zhong and C. M. Lieber (unpublished results).
103. Y. Huang, X. Duan and C. M. Lieber (unpublished results).

Chapter 2

Epitaxial Quantum Wires: Growth, Properties and Applications

Lars Samuelson, B. Jonas Ohlsson,
Mikael T. Björk and Hongqi Xu
Lund University, Solid State Physics/the Nanometer Consortium
Box 118, S-221 00 Lund, Sweden

B. Jonas Ohlsson
QuMat Technologies AB, Malmö, Sweden

1. Introduction

The opportunity to form one-dimensional (1D) materials and devices via epitaxially nucleated growth of whiskers was realized more than 30 years ago by Wagner and others [1]. At that time, technology as well as the general awareness of the technology on small length-scales only permitted whisker dimensions typically down to the micrometer region. It was first via the pioneering work of Hiruma and coworkers [2] at Hitachi during the early part of the 1990s that the nanometer dimension of nanowires were explored and brought to the level where the potential for nanoelectronic and photonic devices was realized. The approach of Hiruma has during the last few years been developed further by several groups, for instance that led by Peydong Yang at UC Berkeley and the one by Lars Samuelson at Lund University. Besides this special version of nanowire formation via epitaxially nucleated growth, various research groups have been developing the method of spontaneous sprouting of nanowires from catalytic nanoparticles, most noticed being the work by the Lieber group at Harvard University. Much of the present excitement in the field stems from the recent breakthrough in forming nanowires containing ideal heterostructure interfaces [3], hence allowing nanoelectronic [4] and nanophotonic heterostructure devices to be formed. In the following we first describe the background in terms of competitive or alternative methods for realization of one-dimensional semiconductor structures, with an emphasis on comparison between conventional top-down as opposed to self-assembly or bottom-up fabrication methods. We then present some recent progress related to the electronic structure of nanowires, including predictions of phenomena that should be feasible using existing technology. This is followed by a rather detailed description of state-of-the-art fabrication of homogeneous nanowires as well as those containing heterostructures along the one-dimensional axis of the wire. Finally we describe some of the technology required for the creation of electrically contacted nanowires with an emphasis on contact technologies and experimental results for simple and more complex 1D heterostructure devices. An outlook on which direction this technology might take then follows and we will give examples of future challenges and opportunities.

2. General Background and Introduction to the Subject

The potential and advantages of going towards one-dimensional structures for nanoelectronics and nanophotonics was realized a long time ago. In 1980, Hiroyuki Sakaki proposed [5] that the restriction to 1D would allow improved ballistic properties related to the dramatic reduction in the possible **k**-space points accessible for carrier scattering in 1D as compared to the 2D or 3D cases. The advantages of quantum size photon emitters were also recognized many years ago and different methods for self-assembly of quantum dots and quantum wires for photonics have been developed. The possibility of combining quantum size effects and single-electron phenomena, or Coulomb blockade effects, has increased the interest in these technologies. One example is the efforts to use single quantum dots as emitters for quantum cryptography, where the goal is to combine resonant tunneling effects and Coulomb blockade in the controlled creation of single-photon-on-demand devices and applications.

Methods used to create low-dimensional materials, such as quantum dots and nanowires, and their use for nanoelectronics and photonics are gaining great interest for science as well as for technological applications. The concept of self-assembly has been applied both to nanowire formation and to one-dimensional stacking of quantum dots. A few examples will be mentioned here, primarily those related to the controlled formation of nanowires using vicinal surfaces or V-grooves etched on the surface of the substrate, and also strain induced self-assembled growth of quantum dots. Takashi Fukui at NTT [6] and Pierre Petroff at UC Santa Barbara [7] pioneered the controlled layer-by-layer epitaxial growth on vicinal surfaces to master how different material compositions could be directed to different stripes, in principle to a fraction of the step-length given by the miss-cut angle of the vicinal surface. This led to the demonstration of vertical quantum wells and to the controlled creation of low bandgap nanowires surrounded by a material of higher bandgap in serpentine superlattice structures.

In the late 80s Eli Kapon at Bellcore [8] demonstrated the use of surface patterning to form quantum wires via a self-assembling process. Due to the differences in surface mobilities of the reactive species in metal-organic vapor phase epitaxy (MOVPE) and the different sticking probabilities on different facets of GaAs and AlGaAs semiconductors, it was possible to controllably develop the facets leading to extremely well controlled V-grooves. In turn, these V-grooves could be used for the formation of GaAs quantum wires at the bottom of the grooves. In this way one could form highly ideal quantum wires and also design heterostructures that assisted charge carrier funneling into the nanowire, leading to high-quality quantum wire lasers [9].

Since the mid 90s great pains have been taken over how growth of a thin strained crystal layer on an underlying crystal with (in most cases) a larger lattice constant may be spontaneously transformed into a large concentration of quantum dots. This mechanism is called Stranski-Krastanow growth [10] and may be used to form single layers of buried quantum dots. Given the optimum conditions, such layers may be stacked to form one-dimensional arrays of quantum dots that are possible to use in one-dimensional multi heterostructure transport devices [11–13].

Top-down fabrication techniques have been used for a long time to form nanowires (or quantum wires) by carving out a narrow wire from an already grown quantum well layer. It has been demonstrated that it is possible to create quantum wires with good optical properties using electron beam lithography followed by etching techniques for lateral wire dimensions as small as 20 nm. Similarly, quantum wire transport devices have been realized in this fashion, leading to considerable progress in electron waveguide-like devices and circuits. Of special interest may be the way ballistic multi-terminal waveguide devices [14] have been shown to offer new circuit opportunities for logics as well as high-speed electronics applications [15]. Although such devices have been limited to top-down fabrication, three-terminal waveguide structures have been demonstrated based on carbon nanotube technology and very similar device performance is expected in these cases [16].

One of the most intensively studied top-down approaches has been the challenge to develop 1D-0D-1D resonant tunneling devices. Attempts to fabricate vertical 1D (nanowire) devices have been made since the late 1980s, as pioneered by Randall, Reed and co-workers at Texas Instruments [17]. The target for this 1D device project and many to follow, has been to create resonant tunneling devices in which the emitter and collector have one-dimensional character while the active or central part of the device is a single quantum dot separated from the emitter and collector regions by tunnel barriers. The top-down approach used at Texas Instruments, which still represents state of the art for this family of quantum devices, is based on epitaxial growth of multi-layers defining the two barriers and the central quantum well followed by reactive ion etching to form narrow columns, which are contacted from the top. In the studies of devices fabricated by this top-down technique, 100–200 nm diameter columns have been made, however, with rather disappointing electrical characteristics and peak-to-valley currents at best around 1.1:1. Other approaches to obtain 1D resonant tunneling devices have included focused ion-beam definition of the vertical current path [18] and selective wet etching to control the lateral extent to the 100 nm levels [19].

In spite of progress in top-down fabrication of both planar and vertical quantum wires or pillars, there are clear reasons to look for fabrication techniques which are not limited by damage induced by processing, and which are not so demanding in terms of lithographic patterning. Below we will describe the details of how 1D-0D-1D resonant tunneling devices have recently been realized in our laboratory using bottom-up fabrication during VLS-growth of nanowires.

3. Theory on the Band Structure of Semiconductor Nanowires

The most common methods of solving the electronic structure of semiconductor nanowires are effective mass- and $\mathbf{k} \cdot \mathbf{p}$-theory [20–23], tight-binding theory [24–26], and pseudopotential theory [27–30]. All these methods are semi-empirical. First principle methods, which have been successfully applied to three-dimensional bulk

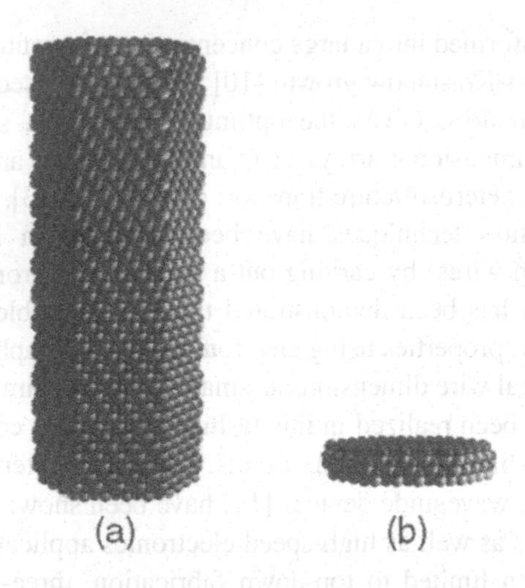

Fig. 1. (a) Schematic view of the atomistic model for a free-standing GaAs nanowire grown along the <1 1 1> direction without including the hydrogen passivation atoms. The nanowire has a hexagonal cross section and the lateral size of the nanowire is defined as the distance between two parallell facets. (b) A unit cell of the nanowire. The pictures are reproduced by courtesy of Persson [26].

semiconductor materials, are computationally too demanding to be used in the electronic structure calculation for nanowires. The difficulty arises from the fact that the semiconductor nanowire structures contain hundreds, or even thousands of atoms per unit cell. Taking the ideal hydrogen-passivated <1 1 1> GaAs nanowire with hexagonal cross section as an example (see Fig. 1 for a schematic view), the unit cell contains 662 host atoms and 126 surface hydrogen atoms for a wire with a lateral size of 4 nm, and 9842 host atoms and 486 surface hydrogen atoms for a 16 nm diameter wire.

A simple view of the electronic structure of semiconductor nanowires is provided by $\mathbf{k}\cdot\mathbf{p}$ theory or effective mass theory [20]. For a semiconductor with a relatively large band gap ($E_g \geqslant 1$ eV), one may assume uncoupled conduction band minimum and valence band maximum and describe the lowest conduction band with an effective mass model and the uppermost valence bands with a four-band or a six-band $\mathbf{k}\cdot\mathbf{p}$ model. These models have been widely used in the study of the electron subband structure with and without electron–electron interaction, and in the study of the properties of excitons in semiconductor nanowires. However, although these models are sometimes successful in the study of large size nanowires, it has been shown that they are inaccurate in the description of the electronic structure of nanometer-sized quantum systems [25, 29].

Tight-binding and pseudopotential theories provide two suitable descriptions for the electronic structure of semiconductor nanowires. In this review, we only present the results of the calculations for semiconductor nanowires with a tight-binding model [26]. For the calculations with pseudopotential models, we refer to the works of Refs. [27–30]. Fig. 2 shows the calculated band structure of freestanding GaAs nanowires,

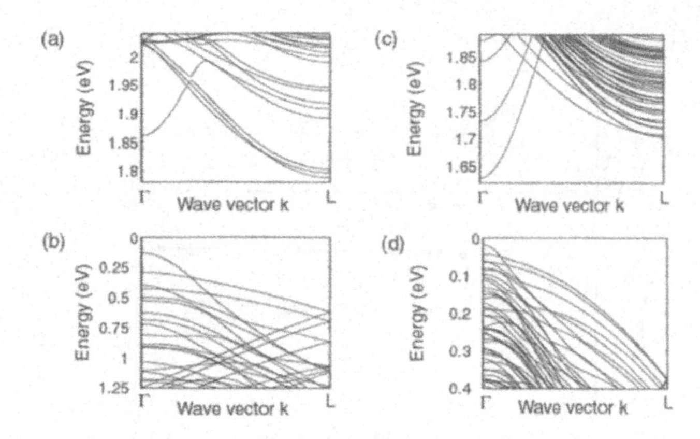

Fig. 2. The band structure of the hydrogen-passivated, free-standing GaAs nanowire, grown along <1 1 1> direction, as shown in Fig. 1(a) and (b) for the nanowire with the size of d = 2.4 nm; (c) and (d) for the nanowire with the size of d = 7.2 nm. Note that the nanowire with the size of d = 7.2 nm has a direct band gap, while that with the size of d = 2.4 nm has an indirect band gap. Reproduced from Ref. [26].

grown along the <1 1 1> direction, with the characteristic lateral size of d = 2.4 nm and d = 7.2 nm, respectively [26]. It is clearly seen that at d = 7.2 nm, the band structure of the nanowires has a direct band gap with the conduction band minimum found at the Γ point. In contrast, the band structure of the nanowire at d = 2.4 nm has an indirect band gap with the conduction band minimum found at the L point. The valence band maximum is in both cases found to remain at the Γ point. Fig. 3 shows the results of tight-binding calculations for the size dependence of the conduction band minima and the valence band maxima at the Γ and L points [26]. These results show that as the characteristic lateral size decreases, the absolute conduction band minimum of the GaAs nanowire shows a Γ→L crossover. In the above given example of the calculations, this crossover takes place at a critical size of $d_{\Gamma \to L}$ of about 4 nm. On the other hand, it is found that the absolute valence band maximum of the GaAs nanowire remains at the Γ point for all lateral sizes. Thus, the Γ→L crossover of the conduction band minimum indicates a direct-to-indirect band gap transition in the band structure. As a result, GaAs nanowires with sufficiently small lateral sizes should show very different optical and electrical properties as compared to large size wires.

Another interesting result predicted by the tight-binding calculations for <1 1 1> GaAs nanowires is that around the Γ point of the Brillouin zone the light-hole band remains the top valence band above the heavy-hole band (see Fig. 2). This result is in strong contrast to the prediction by $\mathbf{k} \cdot \mathbf{p}$ theory, in which the top valence band around the Γ point is found to be the heavy-hole band. However, the tight-binding result is consistent with the fact that the wave functions of the light-hole states in a unit cell have their characteristic extension along the growth direction of the nanowire and are localized deep inside the structure, while the wave functions of the heavy-hole states in a unit cell have their characteristic extension in the transverse direction (see Fig. 4). Thus, the light-hole band has a stronger dispersion relation than the heavy-hole band

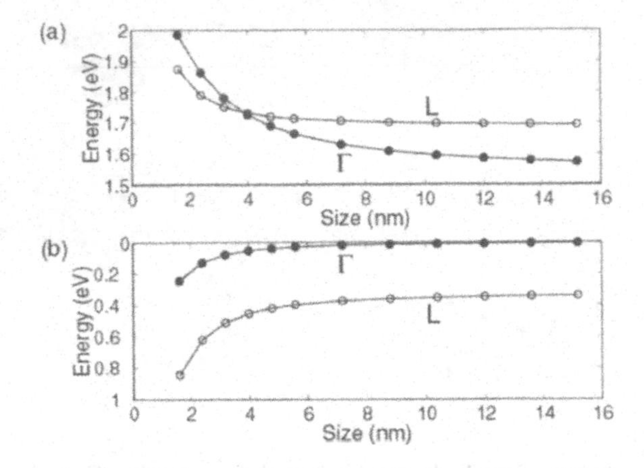

Fig. 3. Calculated energies of (a) the conduction band minima and (b) the valence band maxima at the Γ point (marked by dots) and the L point (marked by circles) as a function of the nanowire size. The marked points present the calculated data and the lines are guides to the eyes. Note that the conduction band minima at the Γ and L points cross over at d ≈ □nm. From Ref. [26].

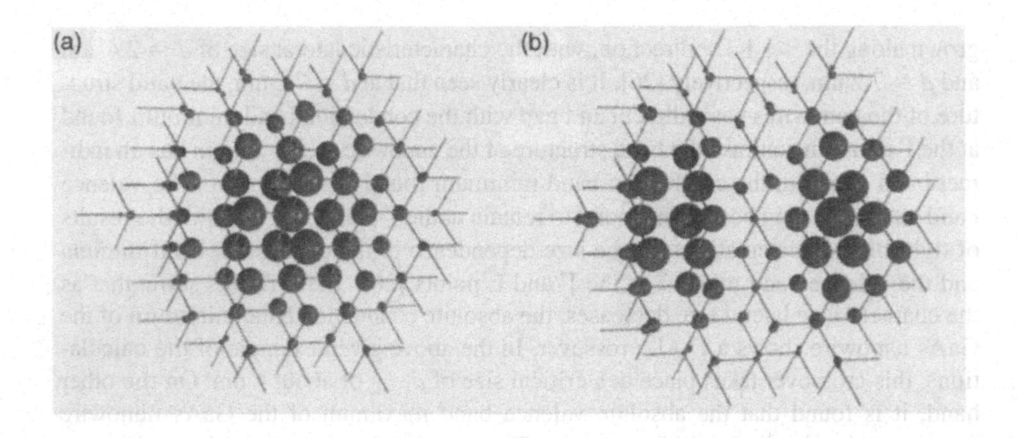

Fig. 4. Probability distributions corresponding to the wave functions of (a) the uppermost light-hole state and (b) the uppermost heavy-hole state of a GaAs nanowire, grown along the <1 1 1> direction, in the tight-binding model representation. The probability distribution on each atomic site is presented by a sphere with the radius proportional to the square root of the probability distribution. The plots show views of the probability distribution in the unit cell when looking at the nanowire from the <1 1 1> direction. Taken from the calculations by Persson with courtesy.

and the effect of the lateral confinement on the light-hole state is smaller than that on the heavy-hole state.

The fact that the absolute conduction band minimum is located at the L point for the smallest diameter GaAs nanowires, for which the electron effective mass is heavy, and that the light hole states stay at the top of the valence bands for all lateral sizes, implies that using holes as majority carriers, as opposed to electrons, in small size nanowire electronic devices may have an advantage in terms of carrier mobility.

4. State of the Art in Epitaxial Growth of Vertical Quantum Wires

In this section we discuss the catalyzed growth of nanowires from a semiconductor substrate. The subjects treated are growth conditions of nanowires, growth of segments in wires (hetero-structures), and designed incorporation of nanowires on chip (positioning and size selection). The nanowires discussed are based on the III–V semiconductors GaAs, InAs and InP, which combine good optical properties with a large span of bandgap variation.

4.1. Vapor–liquid–solid growth

The vapor–liquid–solid (VLS) growth mechanism, which is used for wire growth, is based on local growth catalyzed by a metal droplet [31]. In the growth of III–V and silicon wires, the predominantly used metal is Au, albeit a number of other metals can also be used [32]. In this work Au aerosol particles have been deposited onto GaAs <1 1 1> B substrates [33] but also Au colloids and lithographically positioned Au films has been used. Thorough studies have been made on growth of silicon wires in vapor phase epitaxy (VPE), centering on transport and kinetic mechanisms on the droplet surface and inside the droplet at the liquid–solid interface [1, 31, 32, 34]. The wire growth process also depends on the transport of necessary species to the droplet [32, 35–37].

The growth of III–V wires are further complicated by the binary nature of the material. When heating the substrate, Ga and possibly As in our case is released from the substrate and incorporated into the particles to form eutectic droplets. While the In/Au and the Ga/Au eutectic systems are very similar to the Si/Au system, the addition of As solidifies the melt, hence complicating the picture even further. The eutectic point of GaAs/Au is 630 °C [38] compared to 353 °C for Ga/Au. GaAs nanowires have been grown in different systems between 420 and 580 °C [39–41], suggesting the Ga/Au eutectic point to be the lower limit for wire-growth. The different properties of the column-III source and the column-V source will also have implications for the growth of nanowires and the incorporation of heterostructures in nanowires.

In the following discussion we consider the droplet to be a microscopic liquid-phase epitaxy system, where the inflow of sources comes from the surrounding environment with subsequent precipitation at the droplet/semiconductor surface. The precise composition and transport-ways in and on the droplet are unknown.

4.2. Controlled growth of nanowires in a UHV environment

The wires are grown in a typical chemical-beam epitaxy (CBE) system. In this growth technique, the chemical precursors impinge as molecular beams on a heated substrate for subsequent reaction and epitaxial growth. The growth rate is determined by the arrival of the column-III atoms (Ga or In) while a slight overpressure of the

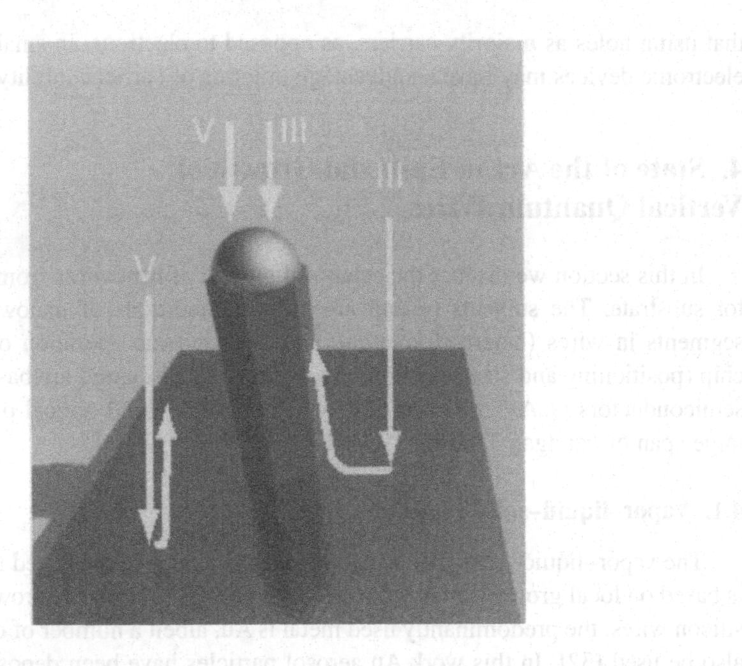

Fig. 5. Different transport paths of the column III- and V-material respectively. The III material either impinges directly on the metal droplet or on the substrate followed by diffusion up the wire sidewalls. The V-material can only reach the droplet by hitting it directly.

column V element, As or P, is maintained. The column-III precursors used are triethylgallium and trimethylindium. The column-V sources are thermally cracked tertiarybutylarsine and tertiarybutylphosphine.

By using CBE we benefit from the same advantages that apply during selective growth [42]. As will be discussed later, the main requirements necessary for incorporation of advanced heterostructures within nanowires are low growth rates and a swift response when changing growth elements. The growth rate of the wires can be lowered with the CBE technology, and will then be comparable to the epitaxial growth outside the droplet. The diameter of the wires does not expand appreciably during growth due to the limited growth of certain types of surfaces, especially the <1 1 0>-type, which constitute the wire sidewalls. Instead, diffusion of the column-III elements will occur from the substrate and up along the wire sidewalls to the droplet (Fig. 5). The inter-diffusion of the column-III element will, however, make the wire growth rate sensitive to the density of wires on the surface and the lateral size of the wires. This results in a decreased growth rate with higher density and thicker wires.

4.3. Growth of hetero-structures: Lattice mismatch and strain release in nanowires

For planar growth of high quality abrupt hetero-structural interfaces, several different growth techniques work excellent. Especially molecular-beam epitaxy (MBE) and CBE, but also low pressure MOVPE produces interfaces with monolayer abruptness. The nature of the source-beams in CBE and MBE guarantees a swift response at the

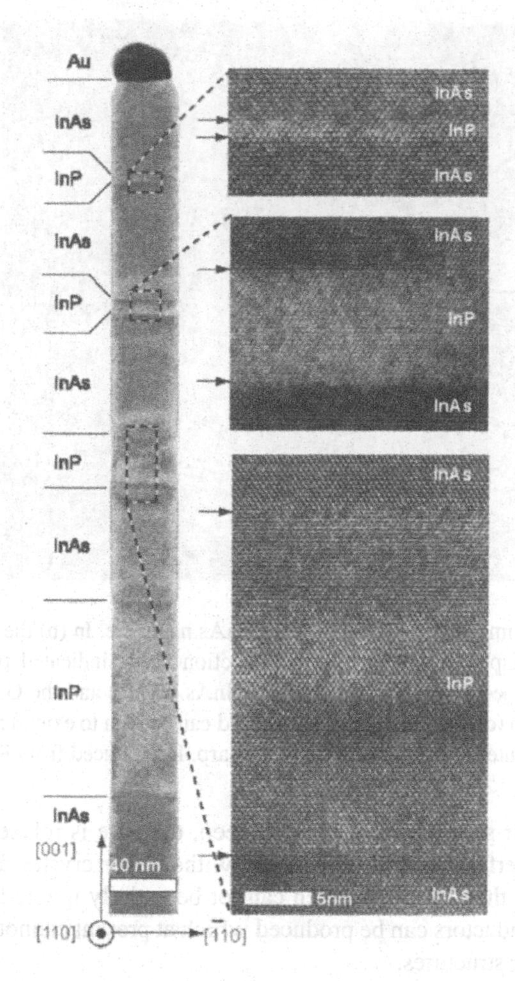

Fig. 6. TEM image of a 40 nm diameter InAs nanowire with four sequential InP segments of lengths (starting from the top) 1.5, 8, 25 and 100 nm respectively. Also shown are magnified views of the three thinnest segments. The interfaces are indicated by arrows as a guide to the eye. Reproduced from Ref. [43].

nucleation-surface when the material sources are switched. For wire growth, the CBE-technology brings down the growth rate to speeds where abrupt hetero-interfaces can be fabricated regularly [3, 43]. In Fig. 6 such interfaces between InP and InAs are shown.

Of particular interest is the fact that the very small cross section of a nanowire allows efficient lateral strain relaxation thereby providing the freedom to combine materials with very different lattice constants to create heterostructures within a nanowire. In ordinary epitaxy, a large lattice mismatch results in Stranski–Krastanow island formation [10] or in the incorporation of misfit dislocations once a critical thickness of a few monolayers is reached.

The interface of InAs/GaAs as shown in Fig. 7 is an example of a highly lattice-mismatched system, with a mismatch Δa of 6.7%. Seen in Fig. 7(c) is the partly

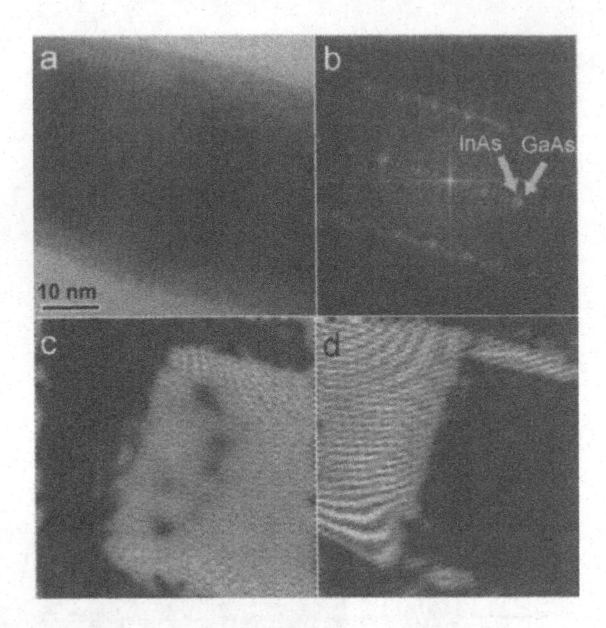

Fig. 7. (a) A HRTEM image of a part of an InAs/GaAs nanowire. In (b) the powerspectrum of the image in (a) is displayed showing split diffraction peaks indicated by arrows. Fourier analysis of the powerspectrum yields a map of the InAs part (c) and the GaAs part (d) of the wire (brighter parts). In (c) the semicircular strainfield can be seen to extend roughly 20 nm into the InAs part and the interface appears to be very sharp. Reproduced from Ref. [53].

exposed semicircular strain field. As can be seen, the wire is relaxed approximately 20 nm from the interface in both directions, without any crystal dislocations. This implies that, even if the lattice mismatch cannot be entirely ignored, structural combinations of semiconductors can be produced which at present cannot be made for any other semiconductor structures.

4.4. Size and position

As mentioned before the column III-element has a high influence on the growth rate. To achieve a similar growth rate for many wires on a substrate we need to have size-homogeneous wires and position them equidistant or isolated. The virtues of these structures would increase abundantly if it were possible to control the size, density, and position in a similar way as with lithographically, etched, and epitaxially grown 1D nanostructures. Implementing the nanowire technology on substrate may be a convenient way to combine traditional, planar semiconductor technology with a bottom up, vertical technology.

In 1970, Wagner showed that macroscopic wires could be positioned by placing the seed particles at desired locations [1]. Some other pioneering work in this field was made in the first half of the 1990s by Hiruma et al. They grew wires in an MOVPE system, by first depositing a thin layer of Au on the substrate. By varying temperature and the thickness of the Au-layer they achieved very high size homogeneity of the wires [44].

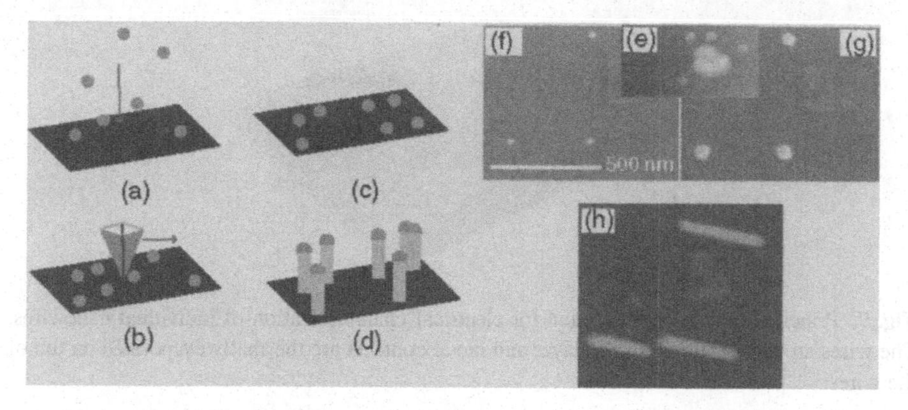

Fig. 8. Nanowire positioning. After Au particle deposition (a) individual particles may be manipulated by an AFM tip (b) to selected positions. After subsequent deoxidation (c) and growth (d) the wires are located at the desired positions. The corresponding experiment is presented in (e) through (h). Image (e) is an AFM image after aerosol deposition showing 40 nm Au particles randomly distributed around a Ti reference marker. After AFM manipulation (f) the particles have been arranged to form a triangle. As seen in the SEM image (g) the grown nanowires are still positioned as a triangle. The tilted view of (h) shows that the wires are homogeneous in diameter but slightly larger than the original Au diameter.

We make use of Au aerosol particles or Au colloids to control the nanowire diameter independently of the density as opposed to Hiruma. The diameter range available is 8–80 nm for the aerosol technique and 2–250 nm using colloids.

The positions of individual particles are usually completely random, however, we have a possibility to position the nanowires by means of manipulation of the Au aerosol particles with an atomic force microscope (AFM) prior to growth (Fig. 8(a)–(d)). In Fig. 8(e)–(g) we present the result of such an experiment, showing that the liquid/solid interface area determines the diameter of the wire, and that the wires can be grown at the exact position of the particle. The diameter of the wire will also depend on the wetting of the particle, but if we assume the temperature and the interface wetting to be constant over the whole substrate, these parameters will give a predictable change in the wire diameter.

5. Processing and Studies of Nanowires as One-Dimensional Devices

To probe the electrical properties of individual nanowires, metallic contacts must be connected to the wires. The basic principle we have adopted is shown schematically in Fig. 9. Wires are deposited on top of a non-conducting surface and the metal interconnects, which are used to probe the electrical properties are placed on top of the wires. In the next part the processes leading to the formation of ohmic contacts to InAs nanowires will be presented.

Fig. 9. Principle device configuration for electrical characterization of individual nanowires. The wires are deposited on a SiO_2 layer and metal contacts are thermally evaporated on top of the wires.

5.1. Samples for characterization

The processing starts from <0 0 1> Si substrates, usually highly p-doped with a resistivity of 0.03 Ωcm. A 1000 Å thermal oxide is then grown on the surface to make it electrically insulating. The substrate is cleaved into 3 by 5 mm samples and on top of each sample macroscopic metal pads for wire bonding is formed by evaporation of Ti and Au through a shadow mask. The resulting pattern will be a set of metal stripes as seen in Fig. 10(a). In between each set of four stripes a gap will be formed. A coordinate system is then created by making regularly spaced reference markers in each of the gaps. Nanowires are removed mechanically from the growth substrate and transferred to one of the samples described above. Wires will be deposited randomly on the SiO_2 surface and can be located by scanning electron microscopy (SEM). The SEM images are used to identify the positions of individual nanowires relative to the prefabricated coordinate system. These coordinates then form the basis for the design of interconnects between a nanowire and the bond pads. Electron beam lithography is used to open up holes in the PMMA resist spin coated on top of the sample.

Ohmic contact technology is one of the cornerstones of semiconductor physics reflected in the wealth of papers published on the subject [45, 46]. If the contact quality is poor the device performance will suffer. A common figure of merit is the specific contact resistivity defined as the resistivity at zero applied bias. A good contact has a value somewhere below 10^{-5} Ωcm^2, depending on the material since some materials are easier to form high quality contacts to. The contact resistance can be calculated given the contact area and the specific contact resistivity. For macroscopic contacts (several μm^2) the contact resistance is most often below 1 Ω but as the contact area is made smaller the resistance will increase rapidly and might finally come to dominate the properties of the device. Thus, the smaller contact area the more important it is to fabricate high quality contacts.

5.2. Resist stripping

The first step towards an ohmic contact is to remove resist residues on the nanowire surface that is exposed to air after lithography. It is known that after development, resist residues are always present [47] to some extent. Since the PMMA resist

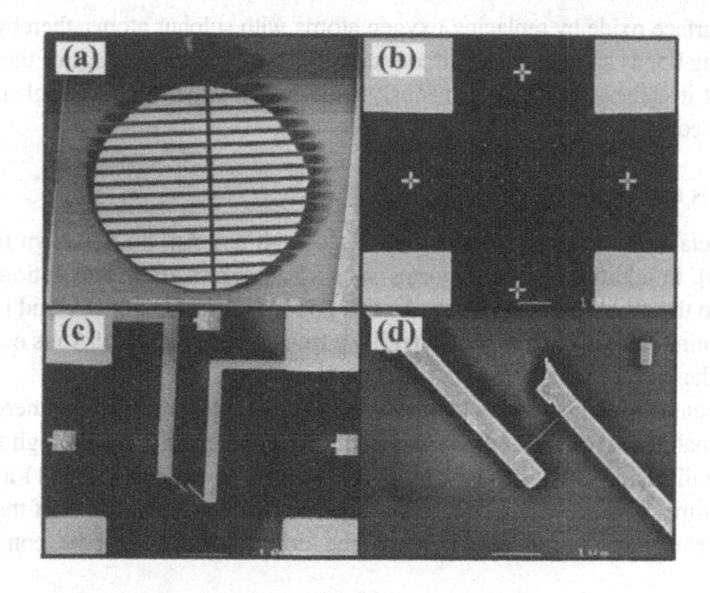

Fig. 10. Sample preparation steps. (a) Macroscopic metal bondpads are fabricated on top of an insulating SiO_2 layer. (b) A lithographically defined coordinate system is fabricated in between each set of four bondpads. These are then used to locate individual wires after the random deposition. (c) Using the coordinates of individual wires, the exposure file containing the interconnects between wire and bondpad are designed and exposed. (d) After metal evaporation and lift-off the wire device is finished. Here is shown a 40 nm diameter InAs wire with Au/In contacts.

consists of polymers, one way of removing residues is oxygen plasma ashing. In this process ionic species of oxygen are created by microwave excitation of oxygen. The organic material is removed by a combustion process with H_2O and CO_2 reaction products [48]. The drawback of this method is the radiation (ions, electrons and UV light), which may inflict damage to the wires. Oxidation of the nanowire surface also takes place. The duration of the ashing is therefore restricted to 15 s. TEM images have shown an oxide thickness after 15 s of oxygen plasma ashing to be roughly 2 nm.

5.3. Oxide removal

Oxide is most often removed by wet etching techniques. For the case of InAs a 5 s HF etch (6%) followed by 60 s in HCl : H_2O, 1 : 1, have given good results. AFM studies of exposed and developed areas of PMMA covered SiO_2 surfaces before and after HF etching has shown that after etching the surfaces are extremely clean. This is probably due to a lift-off process of resist residues and other contaminations on the SiO_2 surface. The HCl does not remove any SiO_2 but etches In- and As-oxides and also the InAs, although at a rather slow rate. A drawback of the HCl etch is the non-uniform etch rate.

An alternative technique is to passivate the surface prior to metal evaporation instead of the HCl etching. The sample is then put in a NH_4S_x : H_2O (1 : 9) solution, which removes surface oxide by replacing oxygen atoms with sulphur atoms, thereby removing dangling bonds and reducing surface states [49]. Another advantage of the passivation is that it etches InAs homogeneously and slowly, which gives a higher process stability as compared to the HCl etch.

5.4. Metals and post-processing

The metals of choice for InAs nanowires are Ni and Au. Ni is known to reduce oxides [45], in addition the metal forms very small grains after evaporation and the adhesion to the sample is also strong. Au is a good electrical conductor and is usually used to minimize heat generation. Generally a thick layer (600 Å) of Au is evaporated on top of the Ni layer and is used for the wire bonding.

If the contacts are not satisfactory after the processing steps described here, a short rapid thermal anneal of the contacts might improve the quality. At too high temperature As out diffusion can destroy the InAs wires and as illustrated in Fig. 11 a too long annealing time causes metal to diffuse through the entire wire. However, if the annealing parameters are chosen appropriately the contact quality can be considerably improved.

5.5. Electrical measurements on homogenous InAs nanowires

The typical current–voltage (*I–V*) characteristic of an InAs nanowire is linear (ohmic) at room temperature. The measured two-point resistance of a good contact varies between tens of kΩ down to a few kΩ. These values might be considered too large for a good contact, however one has to bear in mind that due to the small contact area, the contact resistance can be very high and of course depends on the specific contact resistivity. The wire itself will have a resistance of the order of a few kΩ depending on the dimension (typical lengths are 1 – 2 μm) and doping concentration.

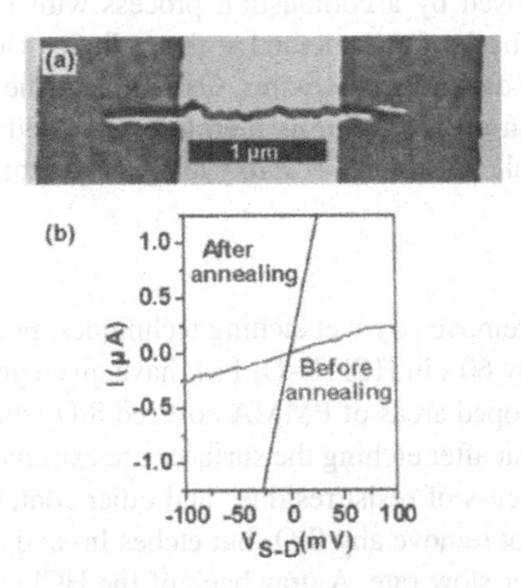

Fig. 11. Rapid thermal annealing processing of nanowire contacts. (a) The result of too high temperature during annealing. Metal has diffused throughout the entire wire. (b) *I–V* characteristics before and after succesful annealing. The resistance has dropped a factor of 40.

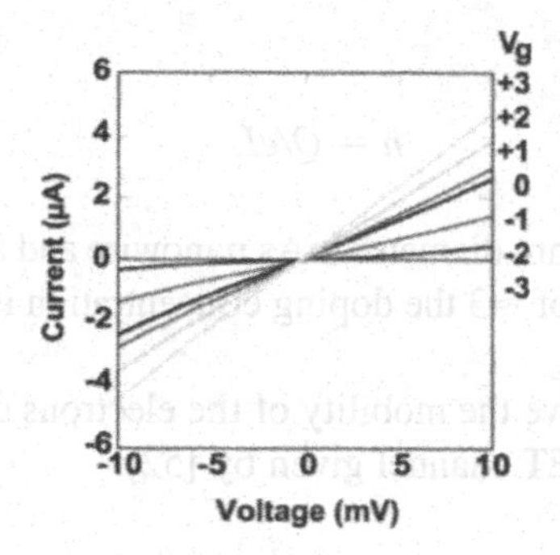

Fig. 12. *I–V* characteristics of a 40 nm diameter InAs nanowire at low temperature. The different curves correspond to different gate voltages applied to the backside Si substrate. The gate effect is consistent with an *n*-type semiconductor.

The linearity of the *I–V* is maintained even if the sample is cooled down to 4.2 K, provided that the resistance is within the limits previously specified. For poorer contacts the *I–V* characteristics will not remain linear at low temperatures most probably because of the existence of a small Schottky barrier between the metal and nanowire due to unsuccessful oxide- or impurity-removal prior to metal evaporation.

In InAs nanowires the main carrier type has been determined to be electrons (i.e. *n*-type) by changing the electrostatic potential on the backside of the underlying Si substrate (Fig. 12). An increasingly negative potential decreases the conductivity of the wire, which is consistent with electron transport. The wires are not intentionally doped making carbon, originating from the methyl-groups of the trimethylindium, the most likely dopant species. Carbon is an amphoteric dopant, which can act as either an acceptor or a donor depending on the semiconductor. It is known that C in InAs is a donor but in GaAs it is an acceptor [50]. Measurements on GaAs indeed show gate dependence consistent with *p*-type material. Changing the gate voltage thus allows one to tune the conductivity as in Si MOSFET's. In the nanowire case the wire itself acts as the inversion layer and the Si substrate as the gate separated from the wire by the SiO_2 layer. Since Hall measurements are extremely difficult to perform on nanowires there is no easy way of measuring mobility and doping concentration. One way of estimating the carrier density is to write down the charge on the nanowire as

$$Q = CV_{g,Th} \tag{1}$$

where C is the nanowire capacitance and $V_{g,Th}$ is the threshold gate voltage needed to completely deplete the wire of carriers. The capacitance per unit length with respect to the gate can be written [51]:

$$C/L = 2\pi \varepsilon \bar{\varepsilon} \ln (2h/r) \tag{2}$$

Here r and L is the wire radius and length respectively and ε is the average dielectric constant and h the thickness of the dielectric layer. The doping concentration is then given by:

$$n = Q/eL \tag{3}$$

For a 1 μm long and 40 nm diameter InAs nanowire and a 900 nm thick gate oxide with a dielectric constant ε of ≈3 the doping concentration is roughly $2*10^7$ cm^{-1} for a threshold voltage of 10 V.

If the transport is diffusive the mobility of the electrons can be estimated from the transconductance g_m of a FET channel given by [52]:

$$g_m = \partial i/\partial V_g = \mu\,(C/L^2)V_{S\text{-}D} \tag{4}$$

C is again the capacitance as given in (2), μ is the mobility and $V_{S\text{-}D}$ is the applied source-drain voltage.

With this method measurements on InAs nanowires yield mobilities μ of a few 100 cm^2/Vs [53], which are one or two orders of magnitude larger than for Si nanowires [54] but roughly two orders of magnitude smaller than for bulk InAs [55]. It should be noted that the calculated values are estimates since the effective radius of the wire is unknown because of surface depletion. In addition the estimates are done without considering the contact resistance. The low mobility is most likely due to surface- and/or impurity scattering and suggests a diffusive transport mechanism.

5.6. Electrical properties of InAs/InP nanowires

InAs nanowires with roughly 80 nm thick InP slices were grown and contacted by the procedures previously described to test the electrical quality of 1D InP/InAs interfaces. The size of the InP piece was chosen so that the strain field around each interface would not overlap (the extent of the strain field is roughly 10 nm for the InAs/InP system as observed from TEM analysis). The central parts of the InP crystal would therefore be totally relaxed and the conduction band off-set between relaxed InAs and InP could be deduced by measuring the thermal excitation of carriers over the InP barrier. The current contribution from the thermal excitation of carriers over a barrier of height $\varnothing_B$, is given by the relation:

$$i = A\,A^*T^2 \exp\{-q\varnothing_B/k_BT\} \tag{5}$$

Here A^* is the effective Richardson coefficient and A is the device area. By varying the temperature and plotting the logarithm of the measured current as a function of inverse temperature, it is possible to calculate the barrier height $\varnothing_B$. In Fig. 13(a) the I–V characteristics of an InAs nanowire containing an 80 nm thick InP barrier at different temperatures is shown. As expected, the current varies nonlinearly as a function of applied voltage. The transport mechanism is governed by thermal excitation of carriers over the barrier according to Equation (5). It is clearly seen in Fig. 13(a) how temperature affects the conductivity of the device. In addition, the effective barrier thickness is reduced as the applied voltage increases. Eventually the barrier thickness will become sufficiently thin for electrons to quantum mechanically tunnel through it

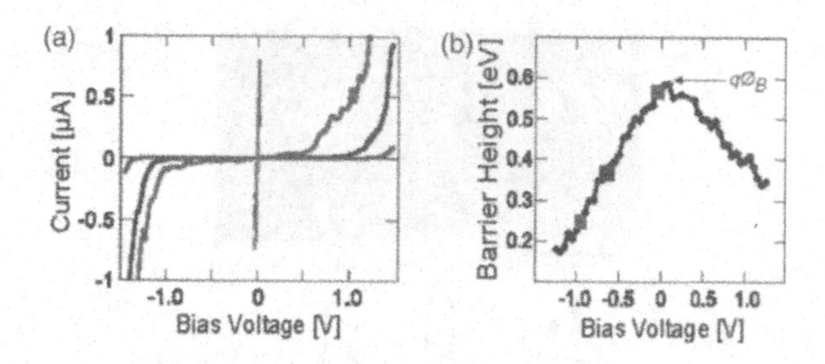

Fig. 13. Single barrier characterization. In (a) is shown the *I–V* curves at three different temperatures for a nanowire containing an 80 nm wide InP barrier. The effect of temperature (230, 270 and 310 K respectively) on the current level is clearly seen. For comparison the *I–V* characteristics of a homogeneous InAs wire is also shown. (b) Effective barrier height as a function of applied voltage. The zero voltage point corresponds to the conduction band off-set between InAs and InP. From the graph the off-set is determined to be 0.6 eV.

and a second transport path is possible. An increase in temperature will shift the tail of the Fermi–Dirac function to higher energies, which will allow more electrons to tunnel through an effectively thinner barrier at higher energies. Thus at large applied voltages the current will have contributions from field-induced tunneling and equation (5) is no longer valid, neither is it valid for very small voltages. At finite voltages the barrier height $\varnothing_B$ should be replaced by an effective, voltage dependent, barrier height $\varnothing_{Beff}(V)$. The conduction band off-set between InAs and InP can be deduced by calculating the barrier height $\varnothing_{Beff}(V)$ for different voltages, see Fig. 13(b), and extrapolate to zero applied voltage. Doing this procedure for several samples yields a mean value of 0.6 ± 0.2 eV[3]. This value is in good agreement with what is expected for the InAs/InP material system [56].

The measurements shown in Fig. 13 is performed in the temperature range of 230–390 K. Below the lower limit of this range the current level drops rapidly. At roughly 150 K the current through the 80 nm barrier is barely measurable with our equipment ($\approx$100 fA at 0.3 V). If thinner barriers are used the current level increases drastically since the tunneling transmission depends exponentially on the barrier thickness.

5.7. Nanowire heterostructure devices

The nanowire double barrier devices consist of an InAs emitter, collector and dot, and InP barriers. In Fig. 14(a) a low resolution TEM image of such a 40 nm diameter nanowire is shown. In this case the barriers are 5 nm thick and the central dot is 15 nm wide. Because of the small size of the dot the energy levels will be quantized in all three direction. For a 15 nm wide potential well 0.6 eV deep, the energy ground state is 40 meV above the conduction band. This value is obtained without consideration of the strain in the dot. The level splitting of the transverse states (due to the lateral size) is of the order of a few meV for a 40 nm diameter wire. Fig. 14(b) shows the approximate band diagram of the structure discussed. The transport mechanism for a double barrier system consists of electrons in the emitter tunneling through the

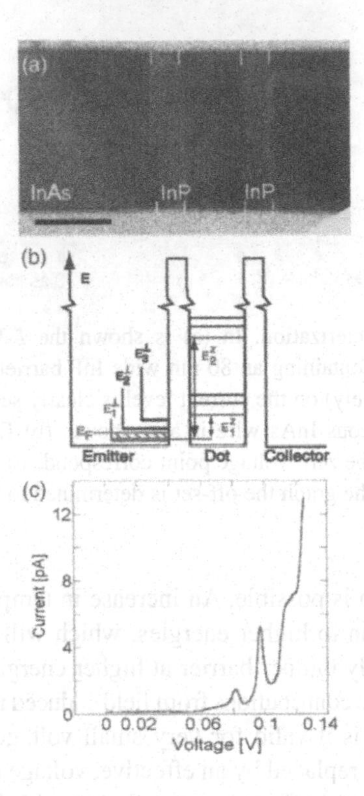

Fig. 14. 1D-0D-1D resonant tunneling structure in a nanowire. (a) TEM image of a InP double barrier structure. The barriers are 5 nm wide and the central quantum dot is 15 nm wide and 40 nm in diameter as determined by the seed particle diameter. (b) The resulting band diagram showing the 1D energy levels of the emitter and the fully quantized energy levels of the quantum dot. (c) 4 K measurements on a nanowire resonant tunneling diode. Energetically sharp peaks with peak-to-valley ratios of 3.5 : 1 is seen in the *I–V* characteristics.

double barriers via available discrete energy states in the InAs quantum dot. As the applied voltage is increased the dot states will move towards lower energy and when the Fermi energy in the emitter lines up with an energy level in the dot the transmission is greatly enhanced. When the level in the dot falls below the bottom of a subband, the transmission again drops to zero and negative differential conductance can be observed. If more than one subband is populated and the level spacing is sufficiently small, a stepwise increase on the rise of the resonance peak is expected. If the lateral quantization is the same in emitter and dot, all levels go out of resonance simultaneously and a sharp drop in current is the outcome. Here it is assumed that no mode mixing occurs.

Experimentally, the *I–V* characteristics of the nanowire devices consist of energetically sharp resonance peaks (FWHM ≈ 5–10 mV) as seen in Fig. 14(c). Virtually no current flows through the device below 70 mV because there are no energy levels in the dot aligned with the Fermi energy of the emitter. Then one or several sharp peaks are seen in the *I–V* characteristics. The current peak-to-valley ratios of the peaks are around 3.5 : 1, however as high as 50 : 1 has been observed [4]. The peak current density is on the order of 1 nA/μm^2 consistent with other similar devices [19]. This value is not corrected for surface depletion making the actual value higher. Negative differential resistance has so far only been observed at temperatures below 40 K, however, the double barrier structures have not yet been optimized.

6. Outlook

As indicated above, the ability to control the location of nucleation of individual nanowires opens the way to parallel positioning and fabrication of millions of identical devices, including advanced heterostructure devices such as resonant tunneling diodes, in just one processing step. Many of the limitations in creating heterostructure devices by conventional epitaxial growth related to lattice mismatch can, as we described in Chapter 4, be avoided by the bottom-up fabrication technique of nanowires. We have so far shown this for combinations of InAs and GaAs, for which the lattice mismatch is almost 7%, and for the InAs and InP combination having about half that lattice mismatch. In both cases we find that the geometrical conditions, with an open surface within some 10 nm distance from any point inside the nanowire, very effectively allows strain relaxation and leads to defect-free nanowire crystals and heterostructures. The importance of ultra-sharp heterostructures is obvious for many heterostructure device applications. The ability of the VLS-technique to offer such ultra-sharp heterointerfaces has clearly been demonstrated primarily in studies from our group, as demonstrated in Chapters 4 and 5. One can predict that the advantages of our nanowire heterostructure technique will lead to a break-through in silicon device and circuit technologies, for instance by incorporating silicon–germanium and silicon–carbon material structures, thus enabling advanced heterostructure devices to be incorporated in and mixed with silicon technology.

There is much interest in silicon transistors reduced to ultra small dimensions, approaching nanowire dimensions. Much of the technological difficulties in these techniques, with the control of the resulting diameters and the absence of pinch-off of localized islands inside these nanowires, would be very easy to avoid by the nanowire fabrication technique that we describe here. The possibilities for accurate lateral positioning of nanowire structures is also very attractive for the realization of interconnects for advanced circuitry in which one would like vertical monolithic nanowires to act as circuit interconnects.

We expect the future use of these parallel fabrication techniques also for photonic devices, such as light emitting devices as well as for photon detection. It is indeed very straight-forward to envisage the realization of arrays of such photonic devices, both acting in vertical mode, e.g. surface emitting devices, and such intended for handling of optical signals in the plane of a wafer. With the technique demonstrated in Chapter 5 where a single quantum dot is effectively wired up and addressed via tunneling through surrounding heterostructure barriers, the basic concepts for a single-photon-on-demand device are realized, along the lines proposed by Imamoglu [57] and Yamamoto [57, 58]. One of the application areas in which our technology will most ideally be implemented is in the formation of one-dimensional superlattices, for which there is a very strong interest for applications as quantum cascade laser structures, as pioneered by Capasso [59] but so far only made in conventional 2D epitaxy methods. Also here, the process-free fabrication via self-assembly will offer important advantages as do the freedom in combining materials of different lattice constants.

A special possible application area for nanowire materials could be for photovoltaic applications, in which "forests" of nanowires may be contacted via polymer filling of the material in between the wires and contacts applied via the substrate as well as from the top surface. Also here, the possibility to design multi-layer structures with great freedom will be of value.

An important area for applications of carbon nanotubes is for their use in field emission, where the very sharp tips of the nanotubes are used to extract and emit electrons that are then accelerated to hit a fluorescent screen. We expect that the possibility with the techniques we have describe here to controllably position individual field-emission tips, be that of silicon or of III–V semiconductors like InAs, will have several advantages compared to the rather uncontrolled formation of trusses of carbon nanotubes as demonstrated hitherto.

We have recently reported the first combination of magnetically active nanowire materials, so far studied for combination of GaAs and MnGaAs in nanowires [60]. We expect that it will be possible to fabricate ferromagnetic semiconductor nanowires and to use these for spin-polarized injection, which would have great application possibilities as read- and write-heads on the nanometer scale for magnetic storage applications. In many applications, like the one mentioned for spin-injection, nanowires can be used as advanced scanning-probe tips formed in a controlled fashion directly on the probe tip. One such application that we suggested more than ten years ago [61] was for the use of designed and very sharp semiconductor tips for injection of monoenergetic charge carriers (electrons or holes) in scanning tunneling spectroscopy applications, with special interest for direct creation of local luminescence by resonant injection of minority carriers as a single-particle approach to luminescence excitation spectroscopy. This was demonstrated in our laboratory using coarse semiconductor tips but this spectroscopy technique would have a significant value could it be implemented in a nanowire scanning probe tip geometry.

It is quite obvious that the nanowire fabrication technique may, in general, become important for scanning probe applications, including that of hugely parallel lithography as pioneered by IBM in the technology called Millipede [62]. We also anticipate the use of electrically contacted and monolithically integrated nanowires for use as ultra-fine probes to be used in life-science applications, e.g. for addressing and for read-out of electrical signals in neuroscience or for studies of individual cells.

With our methods for controlled positioning and structuring of nanowires it will be possible to create photonic bandgap materials by, again, a bottom-up approach as opposed to the top-down fabrication so far used via the drilling of holes in bulk material. This type of work was recently initiated and it is still too early to tell how perfect photonic crystal materials can be obtained this way.

Most of the above-mentioned application areas for our technologies are subject to further investigations and development [63]. Other research labs have pioneered the use of nanowires for other applications, such as for thermoelectrics and as extremely sensitive and selective sensor devices for medical applications. It seems safe to claim that this materials technology is likely to have a profound impact on many areas of technology and on the everyday life of us all.

References

1. R. S. Wagner, in *Whisker Technology*, Edited by A. P. Levitt, Wiley, New York (1970), pp. 47–119.
2. M. Yazawa, M. Koguchi and K. Hiruma, Heteroepitaxial ultrafine wire-like growth of InAs on GaAs substrates, *Appl. Phys. Lett.* **58** (1991) 1080.
3. M. T. Björk, B. J. Ohlsson, T. Sass, A. I. Persson, C. Thelander, M. H. Magnusson, K. Deppert and L. R. Wallenberg, One-Dimensional Steeplechase for Electrons Realized, *Nano Lett.* **2** (2002) 87.
4. M.T. Björk, B. J. Ohlsson, C. Thelander, A. I. Persson, K. Deppert, L. R. Wallenberg and L. Samuelson, Nanowire Resonant Tunneling Diodes, *Appl. Phys. Lett.* **81** (2002) 4458.
5. H. Sakaki, Scattering Suppression and High-Mobility Effect of Size-Quantized Electrons in Ultrafine Semiconductor Wire Structures, *Jpn. J. Appl. Phys.* **19** (1980) L735.
6. S. Hara, J. Ishizaki, J. Motohisa, T. Fukui and H. Hasegawa, Formation and photoluminescence characterization of quantum well wires using multiatomic steps grown by vapor phase epitaxy, *J. Cryst. Growth* **145** (1994) 692.
7. M. S. Miller, C. E. Pryor, H. Weman, L. A. Samoska, H. Kroemer and P. M. Petroff, Serpentine superlattice: concept and first results, *J. Cryst. Growth* **111** (1991) 323.
8. R. Bhat, E. Kapon, D. M. Hwang, M. A. Koza and C. P. Yun, Patterned quantum well heterostructures by OMCVD on non-planar substrates: Applications to extremely narrow SQW lasers, *J. Cryst. Growth.* **93** (1988) 850.
9. E. Kapon, D. M. Hwang and R. Bhat, Stimulated Emission in Semiconductor Quantum Wire Heterostructures, *Phys. Rev. Lett.* **63** (1989) 430.
10. W. Seifert, N. Carlsson, M. Miller, M.-E. Pistol, L. Samuelson and L. R. Wallenberg, In-situ Growth of Quantum Dot Structures by the Stranski-Krastanow Growth Mode, *Prog. Crystal Growth Charact.* **33** (1996) 423.
11. I. E. Itskevich, T. Ihn, A. Thornton, M. Henini, T. J. Foster, P. Moriarty, A. Nogaret, P. Beton, L. Eaves and P. C. Main, Resonant magnetotunneling through individual self-assembled InAs quantum dots, *Phys. Rev. B* **54** (1996) 16401.
12. M. Narihiro, G. Yusa, Y. Nakamura, T. Noda and H. Sakaki, Resonant tunneling of electrons via 20 nm scale InAs quantum dot and magnetotunneling spectroscopy of its electronic states, *Appl. Phys. Lett.* **70** (1996) 105.
13. M. Borgstrom, T. Bryllert, T. Sass, B. Gustafson, L.-E. Wernersson, W. Seifert and L. Samuelson, High peak-to-valley ratios observed in InAs/InP resonant tunneling quantum dot stacks, *Appl. Phys. Lett.* **78** (2001) 3232.
14. I. Shorubalko, H. Q. Xu, I. Maximov, P. Omling, L. Samuelson and W. Seifert, Nonlinear operation of GaInAs/InP-based three-terminal ballistic junctions, *Appl. Phys. Lett.* **79** (2001) 1384.
15. S. Reitzenstein, L. Worschech, P. Hartmann and A. Forschel, Logic AND/NAND gates based on three-terminal ballistic junction, *Electronic Lett.* **38** (2002) 951.
16. C. Papadopoulos, A. Rakitin, J. Li, A. S. Vedeneev and J. M. Xu, Electronic transport in Y-junction carbon nanotubes, *Phys. Rev. Lett.* **85** (2000) 3476.
17. M. A. Reed, J. N. Randall, R. J. Aggarwal, R. J. Matyi, T. M. Moore and A. E. Wetsel, Observation of Discrete Electronic States in Zero-Dimensional Semiconductor Nanostructure, *Phys. Rev. Lett.* **60** (1988) 535.

18. S. Tarucha, Y. Hiarayama, T. Saku and T. Kimura, Resonant tunneling through one- and zero-dimensionalstates constricted by $Al_xGa_{1-x}As/GaAs/Al_xGa_{1-x}As$ heterojunctions and high-resistance regions induced by focused Ga ion-beam implantation *Phys. Rev. B* **41** (1990) 5459.
19. J. Wang, P. H. Beton, N. Mori, L. Eaves, H. Buhmann, L. Monsouri, P. C. Main, T. J. Foster and M. Henini, Resonant Magnetotunneling via One-Dimensional Quantum Confined States, *Phys. Rev. Lett.* **73** (1994) 1146.
20. For a review, see G. Bastard, J. A. Brum and R. Ferreira, in Solid State Physics,Vol. 44, Edited by H. Ehrenreich and D. Turnbull, Academic Press, Boston, (1991), p. 229.
21. C. R. Proetto, Self-consistent electronic structure of a cylindrical quantum wire, *Phys. Rev. B* **45** (1992) 11911.
22. A. Gold and A. Ghazali, Analytical results for semiconductor quantum-well wire: Plasmons, shallow impurity states, and mobility *Phys. Rev. B* **41** (1990) 7626.
23. T. Ogawa and T. Takagahara, Optical absorption and Sommerfeld factors of one-dimensional semiconductors: An exact treatment of excitonic effects, *Phys. Rev. B* **44** (1991) 8138.
24. Y. Arakawa and T. Yamauchi and J. N. Schulman, Tight-binding analysis of energy-band structures in quantum wires, *Phys. Rev. B* **43** (1991) 4732.
25. Y. M. Niquet, C. Delerue, G. Allan and M. Lannoo, Method for tight-binding parametrization: Application to silicon nanostructures, *Phys. Rev. B* **62** (2000) 5109.
26. M. P. Persson and H. Q. Xu, Electronic structure of nanometer-scale GaAs whiskers, *Appl. Phys. Lett.* **81** (2002) 1309.
27. A. Franceschetti and A. Zunger, GaAs quantum structures: Comparison between direct pseudopotential and single-band truncated-crystal calculations, *J. Chem. Phys.* **104** (1996) 5572.
28. A. Franceschetti and A. Zunger, Free-standing versus AlAs-embedded GaAs quantum dots, wires, and films: The emergence of a zero-confinement state, *Appl. Phys. Lett.* **68** (1996) 3455.
29. D. W. Wood and A. Zunger, Successes and failures of the k·p method: A direct assessment for GaAs/AlAs quantum structures, *Phys. Rev. B* **53** (1996) 7949.
30. J.-B. Xia and K. W. Cheah, Quantum confinement effect in thin quantum wires, *Phys. Rev. B* **55** (1997) 15688.
31. R. S. Wagner, W. C. Ellis, Vapor-liquid-solid mechanism of single crystal growth, *Appl. Phys. Lett.* **4** (1964) 89.
32. E. I. Givargizov, in Current Topics in Materials Science, Edited by E. Kaldis, North Holland, Amsterdam (1978), pp. 79–145.
33. M. H. Magnusson, K. Deppert, J.-O. Malm, J.-O. Bovin and L. Samuelson, Gold nanoparticles: Production, reshaping, and thermal charging, *J. Nanopart. Res.* **1** (1999) 243.
34. H. Wang and G. S. Fischman, Role of liquid droplet surface diffusion in the vapor-liquid-solid whisker growth mechanism, *J. Appl. Phys.* **76** (1994) 1557.
35. W. Dittmar and K. Neumann, in: *Growth and Perfection of Crystals*, Edited by R. H. Doremus et al. Wiley, New York (1958), p. 121.
36. J. M. Blakely, K. A. Jackson, Growth of crystal whiskers, *J. Chem. Phys.* **37** (1962) 428.
37. K. Haraguchi, K. Hiruma, K. Hosomi, M. Shirai and T. Katsuyama, Growth mechanism of planar-type GaAs nanowhiskers, *J. Vac. Sci. Technol. B* **15** (1997) 1685.
38. J. Hu, T. W. Odom and C. M. Lieber, Chemistry and Physics in One Dimension: Synthesis and Properties of Nanowires and Nanotubes, *Acc. Chem. Res.* **32** (1999) 435.
39. K. Hiruma, M. Yazawa, T. Katsuyama, K. Ogawa, K. Haraguchi, M. Koguchi and H. Kakibayashi, Growth and optical properties of nanometer-scale GaAs and InAs whiskers, *J. Appl. Phys.* **77** (1995) 447.

40. X. F. Duan, J. F. Wang and C. M. Lieber, Synthesis and optical properties of gallium arsenide nanowires, *Appl. Phys. Lett.* **76** (2000) 1116.
41. B. J. Ohlsson, M. T. Björk, M. H. Magnusson, K. Deppert, L. R. Wallenberg and L. Samuelson, Size-, shape-, and position-controlled GaAs nano-whiskers, *Appl. Phys. Lett.* **79** (2001) 3335.
42. CBE and Related Techniques, Edited by J. S. Foord, G. J. Davies and W. T. Tsang, Wiley, Chichester (1997), pp. 331–393.
43. M. T. Björk, B. J. Ohlsson, T. Sass, A. I. Persson, C. Thelander, M. H. Magnusson, K. Deppert, L. R. Wallenberg and L. Samuelson, One-Dimensional Heterostructures in Semiconductor Nanowhiskers, *Appl. Phys. Lett.* **80** (2002) 1058.
44. K. Hiruma, M. Yazawa, T. Katsuyama, K. Ogawa, K. Haraguchi, M. Koguchi and H. Kakibayashi, Growth and optical properties of nanometer-scale GaAs and InAs whiskers, *J. Appl. Phys.* **77** (1995) 447.
45. T. C. Shen, G. B. Gao, and H. Morkoç, Recent developments in ohmic contacts for III–V compound semiconductors, *J. Vac. Sci. Technol. B* **10** (1992) 2113 and references therein.
46. A. G. Baca, F. Ren, J. C. Zolper, R. D. Briggs and S. J. Pearton, A survey of ohmic contacts to III–V compound semiconductors, *Thin Solid Films* **308–309** (1997) 599 and references therein.
47. I. Maximov, A. A. Zakharov, T. Holmqvist, L. Montelius and I. Lindau, Investigation of polymethylmethacrylate resist residues using photoelectron microscopy, *J. Vac. Sci. Technol. B* **20** (2002) 1139.
48. W. M. Moreau, Semiconductor Lithography, Principles, practices, and materials, Plenum New York (1988), pp. 791–793.
49. V. N. Bessolov and M. V. Lebedev, Chalcogenide passivation of III-V semiconductor surfaces, *Semiconductors* **32** (1998) 1141.
50. M. Kamp, R. Contini, K. Werner, H. Heinecke, M. Weyers, H. Lüth and P. Balk, Carbon incorporation in MOMBE-grown $Ga_{0.47}In_{0.53}As$, *J. Cryst. Growth* **95** (1989) 154.
51. R. Martel, T. Schmidt. H. R. Shea, T. Hertel and Ph. Avouris, Single- and multi-wall carbon nanotubes field-effect transistors, *Appl. Phys. Lett.* **73** (1998) 2447.
52. S. M. Sze, *Semiconductor Devices, Physics and Technology*, Wiley, New York (1985), p. 206.
53. B. J. Ohlsson, M. T. Björk, A. I. Persson, C. Thelander, L. R. Wallenberg, M. H. Magnusson, K. Deppert and L. Samuelson, Growth and characterization of GaAs and InAs nano-whiskers and InAs/GaAs heterostructures, *Physica E* **13** (2002) 1126.
54. Y. Cui, X. Duan, J. Hu and C. M. Lieber, Doping and Electrical Transport in Silicon Nanowires, *J. Phys. Chem. B* **104** (2000) 5213.
55. Physics of Group IV Elements and III–V Compounds, Vol. 17a, Edited by O. Madelung, M. Schulz, H. Weiss and Landholt-Börnstein, Springer, New York (1982), p. 302.
56. I. Vurgaftman, J. R. Meyer and L. R. Ram-Mohan, Band paramters for III/V compound semiconductors and their alloys, *J. Appl. Phys.* **89** (2001) 5815.
57. A. Imamoglu, Y. Yamamoto, Turnstile Device for Heralded Single Photons: Coulomb Blockade of Electron and Hole Tunneling in Quantum Confined p-i-n Heterojunctions, *Phys. Rev. Lett.* **72** (1994) 210.
58. J. Kim, O. Benson, H. Kan, Y. Yamamoto, A single-photon turnstile device, *Nature* **397** (1999) 500.
59. F. Capasso, C. Gmachl, D. L. Sivco, A. Y. Cho, Quantum Cascade Lasers, *Physics Today*, May (2002), p. 34.
60. J. Sadowski, K. Deppert, J. Kanski, J. Ohlsson, A. Persson, L. Samuelson, Migration Enhanced Epitaxial growth (In, Ga, Mn)As magnetic semiconductor nanowhiskers, Proceedings NANO-7/ECOSS-21, Sweden (2002).

61. L. Samuelson, J. Lindahl, L. Montelius, M.-E. Pistol, Tunnel-Induced Photon Emission in Semiconductors using an STM, *Physica Scripta* **T42** (1992) 149.
62. P. Vettiger et al., The "Millipede"— More than one thousand tips for future AFM data storage, *IBM J. Res. Develop.* **44** (2000) 323.
63. More information on the development of these nanowire technologies, including IPR-issues, is available via the web-pages of the Nanometer Consortium, www.nano.ftf.lth.se and QuMat Technologies AB, www.qumat.se.

Part II

Theory of Nanowires

Chapter 3

Theoretical Study of Nanowires

Hatem Mehrez[1] and Hong Guo[2]

[1]*Department of Physics, Harvard University, Cambridge, MA 02138;*
[2]*Center for the Physics of Materials and Department of Physics, McGill University, Montreal, PQ, Canada H3A 2T8*

1. Introduction

Electronic devices are the underpinning of information technology. The recent advent of nano-electronics has opened up a new frontier, which aims at the ultimate miniaturization of electronic devices [1–10]. Indeed, if the Si based micro-electronic technology continues along the Moore's law [11] of size scaling, individual devices will reach molecular scale in a decade. However, it is generally accepted that the micro-electronic technology at its present form will not function well at the nano-scale due to physical limitations. While hybrid micro–nano systems may provide a short "breathing" period of time, it does not provide a fundamental solution for the next generation of smaller, faster, cheaper and massively parallel computer technology. It is these requirements, which drives the nano-electronics research to discover a fundamentally new device paradigm based on quantum phenomena. The recently discovered nano scale systems, including carbon nanotubes [12], fullerenes [13], molecular wires [14], switches [3] and nanoscale magnetic systems [15], as well as new fabrication techniques based on massive self-assembly [16] and/or bio-technology [17], have demonstrated great potential for realizing the new device paradigm in the nano-meter scale.

The most basic component of a nano-electronic system is a nanowire, which conducts electric current. Our interest in this Chapter is on nanowires at the atomic and molecular scale, where details of the microscopic degree of freedom become important. Due to the rapid progress in experimental measurement techniques, it is now possible to ask and even to provide partial answers to questions such as: *what is the conductance of a single atom or a chain of atoms?* Experimental data on nanowires clearly call theoretical analysis in order to understand the basic physics that controls electronic, mechanic, and transport properties of the nanowires. Although these are the simplest nano-electronic systems, complicated issues can however arise. For example, in two dimensions or less, Fermi liquid theory may break down in ordered systems and strong electron correlation emerges [18]. Infinitely long nanowire is quasi-one dimensional, therefore they exhibit non-Fermi liquid behavior as observed recently in carbon nanotube wires [19–21]. On the other hand, for short nanowires connected to external leads, the wire-lead junction can dominate transport features and the non-Fermi liquid property becomes less significant. The electronic property of a nanowire can be drastically different compared to its bulk counterpart. For example, bulk Si is a semiconductor while a Si chain is metallic [22].

The experimental fabrication of molecular scale nanowires has progressed along several lines. A common method of investigation is using the atomic manipulation ability of Scanning Tunneling Microscopy (STM) and/or Atomic Force Microscopy (AFM). When a sharp tip is approaching a substrate, an atomic "neck" can be formed between the tip and the substrate [23–25]. As the tip is pulled away, a wire of atomic scale is formed and electrical conductance is measured [23–34]. During the elongation process of the "neck", the cross section becomes smaller, thereby reducing the number of transport subbands. As a result electrical current between the tip and the substrate is reduced in a stepwise manner: it is "quantized". Similar quantization can also be observed during an indentation process [31, 32]. The wire between the tip and the substrate can be extremely small, involving only a few atoms. Hence the quantized current may be observed at room temperature due to the large energy level spacing in the wires [24, 25, 34, 35]. The high precision measurement can be pushed further to measure the resistance of a single atom sandwiched between a tip and a substrate, as demonstrated by the work of Yazdani et al. [36]. Furthermore, both STM and AFM have been applied to fabricate free standing single electron transistor [37–39] and atomic point contact [40], by using the tip to anodize a thin film.

A remarkably simple technique of fabrication, also based on changing the conductor size while measurement is being carried out, is the mechanically controllable "break-junction" (MCBJ) technique [41, 42]. Here a piece of metal ribbon mounted on a plastic strip is pushed from below until it breaks. A tiny atomic wire is then formed at the break junction that can be controlled by piezoelectric. Good mechanical control can be achieved and detailed charge transport properties can be measured [43].

The above methods are based on clever mechanical techniques, and conductance quantization is owed to the abrupt structural variation of the conductor. From an application point of view, however, substantial effort must be devoted to fabricate "free standing" wires. Several processes based on lithographic techniques have fabricated silicon or polycrystalline Si nanowires [44–48] in the few nanometer range, on top of a substrate. Indications of quantized electrical current as a function of bias voltage were observed [44] at temperatures as high as 170 to 210 K, while Coulomb charging peaks were seen at 75 K. Routkevitch et al. have used an electro-chemical fabrication process to obtain CdS nanowires in the few nanometer range [49, 50]. In addition, there have been experimental evidences that spontaneously formed nano-systems, based on strained island formation on top of a lattice mismatched crystal layer [51, 52] can have transport properties of a point contact [53].

Perhaps the most exciting free-standing nanowire is a carbon nanotube (CNT) [54]. CNTs are either metals or semiconductors depending on their chirality (which is denoted by a pair of integers *(n,m)*). Considerable amount of experimental [7–9, 19–21, 55–69] and theoretical [70–85] work have been devoted to CNTs. Progress has been rapid, and several prototypical devices have already been constructed and demonstrated [7–9, 21, 56–59]. CNTs can be contacted by other materials to achieve nanotube tunnel junctions, for instance magnetic-nanotube tunnel junction [66, 80] and superconductor-nanotube junctions [68, 69, 82] have been experimentally and theoretically explored.

Intensive theoretical investigations on nanowires have already been carried out in order to understand properties of these interesting systems. Phenomenological, semi-classical, and full quantum mechanical, *first principles*, methods have been developed for their investigations. It is the purpose of this Chapter to briefly review these activities. Because the ultimate aim of nanowires is at the nanoelectronic device application, we will focus on theoretical issues of charge transport and electronic property while briefly touch mechanical analysis of the nanowires. Previous works clearly demonstrated that many of the device characteristics of nanowires are directly related to specific atomic scale degrees of freedom and to interactions of the nanowire with device electrodes. Therefore, to be able to make quantitative predictions without phenomenological parameters on quantum charge transport through nanowires (and molecular scale devices), *first principle* formalisms are the best to use.

A powerful technique, which has seen wide-spread application in condensed matter electronic theory, is the density functional theory (DFT) [86, 87]. However, to analyze quantum transport through molecular scale devices under finite bias potentials using DFT type approach, some points of essential importance must be noted. First, one must deal with *open* systems within the DFT formalism. Recall that conventional DFT methods, *for example*, the well-known plane wave methods discussed in Ref. [88] or real-space techniques such as Ref. [89] can treat two kinds of problems: (i) finite systems such as an isolated molecule, as in quantum chemistry; (ii) periodic systems consisting of super-cells, as in solid state physics. In contrast, a nanoelectronic device has *open* boundaries provided by long and different electrodes, which maintain different chemical potential due to an external bias. In other words, typical device geometry is neither isolated nor periodic. Therefore, one must find a *new* technique, beyond the (i,ii) above, to reduce the infinitely large problem of an open device to a finite problem which can be efficiently dealt with. Second, one must calculate the device Hamiltonian, $\boldsymbol{H}[\rho(\boldsymbol{r})]$, within DFT using the correct charge distribution $\rho(\boldsymbol{r})$ which must be constructed under a finite bias with the correct open boundary conditions. For open systems, which extend from $z = -\infty$ to $z = +\infty$, $\rho(\boldsymbol{r})$ is contributed by both scattering states, which connect the reservoirs at $z = \pm\infty$ and crossing the molecular region, and by bound states that may exist inside the molecular region. It is worth noting that when the electro-chemical potentials, $\mu_{l(r)} + \Delta V_{l(r)}$, of the two electrodes are not equal, the device is actually in a nonequilibrium state. Here $\Delta V_{l(r)}$ is the bias voltage applied at the left (right) reservoirs and $\mu_{l(r)}$ is the chemical potentials there. Therefore there is the issue of nonequilibrium distribution in order to construct charge density even if all the scattering and bound states can be obtained. Finally, an efficient and transferable numerical procedure associated with the proper theoretical framework is needed in order to model systems with hundreds of atoms. Fundamental progress has been made on these issues in recent years and will be reviewed in some detail below.

The research of mechanical, electronic, and transport properties of nanowires presents a wide-open field for physicists, chemists, electrical engineers, applied mathematicians, and computer scientists. It is impossible to review the vast existing literature in a short Chapter here. Instead, we will choose to discuss a few important theoretical issues of nanowires, and focus on theoretical approaches for quantitative prediction of transport through nanowires. We organize this Chapter as follows. In the next section, we review important basic concepts of quantum transport theory for nanowires, and briefly present known phenomenology of electronic and mechanical properties of these systems. Following this we review in more detail theoretical tools for predicting charge transport in nano-devices, and apply our new approach to Au nanowires. The last section is reserved for a summary and outlook.

2. Phenomenology

In the rest of this Chapter, we will restrict our definition of a nanowire to those having a strong quantum confinement. This confinement modifies charge transport and other physical properties of the wire in essential ways. We further assume that the linear dimension of the nanowire is less than the phase coherence length so that quantum coherence is maintained.

2.1. Semiconductor quantum wires and sub-bands

Quantum wire research has advanced by semiconductor fabrication technology, which permitted construction of high mobility *two-dimensional electron gas* (2DEG) in semiconductor heterostructure. Using the split gate technique [90, 91], electrons in the 2DEG can be depleted electrostatically by a negative gate voltage (V_g), leaving channels of electrons defining a 2D quantum structure. Fig. 1 schematically shows a 2D quantum wire with width L_x defined electrostatically. The width L_x can be controlled by the value of the gate voltage, V_g.

The main conductance feature of the quantum wire can be obtained from Landauer formula [92]. Since we will consider 3D wires later, let's start by assuming a simple 3D wire along the z-direction while confined in the x–y-directions with a hard wall potential[1]

$$V(x, y, z) = \begin{cases} 0 & \text{for } 0 < x < L_x \text{ and } 0 < y < L_y \\ \infty & \text{otherwise} \end{cases} \tag{1}$$

where L_x, L_y are the widths of the wire along x–y-directions. Again, these parameters can be controlled by V_g. Using effective mass approximation within single particle theory, potential confinement of Eq. (1) allows a separation of the total wave function

[1] In quantitative analysis, the actual confining potential of a system can be calculated numerically. However, in this case we use a simple potential model to solve the problem analytically.

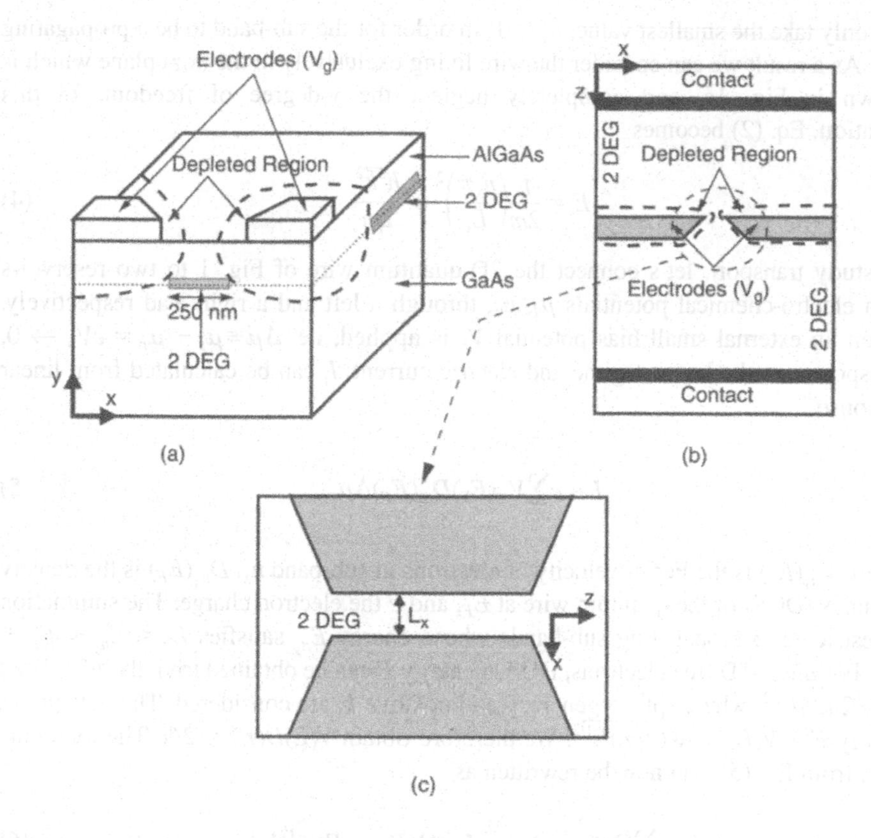

Fig. 1. Schematic diagram of 2DEG quantum point contact. (a) Lateral cross section view and (b) top view. Electrons flow through the undepleted region (enlarged in (c)) which is controlled by V_g, thereby defining a quasi 1D quantum wire with width L_x.

and energy into lateral (x–y-directions) and longitudinal (z-direction) parts,

$$E = E_{x,y} + \frac{\hbar^2 \vec{k}_z^2}{2m} \tag{2}$$

where $\vec{k}_z$ is the momentum vector in the z-direction, m the effective mass, and $E_{x,y}$ the lateral eigen-energy spectrum,

$$E_{x,y} = E_{n_x,n_y} = \frac{\hbar^2}{2m}[(\frac{n_x \pi}{L_x})^2 + (\frac{n_y \pi}{L_y})^2] \quad (n_x, n_y \neq 0) \tag{3}$$

Here integers n_x,n_y define a "sub-band" of the quantum wire. From Eq. (2), for a given electron energy E, only a limited number of sub-bands can propagate inside the quantum wire (see Fig. 1). These propagating sub-bands ($E > E_{x,y}$) have real momentum $\vec{k}_z$. On the other hand, there are an infinite number of "evanescent" modes, for which $\vec{k}_z$ are purely imaginary. For the simpler case of wires with $L_y << L_x$, i.e. when the thickness in the y-direction is very small (Fig. 1a), from Eq. (3) we conclude that n_y

can only take the smallest value, $n_y = 1$, in order for the sub-band to be a propagating one. As a result we can consider the wire living exclusively in the (x,z) plane which is shown in Fig. 1c, and completely neglect the y-degree of freedom. In this situation, Eq. (2) becomes

$$E = \frac{\hbar}{2m}\left(\frac{n_x \pi}{L_x}\right)^2 + \frac{\hbar^2 \vec{k}_z^2}{2m} \tag{4}$$

To study transport, let's connect the 2D-quantum wire of Fig. 1 to two reservoirs with electro-chemical potentials μ_l, μ_r, through a left and a right lead respectively. When an external small bias potential V_b is applied, i.e. $\Delta\mu \equiv \mu_l - \mu_r = eV_b \rightarrow 0$, transport is in the linear regime and electric current, I, can be calculated from linear response,

$$I = e\sum_{n_x} V_{n_x}(E_F) D_{n_x}(E_F) \Delta\mu \tag{5}$$

where $V_{n_x}(E_F)$ is the Fermi velocity of electrons at sub-band n_x. $D_{n_x}(E_F)$ is the density of states (DOS) of the quantum wire at E_F, and e the electron charge. The summation is restricted to propagating sub-bands whose energy E_{n_x} satisfies $E_F < E_{n_x} < E_F + \Delta\mu$. For *quasi*-1D free electrons, DOS at energy E can be obtained trivially [93], $D(E) = \sqrt{2m/hE^{1/2}}$, where spin degeneracy and positive $\vec{k}_z$ are considered. The velocity at energy E is $V(E) = \sqrt{(2/m)E^{1/2}}$. We therefore obtain $V(E)D(E) = 2/h$. The total current, from Eq. (5), can now be rewritten as

$$I = \sum_{n_x} \Theta(E_F + \Delta\mu - E_n) \Theta(E_{n_x} - E_F) \frac{2e}{h} \Delta\mu \tag{6}$$

where we have inserted two step-functions, Θ, which allow us to expand the restricted summation to infinity. Note that only the Θ-functions depend on the sub-band index n_x, and $\sum_{n_x} \Theta_F + \Delta\mu - E_{n_x}) \Theta(E_{n_x} - E_F) = N_c$ which is the total number of propagating (conducting) channels of the quantum wire. We therefore conclude

$$I = \frac{2e}{h} \Delta\mu N_c, \quad \text{hence}$$

$$G = \frac{I}{V_b} = \frac{2e}{h}\frac{\Delta\mu}{V_b} N_c = \frac{2e}{h}\frac{eV_b}{V_b} N_c = \frac{2e^2}{h} N_c \tag{7}$$

Therefore, for a perfect *quasi*-1D quantum wire, the conductance G is an integer multiple of the "quantum conductance" $G_0 \equiv (2e^2/h) = (12.9\ \mathrm{K\Omega})^{-1}$. Importantly, the number of propagating modes, N_c, is controllable through V_g which tunes the width of the wire, L_x (see Fig. 1). As V_g is made more negative, width L_x is decreased, reducing the number of conducting channels N_c. We therefore expect a step-wise, or quantized, decrease of conductance by changing V_g with a G_0 step size. The first experimental results which demonstrated this "ballistic behavior" were provided by van Wees et al. [90] and Wharam et al. [91] more than a decade ago. A schematic description of such experimental results is shown in Fig. 2. These early experiments provided break

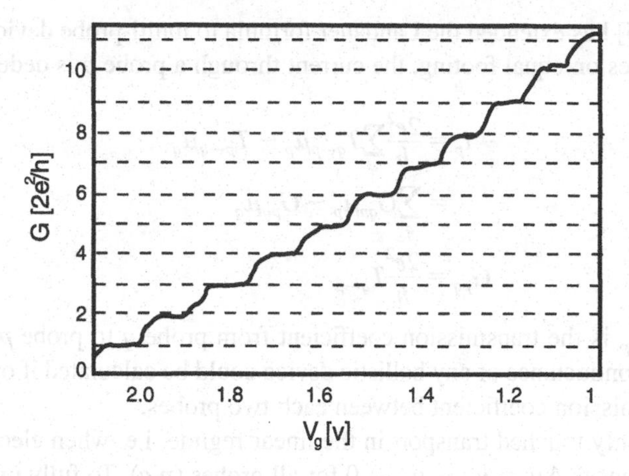

Fig. 2. Schematic diagram of zero-bias conductance of a 2DEG as function of Gate Voltage (V_g) which controls the width of the constriction L_x, or number of sub-bands at Fermi Energy.

through evidence that charge transport in the quantum regime is very different from the classical drift diffusion. Indeed, as Eq. (7) shows, the conductance G is not a function of the nanowire length, which is a nonclassical behavior. Such a property is a direct consequence of the small dimension of the nanowire when its linear dimension is less than the mean-free path of the carriers.

For a quantum wire involving some scattering centers, for instance if an impurity exists inside the wire of Fig. 1, the conductance will reduce from the ideal value of Eq. (7) due to scattering and inter channel mixing occurs. A very similar derivation as above [92, 94, 95] gives:

$$G = \frac{2e^2}{h}T = \frac{2e^2}{h}tr(\tilde{t}^+\tilde{t}) \tag{8}$$

where quantity T is the total transmission coefficient, tr denotes the trace of a matrix, and $\tilde{t}^{(+)}$, is the transmission matrix (and its Hermitian conjugate) through the device between all sub-bands. For the case of independent sub-bands, $\tilde{t}$ is diagonal and Eq. (7) is recovered from Eq. (8). Therefore, the conductance calculation of a device is reduced to the determination of the matrix $\tilde{t}$.

In the Landauer formalism discussed so far, inelastic scattering is neglected in the scattering region, because we assumed $L << L_c$. What is the origin of the measured resistance? For a perfect nanowire, should this resistance approach zero? How can these results be compared to experimental measurements? These questions were among the subtle points for ballistic transport until they were clarified during late 1980s, by the works of Imry and Büttiker. Imry [96] gave an elegant understanding of the origin of resistance in ballistic devices. Although no scattering occurs inside the structure for a perfect nanowire, it is still present at the interface between the reservoir and the device, due to wave function mismatch. Therefore, the measured resistance for a perfect wire, which can be obtained by inverting Eq. (7), is nothing but the contact resistance.

Büttiker [95] has extended the Landauer formula to multi-probe devices. By treating all the probes on equal footing, the current through a probe p is deduced as:

$$\begin{aligned} I_p &= \frac{2e^2}{h}\sum_q T_{q\leftarrow p}\mu_p - T_{p\leftarrow q}\mu_q \\ &= \sum_q G_{qp}\mu_p - G_{pq}\mu_q \\ G_{pq} &= \frac{2e^2}{h}T_{q\leftarrow p} \end{aligned} \tag{9}$$

and $T_{p\leftarrow q} = T_{pq}$, is the transmission coefficient from probe q to probe p. Within this approach, the conductance of any ballistic device could be calculated if one can determine the transmission coefficient between each two probes.

So far we only touched transport in the linear regime, i.e. when electro-chemical potential differences $\Delta\mu = \mu_p - \mu_q \to 0$ for all probes (p,q). To fully understand the operation of a device, one needs to go beyond the linear regime and investigate non-linear current-voltage $(I - V_b)$ characteristics. For a two-probe device, the $I - V_b$ curve can be calculated as:

$$I = \frac{2e^2}{h}\int_{\Delta_1}^{\Delta_2} T(E,V_b)[f_l(E) - f_r(E)]dE \tag{10}$$

where $T(E,V_b)$ is the transmission coefficient at energy E for a given bias voltage V_b, and $f_{l(r)}$ is the Fermi distribution function in the Left (Right) reservoir. It is important to emphasize that $T(E,V_b)$ and the Fermi functions depend on external bias voltage V_b. Although the integration range in Eq. (10) should be from $-\infty \to +\infty$, due to the Fermi functions we only need to worry about a small range, with Δ_1 smaller than $\mu_{\min}$ by a few temperature scale K_bT and Δ_2 greater than $\mu_{\max}$ by a few K_bT. Here $\mu_{\min}$, $\mu_{\max}$ are the minimum and maximum chemical potentials of the two reservoirs, respectively.

2.2. Atomic and molecular wires

The semiconductor quantum wire constructed by 2DEG typically has widths in the range of ≈ 100 nm. Much smaller nanowires can now be fabricated; atomic-wires have cross section area of one or few atoms range [97–99]. Molecular wires are formed by conjugated molecules [2–5, 99–106], or carbon nanotubes [6–8, 19–21, 54, 107–113], their cross section radius is also in the ≈ 1 nm range. These wires have received great attention in recent years. As discussed in the Introduction, several fabrication techniques of atomic wires have been used including scanning probes, mechanical break junctions, lithographic methods, and self-assembly. Because it is rather difficult to precisely determine the positions of each atom in an atomic/molecular wire except in highly nontrivial experiments [98, 114], and it is also very difficult to precisely determine details of the wire-electrode contacts, theoretical analysis of atomic/molecular wires usually start from physically plausible models.

Fig. 3 illustrates the simplest theoretical model of an atomic/molecular wire, where a group of atoms (or conjugated molecules) is sandwiched between two electrodes [22]. In this plot the electrodes are represented by Jellium model where the nuclear charge is uniformly distributed in the volume of the electrodes [22, 115, 116]. Jellium model does

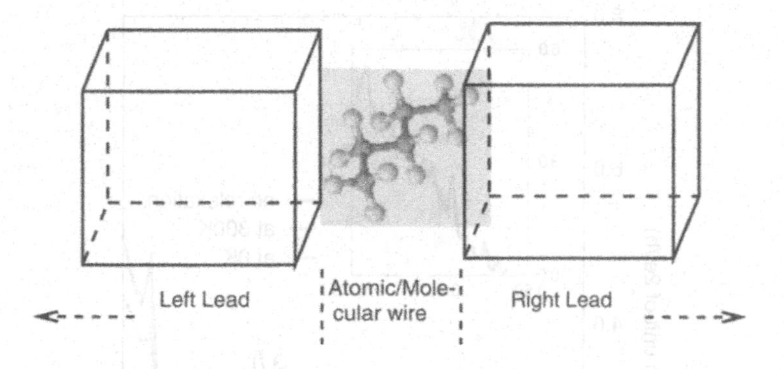

Fig. 3. Schematic plot of a molecular/atomic wire contacted by external leads, which extend from $-\infty \rightarrow +\infty$.

not account fully for the physics and chemistry of the contacts, but it simplifies theoretical and numerical analysis, and gives qualitatively adequate results for many situations. Recent theoretical developments, however, have overcome the short comings of Jellium electrodes by using more realistic electrodes with atomic details and crystalline structure [117–123]. So far, the positions of the atoms in the nanowire section (the middle part of Fig. 3) have been determined by quantum molecular dynamics (QMD) [124] before the transport analysis is carried out [124], but progress has also been made in calculating quantum mechanical forces during charge current flow [125] and therefore, in principle, the nanowire structure can also be determined during transport.

To give an impression of the transport properties of an atomic wire, consider an Al(1 0 0) nano structure arranged with alternating stacks of *four* and *five* atoms in each layer and terminated with the *five*-atom stack at the ends. With *nine* stacks, such a wire has 41 atoms in the molecular section. After full QMD structural relaxation within plane-wave DFT, we found that at temperature $T = 0$ K the final configuration has reached a state which is different from the bulk structure. On the other hand, at $T =$ 300 K the atomic positions are quite disordered. The distorted atomic configuration provides a natural positional disorder to quantum conduction. The QMD simulations not only determined the atomic positions, but also provided interesting information about the elastic properties of the atomic wires [124]. The equilibrium conductance of the atomic wire can be determined by a combination of DFT with scattering matrix theory [22, 115, 116], and in Fig. 4 we plot the conductance $G(E)$ as a function of incoming electron energy, E, where high density Jellium is used to model the two electrodes. The three curves are for wires relaxed at 0, 300 K and no dynamic relaxation. In all cases some degree of conductance "plateau" is observed, as expected from Eq. (7). Here, as the energy, E, is increased, more transmission channels become available in the atomic wire, and conductance increases in step like fashion. However the quantization of $G(E)$ is not perfect, reflecting the fact that the dynamically relaxed atomic wires have some degree of positional disorder that provides scattering to the charge carriers. For the wire without relaxation, the quantization is better due to the crystalline structure of the atomic section. However the atom–electrodes junction still provides scattering to the electrons which gives rise to quantum mode mixing, leading to the

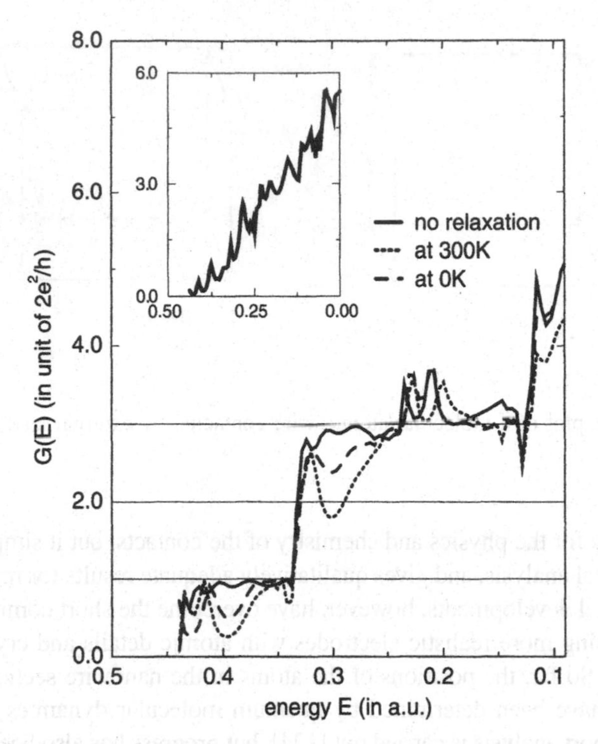

Fig. 4. Conductance, *G(E)*, as a function of incoming electron energy *E* for wires with *41* Al atoms in the atomic section contacted by Jellium leads. Inset: *G(E)* for a wire relaxed at 0K with *free* ends (not pinned by the leads).

imperfections of the quantization and the resonance behavior. Indeed, the experimentally fabricated atomic wires rarely showed perfect quantization [24, 25, 40, 44] due, in part, to the same reason. The inset of 0 plots *G(E)* for a wire dynamically relaxed at 0 K with the ends of the atomic section free to move instead of being pinned by the leads. The QMD simulation showed that in this wire the atomic positions are completely disordered. *G(E)* clearly confirms this fact and shows no quantization at all.

Atomic wires have received great attention because they are, perhaps, the smallest conductors that can be fabricated. Recently, extensive studies are directed to molecular wires [2–5, 99–106] since they have more potential for application. Molecular wires are represented in Fig. 3 with conjugated molecules sandwiched between the metallic leads. The fabrication technique of these devices is known as molecular self-assembly [16, 17]. Molecular wires fabricated in ambient condition are strongly affected by environmental factors, and the outcome of the wire structure is less controllable and their physical property suffer in reproducibility aspect. Although end-to-end control of device fabrication of molecular wires is difficult, these systems are promising as they provide a wide range of electronic responses which makes them good candidates for device application.

Experimentally, conjugated molecules can be deposited onto surfaces [126–128]; they can form a contact structure at a MCBJ [2, 3] junction; or they can form a contact between a STM/AFM tip and a surface [5, 38, 39]. Very different electronic responses have been observed on molecular wires even for seemingly similar structures. Because atomic/molecular details of the nanowires and the contacts play a very important role, results often show some degree of ambiguity, uncertainty, and even contradictory outcome, due to the atomic details of different devices. For example, DNA molecular wires have been reported as both metallic [129] and semiconducting [103] with a gap of $\cong 2$ eV. On the other hand, Ref. [130] reported that a λ–DNA is an insulator, but it can be turned into a conductor if the contact is modified. The uncertainty in the details of the molecular wire also provides difficulty in the interpretation of experimental data. STM experiments on single C_{60} molecular junction reported [38] a conductance $\approx G_0$ for well contacted C_{60} molecule, despite the $\cong 2$ eV HOMO-LUMO gap of C_{60}. This large conductance has been interpreted as due to a structural change of C_{60} when contacted by STM [38]. Recent theoretical analysis [118, 131], on the other hand, have found that there is a charge transfer from the leads to the C_{60} molecule in metal–C_{60}–metal junction, and the transferred charge aligned the C_{60} LUMO to the Fermi level of the leads, causing large conductance to occur. The above analysis of the existing literature on molecular wires clearly demonstrates the need to reach end-to-end control of wire fabrication, as well as extensive theoretical/ numerical analysis to understand the underlying mechanisms for charge transport.

2.3. Carbon nanotube as nanowires

CNTs are fairly new fullerene materials, which have been known for a decade [12]. They are composed of graphite layers (graphene) wrapped to form a tubule. These novel structures have been a focus of research because they have a wide range of physical, chemical and mechanical properties, which make them ideal candidate for nanoelectronics application. CNTs have radius ≈ 10 Å but can have a long length ≈ 1 μm. The large length/radius ratio makes CNTs very good system to analyze 1D-quantum effects. Another most interesting characteristic of CNTs is the dependence of their electronic property, i.e. being metallic or semiconducting, on the tube index *(n,m)* [54]. The intensive research in CNT physics has rapidly advanced our understanding of this class of materials [54], although there still remain many unsolved problems. For example, it is still not yet possible to selectively grow a particular kind of CNT; it is also difficult to fabricate good Ohmic contacts to CNT in a controlled way and to eliminate effects of the surrounding environment of a CNT device.

Fig. 5 schematically shows the extraction of CNT from a graphene layer [54]. The honeycomb structure of graphene is denoted by basis vectors $\vec{a}_1$ and $\vec{a}_2$. A CNT is formed by wrapping a portion of the graphene layer defined by $\vec{C}_h = n\vec{a}_1 + m\vec{a}_2$. This gives an *(n,m)* CNT, formed by the shaded region in Fig. 5.

This section corresponds to the circumference area of the CNT. The translational vector along the tube axis, $\vec{T}$, gives the corresponding number of honeycomb unit cells used to build the CNT unit cell. The chiral angle θ is the angle between $\vec{C}_h$ and the vector which defines a zigzag tube, $(n, 0) = (0, n)$. The circumference of such a zigzag tube is represented by thick dotted line in Fig. 5. Due to the *six* fold symmetry of graphene layer, θ is restricted to $[0, \pi/6]$. This leads to the fact that armchair tubes, *(n,n)*, have $\theta = 0$ chiral angle. In Fig. 5, we illustrate a circumference of one of

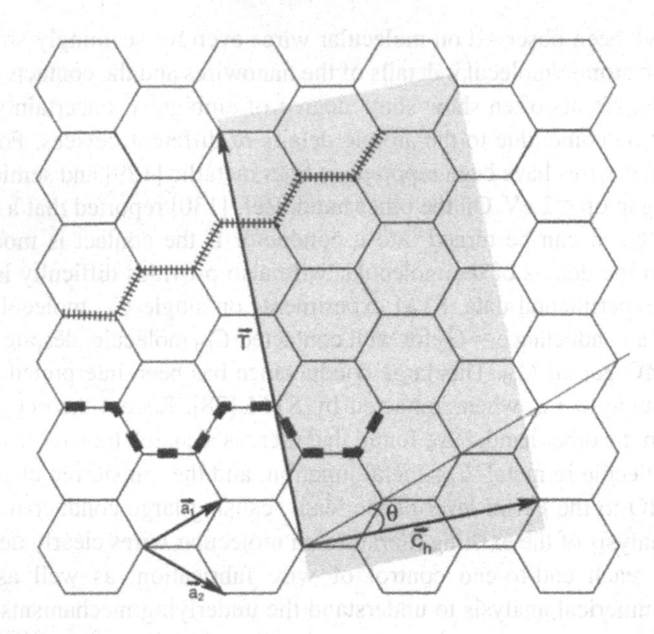

Fig. 5. Formation of CNT from a graphene sheet. $\vec{a}_1$ and $\vec{a}_2$ correspond to basis vectors, and $\vec{C}_h = n\vec{a}_1 + m\vec{a}_2$ defines (n, m) tube. In this case it corresponds to $(n, m) = (2,1)$, and the CNT is formed by folding the shaded region. $(n, 0) = (0, n)$ zigzag tube circumference is represented by thick dotted line and an armchair tube, (n, n), is illustrated by thick dashed line. Angle θ defines the chirality of the tube; due to the *six* fold symmetry of graphene, θ is restricted to $[0, \pi/6]$.

the armchair tubes by thick dashed line. We note that the chiral angle defines the unit cell size of a CNT. Therefore, zigzag and armchair tubes have $\theta = 0$, leading to the smallest translational vector and hence smaller CNT unit cells. By wrapping the graphene in different ways, diverse CNT structures have been obtained. Tubes formed by wrapping a single graphene layer are single walled CNT. But there are abundant multi-walled CNT which are formed from *two* to *few tens* of single wall concentered CNTs with different chiralities and diameters. CNTs tend to bunch together to form bundles or ropes [132]. Typically the inter-tube binding strength is larger than intra-tube interaction; hence multi-wall tubes tend to share common electronic features dominated by single wall CNT. The most important character of CNTs is the dependence of their electronic properties on the tube chirality, (n, m). Zero-gap tubes are found [54] for CNTs satisfying $n-m = 3q$ where q is an integer; whereas other tubes have a moderate gap and they are considered as semiconductors [133–138]. Metallic tubes are expected to have better conductance when compared to graphene sheet. This is mainly due to the carrier concentration in metallic CNT, which is comparable to conductors and is at least *three* orders of magnitude larger than that of graphene sheets. Carbon atoms have valency of *4e* in the *s, p* states, which hybridize in the form of *sp2* as a core state. This leaves a single free electron per atom having a p_z character. Therefore, in graphene sheet, single π orbital nearest neighbor interaction is a reasonable approximation to describe its characteristics. Such a consideration has been generalized to CNTs to construct simple but qualitatively reasonable tight binding (TB)

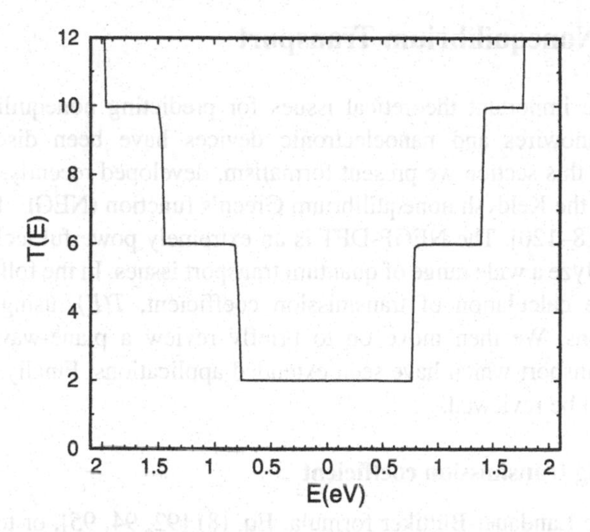

Fig. 6. Transmission coefficient through (10,10) armchair CNT. These calculations are performed with *ab initio* self-consistent calculations within LDA approximations. Energy $E=0$ corresponds to E_F of the system.

models [54, 133]. Indeed, majority theoretical analyses of CNTs use somc version of the TB model. Investigations suggest [135] that at room temperature and for CNTs with radius larger than 3.5 Å (radius of (9,0) CNT), even the nearest neighbor TB model parameterization may provide a qualitatively reasonable picture for CNTs. More recently, transport analysis through CNT by more precise DFT based self-consistent methods have been reported [117, 119, 139]. For a perfect arm-chair CNT, there are two bands crossing the Fermi level, therefore at zero bias and using the Landauer formula, Eq. (7), we expect the equilibrium conductance to be $G = 2G_0$ (resistance $\cong 6.5$ KΩ). Fig. 6 plots the transmission coefficient as a function of the electron energy for an arm-chair CNT using self consistent pseudopotential calculations within LDA. At E_F, the transmission coefficient is 2, and it increases in a step-wise manner as energy is increased so that more subbands participate in transport. Simple TB models give similar results. Experimental measurements of equilibrium conductance G of CNT have given many different values due to the quality of nanotube-metal contacts, but recent study using Ti electrodes [140] has obtained $G \approx 1.8G_0$, close to the theory limit of $2G_0$.

The novel and wide range of electronic properties of CNT have led to fabrication of different nano-devices. CNT intermolecular junctions have been formed [21] and its electronic response measured. These devices are obtained by joining *two* CNTs with different chirality to form metal/metal or metal/semiconductor junction [19]. CNT transistors have also been reported [7, 8, 58] with good device characteristics. *Three*-terminal device based on CNT has been fabricated to form Y-junction [112]. Contacting CNT to ferromagnetic material resulted in molecular scale spin polarized tunnel junction [66], and experimental data showed that nanotubes have very long spin-scattering lengths of at least 1300 Å. There have also been reports of 1D superconductivity in CNT nanowires [141].

3. DFT for Nonequilibrium Transport

Some of the important theoretical issues for predicting nonequilibrium charge transport in nanowires and nanoelectronic devices have been discussed in the Introduction. In this section we present formalism, developed recently, which carries out DFT within the Keldysh nonequilibrium Green's function (NEGF) framework for this purpose [118–120]. The NEGF-DFT is an extremely powerful technique, which allows us to analyze a wide range of quantum transport issues. In the following we will first outline the calculation of transmission coefficient, $T(E)$, using the standard Green's functions. We then move on to briefly review a plane-wave based DFT approach for transport which have seen extended applications. Finally, details of the NEGF-DFT will be reviewed.

3.1. Calculating transmission coefficient

To apply the Landauer-Büttiker formula, Eq. (8) [92, 94, 95], or to calculate the I–V_b curves from Eq. (10), we need the transmission coefficient, $T(E)$. $T(E)$ can be calculated from scattering matrix theory [22, 115], but for nanowires where microscopic details matter, Green's function technique turns out to be more convenient [94]. The equivalence between scattering matrix theory and Green's function formalism has been confirmed by the work of Fisher and Lee [142]. We will calculate $T(E)$ from Green's function. The transmission matrix, introduced in Eq. (8), can be viewed [94] as representing the response of an "out-going" lead to the event of an electron incident from an "in-coming" lead. On the other hand, the Green's function is more powerful as it incorporates the response of the system at any point in space due to any excitation occurring at any other point. It is much easier to include interactions into Green's functions, although the complexity of the problem scales with the complexity of interaction and correlation. Within the single particle picture, Green's function $\boldsymbol{G}$ is simple to calculate from its definition:

$$\boldsymbol{G}(E) = (E\boldsymbol{I} - \boldsymbol{H})^{-1} \tag{11}$$

where $\boldsymbol{H}$ is the Hamiltonian of the system, E is the energy and $\boldsymbol{I}$ is the identity operator.

Therefore, with appropriate basis set to represent the Hamiltonian matrix $\boldsymbol{H}$, we obtain Green's function $\boldsymbol{G}$ by a simple matrix inversion. However, for transport calculations we are dealing with an infinitely large problem because the leads extend to $\pm\infty$, see Fig. 7. In other words, our system has *open* boundaries. Therefore, $\boldsymbol{H}$ and $\boldsymbol{G}$ in Eq. (11) are infinite matrices. An infinite $\boldsymbol{H}$ cannot be directly inverted. Even if it could, we do not need all the information contained in the infinite $\boldsymbol{G}$, and we only need the Green's function near the device scattering region. Hence, we divide the device into three pieces: the device central scattering region, (c), and the two semi-infinite, left (l) and right (r) leads. The three pieces of the Hamiltonian are $\boldsymbol{H}_c$ for the scattering region and $\boldsymbol{H}_{l(r)}$ for the left (right) lead. In addition, the interaction between the scattering region and the leads are defined by Hamiltonians $\boldsymbol{H}_{cl(r)}$. Then it is easy to show [94] that the Green's function $\boldsymbol{G}_c$ of the scattering region, can be written as:

$$\boldsymbol{G}_c(E) = [E\boldsymbol{I} - \boldsymbol{H}_c - \Sigma]^{-1} \tag{12}$$

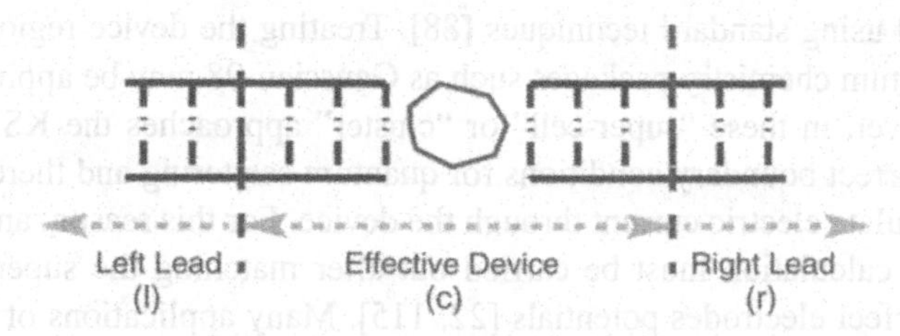

Fig. 7. Schematic description of a real *infinite two* probe device. Perfect lead characteristics (charge density and self-consistent potential) are assumed beyond the effective device region.

where Σ is the self-energy contributed by the leads to the scattering region due to device-lead interactions $\boldsymbol{H}_{cl(r)}$. For a two-lead device, $\Sigma = \Sigma_l + \Sigma_r$, where

$$\Sigma_l = \boldsymbol{H}_{cl}\boldsymbol{G}_l\boldsymbol{H}_{lc} \tag{13}$$

and a similar expression for Σ_r. In Eq. (13), G_l is the Green's function for the semi-infinite left lead, which can be calculated by well-established surface Green's function techniques [143, 144]. Clearly, only the atoms near the scattering region will have significant contribution to self-energy. Therefore, all the matrices in Eq. (12) are finite and can therefore be inverted to obtain G_c. Finally, from the Fisher-Lee relation [142] one can deduce the transmission coefficient [94]

$$\mathbf{T}_{pq}(E) = tr[\Gamma_p \mathbf{G}_c^R \Gamma_q \mathbf{G}_c^A] \tag{14}$$

Here $G_c^{R(A)}$ is the retarded (advanced) Green's functions obtained from Eq. (12) by replacing $\boldsymbol{EI}$ with $(E + (-)i\eta)\boldsymbol{I}$, with $\eta \to 0^+$. The quantity Γ_λ is the line-width function representing the coupling of the scattering region to a lead λ,

$$\Gamma_\lambda = i[\Sigma_\lambda - \Sigma_\lambda^+] \quad (\lambda = p, q). \tag{15}$$

Using Eq. (10) and the transmission coefficient, we easily deduce the I–V_b characteristics. To apply the Green's function analysis of transport, the knowledge of Hamiltonian, $\boldsymbol{H}$, is required. It is clear that the electronic properties of a device such as that of Fig. 7, are strongly influenced by the atomic structure. It is also clear that transport properties should be derived from a quantum mechanical treatment, which at the level of DFT, includes the atomic orbitals, the exchange correlation interactions, the core-valence interactions, the coupling to the electrodes, and the effects of any externally imposed fields. Finally, the microscopic system must be coupled to macroscopic reservoirs in order to investigate nonequilibrium charge transport. These microscopic details can be taken into account within DFT. Approximations to the device Hamiltonian $\boldsymbol{H}[\rho(\boldsymbol{r})]$ may be made by modeling only the central simulation box (the device region, labeled by c) in Fig. 7 as a finite cluster or as a super-cell. If more and more portions of the electrodes are added to the device region, the KS potential near the molecule (see Fig. 3) begins to mimic the correct potential which requires the electrodes to be infinitely long. Treating the device region as a "super-cell", plane-wave basis DFT analysis of the KS potential can be carried out [22, 115] so that

$\boldsymbol{H}$ is calculated using standard techniques [88]. Treating the device region as a finite "cluster", quantum chemistry packages such as Gaussian-98 may be applied to obtain $\boldsymbol{H}[\rho(\boldsymbol{r})]$. However, in these "super-cell" or "cluster" approaches the KS wave functions have incorrect boundary conditions for quantum scattering and therefore cannot be used to calculate electric current through the device. For this reason, an extra quantum scattering calculation must be carried out after matching the super-cell/cluster potential to perfect electrodes potentials [22, 115]. Many applications of these methods have been carried out [22, 115, 116, 124] because DFT provides a self-consistent approach and is parameter free. However, as the effective device potential is derived for a problem with periodic boundary conditions in the super-cell approach, it cannot describe systems with different electrodes. More importantly, it is rather difficult for super-cell and cluster approaches to deal with systems under external bias. Therefore they are best suited for calculating linear response transport coefficients such as the equilibrium conductance. A very important and useful method within DFT formalism is the Lippman–Schwinger (LS) scattering approach pioneered by Lang [116]. In this technique, the self-consistent KS equation is solved for *open* device structure and the charge density is constructed from the *scattering* states of the open device [116], with a plane-wave basis set. Bound states can also be added in an approximate fashion [145]. As such, this method correctly describes the boundary conditions of the device under external bias and, very importantly, I–V_b curves can be obtained from *first principles*. So far, in the LS-DFT approach the electrodes are treated within Jellium approximation, which is adequate for a number of situations. The LS-DFT approach has been applied to several investigations of nanoelectronic devices [116, 145] and provided important insights.

3.2. NEGF-DFT approach

In this subsection we present a novel approach of DFT within the Keldysh NEGF formalism for predicting nonequilibrium charge transport [118–120]. The essential problem is still the evaluation of transmission coefficient $T(E, V_b)$ so that current can be predicted from Eq. (10). Instead of standard DFT methods using super-cells or clusters, we calculate the Hamiltonian $\boldsymbol{H}[\rho(\boldsymbol{r})]$ using the NEGF-DFT. This technique [119] correctly treats the open boundary condition of a nanodevice, accounts for the external bias and gate potentials self-consistently, deals with atoms in the leads and the device scattering region on equal footing, and is computationally efficient so that a large number of atoms can be included in the analysis.

The NEGF–DFT method uses the DFT *ab initio* approach to describe the electronic degrees of freedom. The atomic cores are defined by standard norm-conserving non-local pseudo-potentials [146]. Because transport deals with open boundary conditions, a real space technique is needed to calculate the Hamiltonian and effective potential within DFT. We therefore use an LCAO *s, p, d* orbital basis to expand the KS wave functions. The use of a minimal basis set results in an efficient calculation with an acceptable accuracy as has already been documented in literature [89]. We note that systematic improvements may be achieved in LCAO basis sets so as to give comparable accuracy to large basis set methods such as plane waves, multi-grids, or wavelets. However, these other methods with large basis sets cannot be used in the NEGF–DFT approach: the large basis leads to large Hamiltonian matrix, which cannot be efficiently inverted. The NEGF–DFT constructs charge density from NEGF, instead of the KS wave functions, due to the nonequilibrium transport conditions. But once the charge density is constructed, the rest of the self-consistent iteration procedure is identical to conventional iterative DFT approaches. In what follows, we focus on the two problems, which arise for nonequilibrium transport: How to deal with an open, therefore infinitely large, device system within DFT? How to calculate charge distribution under external bias for such an open system in order to deduce the KS effective potential $V_{eff}(\mathbf{r})$ of the device Hamiltonian?

3.3. Open boundary condition for the Kohn-Sham potential

For the transport problem of Fig. 7, the effective KS potential $V_{eff}[\rho(\mathbf{r})]$ should be constructed by calculating the charge distribution $\rho(\mathbf{r})$ for the ***open*** structure which is, actually, infinitely large because the electrodes extend to $z \rightarrow \pm\infty$. However, we observe that $V_{eff}[\rho(\mathbf{r})]$ deep inside a solid surface (the lead) should be very close to the corresponding bulk KS potential. Therefore, the open boundary condition can be dealt with using a "screening approximation", i.e. we require [119],

$$V_{eff}(\mathbf{r}) = \begin{cases} V_{l,eff}(\mathbf{r}) = V_{l,bulk}(\mathbf{r}), & Z < Z_l \\ V_{c,eff}(\mathbf{r}), & Z_l < Z < Z_r \\ V_{r,eff}(\mathbf{r}) = V_{r,bulk}(\mathbf{r}), & Z > Z_r \end{cases} \quad (16)$$

where the planes $Z = Z_{l(r)}$ are the left (right) limits of the scattering region (indicated by the vertical dashed lines in Fig. 7) which forms our calculation "box". Note that enough lead layers should be included in the calculation box in order for Eq. (16) to hold: this can be checked in the numerical calculation. The potentials $V_{l(r),bulk}(\mathbf{r})$ describe periodic solid structures, and they can be calculated by the conventional super-cell DFT in a separate "electrode" calculation, and stored in a database. The real space potential matching of Eq. (16) allows us to deal with situations where the two electrodes are from different materials. In our numerical code,[2] what matched at the open boundary is actually not the full effective potential, $V_{eff}[\rho(\mathbf{r})]$, because it is sufficient to just match the Hartree potential $V_H[\rho(\mathbf{r})]$ there. The reason is that all other terms in the KS potential $V_{eff}[\rho(\mathbf{r})]$ are functional of the charge density $\rho(\mathbf{r})$. Once the Hartree potential is matched, $\rho(\mathbf{r})$ will automatically match across the boundary, which ensures the equality in Eq. (16) within LDA. Therefore, we solve the Poisson equation in three-dimensional real space of the scattering region using an efficient multi-grid numerical procedure with the Hartree potential of bulk electrodes as boundary condition. This way the external bias and gate potentials can be easily added as additional boundary conditions for the Poisson equation. One can then verify that $\rho_c(\mathbf{r})$ and $\rho_{l(r),bulk}(\mathbf{r})$ become automatically matched perfectly at $Z_{l(r)}$—if $Z_{l(r)}$ is chosen far enough inside the electrode surface. The screening approximation effectively reduces the infinitely large problem to a finite one, corresponding to the scattering region. To apply the screening approximation Eq. (16), we need to calculate the potential of the electrodes $\rho_{l(r),bulk}(\mathbf{r})$. Each electrode consists of a collection of atoms at positions $\vec{R}_1$ in a unit cell, which is repeated to $\pm\infty$. Using LCAO basis, a unit cell should be chosen large enough so that its left and right neighboring cells do not interact. This way, we iterate the KS equation to self-consistency by a standard super-cell

[2]Our NEGF-DFT code, McDCAL, is described in Ref. [119].

DFT approach, in which the charge density is calculated by the KS eigenstates (the Bloch states) $|\Psi_k\rangle$,

$$\hat{\rho} = \int dk |\Psi_k\rangle f_{eq} \langle \Psi_k| \tag{17}$$

where f_{eq} is the Fermi distribution function and the chemical potential is defined such that the total charge in the unit cell is conserved. Finally, the following information is saved to an electrode database: the position of the atoms $\vec{R}_I$ in the unit cell; the effective potential $V_{l(r),bulk}(\mathbf{r})$; the Hartree potential on the boundary of a unit cell, $V_{l(r),bulk}(\mathbf{r})|_{Z=Z_{l(r)}}$; the overlap matrix of the LCAO basis; as well as the chemical potential which ensures charge neutrality.

3.4. Density matrix of an open device system

We now deal with the second problem: the construction of density matrix under external bias for an open system. The density matrix can be constructed in two ways, either by including all the eigenstates of the open device with the proper statistical weight, or from the Keldysh NEGF. The two ways give identical results, but the NEGF is much easier to apply when there is a bias voltage (nonequilibrium situation). It has been shown [119] that for open boundary problems, the bound states, which live inside the scattering region (see Fig. 7) are rather difficult to calculate.[3] NEGF treat bound states and scattering states on equal footing: they are the poles of the Green's function. Furthermore, when there is a bias voltage so that the system is in nonequilibrium, the distribution function in the scattering region is not a Fermi distribution and must be calculated. NEGF naturally takes into account the nonequilibrium distribution [147]. The density matrix is calculated from NEGF $G^<(E)$, as

$$\hat{\rho} = -\frac{i}{2\pi}\int dE G^<(E), \quad where$$
$$G^< = G^R \Sigma^<[f_l, f_r] G^A \tag{18}$$

The above definition of $G^<$ is the Keldysh equation [147] in which $G^{R(A)}$ is the retarded (advanced) Green's function which we have discussed before. The quantity $\Sigma^<[f_l, f_r]$ represents injection of charge from the electrodes. $\Sigma^<[f_l, f_r]$ becomes very simple to calculate in a mean field theory such as the DFT within LDA,

$$\Sigma^<[f_l f_r] = -2i\Im m\left[f_l \Sigma_l^R - f_r \Sigma_r^R\right] \tag{19}$$

where $\Sigma^R_{l(r)}$ is the retarded self-energy of the left (right) electrode. The distribution functions $f_{l(r)}$ of the electrodes, or reservoirs, are taken as Fermi distributions, f_{eq}, but with its chemical potential shifted by the applied bias voltage,

$$f_l(E, \mu_l + \Delta V_{b,l}) = f_{eq}(E, \mu_l + \Delta V_{b,l})$$
$$f_r(E, \mu_r + \Delta V_{b,r}) = f_{eq}(E, \mu_r + \Delta V_{b,r}) \tag{20}$$

[3] The reason that there may exist bound states in the scattering region is that the potential in the scattering region may be deeper than that of the leads.

Once the density matrix is constructed from Eq. (18), the rest of the self-consistent DFT iteration is the same as in the conventional DFT analysis. The use of the NEGF, as opposed to eigenstates of the open system, to construct charge density is the most important and novel point of this *ab initio* technique. Numerically, an efficient procedure is needed for evaluating the energy integration in Eq. (18): too few points in energy will not produce accurate density because of the van Hove singularities at band edges of DOS (which is the integrand), while too many points make computation very expensive because for each energy E one must evaluate the NEGF $\boldsymbol{G}^{<}$.

From the point of view of constructing charge density $\rho_c(r)$, the essential difference between $\boldsymbol{G}^{<}$ and the retarded Green's function $\boldsymbol{G}^{R}$, is that $\boldsymbol{G}^{<}$ contains information about the distribution function through the quantity $\Sigma^{<}[f_l, f_r]$. In other words, the NEGF tells us how to fill the states of an open device system under bias, according to Eq. (18). In equilibrium without bias, i.e. when the chemical potentials of the electrodes are equal, $\boldsymbol{G}^{<}$ is reduced to a simple form:

$$\Im m G^{<} = -2 f_{eq} \Im m[G^{R}] \tag{21}$$

Importantly, this expression remains true even for situations where the electrochemical potentials are different as long as $f_l(E) = f_r(E) = 1$. Therefore, we can split the integral in Eq. (18) into two terms: an "equilibrium" contribution, ρ_{eq}, where the Fermi distribution functions are unity and a "non-equilibrium" contribution ρ_{neq} corresponding to the energy window between both chemical potentials. Hence, the density matrix $\rho = \rho_{eq} + \rho_{neq}$ where:

$$\begin{aligned} \rho_{eq} &= -\frac{1}{\pi}\Im m \int_{-\infty}^{\mu_{\min}} dE G^{R}(E) \\ \rho_{neq} &= -\frac{1}{\pi}\Im m \int_{\mu_{\min}}^{\mu_{\max}} dE G^{<}(E) \end{aligned} \tag{22}$$

where $\mu_{\min(\max)} = \min(\max)(\mu_l + eV_{b,l}, \mu_r + eV_{b,r})$. In the above equation, temperature is set to zero so that the distribution functions are step functions. The extension to finite temperature is straightforward.

The equilibrium charge contribution ρ_{eq} can be calculated by a contour integral because the retarded Green's function G^{R} has no poles in the upper half complex energy plane. A semi-circular contour illustrated in Fig. 8 is chosen [119], where $E_{\min}$ is fixed at sufficiently low value so that it is lower than the lowest eigenvalue of the Hamiltonian. The integration itself is accomplished using a Gaussian quadrature and the result converges rapidly as the number of quadrature points is increased. Typically *thirty* quadrature points are enough to evaluate Eq. (22). We note that such a complex contour automatically includes the charge contribution from any bound state below $\mu_{\min}$, which would appear as poles of $\boldsymbol{G}^{R}$ along the real energy axis. The contour integration is very important for a self-consistent analysis. On the other hand, a direct integration of Eq. (18) would normally require a prohibitively large number of integration points to reach a reasonable accuracy. The non-equilibrium charge, Eq. (22), cannot be calculated by the contour trick because for $\mu_{\min} < E < \mu_{\max}$, Eq. (21) does not hold. Therefore we evaluate Eq. (22) directly on the real axis,

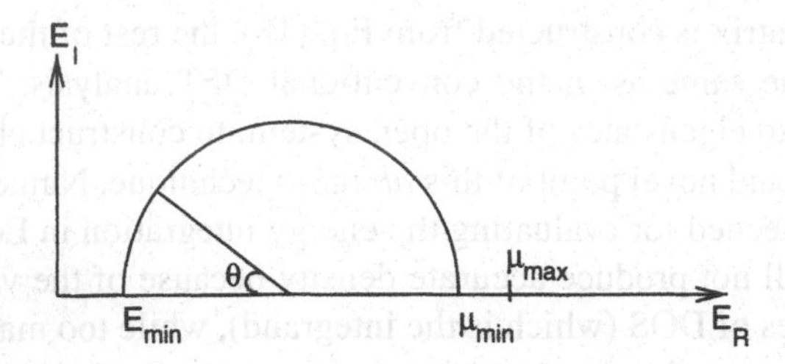

Fig. 8. Analytical continuation of G^R into the complex energy plane, (E_R, E_I). The equilibrium charge distribution is calculated by exploiting the analytical property of G^R. The integral around the semicircle is equal to the integral on the real axis from $E_{min} \rightarrow \mu_{min}$. The equilibrium charge contribution is calculated by choosing a sufficiently low value of E_{min} and taking $\mu_{min} = \min(\mu_l + eV_{b,l}, \mu_r + eV_{b,r})$.

$$\rho_{neq} = -\frac{i}{2\pi}\int_{\mu_{min}}^{\mu_{max}} dEG^R\Sigma[f_l, f_r]G^A(E) \tag{23}$$

As long as there are no band edges between the chemical potentials of the reservoirs, the energy integration will be smooth along the real axis and a Gaussian quadrature is found to converge with a small number of evaluations of $G^<$. Since the semi-infinite electrodes are taken into account through self-energies Eq. (12), we only need to evaluate density matrix in the scattering region (see Fig. 7). Changes in extensive physical quantities such as the band-structure energy or number of electrons are evaluated in the scattering region by tracing over the indices of density matrix corresponding to that region. Once the KS self-consistent potential $V_{eff}(r)$ of the scattering region is iterated to convergence, the transport properties of the device are calculated by the Green's function technique discussed above.

4. Nonequilibrium Transport in Au Atomic Wire

To illustrate the NEGF-DFT formalism for nonequilibrium transport, in this section we present an analysis of transport properties of gold nanowires under external bias. The experimental results [148, 149] of Au nanowires showed linear response for gold structures formed using RC [150] and/or STM under UHV conditions [151]. However, other measurements [151, 152], using STM in air, resulted in non-linear I–V_b behavior. To understand and interpret these contradictory data, extensive *ab initio* theoretical work is required. In fact, current-voltage characteristics are so sensitive to ambient conditions that, perhaps, any desired experimental characteristic could be fitted by theories using empirical parameters. From an experimental point view, the measured transport response could be either due to structural or to electronic origins. It is therefore necessary to resolve these factors from each other. Recently, Wlasenko and Grütter [149] developed a useful experimental approach which can, possibly, separate electronic response from atomic rearrangement in Gold nano wires

[148, 149]. In this section we investigate Au atomic wires and develop a possible physical picture capable of explaining the contradictory data [148].

4.1. Clean Au atomic wires

Consider a clean Au contact composed of four Au atoms in contact with Au(1 0 0) leads, as illustrated in Fig. 9c. The scattering region is bonded by two semi-infinite Au leads, which extend to electron reservoirs at $\pm\infty$, where bias voltage is applied and current is collected. The device scattering region, indicated by c, is described by *three* Au layers from the left lead, *four* Au atoms in a chain, and *two* layers of Au from the right lead. We have also increased the *two* Au layers on the right side of the chain to *four* to ensure that convergence is reached with respect to the screening length. In this structure, the registry of the atomic chain with respect to the lead surface layer can be different. The most common structures that we analyze are the hollow and top sites. In the former configuration, the atomic chain faces the vacant position in the lead layer as shown in Fig. 9d, and in the latter structure, the atomic chain and an atom from the lead surface layer face each other as illustrated in Fig. 9e. Our calculations show that in all cases charge transfer between the Au chain and the leads is not important, being only 0.07–0.1 electrons per atom to the Au chain at different bias voltages. This corresponds to less than %1 difference in electron population per atom and, therefore, does not play any significant role in the I–V_b curves. In addition, solving for energy eigenvalues of the four-atom chain gives a HOMO–LUMO gap of 0.68 eV, indicating that the molecule eigen-states should only have a secondary effect on the transport properties. Therefore, the major effect on the I–V_b curve is due to the character of electronic states in the leads and their couplings to the wire at the wire-lead interface. In Fig. 9a, we show results for a system with atoms in the hollow site. At small bias V_b (applied to the right lead), we note that current is a linear function of V_b with a slope $G \cong 0.94G_0$. The linear function suggests that $T(E, V_b) = T_0$ with only weak voltage dependence. By carefully investigating the transmission function $T(E, V_b)$, we found [148] that $T(E, V_b)$ in general has a nonlinear dependence on E for a given bias voltage, which increases faster than linear behavior. However, for $V_b >$ 100 meV, $T(E, V_b)$ reduces as V_b increases. Therefore, there is this compensating effect which produced the apparent linear I–V_b curves. Hence, our self-consistent calculation gives a result apparently similar to previous theoretical work in the low-bias regime in which the transmission coefficient is independent from the bias voltage [153, 154]; the physical origin is rather different. Another major feature of Fig. 9a is the huge plateau in the I–V_b curve, as well as the fine structures observed for larger voltages. In Fig. 9f, we show the band structure of the Au(1 0 0) lead along the z-direction (transport direction). Even though at a given energy E there are many electronic states that are potential candidates for transporting current, our investigation found that for $E < 0.5$ eV, there is only one state that is actually conducting, presented by a continuous line in the band diagram. Once this state is terminated at $E \approx 0.5$ eV, the current is saturated resulting in a large plateau until new conducting states emerge at higher bias voltages. For $E > 0.9$ eV, transport properties are more complex since more states contribute to transmission. Under these circumstances, band crossing occurs more frequently, and $T(E, V_b)$ changes over small ranges of bias leading to the small structures

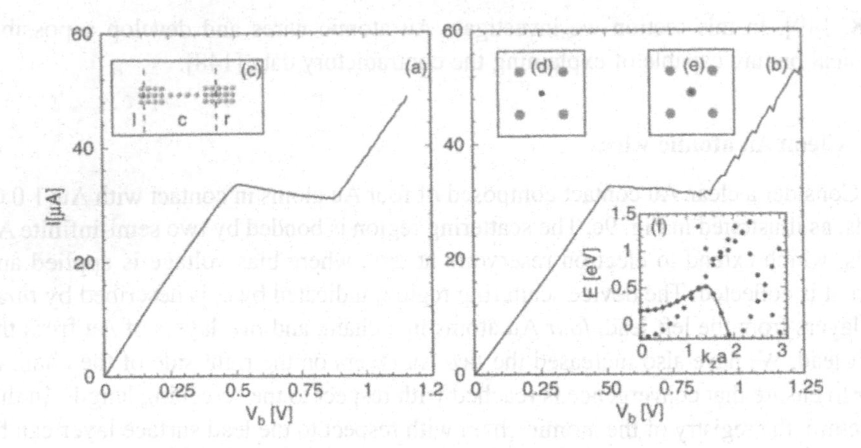

Fig. 9. I–V_b characteristics of Au contacts with hollow site registry (a) and Top site registry (b). The structure of the device is illustrated in (c) where *l*, *c* and *r* correspond to Left Lead, effective Device and Right Lead, respectively. The Hollow site is shown in (d) and Top site (e) where chain atoms are illustrated by dark circles. (f) shows the band structure of the lead along the transport direction, the conducting band is shown by solid line.

seen in the I–V_b characteristics. The fine structures have been observed in the experiments [148].

The effect of site registry is studied by placing the end atom of the Au chain at the top site of the leads. In this situation the chain atoms are facing one atom of the lead surface layers. This analysis is quite important, because it was shown that atoms at the junction change registry from hollow to top sites resulting in bundle formation just before the nano-structure breaks [155]. The I–V_b characteristics of this system (shown in Fig. 9b) are similar to the previous results, therefore no change is observed in the transport properties for the top-site registry. However, we note the difference of zero-bias conductance, $G \approx 0.8G_0$ for the top-site device and $G \approx 0.94G_0$ for the hollow-site device. This difference is due to a change in the coupling between the end atoms of the wire and the surface of the leads. To ensure the same nearest-neighbor separation distance for both cases, we end up with *four* nearest neighbors for the hollow-site registry and only one nearest neighbor for the top-site registry. Under these circumstances, the hollow site has a better coupling to the nanowire and hence a larger conductance. A pure and perfect Au nano-contact, as studied in this subsection, shows rich and interesting transport properties. However, it has a linear I–V_b curve at small bias voltage, $V_b < 0.5$V, rather than the experimentally observed nonlinear I–V_b characteristics of STM [151, 152] in air and of MCBJ contacts [149].

4.2. Au nano-contacts with S impurity

How can a nanowire produce a nonlinear I–V_b curve? The simplest possibility to observe such a phenomenon is to generate a tunneling barrier at the wire-lead junction whose effect gradually collapses as a function of increasing bias voltage. While a tunneling barrier can be established by several means, we investigate a model where it is a result of impurities. Indeed, this is supported by the STM experimental findings

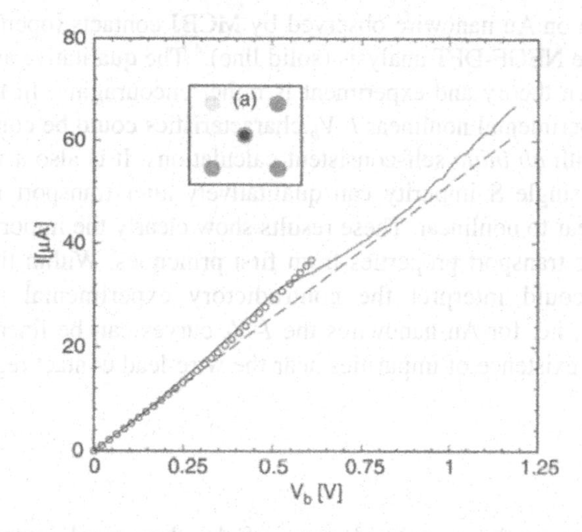

Fig. 10. $I-V_b$ characteristics of Au contact doped with S impurity (Solid line). The dashed line has a slope $\cong 0.67G_0$ and is shown to help view the onset of non-linearity; Circles correspond to the experimental results. The inset (a) shows the atoms registry, where dark circles show the atomic chain, grey ones are Au atoms in the leads' surface and light grey is the S atom.

[151] which indicate that perfectly linear I–V_b characteristics were reproducibly found in gold–gold nano-contacts under UHV, while nonlinear effects emerge when the experiment was performed in the air. This suggests that impurities play an important factor. To simulate the effect of an impurity at the contact, we have replaced one of the Au atoms at the interface layer with a sulfur atom. This is presented in the inset of Fig. 10. The choice of sulfur is motivated by the fact that in experimental labs, sulfur is a non-negligible airborne pollutant (diesel exhausts) and sulfur atoms bond actively with Au. We also note that the bandwidth of sulfur, ≈ 10 eV, is much higher than that of Au, ≈ 1 eV, thus a tunneling barrier is expected to be provided by the presence of S atoms. In this system, charge transfer to the atomic chain is still small, and thus inadequate in explaining the experimentally observed I–V_b characteristics. We note that the S atom suffers from an electron deficiency of $\approx$%4. Also due to the presence of the S atom, the coupling of the electronic states in the leads to the device scattering region is quite different as compared to the mono-atomic gold structure. When the S atom is present, our calculations found that all Bloch states are coupled to the scattering region, with the highest transmitting mode still being that corresponding to the conducting mode of the mono-atomic gold system. The I–V_b characteristics of the S doped Au nano-contacts are shown in Fig. 10. We note that the I–V_b curve for voltages up to 0.5V exhibit the experimental nonlinearity which is set at a nonzero bias voltage. We also note that the huge current plateau of the pure Au device has now diminished because more states contribute to electronic transport. We can compare these results with the experimental measurements after eliminating effects due to atomic rearrangements. For some Au nanowires the elimination can actually be done as reported in recent experimental measurements [148, 149]. Fig. 10 shows the

experimental data on Au nanowire observed by MCBJ contacts (open circles), to be compared with the NEGF-DFT analysis (solid line).[4] The qualitative and quantitative agreement between theory and experiment is rather encouraging. In fact, this is the first time that experimental nonlinear I–V_b characteristics could be compared so well and so directly with *ab initio* self-consistent calculations. It is also a very surprising result because a single S impurity can qualitatively alter transport in these nano-contacts from linear to nonlinear. These results show clearly the importance of investigating electronic transport properties from first principles. Within the NEGF-DFT formalism, we could interpret the contradictory experimental data within a single framework, i.e. for Au nanowires the I–V_b curves can be linear or nonlinear depending on the existence of impurities near the wire-lead contact region.

5. Summary

In this chapter we have outlined some of the theoretical issues of modeling molecular scale nanowires from *first principles*. We have focused on the NEGF–DFT formalism [119] which is suitable for analyzing nonequilibrium charge transport properties of molecular devices self-consistently. The novelty of this technique lies in constructing the electronic charge density via the Keldysh nonequilibrium Green's function, instead of using eigenstates of the conductor, with the help of a screening approximation which reduces the infinitely large problem to a finite calculation in the scattering region. Because our DFT analysis is based on a real space technique with the Hartree potential solved on a real space grid, this allows simulations of devices with different electrodes, *for example*, one carbon and the other Au, with bias as well as gate potentials. As a simple example we predicted the nonlinear I–V_b characteristics of Au nanowires and compared the results with the experimental data. Consistency is reached without using any parameter, and our analysis has lifted off some of the contradictory results of recent experimental measurements. The NEGF-DFT approach has already seen a wide range of applications, including the investigations of C_{60} molecular tunnel junctions [118]; the nonlinear transport properties of carbon nanowires [156]; Impurity effects on conduction of carbon nanotubes [139]; C_{20} molecular wires [157]; Si atomic devices [158]; transport in nanotube-metal interfaces [119]; current-triggered molecular dynamics [159], and molecular wires [160], etc.

The NEGF–DFT *first principles* formalism presented here deals only with situations where electron correlation is weak so that DFT within LDA can be adequately applied. There are however certain problems involving strong correlation for which the mean field theories fail. New and more powerful numerical techniques are needed for these situations. Electron-phonon effect has so far been neglected but it could play an important role for electronic transport in nanowires. Progress along this direction has already been seen [161, 162], but numeric are limited due to the large phonon basis set. A very important issue, which has started to receive attention, is the current induced quantum mechanic forces [125, 159, 163, 164]. Our analysis presented here

[4] In order to make the quantitative comparison, we have normalized the experimental data to the corresponding theoretical zero bias conductance.

did not include atomic relaxation *during* charge transport, but such effect could be very important in real devices. However, in order to carry out quantum molecular dynamics simulation during transport, improvement in numerical efficiency and parallel computing should be developed. Experimental advances in atomic/molecular devices have been quite rapid in recent years. However, although various prototype devices exist, end-to-end control for systems design and fabrication has so far not been systematically achieved. As we have shown with the Au nanowire example, the charge transport properties are sensitive to microscopic details including chemistry and structure of the contacts, in addition to the molecules in the scattering region. Experimentally it is very challenging to determine these details. Therefore, more extensive and combined experimental, numerical and theoretical analyses are required before a viable molecular electronics can be developed. Because nanoelectronics is such an interesting possibility, solid steps toward its realization are expected.

References

1. C. P. Collier, E. W. Wong, M. Belohradsky, F. M. Raymo, J. F. Stoddart, P. J. Kuekes, R. S. Williams and J. R. Heath, *Science* **285** (1999) 391; C. P. Collier, G. Mattersteig, E.W. Wong, Y. Luo, K. Beverly, J. Sampaio, F. M. Raymo, J. F. Stoddart and J. R. Heath, *ibid.* **289** (2000) 1172.
2. M. A. Reed, C. Zhou, C. J. Muller, T. P. Burgin and J. M. Tour, *Science* **278** (1997) 252.
3. J. Chen, M. A. Reed, A. M. Rawlett and J. M. Tour, *Science* **286** (1999) 1550.
4. C. Joachim, J. K. Gimzewski, R. R. Chlitter and C. Chavy, *Phys. Rev. Lett.* **74** (1995) 2102.
5. J. K. Gimzewski and C. Joachim, *Science* **283** (1999) 1683.
6. Thomas Rueckes, Kyoungha Kim, Ernesto Joselevich, Greg Y. Tseng, Chin-Li Cheung and Charles M. Lieber, *Science* **289** (2000) 94.
7. S. J. Tans, M. H. Devoret, R. M. Alwin and H. Dai, *Nature* (London) **386** (1997) 474.
8. S. J. Tans, A. R. M. Verschueren and C. Dekker, *Nature* (London) **393** (1998) 49.
9. J. W. G. Wildoeer, L. C. Venema and A. G. Rinzler, *Nature* (London) **391** (1998) 59.
10. Y. Xue, S. Datta, S. Hong, R. Reifenberger, J. L. Henderson and C. P. Kubiak, *Phys. Rev. B* **59** (1999) 7852.
11. G. E. Moore, *Electronics* **38** (1965) 114.
12. S. Ijima, *Nature* (London) **354** (1991) 56.
13. H. W. Kroto, J. R. Heath, S. C. O'Brien, R. F. Curl and R. E. Smalley, *Nature* (London) **318** (1985) 6042.
14. A. Aviram and M. Ratner, *Chem. Phys. Lett.* **29** (1974) 277.
15. G. A. Prinz, *Science* **282** (1998) 1660.
16. K. K. Berggren, A. Bard, J. L. Wilbur, J. D. Gillaspy, A. G. Helg, J. J. McClelland, S. L. Rolston, W. D. Phillips, M. Prentiss and G. M. Whitesides, *Science* **269** (1995) 1255.
17. B. C. Bunker, P. C. Rieke, B. J. Tarasevich, A. A. Campbell, G. E. Fryxell, G. L. Graff, L. Song, J. Liu, J. W. Virden and G. L. McVay, *Science* **264** (1994) 48.
18. For a review, see, for example, J. Voit, *Rep. Prog. Phys.* **58** (1995) 977.
19. S. J. Tans, M. H. Devoret, R. J. A. Groeneveld and C. Dekker, *Nature* (London) **394** (1998) 761.
20. Marc Bockrath, David H. Cobden, Jia Lu, Andrew G. Rinzler, Richard E. Smalley, Leon Balents and Paul L. McEuen, *Nature* (London) **397** (1999) 598.
21. Zhen Yao, Henk W. Ch. Postma, Leon Balents and Cees Dekker, *Nature* (London) **402** (1999) 273.

22. José-Luis Mozos, C. C. Wan, Gianni Taraschi, Jian Wang and Hong Guo, *Phys. Rev. B* **56** (1997) R4351.
23. J. K. Gimzewski and R. Möller, *Phys. Rev. B* **36** (1987) 1284.
24. J. I. Pascual, J. Méndez, J. Gómez-Herrero, A. M. Baró, N. García and Vu Thien Binh, *Phys. Rev. Lett.* **71** (1993) 1852.
25. J. I. Pascual, J. Méndez, J. Gómez-Herrero, A. M. Baro, N. García, Uzi Landman, W. D. Luedtke, E. N. Bogachek and H.-P. Cheng, *Science* **267** (1995) 1793.
26. N. García and L. Escapa, *Appl. Phys. Lett.* **54** (1989) 1418.
27. L. I. Glazman, G. B. Lesovik, D. E. Khmelnitskii and R. I. Shekhter, *JETP Lett.* **48** (1988) 238.
28. E.N. Bogachek, M. Jonson, R. I. Shekhter and T. Swahn, *Phys. Rev. B* **47** (1993) 16635.
29. M. Ogata and H. Fukuyama, *Phys. Rev. Lett.* **73** (1994) 468.
30. Uzi Landman, W. D. Luedtke, Brian E. Salisbury and Robert L. Whetten, *Phys. Rev. Lett.* **77** (1996) 1362.
31. N. Agraït, J. G. Rodrigo and S. Vieira, *Phys. Rev. B* **47** (1993) 12345.
32. L. Olesen, E. Lægsgaard, I. Stensgaard, F. Besenbacher, J. Schiøtz, P. Stoltze, K. W. Jacobsen and J. K. Nørskov, *Phys. Rev. Lett.* **72** (1994) 2251.
33. M. Brandbyge, J. Schiøtz, M. R. Sørensen, P. Stoltze, K. W. Jacobsen, J. K. Nørskov, L. Olesen, E. Lægsgaard, I. Stensgaard and F. Besenbacher, *Phys. Rev. B* **52** (1995) 8499.
34. Ronald P. Andres, Thomas Bein, Matt Dorogi, Sue Feng, Jason I. Henderson, Clifford P. Kubiak, William Mahoney, Richard G. Osifchin and R. Reifenberger, *Science* **272** (1996) 1323.
35. Tekman and S. Ciraci, *Phys. Rev. B* **43** (1991) 7145.
36. Ali Yazdani, D. M. Eigler and N. D. Lang, *Science* **272** (1996) 1921.
37. K. Matsumoto, M. Ishii, K. Segawa, Y. Oka, B. J. Vartanian and J. S. Harris, *Phys. Rev. Lett.* **68** (1996) 34.
38. C. Joachim and J. K. Gimzewski, *Proc. IEEE* **84** (1998) 184.
39. C. Joachim, J. K. Gimzewski and H. Tang, *Phys. Rev. B* **58** (1998) 16407.
40. E. S. Snow, D. Park and P. M. Campbell, *Appl. Phys. Lett.* **69** (1996) 269.
41. J. Moreland and P. K. Hansma, *Rev. Sci. Instrum.* **55** (1984) 399.
42. C. J. Muller, J. M. van Ruitenbeek and L. J. de Jongh, *Phys. Rev. Lett.* **69** (1992) 140.
43. For a recent review, see, for example, "Quantum properties of atomic-size conductors", by Nicolas Agraït, Alfredo Levy Yeyati and Jan M. van Ruitenbeek, *Phys. Rep.* **377** (2003) 81.
44. H. Namatsu, Y. Takahashi, M. Nagase and K. Murase, *J. Vac. Sci. Technol. B* **13** (1995) 2166.
45. Yasuo Wada, Tokuo Kure, Toshiyuki Yoshimura, Yoshimi Sudo, Takashi Kobayashi, Yasushi Goto and Seiichi Kondo, *Jpn. J. Appl. Phys.* **33** (1994) 905.
46. Y. Nakajima, Y. Takahashi, S. Horiguchi, K. Iwadate, H. Namatsu, K. Kurihara and M. Tabe, *Appl. Phys. Lett.* **65** (1994) 2833.
47. Yasuyuki Nakajima, Yasuo Takahashi, Seiji Horiguchi, Kazumi Iwadate, Hideo Namatsu, Kenji Kurihara and Michiharu Tabe, *Jpn. J. Appl. Phys.* **34** (1995) 1309.
48. H. I. Liu, D. K. Biegelsen, F. A. Ponce, N. M. Johnson and R. F. W. Pease, *Appl. Phys. Lett.* **64** (1994) 1383.
49. D. Routkevitch, T. L Haslett, L. Ryan, T. Bigioni, C. Douketis and M. Moskovits, *Chem. Phys.* **210** (1996) 343.
50. D. Routkevitch, T. Bigioni, M. Moskovits and J. M. Xu, *J. Phys. Chem.* **100** (1996) 14037.
51. D. Leonard, M. Krishnamurthy, C. M. Reaves, S. P. Denbaars and P. M. Petroff, *Appl. Phys. Lett.* **63** (1993) 3203.
52. J. Tersoff and R. M. Tromp, *Phys. Rev. Lett.* **70** (1993) 2782.

53. Masanobu Miyao, Kiyokazu Nakagawa, Masakazu Ichikawa, Kenji Hiruma and Kazuo Nakazato, *Jpn. J. Appl. Phys.* **33** (1994) 7214.
54. For reviews of nanotube physics, see, for example: M. S. Dresselhaus, G. Dresselhaus, and P. C. Eklund, *Science of Fullerenes and Carbon Nanotubes*, Academic Press, San Diego (1996); R. Saito, G. Dresselhaus, M. S. Dresselhaus, *Physical Properties of Carbon Nanotubes*, ICP Press World Scientific, Singapore (1998); T. W. Ebbesen, *Carbon Nanotubes: Preparation and Properties*, CRC press (1997); J. Bernholc, C. Roland and B. I. Yakobson, *Crit. Rev. Sol. Mat. Sci.* **2** (1997) 706.
55. S. Frank, P. Poncharail, Z. L. Wang and W. A. de Heer, *Science* **280** (1998) 1744.
56. J. W. Odom, J.-L. Huang, P. Kim and C. M. Lieber, *Nature* (London) **391** (1998) 62.
57. R. D. Antonov and A.T. Johnson, *Phys. Rev. Lett.* **83** (1999) 3274.
58. R. Martel, T. Schmidt, H. R. Shea, T. Hertel and Ph. Avouris, *Appl. Phys. Lett.* **73** (1998) 2447.
59. S. Paulson, M. R. Falvo, N. Snider, A. Helser, T. Hudson, A. Seeger, R. M. Taylor, R. Superfine and S. Washburn, *Appl. Phys. Lett.* **75** (1999) 2936.
60. P. G. Collins, Z. Zettl, H. Bando, A. Thess and R.S. Smalley, *Science* **278** (1996) 100; S. N. Song, X. K. Wang, R. P. Chang and J. B. Ketterson, *Phys. Rev. Lett.* **72** (1998) 697; L. Langer, L. Stockman, J. P. Heremans, V. Bayot, C. H. Olk, C. Van Haesendonck, Y. Bruynseraede and J.-P. Issi, *J. Mater. Res.* **9** (1994) 927; L. Langer, V. Bayot, E. Grivei, J.-P. Issi, J. P. Heremans, C. H. Olk, L. Stockman, C. Van Haesendonck and Y. Bruynseraede, *Phys. Rev. Lett.* **76** (1996) 479.
61. A. Bezryadin, A. R. M. Verschueren, S. J. Tans and C. Dekker, *Phys. Rev. Lett.* **80** (1998) 4036.
62. D. H. Cobden, M. Bochkrath, P. L. McEuen, A. G. Rinzler and R. E. Smalley, *Phys. Rev. Lett.* **81** (1998) 681.
63. M. Brockrath, D. H. Cobden, P. McEuen, N. Chopra, A. Zettl, A. Thess and R. E. Smalley, *Science* **275** (1997) 1922; A. Bachtold, C. Strunk, J. P. Salvetat, J.-M. Bonnard, L. Forro, T. Nussbaumer, and C. Schoenenberger, *Nature* (London) **397** (1999) 673.
64. L. C. Venema, J. Wildoeer, J. Janssen, S. J. Tans, H. Temminck Tuinstra, L. Kouewenhovem and C. Dekker, *Science* **283** (1999) 52.
65. A. Bachtold, M. S. Fuhrer, S. Plyasunov, E. H. Anderson, A. Zettl and P. McEuen, *Phys. Rev. Lett.* **84** (2000) 6082; M. Bockrath, J. Horne, A. Zettl, P. McEuen, A. G. Rinzler and R. E. Smalley, *Phys. Rev. B* **61** (2000) R10606.
66. K. Tsukagoshi, B. W. Alpenaar and H. Ago, *Nature* (London) **401** (1999) 572.
67. Y. Zhang, T. Ichihashi, E. Landree, F. Nihey and S. Iijima, *Science* **285** (1999) 1719.
68. A. F. Morpurgo, J. Kong, C. M. Marcus and H. Dai, *Science* **286** (1999) 263.
69. A. Yu. Kasumov, R. Deblock, M. Kociak, B. Reulet, H. Bouchiat, I. I. Khodos, Yu. B. Gorbatov, V. T. Volkov, C. Journet and M. Burghard, *Science* **284** (1999) 1508.
70. X. Blase, L. X. Benedict, E. L. Shirley and S. G. Louie, *Phys. Rev. Lett.* **72** (1994) 1878; Y. A. Krotov, D.-H. Lee and S. G. Louie, *ibid.* **78** (1997) 4245; L. Chico, V. Crespi, L. X. Benedict, S. G. Louie and M. L. Cohen, *ibid.* **76** (1996) 971; L. Chico, L. X. Benedict, S. G. Louie and M. L. Cohen, *Phys. Rev. B* **54** (1996) 2600; V. H. Crespi, M. L. Cohen and A. Rubio, *Phys. Rev. Lett.* **79** (1998) 2093; P. McEuen, M. Bockrath, D. Cobden, Y. G. Yoon and S. G. Louie, *ibid.* **83** (1999) 5098; H. J. Choi, J. Ihm, S. G. Louie and M. L. Cohen, *ibid.* **84** (2000) 2917.
71. H. J. Choi, J. Ihm, Y. G. Yoon and S. G. Louie, *Phys. Rev. B* **60** (1999) R14009.
72. L. Chico, M. P. L. Sancho and M. C. Munoz, *Phys. Rev. Lett.* **81** (1998) 1278.
73. W. Tian and S. Datta, *Phys. Rev. B* **49** (1994) 5097.
74. R. Tamura and M. Tsukada, *Phys. Rev. B* **55** (1997) 4991.
75. M. P. Anantram and T. R. Govindan, *Phys. Rev. B* **58** (1998) 4882.

76. A. A. Farajian, K. Esfarjani and Y. Kawazoe, *Phys. Rev. Lett.* **82** (1998) 5084.
77. M. Buongiorno Nardelli, *Phys. Rev. B* **60** (1999) 7228; M. Buongiorno Nardelli and J. Bernholc, *Phys. Rev. B* **60** (1999) 16338.
78. A. Rochefort, Ph. Avouris, F. Lesage and D. Salahub, *Phys. Rev. B* **60** (1999) 13824; A. Rochefort, D. Salahub and Ph. Avouris, *Chem. Phys. Lett.* **297** (1998) 45.
79. D. Orlikowski, M. Buongiorno Nardelli, J. Bernholc and C. Roland, *Phys. Rev. Lett.* **83** (1999) 4132; *Phys. Rev. B* **61** (2000) 14194.
80. H. Mehrez, J. Taylor, H. Guo, J. Wang and C. Roland, *Phys. Rev. Lett.* **84** (2000) 2682.
81. C. Roland, M. Buongiorno Nardelli, J. Wang and H. Guo, *Phys. Rev. Lett.* **84** (2000) 2921.
82. Y. Wei, J. Wang, H. Guo, H. Mehrez and C. Roland, *Phys. Rev. B* **63** (2001) 195412.
83. R. Saito, G. Dresselhaus and M. S. Dresselhaus, *Phys. Rev. B* **50** (1994) R14698.
84. H. Ajiki and T. Ando, *J. Phys. Soc. Jpn.* **62** (1993) 1255; *ibid.* **62** (1993) 2470.
85. S. Roche and R. Saito, *Phys. Rev. B* **59** (1999) 5242.
86. P. Hohenberg and W. Khon, *Phys. Rev.* **136** (1964) 864.
87. W. Khon and L. J. Sham, *Phys. Rev.* **140** (1965) 1133.
88. M. C. Payne, M. P. Teter, D. C. Allan, T. A. Arias and J. D. Joannopoulos, *Rev. Mod. Phys.* **64** (1992) 1045.
89. P. Ordejón, E. Artacho and José M. Soler, *Phys. Rev. B* **53** (1996) R104441.
90. B. J. van Wees, H. van Houten, C. W. J. Beenakker, J. G. Williamson, L. P. Kouwenhoven, D. van der Marel and C. T. Foxon, *Phys. Rev. Lett.* **60** (1988) 848.
91. D. A. Wharam, T. J. Thornton, R. Newbury, M. Pepper, H. Ahmed, J. E. F. Frost, D. G. Hasko, D. C. Peacock, D A Ritchie and G A C Jones, *J. Phys. C* **21** (1988) L209.
92. R. Landauer, *IBM J. Res. Dev.* 1 (1957) 223; *Phys. Rev. B* **16** (1977) 4698; *IBM J. Res. Dev.* **32** (1988) 306.
93. Charles Kittel, *Introduction to Solid State Physics* (6th ed.), John Wiley and Sons, Inc. (1986).
94. Supriyo Datta, *Electronic Transport in Mesoscopic Systems*, Cambridge University Press (1997).
95. M. Büttiker, *IBM J. Res. Dev.* **32** (1988) 317.
96. Y. Imry, *In directions in condensed matter physics*, Wold Scientific Press, Singapore (1986) 101.
97. J. Nogami, *In Atomic and Molecular Wires*, vol. **341**, Edited by C. Joachim and S. Roth, Kluwer, Dordrecht, (1997) p. 11.
98. Hideaki Ohnishi, Yukihito Kondo and Kunio Takayanagi, *Nature* (London) **395** (1998) 780.
99. A. I. Yanson, G. Rubio Bollinger, H. E. van den Brom, N. Agraït and J. M. van Ruitenbeek, *Nature* (London) **395** (1998) 783.
100. L. A. Bumm, J. J. Arnold, M. T. Cygan, T. D. Dunbar, T. P. Burgin, L. Jones II, D. L. Allara, J. M. Tour and P. S. Weiss, *Science* **271** (1996) 1705.
101. Erez Braun, Yoav Eichen, Uri Sivan and Gdalyahu Ben-Yoseph, *Nature* (London) **391** (1998) 775.
102. W. B. Davis, W. A. Svec, M. A. Ratner and M. R. Wasielewski, *Nature* (London) **396** (1998) 60.
103. C. Kergueris, J.-P. Bourgoin, S. Palacin, D. Esteve, C. Urbina, M. Magoga and C. Joachim, *Phys. Rev. B* **59** (1999) 12505.
104. D. Porath, A. Bezryadin, S. de Vries and C. Dekker, *Nature* (London) **403** (2000) 635.
105. V. Mujica, M. Kemp, A. Roitberg and M. A. Ratner, *J. Chem. Phys.* **104** (1996) 7296.
106. M. P. Samanta, W. Tian, S. Datta, J. I. Henderson and C. P. Kubiak, *Phys. Rev. B* **53** (1996) R7626.
107. C. Joachim, J. K. Gimzewski and A. Aviram, *Nature* (London) **408** (2000) 541.
108. Jing Kong, Nathan R. Franklin, Chongwu Zhou, Michael G. Chapline, Shu Peng, Kyeongjae Cho and Hongjie Dai, *Science* **287** (2000) 622.

109. T. W. Ebbesen, H. J. Lezec, H. Hiura, J. W. Bennett, H. F. Ghaemi and T. Thio, *Nature* (London) **382** (1996) 54.
110. A. Bachtold, M. Henny, C. Terrier, C. Strunk, C. Sch\"onenberger, J.-P. Salvetat, J.-M. Bonard and L. Forró, *Appl. Phys. Lett.* **73** (1998) 274.
111. P. G. Collins, A. Zetti, H. Bando, A. Thess and R. E. Smalley, *Science* **278** (1997) 100.
112. J. T. Hu, O. Y. Min, P. D. Yang and C. M. Lieber, *Nature* (London) **399** (1999) 48.
113. Jing Li, Chris Papadopoulos and Jimmy Xu, *Nature* (London) **402** (1999) 253.
114. J. Lefebvre, J. F. Lynch, M. Llaguno, M. Radosavljevic and A. T. Johnson, *Appl. Phys. Lett.* **75** (1999) 3014.
115. Daniel Sánchez-Portal, Emilio Artacho, Javier Junquera, Pablo Ordejón, Alberto García and José M. Soler, *Phys. Rev. Lett.* **83** (1999) 3884.
116. Kenji Hirose and Masaru Tsukada, *Phys. Rev. Lett.* **73** (1994) 150; *Phys. Rev. B* **51** (1995) 5278.
117. N. D. Lang, *Phys. Rev. B* **52** (1995) 5335; N. D. Lang and Ph. Avouris, *Phys. Rev. Lett.* **81** (1998) 3515.
118. Hyoung Joon Choi and Jisoon Ihm, *Phys. Rev. B* **59** (1999) 2267.
119. Jeremy Taylor, Hong Guo and Jian Wang, *Phys. Rev. B* **63** (2001) R121104.
120. Jeremy Taylor, Hong Guo and Jian Wang, *Phys. Rev. B* **63** (2001) 245407.
121. Mads Brandbyge, José-Luis Mozos, Pablo Ordejón, Jeremy Taylor and Kurt Stokbro, *Phys. Rev. B* **65** (2002) 165401.
122. P. S. Damle, A. W. Ghosh and S. Datta, *Phys. Rev. B* **64** (2001) R201403.
123. J. Heurich, J. C. Cuevas, W. Wenzel and G. Schön, *Phys. Rev. Lett.* **88** (2002) 256803.
124. J. J. Palacios, A. J. Pérez-Jiménez, E. Louis, E. SanFabiàn and J. A. Vergés, *Phys. Rev. B* **66** (2002) 035322.
125. Gianni Taraschi, José-Luis Mozos, C. C. Wan, Hong Guo and Jian Wang, *Phys. Rev. B* **58** (1998) 13138.
126. Massimiliano Di Ventra and Sokrates T. Pantelides, *Phys. Rev. B* **61** (2000) 16207.
127. J. H. Schön, Hong Meng and Zhenan Bao, *Nature* (London) **413** (2001) 713.
128. T. Ishida, W. Mizutani, Y. Aya, H. Ogiso, S. Sasaki and H. Tokumoto, *J. Phys. Chem. B* **106** (2992) 5886.
129. N.B. Zhitenev, H. Meng and Z. Bao, *Phys. Rev. Lett.* **88** (2002) 226801.
130. H. W. Fink and C. Schönenberger, *Nature* (London) **398** (1999) 407.
131. P. J. de Pablo, F. Moreno-Herrero, J. Colchero, J. Gómez Herrero, P. Herrero, A. M. Baró Pablo Ordejón, José M. Soler and Emilio Artacho, *Phys. Rev. Lett.* **85** (2000) 4992.
132. J. J. Palacios, A. J. Pérez-Jiménez, E. Louis and J. A. Vergés *Phys. Rev. B* **64** (2001) 115411.
133. A. Thess, R. Lee, P. Nikolaev, H. Dai, P. Petit, J. Robert, C. Xu, Y. H. Lee, S. G. Kim, A. G. Rinzler, D. T. Colbert, G. E. Scuseria, D. Tomanek, J. E. Fischer and R. E. Smalley, *Science* **483** (1996) 5274.
134. N. Hamada, S. I. Sawada and A. Oshiyama, *Phys. Rev. Lett.* **68** (1992) 1579.
135. R. Saito, M. Fujita, G. Dresselhaus and M. S. Dresselhaus, *Appl. Phys. Lett.* **60** (1992) 2204.
136. R. Saito, M. Fujita, G. Dresselhaus and M. S. Dresselhaus, *Phys. Rev. B* **46** (1992) 1804.
137. J. W. Mintmire, B. I. Dunlup and C. T. White, *Phys. Rev. Lett.* **68** (1992) 631.
138. J. W. Mintmire, D. H. Robertson and C. T. White, *J. Phys. Chem. Solids* **54** (1993) 1835.
139. R. Saito, G. Dresselhaus and M. S. Dresselhaus, *J. Appl. Phys.* **73** (1993) 494.
140. C.C. Kaun, B. Larade, H. Mehrez, J. Taylor and H. Guo, *Phys. Rev. B* **65** (2002) 205416.
141. Jing Kong, Erhan Yenilmez, Thomas W. Tombler, Woong Kim, Hongjie Dai, Robert B. Laughlin, Lei Liu, C. S. Jayanthix and S. Y. Wu, *Phys. Rev. Lett.* **87** (2001) 106801.
142. Z. K. Tang, Lingyun Zhang N. Wang, X. X. Zhang, G. H. Wen, G. D. Li, J. N. Wang, C. T. Chan and Ping Sheng, *Science* **292** (2001) 2462.

143. D. S. Fisher and P. A. Lee, *Phys. Rev. B* **23** (1981) 6851.
144. M. P. López Sancho, J. M. López Sancho and J. Rubio, *J. Phys. F: Met. Phys.* **14** (1984) 1205; *ibid.* **15** (1985) 851.
145. S. Sanvito, C. J. Lambert, J. H. Jefferson and A. M. Bratkovsky, *Phys. Rev. B* **59** (1999) 11936.
146. M. Di Ventra, S. T. Pantelides and N. D. Lang, *Phys. Rev. Lett.* **84** (2000) 979.
147. G. B. Bachelet, D. R. Hamann and M. Schlüter, *Phys. Rev. B* **26** (1982) 4199.
148. H. Haug and A.-P. Jauho, *Quantum Kinetics in Transport and Optics of Semi-conductors*, Springer-Verlag, New York (1998).
149. H. Mehrez, Alex Wlasenko, Brian Larade, Jeremy Taylor, Peter Grütter and Hong Guo, *Phys. Rev. B* **65** (2002) 195419.
150. Alex Wlasenko and Peter Grütter, *Rev. Sci. Instrum.* **73** (2002) 3324.
151. Katsuhiro Itakura, Kenji Yuki, Shu Kurokawa, Hiroshi Yasuda and Akira Sakai, *Phys. Rev. B* **60** (1999) 11163.
152. K. Hansen, S. K. Nielsen, M. Brandbyge, E. Lægsgaard, I. Stensgaard and F. Besenbacher, *Appl. Phys. Lett.* **77** (2000) 708.
153. J. L. Costa-Krämer, N. García, M. Jonson, I.V. Krive, H. Olin, P. A. Serna and R. I. Shekhter, *Coulomb Blockade Effect, Nanoscale Science and Technology*, vol. **348** of NATO Advanced Study Institute Series E: Applied Sciences, Edited by N. García, M. Nieto-Vesperinas and H. Rohree, Kluwer Academic, Dordrecht (1998) p. 1.
154. T. N. Todorov and A. P. Sutton, *Phys. Rev. Lett.* **70** (1993) 2138; *Phys. Rev. B* **54** (1996) R14 234.
155. A. Levy Yeyati, A. Martín-Rodero and F. Flores, *Phys. Rev. B* **56** (1997) 10369.
156. H. Mehrez, S. Ciraci, C. Y. Fong and Ş. Erkoç, *J. Phys.: Condensed Matter* **9** (1997) 10843; H. Mehrez and S. Ciraci, *Phys. Rev. B* **56** (1997) 12632.
157. Brian Larade, Jeremy Taylor, H. Mehrez and Hong Guo, *Phys. Rev. B* **64** (2001) 75420.
158. Christopher Roland, Brian Larade, Jeremy Taylor and Hong Guo, *Phys. Rev. B* **65** (2002) R041401.
159. Christopher Roland, Vincent Meunier, Brian Larade and Hong Guo, *Phys. Rev. B* **66** (2002) 035332.
160. Saman Alavi, Brian Larade, Jeremy Taylor, Hong Guo and Tamar Seideman, to appear in a special issue of *Molecular Electronics* in Chemical Physics, Elsvier (2002).
161. B. Larade, C. Kaun, H. Guo, P. Grütter and B. Lennox, McGill University (preprint).
162. H. Ness, S. A. Shevlin and A. J. Fisher, *Phys. Rev. B* **63** (2001) 125422.
163. H. Mehrez, *Phys. Rev. B* (submitted).
164. T. N. Todorov, J. Hoekstra and A. P. Sutton, *Phil. Mag. B* **80** (2000) 421.
165. B. Larade, Ph.D. Thesis, McGill University (2002).

Chapter 4

Modeling and Simulation of the Mechanical Response of Nanowires

Wuwei Liang, Vikas Tomar and Min Zhou
The George W. Woodruff School of Mechanical Engineering, Georgia Institute of Technology, Atlanta, GA 30332-0405, USA

We present here a contemporary review of the hitherto computational analyses of the mechanical deformation of nanowires. The bulk of the research reported in the literature concern metallic nanowires made of copper, gold, nickel, and their alloys and carbon nanotubes. Research has also been reported for nanowires with molecules having long chain structures (e.g., Silicon Diselenide ($SiSe_2$)). Calculations have primarily focused on discrete simulations using molecular dynamics (MD). In some cases, *ab initio* and first principle calculations have also been carried out [1, 2]. Recently, some researchers have attempted to use continuum formulations to obtain macroscale interpretations of the results of molecular simulations [3, 4]. Interatomic potentials used for modeling interactions between the atoms/molecules in these nanowires include empirical two-body potentials such as the Lennard-Jones potential and many-body potentials such as the embedded atom method (EAM) potential [5, 6]. Issues analyzed include structural changes under loading [7–10], the formation and propagation of defects [11–13], and the effect of the magnitude of applied loading on deformation mechanisms [2, 8, 10]. Efforts have also been made to correlate results of simulations with experimental observations. However, direct comparisons are difficult since most simulations are carried out under conditions of extremely high strain rates due to computational limitations. In some cases, excellent results have been reported and clear understandings have been obtained. However, this area of research is still in a nascent stage and significant work lies ahead in terms of problem formulation, interpretation of results, identification and delineation of deformation mechanisms, and constitutive characterization of behavior. We first present an introduction to commonly adopted methods in the studies of the deformation of nanowires, followed by an overview of findings concerning the mechanics of nanowires. We will also present some recent results from our own work with an emphasis on the effects of strain rate and wire size on the stress–strain relations of nanowires. One interest is to reach down to lower strain rates (10^7 s^{-1} at this time) and avoid artificial schemes necessitated by computational limitations. We will conclude by offering some thoughts on future directions in the rich and largely under-explored territory of the computational mechanics of nanowires.

1. Introduction

Metallic nanowires have great potentials for important applications in microdevices. For example, gold nanowires are already used as interconnects in chips. Nanowires are most commonly assumed to have rectangular cross-sections in simulations and maintain such cross-sectional shapes as deformation progresses until late stages of the process when necking occurs. Such assumptions do not account for all cases, however. For example, multi-shell structures have been both observed in experiments [14, 15] and predicted by molecular dynamics (MD) simulations [16, 17] for gold nanowires. The formation of cylindrical shells in FCC gold nanowires is most pronounced for wires with an initial [111] orientation. MD simulations have also predicted multi-shell and filled structures in aluminum and copper nanowires at room temperature [18].

Depending on its diameter or cross-sectional area and shape, a nanowire can yield by different mechanisms. If the wire maintains an ordered (crystalline) structure and has a relatively larger cross-section, dislocation motion and/or slip on glide planes are the primary mechanisms [19]. On the other hand, if the cross-section is small, order-disorder transformations (amorphization) and single-atom processes lead to large deformation [20]. Techniques used to characterize the deformed structures include centrosymmetry parameters [21], density contours, temperature contours and the radial distribution function (RDF). In many simulations, periodic boundary conditions (PBCs) are imposed in the longitudinal direction to account for infinite lengths. In some other cases, PBCs are used along the lateral directions [1]. However, such treatments restrict lateral deformation and eliminate surface effects. It represents an infinite dimensional extent and is not a valid treatment of the deformation of nanowires. Another issue is relaxation and its relation to applied loading. Some simulations have included computational steps to allow relaxation (thermal equilibration under constant external load or with no progression of deformation) under conditions of high strain rate deformation where the process is primarily adiabatic. This is necessitated, for example, by the interest to avoid shear localization [8]. It should be noted that this is purely a computational necessity and is not part of any realistic physical deformation process. However, it is useful in that it allows physically interpretable results to be obtained using available computer resources through the consideration of artificially high rates of deformation.

An important part of the numerical analyses of the mechanical response of nanocomponents has been directed towards carbon nanotubes. Depending on the diameter and the helicity, carbon nanotubes can be either metallic or semiconducting. They can also possess quantum wire properties. Because of these reasons, they are important for nanoelectronic devices. Some potential applications of carbon nanotubes are based on their extraordinary mechanical properties, including the ability to recover original shape after extremely large deformations and high strength. It has been reported that carbon nanotubes have very high strengths [22, 23]. However, such suggestions are tempered by the realization that some of the high values are due to the use of impractical reference areas for stress calculation. Specifically, strength should be reckoned in terms of macroscopically accessible, overall size (large) of a nanotube rather than the atomic wall size (small) of a nanotube. The deformation and generation of defects during deformation have been studied for single-wall and multi-wall tubes as well as for ropes composed of carbon nanotubes [11–13, 24]. It is found that highly distorted carbon nanotubes return to their original shape when loading is released. As a consequence, carbon nanotubes are useful in development of high-strength fibers, composite materials, and probe tips in scanning-probe microscopes [11–13, 24]. The properties of carbon nanotubes are in part the manifestation of the strength and the rigidity of the graphitic bond. The cylindrical structure of carbon nanotubes also increases their elasticity and strength.

2. Mechanical Response of Nanowires

Most MD simulations have focused on deformations at strain rates at or above 10^9 s^{-1}. This is primarily due to the fact that MD calculations require small time steps and only short physical time spans can practically be considered on computers available until recently. Issues analyzed so far are strain-induced structural changes, strain rate dependent deformation mechanisms, tensile rupture, constitutive relations, and correlation between simulations and experiments.

2.1. Strain-induced structural changes

Nanowires can undergo a crystalline-to-amorphous structural change during deformations at ultra high strain rates (above 10^{10} s^{-1}) [8]. Li et al. [7] investigated amorphization and fracture in Silicon Diselenide ($SiSe_2$) nanowires which have a very special crystalline structure consisting of nonintersecting chains of edge-sharing tetrahedrons. Theoretically, a nanowire containing a finite number of chains can be obtained readily from the crystalline structure of Silicon Diselenide ($SiSe_2$) molecules. The MD simulations are based on an effective interatomic potential which accounts for two-body steric repulsion, three-body covalent interaction, screened Coulomb interaction due to charge transfer, and charge-dipole interaction due to large electronic polarizability of selenium ions. The length of the cylindrical nanowires considered is 3510 Å and the diameter is 10–60 Å. The number of particles in these nanowires ranges from 3600 to 230400. Periodic boundary conditions are applied only along the *c* crystalline axis. The nanowires are thermalized (thermally relaxed through atomic oscillations without progression of deformation) for 4.5 picoseconds after a tensile strain of 1%. Stable configurations of the nanowires are determined by the steepest descent quench scheme [25]. A constant temperature of 100 K is maintained throughout the calculations. The authors reported a fracture strain of 15% for all cases analyzed, independent of the number of chains in the nanowires or cross-sectional sizes of the nanowires.

As the nanowires are stretched, disordered structures first form locally along the *c* crystalline axis and then expand longitudinally with time, see Fig. 1. This disordered

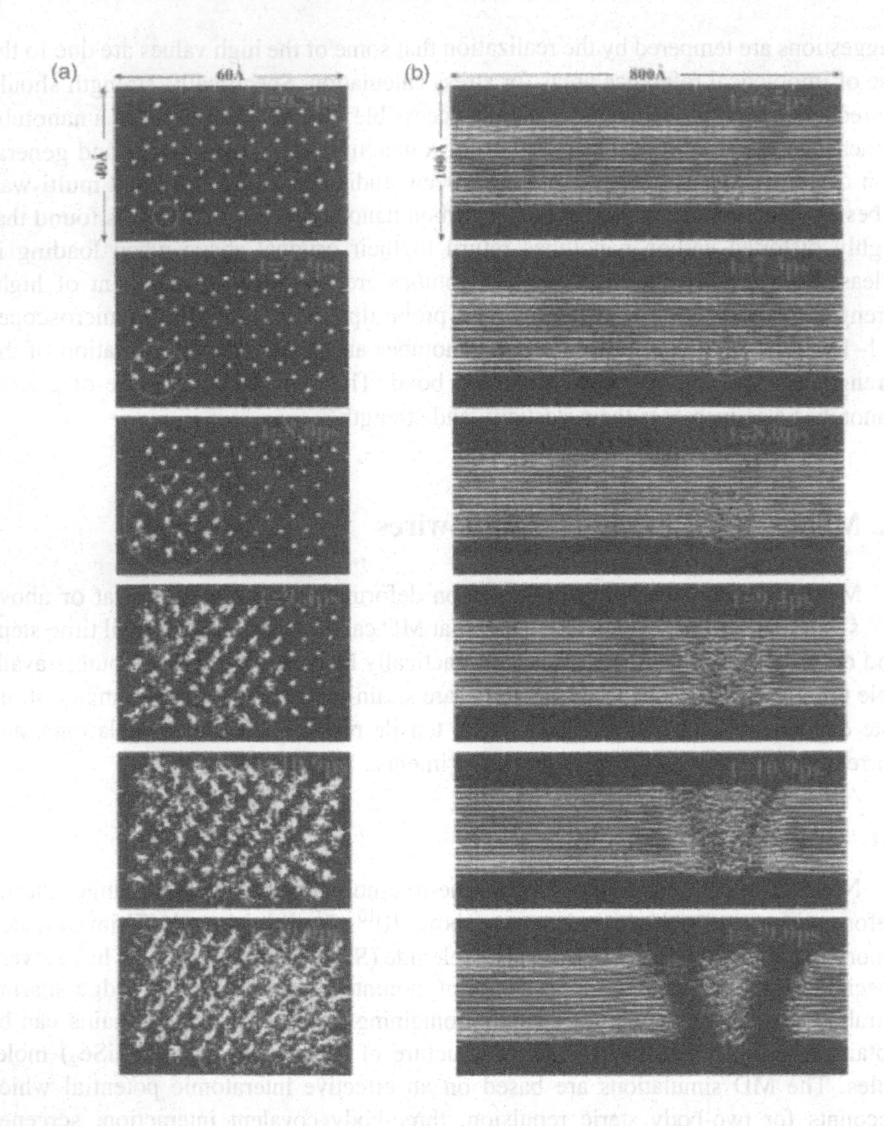

Fig. 1. Amorphization and fracture in a 64-chain nanowire. Snapshots of Si atoms projected onto (a) the a-b plane and (b) the b-c plane. Only a segment of the nanowire is shown. (From [7], Courtesy APS)

structure represents local amorphization and the rest of the nanowire remains highly crystalline. Fracture occurs at one of the two boundaries of this amorphous region. Temperature contours show that high temperatures are localized at the fracture site. Local density contours show that all cracks nucleate at the boundaries of the amorphous region. Amorphization precedes fracture.

Ikeda et al. [8] used Finnis-Sinclair potentials to study strain rate induced amorphization in Ni and Ni-Cu alloy nanowires [8]. The specimens considered are five unit cells wide (≈2 nm) in the x- and y-directions and 10 unit cells long (≈4 nm) in the z-direction. A 1-D periodic boundary condition is used in the *c*-direction. Deformation at strain rates between 0.5% ps^{-1} and 5% ps^{-1} (5×10^9 s^{-1} and 5×10^{10} s^{-1}) are considered. The calculations are isothermal, at a constant temperature of 300 K. The radial distribution function (RDF) is used to determine transformation from an FCC structure to an amorphous state. Stress distributions between strain increments are allowed to relax to homogeneous thermodynamic equilibrium under a nominally uniaxial tensile stress state. This constant temperature relaxation scheme is used in order to avoid the possibility of shear localization arising from adiabatic heating and local thermal softening at such artificially high strain rates. Such highly non-equilibrium, dynamic processes are quite difficult to obtain experimentally for nanowires. Such high strain rates are only observed experimentally in shock waves resulting from high velocity impact. Since such high strain rate processes are more adiabatic than isothermal in nature, the stress relaxation and temperature control are more numerical necessities than realistic treatments. Indeed, at the size scale of individual nanowires, there is no effective means for a heat sink at time scales of the order of a small fraction of a picosecond.

Ikeda et al. [8] also found that at strain rates between 0.5% s^{-1} and 5% s^{-1} Ni-Cu nanowires can be deformed to a strain of 100% without fracture, see Fig. 3. Additionally, yield stress increases with strain rate. The maximum stress at a strain rate of $\dot{\varepsilon} = 0.5\%$ s^{-1} reaches 5.5 GPa which occurs at a strain of 0.08. At this strain rate, cooperative shear within the crystal produces coherent shear bands which are often "twins". Multiple coherent shearing events lead to necking and eventual failure, see Fig. 2. However, for $\dot{\varepsilon} = 5\%$ s^{-1}, the behavior is fundamentally different and no shear bands or twins are seen. Instead, the specimen transforms homogeneously to an amorphous state at a strain of only 0.15 where a maximum stress of 9.5 GPa is observed, see Fig. 3. For pure Ni specimens at 300 K, the critical strain rate at which transformation to the amorphous state observed is also close to 5% s^{-1}.

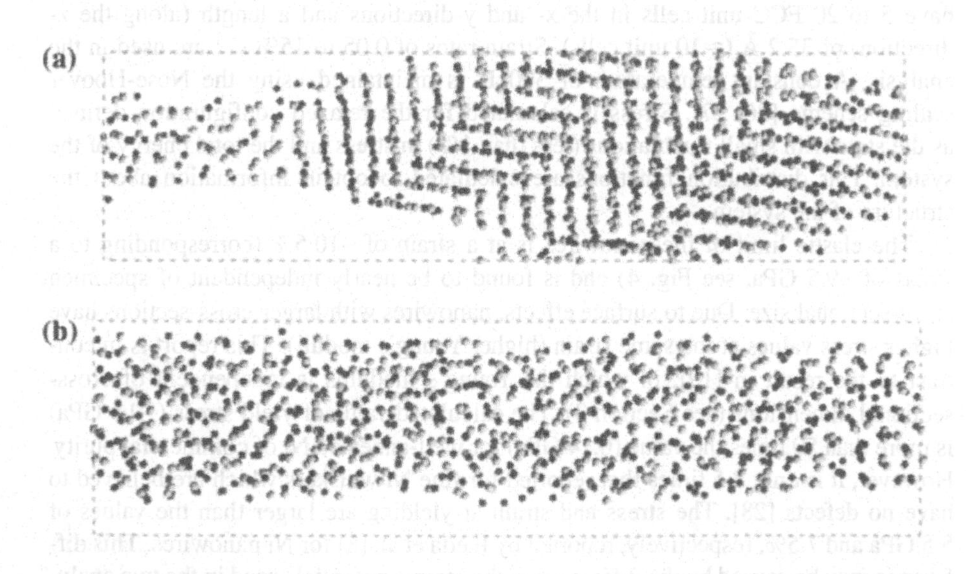

Fig. 2. Snapshots of an FCC Ni-Cu alloy nanowire at 100% strain, deformed at constant strain rates starting with a random alloy single crystalline state at 300 K. (a) 0.5% ps^{-1}; (b) 5.0% ps^{-1}. (From [8], Courtesy APS)

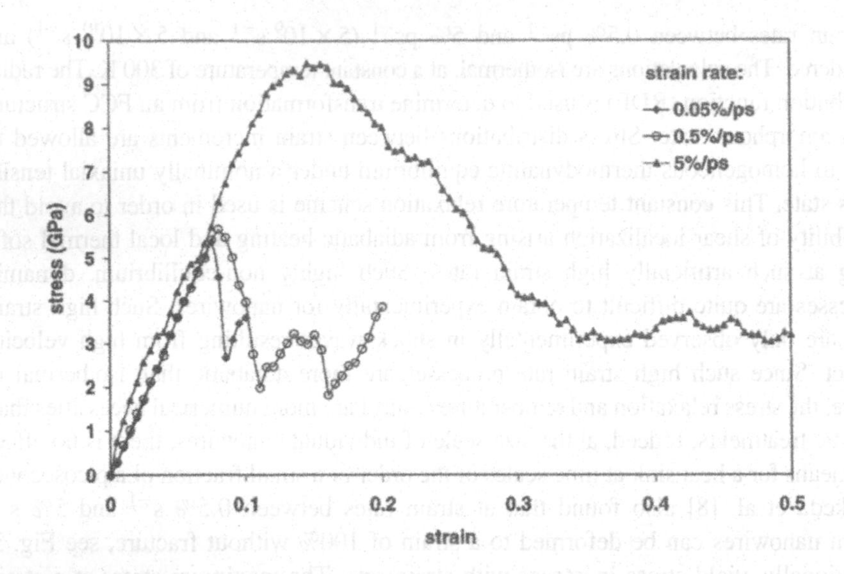

Fig. 3. Stress–strain curves at strain rates between 0.05% and 5% ps^{-1} for Ni-Cu nanowires at T = 300 K. Curves at all three strain rates show elastic response up to a strain of 7.5%. At the rates of 0.05% and 0.5% ps^{-1}, stress relaxation and subsequent hardening due to twin formation occur, leading to zigzag stress–strain curves. At a strain rate of 5% ps^{-1}, a continuous transformation to the amorphous phase is observed. (From [8], Courtesy APS)

Branício and Rino [9] studied the deformation and amorphization of Ni nanowires under conditions of uniaxial strain using an EAM potential. The nanowires studied have 5 to 20 FCC unit cells in the x- and y-directions and a length (along the z-direction) of 35.2 Å (≈10 unit cells). Strain rates of 0.05 to 15% s^{-1} are used in the analysis. A constant temperature of 300 K is maintained using the Nose-Hoover scaling scheme [26, 27]. Stress is calculated for the relaxed configuration defined as the state with small fluctuations (less than 1%) in stress and the total energy of the system. Pair distribution functions are calculated to obtain information about the structure of the system.

The elastic limit of the nanowires is at a strain of ~10.5% (corresponding to a stress of ~9.5 GPa, see Fig. 4) and is found to be nearly independent of specimen cross-sectional size. Due to surface effects, nanowires with larger cross-sections have higher stress values at the same strain (higher Young's moduli). This report is in contrast to the result in [19], in which the Young's modulus is independent of cross-sectional dimensions (see Section 3). The calculated value of yield stress (~9.5 GPa) is more than 50 times the value (0.14 GPa) for well-annealed Ni of commercial purity. However, it is only 2.5 times that reported for fine Ni whiskers which are believed to have no defects [28]. The stress and strain at yielding are larger than the values of 5.5 GPa and 7.5%, respectively, reported by Ikeda et al. [8] for Ni nanowires. This difference may be caused by the difference in the atomic potentials used in the two analyses. The cross-sectional area of the neck goes to nearly 0 at final separation, a behavior commonly associated with superplasticity. Amorphization is observed to occur at

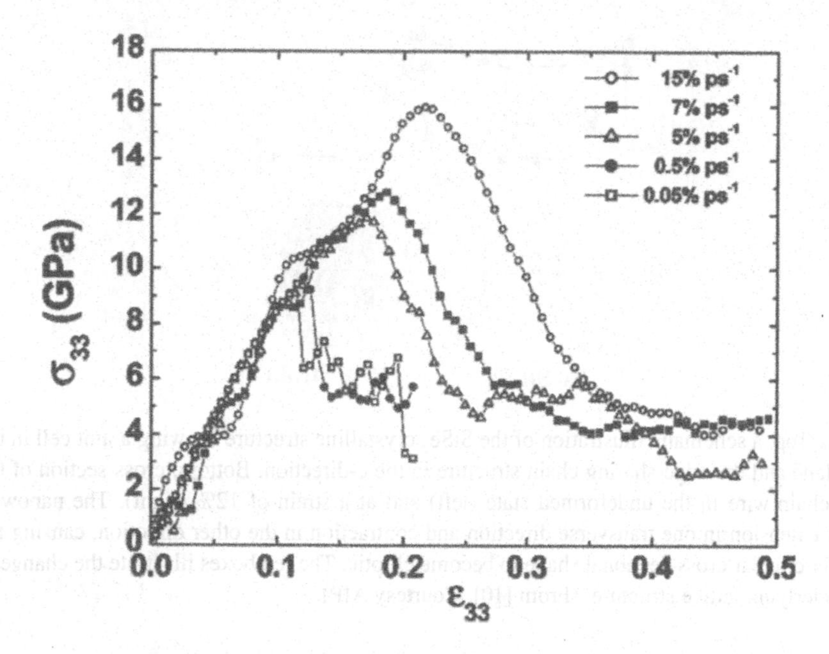

Fig. 4. Stress–strain curves for strain rates of 0.05 to 15% ps^{-1} (specimens are a single crystal Ni initially, temperature is constant at 300 K, the embedded-atom model (EAM) is used). Elastic response is seen for strains up to 11% at all strain rates. For 0.05 to 5% ps^{-1}, stress relaxation due to plastic deformation is observed, leading to zigzag stress–strain curves. For strain rates at or above 7% ps^{-1}, a continuous transformation to the amorphous phase is observed. (From [9], Courtesy AIP)

strain rates above 5% s^{-1}. This observation is slightly different from that in [8] in that a strain rate between 5% and 7% s^{-1} is not sufficient to prevent the amorphous phase from recrystallizing between increments of strain. In contrast, Ikeda et al. [8] reported that a strain rate of 5% s^{-1} or higher is sufficient to both induce amorphization and prevent crystallization.

Walsh et al. [7] studied structural transformation, amorphization, and fracture in $SiSe_2$ nanowires [10] using an empirical MD method based on the effective interatomic potential. Two types of nanowires are used in their study. One type has 128 nonintersecting chains of edge-sharing tetrahedra with a diameter of 65 Å and an initial length of 895 Å (a total of 117504 atoms). The other type has 1204 nonintersecting chains of edge-sharing tetrahedra with a diameter of 210 Å and an initial length of 3580 Å (a total of 4465152 atoms). Periodic boundary conditions are applied only along the *c* crystalline axis (Fig. 5). The nanowires are thermalized for 4.5 ps after a stretch of 1%. Stable configurations are determined by the steepest descent quench scheme. It is observed that the peaks in the Si-Si pair distribution function corresponding to the lattice parameters ($a = 9.669$ Å and $b = 5.998$ Å) merge into one peak when a strain of 15% is attained. This essentially indicates a shift from an orthorhombic to a tetragonal structure. As a result of this transformation, the initially circular cross-sectional shape becomes elliptical, see Fig. 5. This transformation occurs at

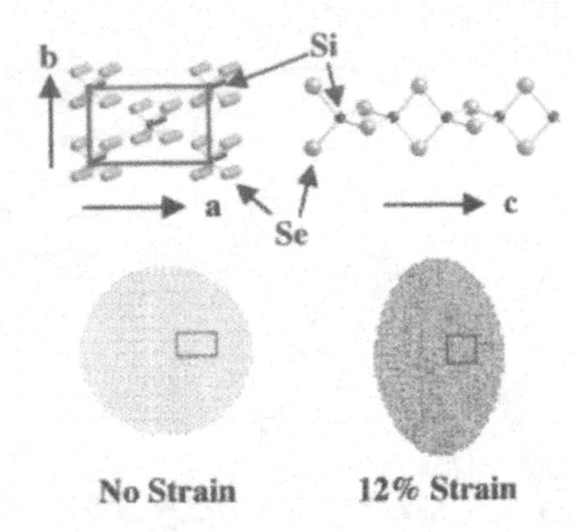

Fig. 5. Top: a schematic illustration of the $SiSe_2$ crystalline structure showing a unit cell in the a–b plane and the edge-sharing chain structure in the c-direction. Bottom: cross-section of the 1204-chain wire in the undeformed state (left) and at a strain of 12% (right). The nanowire shows extension in one transverse direction and contraction in the other direction, causing the initially circular cross-sectional shape to become elliptic. The red boxes illustrate the change in the underlying lattice structure. (From [10], Courtesy AIP)

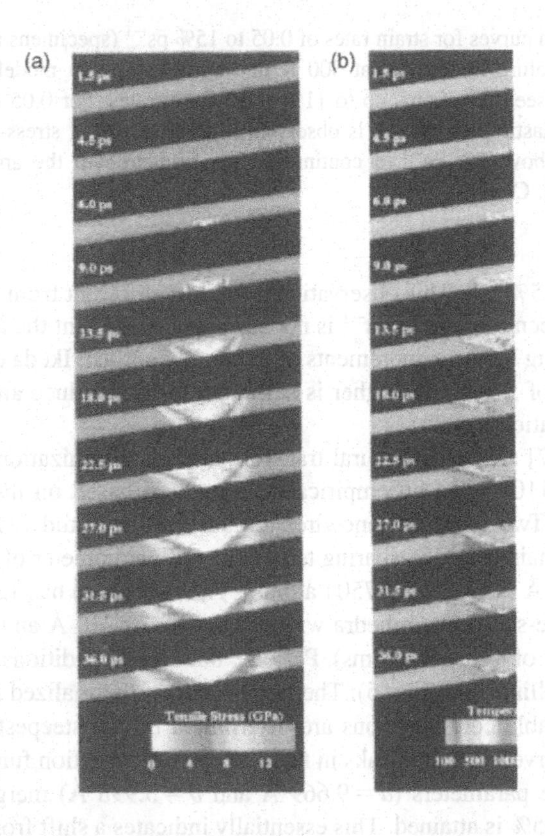

Fig. 6. (a) Time progression of local tensile stress of a 1204-chain nanowire; (b) time progression of local temperature of the same nanowire. (From [10], Courtesy AIP)

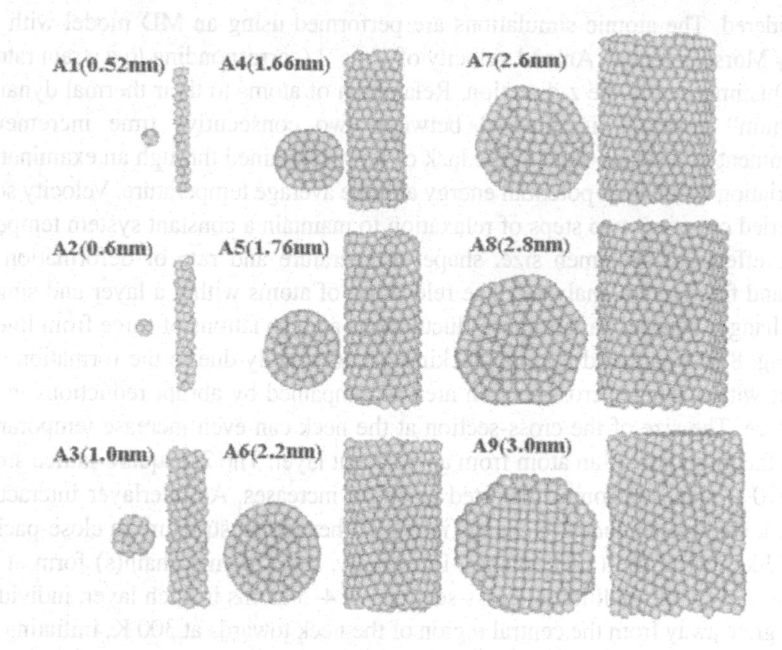

Fig. 7. Morphology of selected gold nanowires with diamcters from 0.5 to 3.0 nm. (From [29], Courtesy APS)

different strains under different strain rates. The critical strain at which the structural transformation occurs at each strain rate is yet to be determined. The nanowires remain crystalline and elastic for strains up to 15%. A Young's modulus of ~130 GPa is obtained from the stress–strain curves. Amorphization occurs before fracture. The amorphous region begins to form at a strain of 15%, see Fig. 6.

Wang et al. [29] used genetic algorithms based on a glue potential to study the response of structures of free standing gold nanowires. Helical and multiwalled cylindrical structures are found for thin nanowires of diameters up to 2.2 nm and bulk-like FCC structures are observed in thicker nanowires with diameters of up to 3 nm, see Fig. 7. They reported that this noncrystalline to crystalline transition starts from the core region of the specimens. A bulk-like behavior is found for the vibrational and electronic properties of wires with the FCC crystalline structure. When the diameter is below 0.6 nm, the helical packing composed of several identical spiral strands is the most stable structure. When the diameter is above 0.6 nm, the structure takes on a multiwalled cylindrical form.

2.2. Deformation and failure

The localized deformation and failure of nanowires has been another area of active investigations, as it is related to fundamental deformation mechanisms. Mehrez et al. [2] presented an atomic-scale study of the elastic and plastic deformations of Cu nanowires under tension. The atomic layers are taken as quasi-circular, with (001) planes being the x-y cross-sectional plane and the [001] direction being the z-axis. Three types of specimens (wide-neck, thin-neck, and wires with tapered thin necks) are considered. The atomic simulations are performed using an MD model with the two-body Morse potential. An end velocity of 1 ms^{-1} (corresponding to a strain rate of ~10^9 s^{-1}) is imposed in the z-direction. Relaxation of atoms to their thermal dynamic "equilibrium" positions is allowed between two consecutive time increments. The attainment of a steady state or the lack of it is ascertained through an examination of the variation of the total potential energy and the average temperature. Velocity scaling is carried out every two steps of relaxation to maintain a constant system temperature. The effects of specimen size, shape, temperature and rate of deformation on necking and fracture are analyzed. The relocation of atoms within a layer and single-atom exchanges between layers cause fluctuations and deviations of force from linearity, see Fig. 8. It was found that the necking occurs mainly due to the formation of a new layer with a smaller cross-section area accompanied by abrupt reductions in the tensile force. The size of the cross-section at the neck can even increase temporarily, owing to the migration of an atom from an adjacent layer. The 2D square-lattice structure of (0 0 1) planes becomes distorted as strain increases. As interlayer interaction weakens, it shows a tendency to change into a 2D hexagonal structure (a close-packed structure like that of the (1 1 1) plane). Eventually, short atomic chain(s) form at the neck. For wires with small initial cross-sections of 4–5 atoms in each layer, individual atoms migrate away from the central region of the neck towards at 300 K, initiating the neck formation at small strains. The migration of atoms occurs more easily at higher temperatures. At the very low temperature of 1 K, the applied force exhibits sharp jumps and amorphization is observed at the neck. Similarly, at lower strain rates, the force-stretch curves exhibit well-defined jumps. At higher strain rates, the wire breaks before a new layer can form. In a related paper, Mehrez and Ciraci [1] presented a

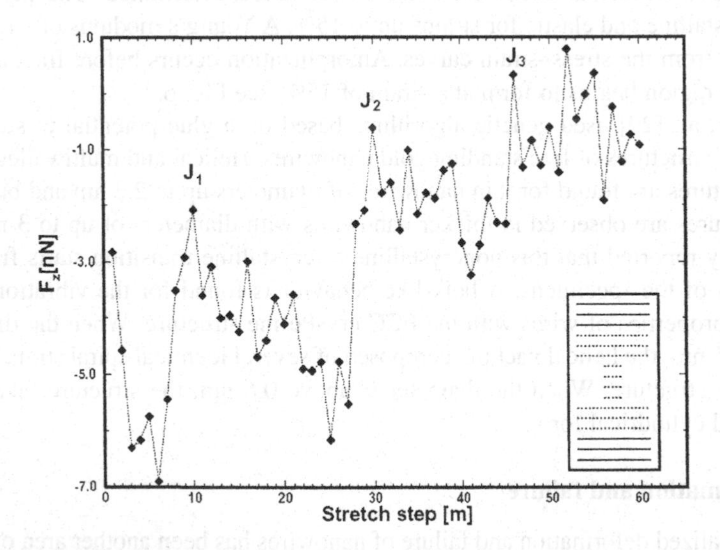

Fig. 8. The tensile force as a function of the number of stretching steps **m** (the strain increment of each stretching step is 0.55%) for the a (001) Cu nanowire at 300 K. J1, J2 and J3 indicate abrupt reductions in the neck size. The inset illustrates the initial, relaxed structure of the wide neck as projected onto the xz-plane. The continuous and dotted lines correspond to robust and dynamic layers, respectively. (From [2], Courtesy IOP)

detailed analysis of the atomic structure of and force variations in Cu nanowires under tensile loading. In addition to MD simulations with the two-body Morse potential, *ab initio* self-consistent field calculations using the local density approximation are also used. The atomic layers are taken as quasi-circular. One type is made of (0 0 1) planes and the other type is made of (1 1 1) planes. The calculations are conducted for 300 K and 150 K. Periodical boundary conditions are specified in the x-y plane. Nose-Hoover drag temperature control is used. The authors found that yielding and fracture mechanisms depend on size, atomic arrangement, and temperature.

The elongation under uniaxial stress shows alternating quasi-elastic and yielding stages, see Fig. 9. In a quasi-elastic stage, the stored strain energy and tensile force increase with deformation while the layer structure remains unchanged. Fluctuations occur due to the movement and relocation of atoms within each layer or due to atomic exchanges between adjacent layers. Tensile force varies approximately linearly in such a stage and deviations from linearity are seen as the number of atoms in the neck becomes smaller at large strains. Each quasi-elastic stage is followed by a yielding stage during which the tensile force decreases rapidly. This is associated with the abrupt reduction in the size of the narrowest neck layers. When the neck became very narrow (having 3–4 atoms), the yielding was realized, however, by single atom jumping from one of the adjacent layers to interlayer space. Neck development is through the migration of atoms, mainly by the sequential insertion of new layers with smaller cross-sections accompanied by abrupt decreases in the tensile force. The large strains in the neck region cause atomic rearrangement. In certain instances, a bundle of atomic chains or a single atomic chain forms shortly before final breakage.

The stages of plastic deformation are shorter at low temperatures than they are at high temperatures, due to limited mobility of the atoms at lower temperatures. The tendency to minimize surface area and system energy is found to be the main driving force for the necking process. Wires with very small cross-sectional areas show

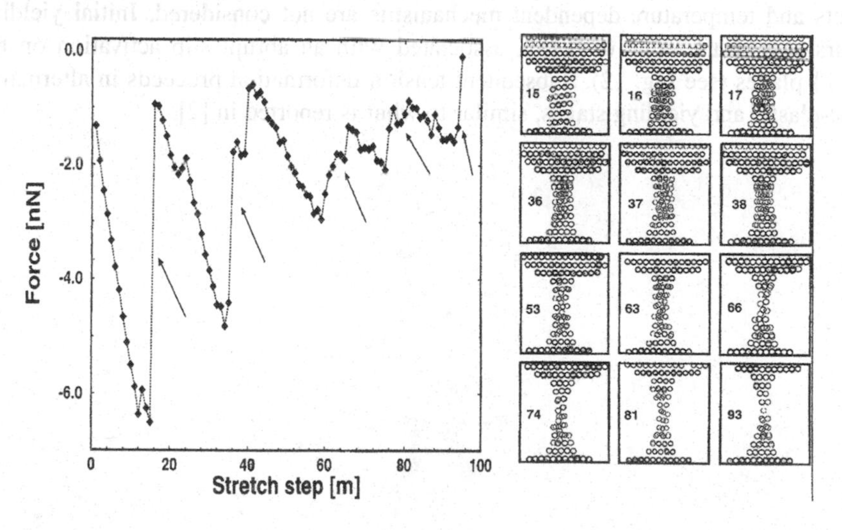

Fig. 9. Tensile force as a function of deformation. The side views show atomic positions at relevant stretch steps (m) for the Cu (1 1 1) nanowire. The strain increment of each stretching step is 0.55%. The MD simulations are performed at 300 K with $\Delta t = 10^{-13}$ s. (From [1], Courtesy APS)

further progression of necking even without increase in the overall tensile strain. This result is consistent with the experimental observation of the breaking of a neck in a gold nanowire at room temperature [30].

Silva et al. [31] analyzed the tensile rupture of gold nanowires using a tight binding molecular dynamics (TBMD) model. They studied structural evolution the formation of necks induced by defects that eventually results in one-atom chains. The structure of gold nanowire tips after rupture is also analyzed. The authors used periodic supercells with the dimensions of $20 \times 20 \times (25.5–41)$ Å. Although the strain rates are much higher than those in experiments, the tensile forces are between 1.0 and 2.5 nN, quite similar to measured values. The applied force right before rupture is around 1.8 nN, in very good agreement with the experimental value of 1.5 ± 0.3 nN. The authors reported that the tensile deformation causes the cylindrical wire to become hollow, with the central atoms of the seven-atom planes migrating outward toward the surface, resulting in six-atom rings. Further deformation leads to thinning of the wall of the deformation-produced "nanotube" a one-atom neck. This one-atom neck continued to grow to a five-atom chain before breaking, (see Fig. 10). It was found that the wire breaks only after the tips attain a rather stable structure. Before this configuration is reached, the system prefers to move the atoms from the less stable positions in the tip towards the neck, rather than breaking the wire. After breaking, the gold nanowire tip resembles a child's French hat being formed basically by two hexagonal structures that share four atoms among themselves, (see Fig. 11).

Kang and Hwang [32] studied the deformation of Cu nanowires using the steepest–descent MD method. A potential function of the second-moment approximation of the tight-binding (SMA-TB) scheme is used [33, 34]. The nanowire analyzed has 10 atomic layers (each of 18 atoms) perpendicular to the wire axis ([0 0 1] direction). The authors considered tension, shear, torsion, and combined torsion/tension deformations. The nanowires are relaxed between time increments, therefore thermal effects and temperature-dependent mechanisms are not considered. Initial yielding occurs at a strain value of 0.133, associated with an abrupt slip activation on the {1 1 1} planes (see Fig. 12). Subsequent tension deformation proceeds in alternating quasi-elastic and yielding stages, similar to what is reported in [2].

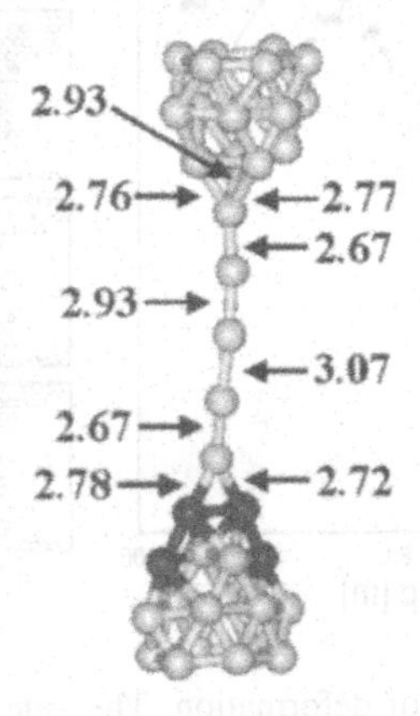

Fig. 10. The structure of a nanowire just before breaking with a single-atom chain comprising the neck. The bond distances are in Å. (From [31], Courtesy APS)

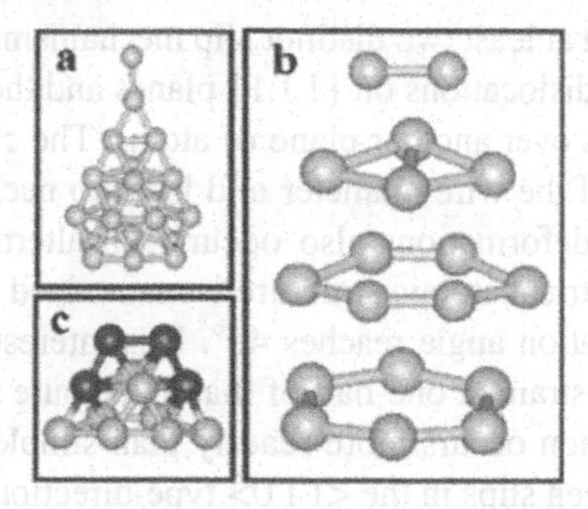

Fig. 11. Stability of the nanowires tip. The tip, before and after the breaking, shows a highly symmetric and stable structure. In (a), the whole tip is shown. In (b), the layers forming the tip are illustrated. In (c), the French hat structure is shown. (From [31], Courtesy APS)

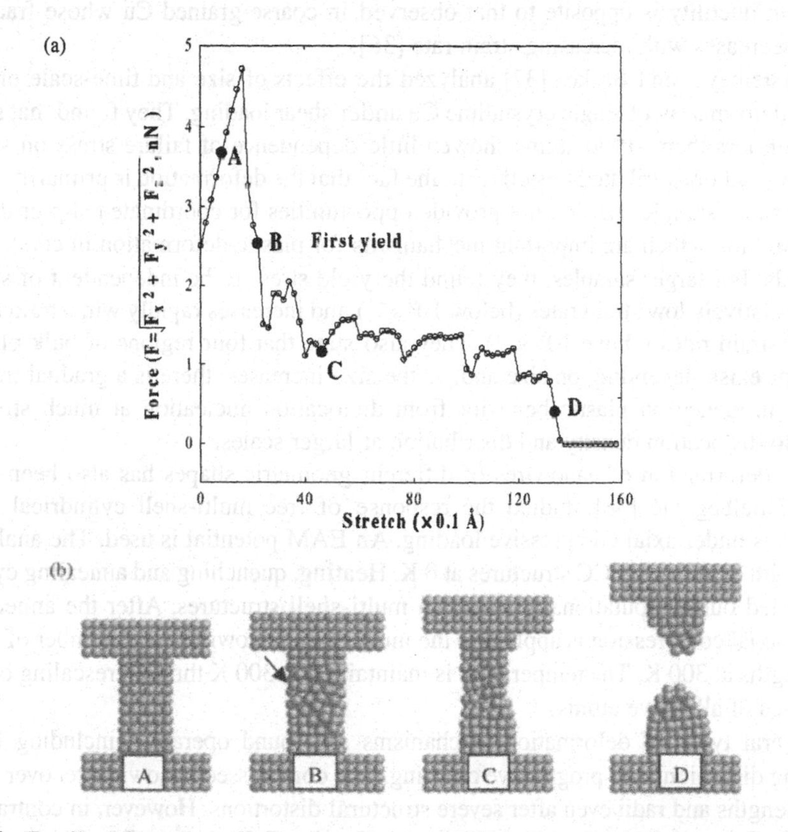

Fig. 12. Tensile deformation of a Cu nanowire comprising of ten atomic layers, each of which has 18 atoms. A stable atomic configuration predicted by the SD method. (a) The variation of tensile force with elongation. (b) Atomic arrangements at points A–D in (a). The arrow denotes slip in the [1 1 0]-direction. (From [32], Courtesy IOP)

The authors found that at least two distinct slip mechanisms affect nanowire deformation. One is a glide of dislocations on {1 1 1} planes and the other is homogeneous slip of one plane of atoms over another plane of atoms. The crossover of the two slip events causes reduction of the wire diameter and leads to necking. Rupture occurs at a strain of 0.752. Shear deformations also occur with alternating quasi-elastic and yielding stages. A symmetrical atomic structure is maintained during rotational deformations until the end rotation angle reaches 45°. It is interesting to note that, initial yielding occurs at a shear strain of one-half of that in the pure tensile case. Elongation with a superimposed torsion occurs more readily than simple tension or simple torsion. The crossover between slips in the <1 1 0> type-directions on {1 1 1} planes and homogeneous slips are observed under conditions of pure tension, shear and combined torsion/tension.

The dependence of deformation mechanisms on strain rates has also been analyzed. For example, Lu et al. (2001) observed that the ductility of nano-crystalline Cu specimens increases as strain rate is increased in experiments [35]. This strain rate effect on ductility is opposite to that observed in coarse-grained Cu whose fracture strain decreases with increasing strain rate [36].

Horstemeyer and Baskes [37] analyzed the effects of size and time-scale on the plastic deformation of single crystalline Cu under shear loading. They found that samples with less than ~1000 atoms showed little dependence of failure stress on strain rate. The authors attributed this effect to the fact that the deformation is primarily elastic and small sample sizes do not provide opportunities for coordinated slip or dislocation motion which are important mechanisms for plastic deformation in crystalline materials. For larger samples, they found the yield stress to be independent of strain rate at relatively low strain rates (below 10^9 s^{-1}) and increases rapidly with strain rates at high strain rates (above 10^9 s^{-1}). They also state that four regions of bulk plastic behavior exist, depending on size and, as the size increases, there is a gradual transition of influence on plastic behavior from dislocation nucleation at much smaller scales to dislocation density and distribution at larger scales.

The deformation of nanowires of different geometric shapes has also been analyzed. Bilalbegović [38] studied the response of free multi-shell cylindrical gold nanowires under axial compressive loading. An EAM potential is used. The analyses begin with cylindrical FCC structures at 0 K. Heating, quenching and annealing cycles are carried out computationally to obtain multi-shell structures. After the annealing cycles, axial compression is applied to the multi-shell nanowires for a number of radii and lengths at 300 K. The temperature is maintained at 300 K through rescaling of the velocities of all active atoms.

Several types of deformation mechanisms are found operative, including large buckling distortions and progressive crushing. The compressed nanowires recover their initial lengths and radii even after severe structural distortions. However, in contrast to what is observed for carbon nanotubes (see Section 2.3), irreversible local atomic rearrangements occur even under small compressive strains. The analysis also revealed the role of defects in inducing the formation of necks that eventually lead to one-atom chains. The authors also characterized the structure of the nanowire tip after rupture.

Short nanowires (e.g., $R_0 = 0.9$ nm, $L_0 = 4$ nm) exhibit only one morphological pattern under compression. They progressively shorten but remain straight when compressed. Long and thin nanowires (e.g., $R_0 = 0.5$ nm, $L_0 = 12$ nm) crush into irregular shapes. On the other hand, thick and short nanowires (e.g., $R_0 = 1.2$ nm, $L_0 = 8$ nm) show three morphological patterns. Under small compressive stresses, ripples develop. For stresses up to 16 GPa, one end of the wire shows more severe deformation than the other end (see Fig. 13). When the stresses approach 32 GPa, crushing and flattening are observed. The most interesting behavior is exhibited by long and slender wires (e.g., $R_0 = 0.9$ nm, $L_0 = 12$ nm), see Fig. 14. Buckling and substantial distortion occur. Fig. 14(b) shows the deformation of the same nanowire in Fig. 14(a) under a stress of 4.81 GPa. The deformed shape is essentially symmetric. Although they largely recover their initial shapes upon unloading, gold nanowires, retain permanent changes in local atomic arrangements. This results from the nature of the atomic bond in gold and the structure of the gold nanowires. Note that

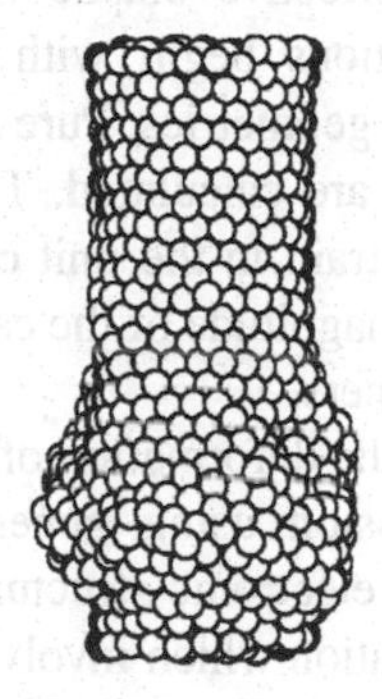

Fig. 13. Deformation of a nanowire with $R_0 = 1.2$ nm and $L_0 = 8$ nm, under a compressive stress of 19.23 GPa. (From [38], Courtesy IOP)

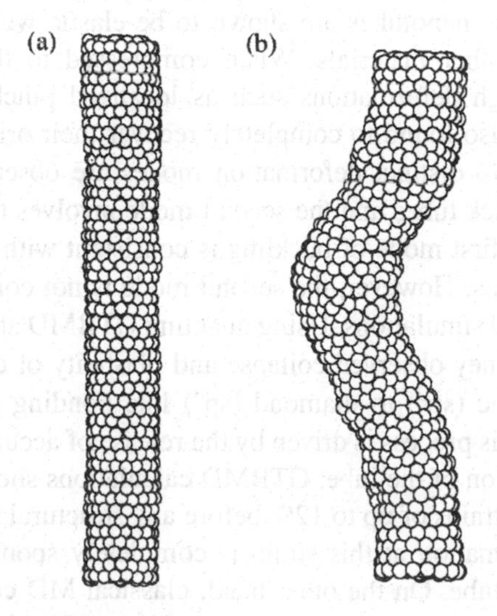

Fig. 14. A nanowire with $R_0 = 0.9$ nm and $L_0 = 12$ nm: (a) the uncompressed configuration; (b) configuration under a stress of 4.81 GPa. (From [38], Courtesy IOP)

virtually no defects are found in carbon nanotubes after the compressive load is released [11–13, 24]. The ability of carbon nanotubes to recover original shapes without damage after large deformations is a consequence of the carbon bonding in graphite layers and their cylindrical structure. However, other than the different lattice structures a detailed explanation as to what causes the differences in behavior between the gold nanowires and carbon nanotubes has not been given at this time. The ability of nanowires and nanotubes to return to original shapes after severe deformations enables them to store large amounts of strain energy and release it upon unloading.

2.3. Carbon nanotubes

Saether et al. [39] formulated an MD model with the Lennard-Jones potential to analyze the elastic behavior of carbon nanotube bundles. They used a consistent unit cell approach to determine the effective elastic moduli of carbon nanotubes in a bundled configuration. Calculations begin with equilibrium configurations for both square and hexagonal packing geometries. Pure shear deformation is imposed on a unit cell and interatomic forces are calculated. The forces and deformations are resolved into shearing stress and strain in the unit cell, yielding an effective virtual transverse shearing modulus. The magnitude of the calculated shear modulus is on the same order as the moduli for polymers.

Srivastava et al. [12] analyzed the deformation of single-walled carbon nanotubes (SWNT) under uniaxial compression using generalized tight-binding molecular dynamics (GTBMD) and *ab initio* electronic structure calculations. They observed a novel mechanism for nano-deformation which involves a collapse of bonding geometry from that of graphite (sp^2) to a localized diamond-like (sp^3) reconstruction (see Fig. 15). The calculated critical stress for the transition (~153 GPa) and the shape of the deformed specimens are in good agreement with recent experimental observations of the collapse and fracture of carbon nanotubes in polymer composites. Both single and multi-walled carbon nanotubes are shown to be elastic well-beyond the typical yield strain for most other materials. When compressed to the non-linear elastic response regime through deformations such as localized pinches and kinks, these carbon nanotubes are also found to completely recover their original shape. At large compressive strains, two distinct deformation modes are observed. The first mode involves buckling of thick tubes and the second mode involves the collapse and fracture of thin tubes. The first mode of buckling is consistent with what is predicted by classical MD simulations. However, the second mode is not consistent with the predictions of classical MD simulations. Using quantum GTBMD and *ab initio* electronic structure calculations, they observed collapse and plasticity of compressed thin nanotubes through graphitic (sp^2) to diamond (sp^3) like bonding reconstruction at the location of collapse. This process is driven by the release of accumulated strain energy in the uncollapsed section of the tube. GTBMD calculations show that the nanotubes can be compressed to strains of up to 12% before any structural transformation starts. The structural transformation at this strain is completely spontaneous and leads to plastic collapse of the tube. On the other hand, classical MD calculations show that structural transformation starts at strains between 8% and 9%. Additionally, the structural transformation results in symmetric-pinching and is fully elastic.

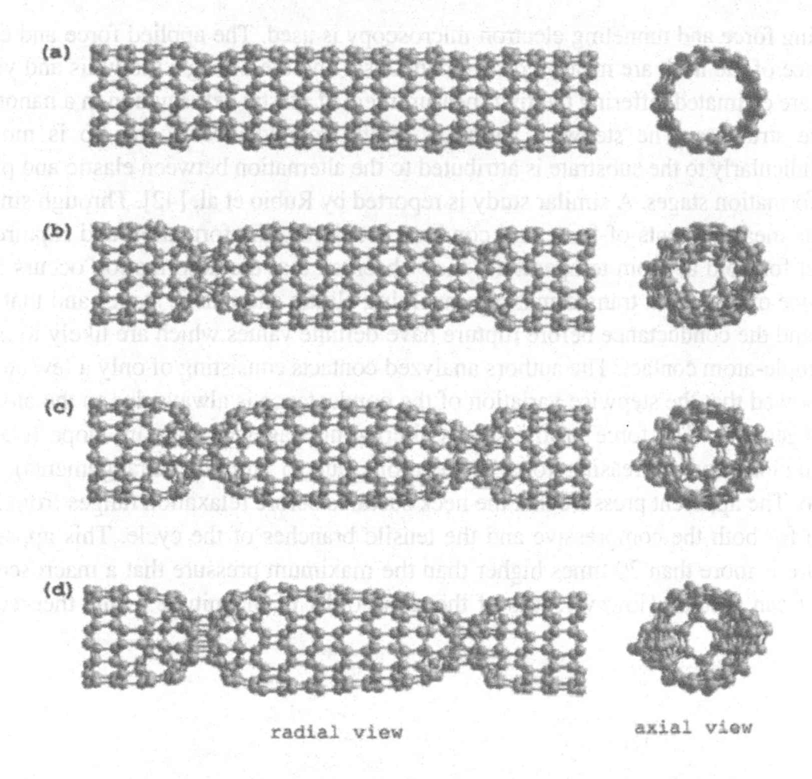

Fig. 15. The four stages of spontaneous plastic collapse of a carbon nanotube with a compressive strain of 12%; (a) nucleation of local deformation; (b) and (c) inward collapse; and (d) graphitic to diamond-like structural transition at locations of the collapse. (From [12], Courtesy APS)

2.4. Correlations with experiments

Stalder and Durig [40] studied the yielding of nanometer-scale gold contact points using a scanning tunneling microscope supplemented by a force sensor. The contact is between an indenting tip and a substrate. Indentation of the substrate is typically 10 nm. An adhesion neck is formed in the tip during retraction of the tip. The neck is found to undergo three consecutive phases of deformation: (1) buildup of tensile stress, (2) incomplete fracture, and (3) quasicontinuous plastic flow. Separation occurs at the neck when only three to four atoms are left in the contact region. In the plastic flow regime, the conductance and thus the contact area shrink exponentially with the elongation of the neck, suggesting that plastic deformation occurs locally within 5 to 6 atomic layers. The stress during initial plastic flow of the order of 10 GPa and gradually increases to 20 GPa shortly before breaking of the neck. The yield strength is estimated to be of the order of 5–8 GPa, more than an order of magnitude higher than the macroscopic yield strength of gold.

Agräit et al. [41] studied the plastic deformation of connective necks in gold nanowires formed by cohesive bonding between a metallic tip and a substrate after contact. Experiments show force oscillations as a result of successive stages of elastic and plastic deformation, as predicted by the MD calculations [1]. A combination of scanning force and tunneling electron microscopy is used. The applied force and conductance of the neck are measured simultaneously and the Young's modulus and yield stress are estimated, offering the first measurement of plastic deformation in a nanometer-size structure. The stepwise variation of the conductance as the tip is moved perpendicularly to the substrate is attributed to the alternation between elastic and plastic deformation stages. A similar study is reported by Rubio et al. [42]. Through simultaneous measurements of force and conductance during the formation and rupture of contact for gold at room temperature, they observed that the deformation occurs as a sequence of structural transformations involving elastic and plastic stages and that the force and the conductance before rupture have definite values which are likely to indicate single-atom contact. The authors analyzed contacts consisting of only a few atoms and showed that the stepwise variation of the conductance is always due to the atomic rearrangements. The force profile shows alternating stages of constant slope (elastic deformation) and decreasing force (relaxations due to atomic rearrangements), see Fig. 16. The apparent pressure that the neck sustains before relaxation ranges from 3 to 6 GPa for both the compressive and the tensile branches of the cycle. This apparent pressure is more than 20 times higher than the maximum pressure that a macroscopic contact can sustain. However, it is of the same order of magnitude as the theoretical

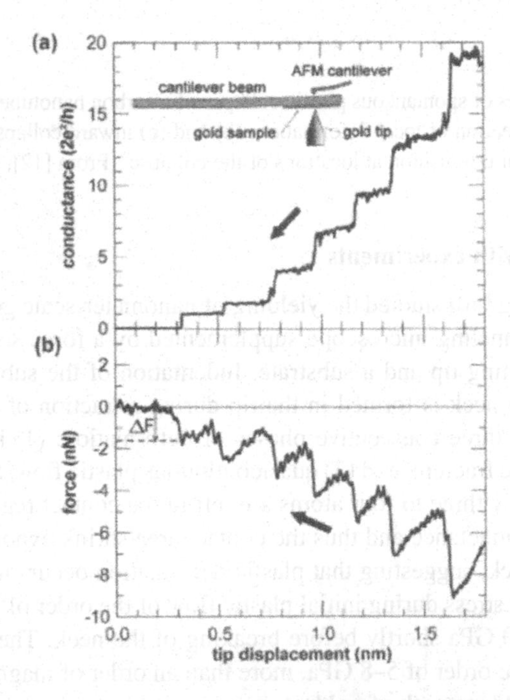

Fig. 16. Simultaneous recording of conductance and force during the elongation of an atomic-sized constriction at 300 K. This constriction is stretched until rupture by retracting the tip by 1.8 nm (x axis). (a) Variation of conductance during the deformation in conductance quantum (2e2yh). (b) Force history measured with a cantilever beam of an effective spring constant of 25 Nym. The inset shows the experimental setup. (From (42), Courtesy APS)

value in the absence of dislocations [43, 44]. This pressure is up to 13 GPa for the smallest contact analyzed (of one quantum conductance unit).

The authors obtained the effective spring constant K_{eff} of the neck from the slope of the force curve in the elastic stage, using the continuum contact mechanics relation of

$$K_{eff} = BEa/(1 - v^2), \tag{1}$$

where, a is the radius of the contact, v is the Poisson's ratio, E is the Young's modulus, and B is a factor that depends on the geometry of the contact. The Young's modulus of gold nanowires ranges from 43 to 117 GPa, depending on the crystalline orientations. The experiments show that continuum contact mechanics can be used to estimate the elastic properties of this type of nanostructures down to contact areas whose conductance is 10 quantum units.

Following the work of Rubio et al. [42], Kassubek et al. [45] derived a semi-classical trace formula for the mesoscopic oscillations of the cohesive force in metallic nanowires of constant cross-sections. They showed that the mesoscopic oscillations of the cohesive force are *universal*, in the sense that their root mean square (RMS) amplitude is independent of the area of the cross-section, provided that the cross-section possesses a continuous one-dimensional symmetry. The fact that axial symmetry is favored by surface tension may be explained by the universality of the force oscillations observed in gold nanowires. For a wire with a rectangular cross-section, the RMS amplitude of force oscillations is shown to be proportional to the aspect ratio of the cross-section.

Rubio-Bollinger et al. [46] used a scanning tunneling microscope supplemented with a force sensor to study the mechanical properties of a freely suspended chain of single gold atoms and found that the bond strength of the nanowire is about twice that of a bulk metallic bond. *Ab initio* calculations of the force at chain fracture are performed and compared quantitatively with experimental measurements. The mechanical failure and nanoelastic processes during atomic wire synthesis are investigated using MD simulations. It is found that the total effective stiffness of the nanostructure is strongly affected by the detailed local atomic arrangement at the chain bases. Upon stretching, a chain of single gold atoms is, in some cases, formed by extraction of atoms from the neighboring electrodes. The force profile has a saw tooth shape consistent with the experimental measurement, see Fig. 17. Experiments show that the force at which atomic chains break is around 1.5 nN and is independent of the length of the chain. This value is considerably higher than the strength of individual bonds in bulk gold (0.8–0.9 nN). The largest atomic displacements occur in the electrodes, not in the chain itself. This is due to a combination of two effects. One is the electronic effect which makes the bonds in the chain stronger than the bonds in the more bulk-like electrodes. The other effect is geometrical. The atoms in the electrodes are in a structure in which the breaking of bonds can proceed through a more concerted motion of atoms, giving rise to longer paths and hence requiring smaller forces.

2.5. Characterization of constitutive behavior

Odegard et al. [4] recently proposed a technique for developing constitutive models for polymer composite systems reinforced with SWNTs. The main emphasis

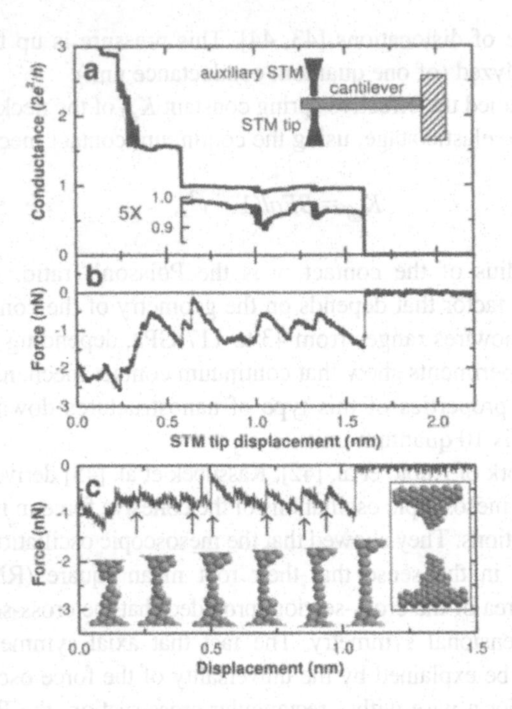

Fig. 17. Simultaneous measurements of (a) conductance and (b) force during chain formation and breaking. The inset shows a schematic illustration of the experimental setup. (c) Calculated force history. The arrows indicate locations where new atoms pop into the chain and snapshots of the structure at these positions. (From [46], Courtesy APS)

is on developing structure-property relationships for nano-structured materials. Since polymer molecules are of the same size scale as nanotubes, the interaction at the polymer/nanotube interface is highly dependent on local molecular structure and bonding. At these small length scales, the lattice structures of the nanotubes and the polymer chains cannot be considered to be continuous and the bulk mechanical properties of the SWNT/polymer composites can no longer be determined through traditional micromechanical approaches that are formulated using only continuum mechanics considerations.

The proposed framework assumes that the nanotube, the local polymer near the nanotube, and the nanotube/polymer interface can be modeled as an effective continuum fiber using an equivalent-continuum model. Such a method provides a link between computational chemistry and solid mechanics by substituting discrete molecular structures with an equivalent continuum model. The equivalence here, however, is only based on energy. Intrinsic dynamic effects inherent in the nonequilibrium molecular system are neglected since work rates and momentum are not considered.

The effective fiber retains the local molecular structure and bonding information and served as a means for incorporating micromechanical analyses for the prediction of bulk properties of SWNT/polymer composites with different nanotube sizes and orientations. As an example, the authors used the proposed approach to model the constitutive behavior of two SWNT/polyethylene composite systems, one with continuous and aligned SWNT and the other with discontinuous and randomly oriented nanotubes. Molecular simulations are conducted to obtain the equilibrium structure of a SWNT surrounded by polyethylene molecules. The simulation is performed for a rectangular box (volume ~36 nm^3) at a constant temperature of 300 K and the atmospheric pressure. The representative volume element (RVE) is developed using an equivalent truss model in which each truss element represents an atomically bonded or non-bonded interaction. The moduli of the truss elements are specified based on the molecular mechanics force constants that describe the contribution of each bonded or non-bonded interaction to the total vibrational potential energy. The continuum RVE was developed by equating the total strain energies of the truss and continuum models under identical loading conditions. Constitutive models of SWNT/polymer composites are developed with a micromechanical analysis using the mechanical properties of the effective fiber and the bulk matrix material. However, it must be pointed out that the "equivalence" here is only in a very limited sense. It is not the same as the fully dynamic, time-resolved continuum/molecular system equivalence represented in the newly developed theoretical framework [47, 48].

3. Size and Rate Effects in Nanowires

The deformation of nanowires and atomic systems at finite temperatures in general is an intrinsically dynamic process. Size and strain rate effects arise out of several factors and play important roles in determining the response of nanostructures. For example, the behavior and properties of nanowires are size-dependent due to the discreteness of atomic structures, the length scale (or spatial range) of the atomic interactions, lattice structure, and lattice size scale. The dynamic inertia effect and the finite speeds at which lattice waves propagate also introduce length scales to the problem and contribute to size-dependence of atomic behavior. The inertia effect and finite wave speeds, along with phonon effects, also cause the response of nanostructures to be deformation-rate dependent. Historically and even presently, MD calculations of the mechanical response of atomic systems have been almost exclusively carried out at very high strain rates which are above 10^9 s^{-1}. This is primarily out of necessity. Specifically, the time steps allowable in MD calculations are limited by the need to resolve high frequency thermal oscillations for atoms and are quite small (on the order of 1 picosecond). High rates of deformation allow high levels of strain to be reached with practically available computer resources. The use of high deformation rates introduces several issues. First, direct comparisons with experiments are extremely difficult to justify since it is so far not possible to conduct controlled laboratory experiments at high strain rates on nanowires or nano-structures. The artificially high rates also have necessitated computational schemes that allow computations to proceed. One issue is temperature control. Many authors have carried MD simulations of nanowires using the Nose-Hoover thermostat scheme or the velocity scaling scheme which keep the temperature at constant values. If such schemes were not used, the artificially high strain rates would lead to temperatures over the melting point of the system under consideration, invalidating the results and preventing analyses to be carried out. The use of such schemes has allowed results to be obtained and important understandings to be arrived at. However, we note that at the size and time scales of the dynamic deformation of nanowires, there is usually no effective mechanisms for heat to be conducted, convected or radiated out of the system. It is clearly desirable and important to conduct numerical simulations under conditions that do not necessitate artificial schemes for pure numerical reasons. There has been an effort in carrying out simulations at lower strain rates. We describe here some of our recent calculations at strain rates between 10^7 and $10^9\ s^{-1}$. No temperature controlling algorithms are used, providing a more realistic account of the conditions of the dynamic deformation of nanowires. The focus of the analyses is on the size and rate effects on the constitutive response of Cu nanowires. Parameters varied include loading rate and specimen size.

These artificially high rates are, to a degree, necessitated by speeds of computers available at the time. However, it is important to note that this treatment is artificial and is inconsistent with the dynamic nature of the deformation. At strain rates on the order of $10^7\ s^{-1}$ or higher, there is not sufficient time for the specimens to exchange heat with the environment. Also, at such small size scales, efficient heat exchange mechanisms are not easily available. The deformation of nanowires is more close to an adiabatic process rather than an isothermal process. Consequently, more realistic treatments of the issue should entail proper account of temperature increases and should avoid artificial or arbitrary numerical schemes.

Since the deformation processes are fundamentally dynamic, proper distinction between internal stress and externally applied stress (traction) must be made. In our analysis, both the internal stress and externally applied traction are tracked separately. This allowed the fully dynamic nature of the deformation process to be quantified. The average internal stress [47, 49] in the specimen is

$$\overline{\sigma} = \frac{1}{2V}\sum_{i}\sum_{j(\neq i)} \mathbf{r}_{ij} \otimes \boldsymbol{f}_{ij}, \tag{2}$$

where V is the volume of the specimen, $\mathbf{r}_{ij} = \mathbf{r}_j - \mathbf{r}_i$ with $\mathbf{r}_i$ being the position of particle i and $\otimes$ denotes the tensor product of two vectors with $(\mathbf{a} \otimes \mathbf{b})_{\alpha\beta} = a_\alpha b_\beta$ (a_α and b_β are Cartesian components of $\mathbf{a}$ and $\mathbf{b}$ respectively, $\alpha,\beta = 1,2,3$). $\boldsymbol{f}_{ij}$ is the interatomic forces applied on atom i by atom j. The externally applied traction (stress) is

$$T = F/A, \tag{3}$$

where, T is the average externally applied stress, F is the total applied force, and A is the cross-sectional area of the specimen. Note that in general, $T \neq \overline{\sigma}_{33}$ under conditions of finite temperature. Further, the concept of virial stress [50–52] does not apply here, since the nanowire is in fully dynamic deformation and is not in a statistically non-deforming or steady state [47, 49].

In order to analyze the plastic deformation in detail, the locations and types of defects in the specimens must be identified. Several techniques can be used for this purpose. For example, techniques relying on the electron density, the potential energy,

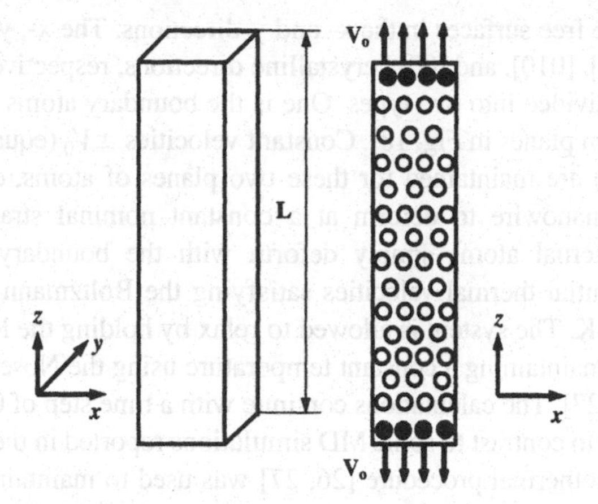

Fig. 18. A schematic illustration of the computational model for Cu nanowires. Open circles represent the active atoms and the dark circles represent boundary atoms.

the dislocation density tensor [53], or the atomic level stress tensor [54] are available. Kelchner et al. [21] defined a centrosymmetry parameter and used it to identify defects. This technique is based on the fact that a centrosymmetric material (such as FCC metals) will remain centrosymmetric under homogeneous elastic deformations. The concept uses the fact that in centrosymmetric materials each atom has pairs of equal and opposite bonds to its nearest neighbors. As lattice is distorted, these bonds will change direction and/or length, but they will remain equal and opposite under uniform deformations. When a defect is introduced nearby, this centrosymmetric characteristic is lost, providing a mechanism for the defect to be identified. For FCC structures, the centrosymmetry parameter is defined by

$$P = \sum_{i=1,6} |\mathbf{R}_i + \mathbf{R}_{i+6}|^2, \tag{4}$$

where $\mathbf{R}_i$ and $\mathbf{R}_{i+6}$ are the location vectors for the six pairs of opposite neighbors in an FCC lattice. The centrosymmetry is zero for undeformed lattices and for homogeneously (elastically) deformed lattices. It is nonzero under any plastic deformation or if the elastic deformation is inhomegeneous (giving rise to lattice distortion). In the above definition, the value of the centrosymmetry depends not only on the amount of the plastic deformation but also on the lattice constant, which is different from material to material. We normalize the centrosymmetry by the lattice constant, i.e.

$$C = \sqrt{P}/a, \tag{5}$$

where, a is the lattice constant. The normalized centrosymmetry C depends only on the amount of plastic deformation and is material-independent. We carried out MD simulations of the simple tension of single crystal Cu nanowires, see Fig. 18. The length of all specimens considered is 60 lattice constants (or 21.68 nm). The cross-sectional dimensions of the nanowires vary from 5 to 20 lattice constants (1.8 nm–7.2 nm).

The nanowires have free surfaces in the x- and y-directions. The x-, y-, and z-axes are oriented in the [100], [010], and [001] crystalline directions, respectively. The atoms in the specimens are divided into two types. One is the boundary atoms (the black atoms at the top and bottom planes in Fig. 18). Constant velocities $\pm V_0$ (equal magnitude and opposite directions) are maintained for these two planes of atoms, effecting loading necessary for the nanowire to deform at a constant nominal strain rate equal to $\dot{\varepsilon} = 2V_0/L$. The internal atoms simply deform with the boundary atoms and are assigned random initial thermal velocities satisfying the Boltzmann distribution at a temperature of 300 K. The system is allowed to relax by holding the length of the wire unchanged and by maintaining a constant temperature using the Nose-Hoover thermostat procedure [26, 27]. The calculations continue with a time step of 0.001 ps until the nanowire fractures. In contrast to some MD simulations reported in the literature where the Nose-Hoover isothermal procedure [26, 27] was used to maintain a constant temperature, no thermal constraints are applied to the specimen during the calculations here, therefore, the temperature rises adiabatically. This more closely simulates the tensile deformation of a nanowire undergoing high rate deformation. The nanowires are not relaxed to a zero stress state and the beginning of deformation is at stress levels of 0.4–1.6 GPa. Fig. 19 shows the stress–strain relations of a nanowire at three different strain rates, between 1.67×10^7 s^{-1} and 1.67×10^9 s^{-1}. The cross-sectional dimensions of the nanowire are 5×5 lattice constants (1.8×1.8 nm). The stress–strain relations are essentially linear at small strains. The yield stress is the maximum stress in these cases. The curves for different strain rates coincide during the elastic part of the deformation, indicating rate-independence of elastic deformations which is expected. The curves show a rate-independent Young's modulus of 70 GPa for <001> type crystalline directions. The strain at which yielding occurs increases with strain rate. The yield stress increases from 6.8 GPa to 7.7 GPa as the strain rate increases from 1.67 $\times 10^7$ s^{-1} to 1.67×10^9 s^{-1}. This dependence of yielding on strain rate is due to the dynamic wave effect or phonon drag that impedes the motion of dislocations.

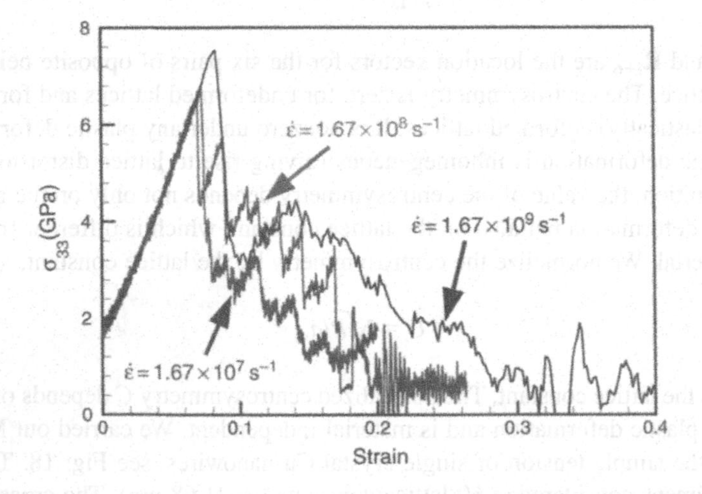

Fig. 19. Stress–strain curves of $5 \times 5 \times 60$ Cu nanowires at three different strain rates.

The strain at which ductile rupture occurs increases from 0.2 at $1.67 \times 10^7\ s^{-1}$ to 0.32 at $1.67 \times 10^9\ s^{-1}$. This strain rate dependence of ductility is consistent with the experimental results reported by Lu et al. [35].

Due to the lack of defects and high strain rates, the yield stress for nanowires undergoing dynamic deformation can far exceed that of bulk Cu. The stress–strain curves in Fig. 19 show precipitous drops in stress after yielding. This sharp drop is caused by the initiation of plastic deformation which occurs at different levels of stress for different strain rates. Crystalline slip along {1 1 1} planes clearly provides the mechanism for the plastic deformation. Activation of multiple slip planes and cross slip are responsible for the formation of the neck in the specimen (Fig. 20(c)). At the strain rate $1.67 \times 10^8\ s^{-1}$, regularly distributed, alternating slip bands are clearly observed throughout the specimen (Fig. 20(a, b)). These slip bands are not simultaneously activated. They propagate from one end to the other within a certain period of time (Fig. 21). This novel phenomenon can be explained by the initially uniform but high strain energy state of the nanowire since it is initially pre-strained with fixed ends. At such state, slip planes are likely to be activated by small thermal or mechanical perturbations. The activation

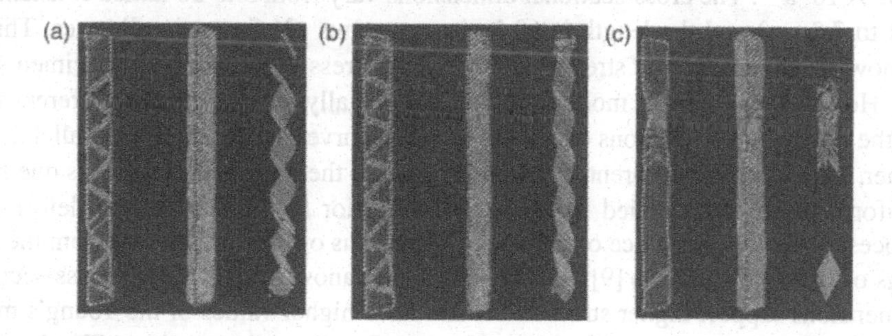

Fig. 20. Deformed configurations of Cu nanowires at a strain of 10%; (a) $\dot{\varepsilon} = 1.67 \times 10^7\ s^{-1}$; (b) $\dot{\varepsilon} = 1.67 \times 10^8\ s^{-1}$ and (c) $\dot{\varepsilon} = 1.67 \times 10^9\ s^{-1}$. In each picture, the left image shows an internal cross-section, the center images shows a solid view of the wire with all atoms, and the right image shows only atoms involved in defects. Graphics generated using VMD1.7 [57]. See color plate 4.

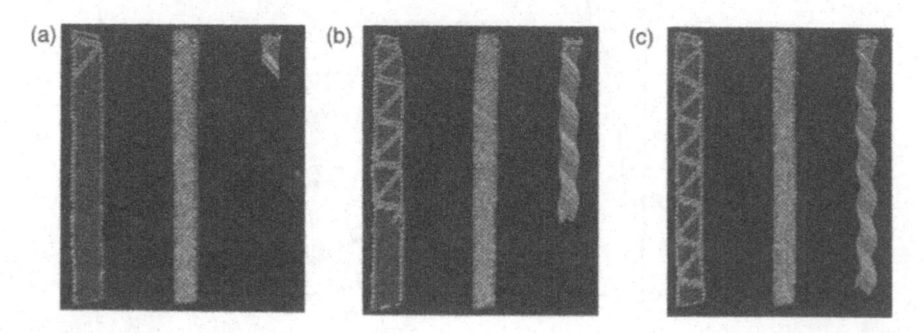

Fig. 21. The propagation of slip bands at a strain rate of $1.67 \times 10^8\ s^{-1}$, (a) $\varepsilon = 4.67\%$, (b) $\varepsilon = 5.33\%$, and (c) $\varepsilon = 6.0\%$. The specimen size is $5 \times 5 \times 60$. See color plate 5.

of a slip plane at one location can set off a successive, chain-reaction type activation of slip planes down the specimen. This mechanism of plasticity activation is observed for the strain rates of 1.67×10^7 s^{-1} and 1.67×10^8 s^{-1}. The speed at which the active front of the planes propagates along the wire is very high for $\dot{\varepsilon} = 1.67 \times 10^7$ s^{-1} and is 355 ms^{-1} for $\dot{\varepsilon} = 1.67 \times 10^8$ s^{-1}. The rapid activation of slip planes across the specimen is responsible for the nearly vertical drop of stress seen following yielding in Fig. 19. The stress–strain curves show oscillatory decreases of stress after the onset of plastic deformation. These oscillations are likely due to successive stages of gradual elastic stretching of temporarily "stationary" lattice structures and rapid plastic slip along the well-defines slip planes. During the elastic stages, straining of the lattice allows strain energy to be accumulated and stored. The stored strain energy allows the activation of slip planes in short, "quick-fire" bursts, leading to relaxation and drop in stress as strain increases. Since specimens deforming at lower strain rates need longer times to "catch up" through further elastic straining, stress decreases are sharper and last longer in time.

The effect of specimen size on the deformation is also analyzed. Fig. 22 shows stress–strain curves for specimens of three different sizes at a strain rate of 1.67×10^9 s^{-1}. The cross-sectional dimensions vary from 5 to 20 lattice constants (or 1.8 to 7.2 nm) and the length is 60 lattice constants (21.7 nm) in all cases. Thinner nanowires support higher stresses and the yield stress decreases with specimen size.

However, the Young's modulus remains essentially the same for the different sizes, as the initial, elastic portions of the stress–strain curves are essentially parallel to each other, except for the different amounts of offset at the origin. This offset is due to the uniform pre-stretch applied to the specimens prior to the start of the deformation process. This independence of the Young's modulus on size is different from the findings of Branício and Rino [9], who reported that nanowires with larger cross-sectional dimensions support higher stress levels and have higher values of the Young's modulus. The calculations here also show enhanced ductility at larger sizes. Clearly, this is due to the fact that smaller samples offer fewer opportunities for slip and dislocation motion and larger specimens offer more opportunities for crystalline slip. This effect

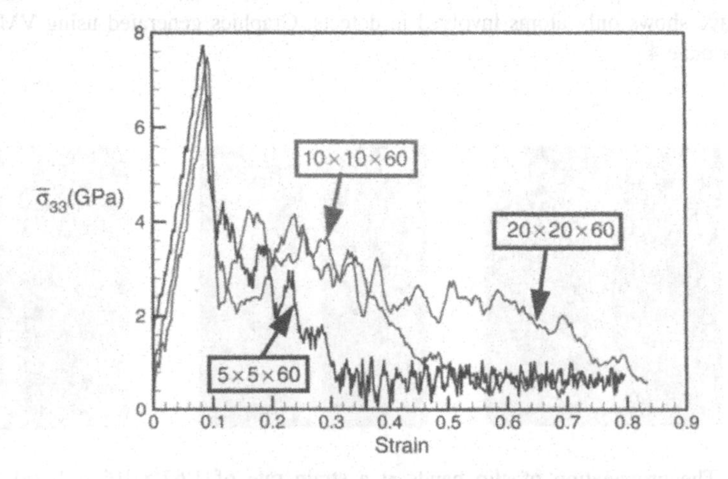

Fig. 22. Stress–strain relations for specimens of different sizes, 1.67×10^9 s^{-1}.

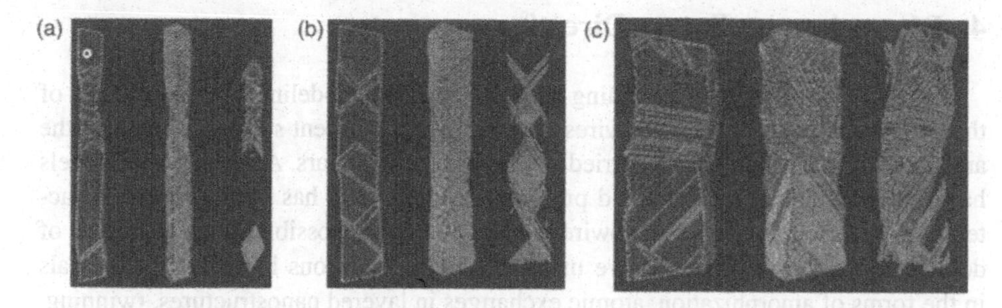

Fig. 23. Deformed configurations of Cu nanowires of different sizes ($\dot{\varepsilon} = 1.67 \times 10^9\ \mathrm{s}^{-1}$); (a) specimen with a cross-sectional size of 5 lattice spacings ($\varepsilon = 9.17\%$); (b) specimen with a cross-sectional size of 10 lattice spacings, ($\varepsilon = 18.3\%$); and (c) specimen with a cross-sectional size of 20 lattice spacings ($\varepsilon = 23.3\%$). In each picture, the left image shows an internal csection, the middle image shows an external view, and the right image shows only atoms involved in defects. See color plate 6.

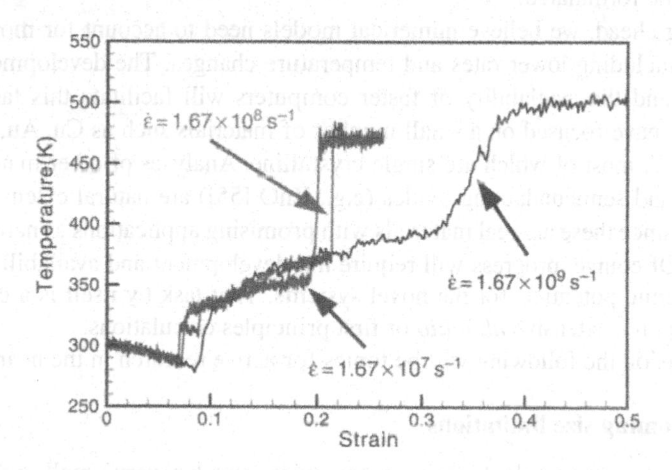

Fig. 24. Temperature changes in Cu nanowires of different sizes during the deforming process.

is clearly seen in Fig. 23. The larger specimen in Fig. 23(c) shows more extensive slip activation and cross slip.

The variation of temperature as a function of strain is shown in Fig. 24 for $5 \times 5 \times 60$ specimens deformed at different strain rates. The thermal behavior of specimens is similar at the different strain rates. In the elastic deformation stage, temperature decreases slightly as part of the kinetic energy is transformed into potential energy (or strain energy). Subsequently upon yielding, the temperature begins to increase abruptly and continues the upward trend until the nanowires rupture. This temperature increase is primarily due to plastic dissipation but is also due to thermoelastic dissipation at the atomic level. Sharp temperature rises are observed at fracture for $\dot{\varepsilon} = 1.67 \times 10^7\ \mathrm{s}^{-1}$ and $\dot{\varepsilon} = 1.67 \times 10^8\ \mathrm{s}^{-1}$, primarily because of the conversion of external work to kinetic energy. The gradual increase of temperature for $\dot{\varepsilon} = 1.67 \times 10^9\ \mathrm{s}^{-1}$ at late stages of deformation echoes the more ductile and prolonged deformation at this strain rate.

4. Discussions on Future Directions

It is obvious from the preceding sections that the modeling and simulation of the mechanical response of nanowires are still in their nascent stage. The bulk of the analyses so far have only been carried out in the last few years. Although MD models have been developed and used and preliminary knowledge has been gained, characterization of the behavior of nanowires is just becoming possible. A clear picture of deformation mechanisms operative under different conditions in different materials in the forms of amorphization, atomic exchanges in layered nanostructures, twinning, slip, cross slip, and their interactions has yet to emerge. We are still at a stage of qualitative discovery of deformation mechanisms and understanding of associated phenomena. Quantification of strength, ductility, and size and rate effects is clearly absent. It is fair to say that the area has just developed to a state where quantification can be obtained. Furthermore, conditions for the activation of deformation mechanisms have yet to be delineated and criteria for transitions among these mechanisms have yet to be formulated.

Looking ahead, we believe numerical models need to account for more realistic conditions including lower rates and temperature changes. The development of new algorithms and the availability of faster computers will facilitate this task. So far, calculations have focused on a small number of materials such as Cu, Au, Ni, $SiSe_2$, Ni-Cu, and C, most of which are single crystalline. Analyses of other materials such as Si, SiC, and semiconducting oxides (e.g., ZnO [55]) are natural extensions in the near future since these are real materials with promising applications as nanowires and nanobelts. Of course, progress will require the development and availability of accurate interatomic potentials for the novel systems. That task by itself is a challenging one and requires extensive *ab initio* or first principles calculations.

We envision the following will be topics for active research in the near future.

4.1. Overcoming size limitations

Until so far, most calculations on nanowires involve very small systems, with diameters less than 5 nm. The behavior of bulk materials can be quite different form those at the nanoscale. At the nanoscale, the behavior can change dramatically when lateral dimensions vary from 1 to 20 nanometers. It is essential to quantify the behavior of nanowires at different sizes. Changes in diameter and length must be considered. Some calculations reported in the literature concern very short wires and periodic boundary conditions have been used in the longitudinal direction to equate them to infinitely long wires. More realistic handling of this issue can and should be pursued. Direct account of the full length of nanowires with load or displacement boundary conditions should be used with more and more powerful computers.

4.2. Lower strain rates and time scale

Strain rate plays an important role in affecting the behavior of nanowires. The strain rates considered in simulations so far (at or above 10^7 s^{-1}) are much higher than what can be obtained in controlled experiments. These artificially high strain rates are, to a degree, necessitated by the speed of computers available. The time scale that can be reached in many simulations is of the order of 100 ns, partly limited by the time step needed to resolve the high frequency thermal vibrations of atoms in MD calculations which is of the order of 0.001 ps. To achieve significant deformation within such a short time, high strain rates are needed. High deformation rates cause rapid and extremely high temperature increases, often necessitating Nose-Hoover type isothermal procedures [26, 27] to maintain a constant temperature or limit temperature increase. These are largely numerical schemes aimed at facilitating calculations. Such treatments are nonphysical and inconsistent with the dynamic nature of the deformation. To bring the strain rates in computations to levels in line with those in experiments, more efficient algorithms or faster computers are necessary. So far, we have been able to consider strain rates on the order of 10^7 s^{-1}.

4.3. Equivalent continuum and constitutive characterization

Model characterizations for the behavior of nanowires are needed. Constitutive and failure characterization must describe and quantify the behavior over a range of conditions including temperature, loading (or deformation) rate, and size. Deformation mechanisms for elasticity, plasticity, viscoplasticity, creep, and failure must be accounted for. Quantification and evolution description require a fundamentally sound approach for transitioning from discrete molecular descriptions to continuum descriptions. The work of Odegard et al. [3, 4] represents a worthwhile and important attempt. However, the energy-based equivalence does not provide consistent transition for fully dynamic conditions. Full dynamic equivalence requires time-resolved temporal account of momentum, work rates, mass and deformation. The equivalent continuum (EC) approach [47, 48, 56] offers a systematic and general framework.

Acknowledgement

This work is supported by grant NAG-1-02054 from the NASA Langley Research Center. Research conducted under NSF CAREER grant CMS9984298 and an AFOSR MURI grant at Georgia Tech benefited this work. We would like to thank Dr. S. Plimpton for sharing his MD code and for helpful discussions on the code. Some of the figures are generated using VMD 1.7 provided by the University of Illinois at Urbana-Champaign.

References

1. H. Mehrez and S. Ciraci, *Phys. Rev. B* **56** (1997) 12632.
2. H. Mehrez, S. Ciraci, C. Y. Fong and S. Erkoc, *J. Phys.: Condens. Matter* **9** (1997) 10843.
3. G. M. Odegard, T. S. Gates, L. M. Nicholson and K. E. Wise, *Equivalent-Continuum Modeling of Nano-structured Materials*, NASA Langley Research Center, Hampton, VA (2001).
4. G. M. Odegard, V. M. Harik, K. E. Wise and T. S. Gates, *Constitutive Modeling of Nanotube-Reinforced Polymer Composite Systems*, NASA Langley Research Center, Hampton, VA (2001).

5. M. S. Daw and M. I. Baskes, *Phys. Rev. B (Condens. Matter)* **29** (1984) 6443.
6. M. S. Daw, S. M. Foiles and M. I. Baskes, *Mater. Sci. Rep.* **9** (1993) 251.
7. W. Li, R. K. Kalia and P. Vashishta, *Phys. Rev. Lett.* **77** (1996) 2241.
8. H. Ikeda, Y. Qi, T. Cagin, K. Samwer, W. L. Johnson and W. A. G. III, *Phys. Rev. Lett.* **82** (1999) 2900.
9. P. S. Branício and J.-P. Rino, *Phys. Rev. B: Condens. Matter* **62** (2000) 16950.
10. P. Walsh, W. Li, R. K. Kalia, A. Nakano and P. Vashishta, *Appl. Phys. Lett.* **78** (2001) 3328.
11. B. I. Yakobson, C. J. Brabec and J. Bernholc, *Phys. Rev. B* **76** (1996) 2511.
12. D. Srivastava, M. Menon and K. Cho, *Phys. Rev. B* **83** (1999) 2973.
13. T. Ozaki, Y. Iwasa and T. Mitani, *Phys. Rev. B* **84** (1999) 1712.
14. V. Rodrigues, T. Fuhrer and D. Ugarte, *Phys. Rev. Lett.* **85** (2000) 4124.
15. H. Ohnishi, Y. Kondo and K. Takayanagi, *Nature* **395** (1998) 781.
16. G. Bilalbegović, *Phys. Rev. B* **58** (1998) 15412.
17. G. Bilalbegović, *Solid State Commun.* **115** (2000) 73.
18. G. Bilalbegović, *Comput. Mater. Sci.* **18** (2000) 333.
19. W. Liang and M. Zhou, to appear in the Proceedings of the 2003 Nanotechnology Conference and Trade Show (Nanotech 2003), San Francisco, CA (2003).
20. Y. Qi, *Phys. Rev. B: Condens. Matter* (2002) (submitted).
21. C. L. Kelchner, S. J. Plimpton and J. C. Hamilton, *Phys. Rev. B (Condens. Matter)* **58** (1998) 11085.
22. P. L. McEuen, *Phys. World* **13** (2000) 31.
23. H. Dai, *Phys. World* **13** (2000) 43.
24. C. F. Cornwell and L. T. Wille, *Solid State Commun.* **101** (1997) 555.
25. F. H. Stillinger and T. A. Weber, *Phys. Rev. A* **28** (1983) 2408.
26. S. Nose, *Mol. Phy.* **52** (1984) 255.
27. W. G. Hoover, *Phys. Rev. A* **31** (1985).
28. R. W. Hertzberg, *Deformation and Fracture Mechanics of Engineering Materials*, John Wiley & Sons, New York (1976).
29. B. Wang, S. Yin, G. Wang, A. Buldum and J. Zhao, *Phys. Rev. Lett.* **86** (2001) 2046.
30. C. J. Muller, J. M. Krans, T. N. Todorov and M. A. Reed, *Phys. Rev. B* **53** (1996) 1022.
31. E. Z. d. Silva, A. J. R. d. Silva and A. Fazzio, *Phys. Rev. Lett.* **87** (2001) 256102.
32. J. W. Kang and H. J. Hwang, *Nanotechnology* **12** (2001) 295.
33. F. Cleri and V. Rosato, *Phys. Rev. B* **48** (1993) 22.
34. D. Tománek, A. A. Aligia and C. A. Balseiro, *Phys. Rev. B* **32** (1985) 5051.
35. L. Lu, S. X. Li and K. Lu, *Scripta Mater.* **45** (2001) 1163.
36. P. I. Polikin, G. Y. Gun and A. M. Galkin, *Handbook of Resistance to Plastic Deformation of Metals and Alloys*, Metallurgy Press, Moscow (1976).
37. M. F. Horstemeyer and M. I. Baskes, *Trans. of the ASME J. Eng. Mater. Technol.* **121** (1999) 114.
38. G. Bilalbegović, *J. Phys.: Condens. Matter* **13** (2001) 11531.
39. E. Saether, R. B. Pipes and T. Gates, Extended Abstract for the 43rd AIAA/ASME/ASCE/AHS/ASC Structures, Structural Dynamics and Materials Conference and Exhibit, Denver, CO (2002)
40. A. Stalder and U. Durig, *Appl. Phys. Lett.* **68** (1995) 637.
41. N. Agräit, G. Rubio and S. Vieira, *Phys. Rev. Lett.* **74** (1995) 3995.
42. G. Rubio, N. Agrait and S. Vieira, *Phys. Rev. Lett.* **76** (1996) 2302.
43. C. Kittel, *Introduction to Solid State Physics*, Wiley, New York (1971).
44. A. Kelly and N. H. Macmillan, *Strong Solids*, Clarendon, Oxford (1986).
45. F. Kassubek, C. A. Stafford and H. Grabert, *Physica B* **280** (2000) 438.

46. G. Rubio-Bollinger, S. R. Bahn, N. Agrait, K. W. Jacobsen and S. Vieira, *Phys. Rev. Lett.* **87** (2001) 026101.
47. M. Zhou, *Proc. R. Soc. A* (2003) (to appear).
48. M. Zhou and D. L. McDowell, *Phil. Mag. A* **82** (2002) 2547.
49. M. Zhou, *Phys. Rev. B*, (2002) (submitted).
50. D. H. Tsai, *J. Chem. Phys.* **70** (1979) 1375.
51. J. S. Rowlinson and B. Widom, *Molecular Theory of Capillarity*, Clarendon Press, Oxford (1982).
52. R. J. Swenson, *Am. J. Phys.* **51** (1983) 940.
53. E. Kröner (ed.), *Physics of Defects*, North-Holland, Amsterdam (1981).
54. D. T. Kulp, G. J. Ackland, M. Sob, V. Vitek and T. Egami, *Model. Simul. Mater. Sci. Eng.* **1** (1993) 315.
55. Z. W. Pan, Z. R. Dai and Z. L. Wang, *Science* **9** (2001) 1947.
56. M. Zhou, *Key Eng. Mater.* **233–236** (2003) 597.
57. W. Humphrey, A. Dalke and K. Schulten, *J. Molec. Graphics* **14** (1996) 33.

Part III

Molecular Nanowires and Metallic Nanowires

Chapter 5

Molecular and Ionic Adsorption onto Atomic-Scale Metal Wires

H. X. He, S. Boussaad, B. Q. Xu and N. J. Tao

Department of Electrical Engineering & The Center for Solid State Electronics Research, Arizona State University, Tempe, AZ 85287

1. Introduction

The recent surge of interest in nanostructured materials and devices is triggered not only by the need of device miniaturization but also by the appearance of many novel quantum phenomena at the nanometer scale. These phenomena may lead to new applications from electronic and optical devices to chemical and biological sensors. One interesting example that will be discussed in this chapter is conductance quantization in atomic-scale metal wires [1–5].

In a classical (macroscopic) metal wire, the conduction electrons experience multiple diffusive scatterings when they traverse through the wire. The resistance, R, of the wire is given by,

$$R = \rho \frac{L}{A} \tag{1}$$

where ρ is resistivity, L and A are the length and the cross sectional area of the wire, respectively. The inverse of R in Eq. 1 is defined as the conductance which is related to conductivity $\sigma = 1/\rho$ by

$$G = \sigma \frac{A}{L}. \tag{2}$$

In the classical wire, the conductivity, σ, is a physical property that depends on the material of the wire. When decreasing the length below the electron mean free path (~30 nm for Cu at room temperature), the electron transport is ballistic, i.e. without collisions with impurities and defects, along the wire. If, in addition, the diameter of the wire is shrunk to the order of the electron wavelength, the electrons in the transverse direction form well-defined quantum modes or standing waves. When such a wire is connected to the outside world via two bulk electrodes (electron reservoirs), one on each side of the wire, the conductance of the system is described by the Landauer formula [6],

$$G = \frac{2e^2}{h} \sum_{i=1}^{N} T_i \tag{3}$$

where e is the electron charge, h is Planck's constant and T_i is the transmission probability of each mode. The summation is over all the quantum modes, so N is the total number of modes with non-zero T_i values, which is determined by number of standing waves at the narrowest portion of the wire (Fig. 1). In the ideal case, T_i is either 0 or 1 and Eq. 3 is simplified as $G = NG_0$, where $G_0 = 2e^2/h \sim 1/13\ k\Omega^{-1}$ is the conductance quantum, so the conductance is quantized. When one increases the diameter of the wire, N increases by one each time a new standing wave or mode fits into the narrowest cross section. For a two-dimensional wire, N is simply determined by the ratio of the width (W) of the wire and half the Fermi wavelength (λ_F), $N = \mathrm{INT}(2W/\lambda_F)$. For a three-dimensional wire, $N \sim (\pi R/\lambda_F)^2$, R is the radius of the circular cross section of the narrowest portion.

This quantization of conductance was first clearly demonstrated in semiconductor devices containing a two-dimensional electron gas confined into a narrow constriction by the gate voltage, where $\lambda_F \sim 400$ Å is much larger than the atomic scale [1, 2]. A similar conductance quantization has been observed in three-dimensional metallic wires created by mechanically breaking a fine metal wire [7] or separating two electrodes in contact [4, 8–11]. Since the wavelength of conduction electrons in a typical metal is only a few Å, comparable to the size of an atom, a metallic wire with conductance quantized at the lowest steps must be atomically thin. This conclusion has been directly confirmed by high resolution transmission electron microscopy [12]. Because the metal wires that exhibit conductance quantization are usually very short, they are often called atomic-scale contacts or simply nanocontacts. They are also referred to as metallic quantum wires and nanowires, or atomic-scale metal wires. Here we use atomic-scale metal wires or atomic-scale contacts interchangeably to emphasize the dimension of the structure.

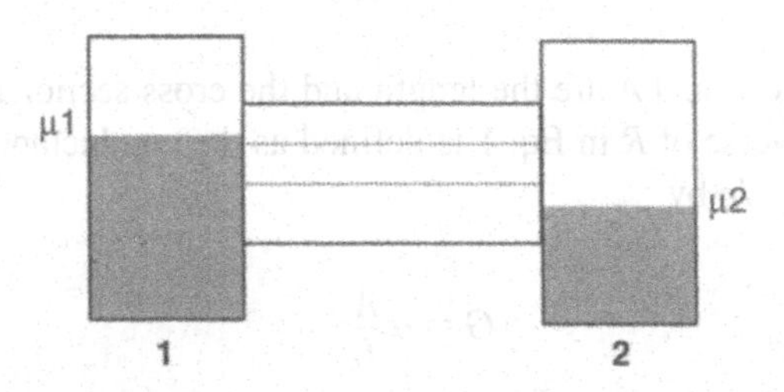

Fig. 1. A quantum wire is connected to two electrodes 1 and 2, whose chemical potentials are μ1 and μ2, respectively. The difference, μ1 − μ2 is maintained by an external voltage source. The dashed lines represent the energy levels of the quantum modes formed in the transverse direction. Each of the mode lying between the μ1 and μ2 contributes to the total conductance of the wire system.

Since the discovery of conductance quantization, various studies have been conducted for better understanding of the phenomenon and for possible applications based on the phenomenon. Motivated by the possibility for chemical sensing applications, molecular and ionic adsorption onto atomic-scale metal wires have been studied [13, 14], which is the focus of this chapter. The remaining part of the chapter is arranged as follows. In Section 2, we describe the fabrication methods of the atomic-scale wires. We then discuss molecular adsorption and anionic adsorption in Sections 3 and 4, respectively. In Section 5, we provide a summary.

2. Fabrication of Atomic-Scale Metal Wires

To date two approaches have been developed to fabricate atomic-scale metal wires that exhibit conductance quantization, one is mechanical methods and the other one is based on electrochemical etching and deposition. Each of the two approaches is described below.

2.1. Mechanical methods

Mechanical methods create an atomic-scale wire by mechanically separating two electrodes in contact. During the separation process, a metal neck is formed between the electrodes due to strong cohesive energy of the metal, which is stretched into an atomically thin wire before breaking. One such method is based on Scanning Tunneling Microscope (STM) (Fig. 2) [4, 8, 9, 15–18], in which the STM tip is driven into the substrate and the conductance is recorded while the tip is gradually pulled out of the contact with the substrate. The first experiment of this type was reported by Gimzewski et al. to study the transition in the electron transport between tunneling and ballistic regimes [4]. Fig. 3a shows several conductance traces during stretching of a Au wire with a STM. The conductance decreases in a stepwise fashion. The steps tend to occur near the integer multiples of G_0, indicating the origin of conductance quantization. However, they can deviate significantly from the simple integer multiples. For this reason, conductance histogram is often constructed from thousands of

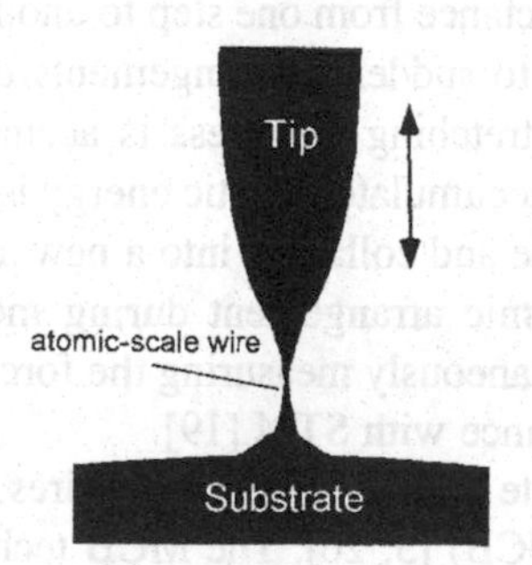

Fig. 2. Mechanical fabrication of atomically thin metallic wire with a STM setup. A STM tip is first pressed into a metal substrate and then pulled out of contact during which an atomically thin wire is formed before breaking.

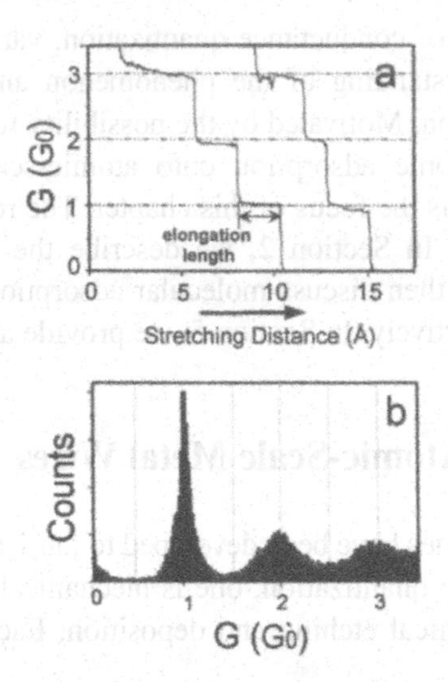

Fig. 3. (a) Typical conductance vs. stretching distance traces that show the quantized variation in the wire conductance. An important quantity, elongation length, is defined as the distance over which a wire can be stretched before its conductance jumps to a lower step. (b) Conductance histogram of Au wires in 0.1 M $NaClO_4$. The well-defined peaks near integer multiples of $G_0 = 2e^2/h$ have been attributed to conductance quantization. Each histogram was constructed from over one thousand individual conductance traces like the ones shown in (a).

the conductance traces to show the occurrence of the steps at different conductance values (Fig. 3b). For metals such as Au, Na and K, well-defined peaks near the integer multiples are found. For Na and K, the peaks near 2 G_0 and 4 G_0 are missing, which is expected because of the mode degeneracy for a cylindrical symmetry of a wire in free electron gas model. The observation, therefore, demonstrates beautifully that these metal wires can be described as free electron gas model in cylindrical wires. Not all metals show pronounced histogram peaks near the integer conductance values. Most other metals show only a broad first peak that is not an integer value. For example, niobium has a first peak near 2.5 G_0.

The transition in conductance from one step to another during mechanical stretching is usually abrupt, due to sudden rearrangements of the atoms in the wire upon stretching. That is, upon stretching the stress is accumulated in the metallic bonds between atoms. When the accumulated elastic energy is large enough, the atomic configuration becomes unstable and collapses into a new atomic configuration that has a thinner diameter. This atomic arrangement during mechanical stretching has been directly observed by simultaneously measuring the force with the atomic force microscope (AFM) and conductance with STM [19].

Another tool to fabricate atomic-scale metal wires is to use a mechanically controllable break-junction (MCB) [5, 20]. The MCB technique creates a clean fracture surface by mechanically breaking a metal and can provide better stability than the STM method. Other methods include the use of a tip and plate [21], two macroscopic electrodes in contact [10], mechanical relay [11]. The mechanical methods can repeatedly fabricate (and break) a large number of wires within a relatively short time, which is particularly convenient for statistical analysis, such as conductance histograms. However, the methods form one wire at a time and use piezoelectric or other mechanical transducers, which are not desirable for device applications. It has been recently demonstrated that metallic wires with quantized conductance can be fabricated with electrochemical etching and deposition [22–27]. In contrast to the mechanical methods, the electrochemical method can fabricate arrays of atomic-scale wires supported on solid substrates.

2.2. Electrochemical fabrications

One simple electrochemical method starts with a single or an array of relatively thick metal wires supported on a solid substrate such as glass slide or silicon chip (Fig. 4) [23]. It then reduces the diameter of each wire by electrochemical etching until the conductance becomes quantized. The process is controlled with a bipotentiostat that can simultaneously control the etching rate and measure the conductance of the wire. In order to form a wire with a desired conductance, a feedback loop is used, which works by comparing the measured conductance with a preset value. When the conductance is greater than the preset value, etching is activated to reduce the diameter of the wire. When the conductance is smaller than the preset value, deposition is turned on to increase the diameter. These atomic-scale wires can last from minutes to days. Further improvement of the stability may be achieved by "coating" the wire with appropriate molecules. The method has been used to fabricate atomic-scale Au, Ag,

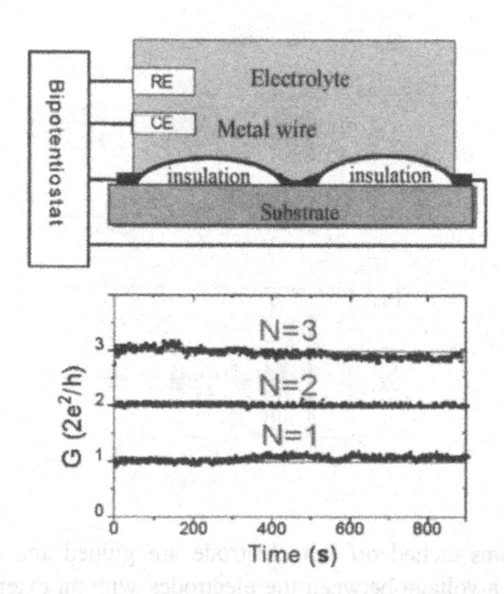

Fig. 4. Top: Electrochemical fabrication of atomic-scale metal wires. A metal wire is attached to the bottom surface of a Plexiglas solution cell. The wire is coated with wax except for a small section that is exposed to electrolyte for etching. The electrochemical potential of the wire is controlled by a bipotentiostat with a reference electrode (RE) and a counter electrode (CE). Bottom: Electrochemically fabricated atomic-scale Cu wires with conductance at $N = 1$, 2 and 3 quantum steps.

Cu and Ni wires with conductance quantized at different steps. The method described above has been used to fabricate atomic-scale wires for many measurements, but a simpler method has been demonstrated [28]. The method does not require a bipotentiostat and feedback control, and it can *simultaneously* fabricate a large array of wires.

The principle is illustrated in Fig. 5a. It starts with a pair of electrodes separated with a relative large gap in an electrolyte. When applying a bias voltage between the two electrodes, metal atoms will be etched off from the anode and deposited onto the cathode. As we have found experimentally, the etching takes place all over the anode surface, but the deposition is localized to the sharpest point on the cathode, due to high electric field and metal ion concentration near the sharpest point. Consequently, the gap narrows and disappears eventually when the two electrodes connect. In order to control the formation of the contact between the electrodes, the etching and deposition processes must be terminated once a desired contact is formed. While a feedback mechanism may be used for this purpose, we introduce a much simpler self-termination mechanism by connecting one electrode in series with an external resistor (R_{ext}). So the effective voltage (or overpotential) for etching and deposition is given by

$$V_{gap} = \frac{R_{gap}}{R_{gap} + R_{ext}} V_0 \quad (4)$$

where R_{gap} is the resistance between the two electrodes, and V_0 is the total applied bias voltage. Initially, the gap is very large and R_{gap} is determined by ionic conduction (leakage current) between the electrodes. By coating the electrodes with a SiN

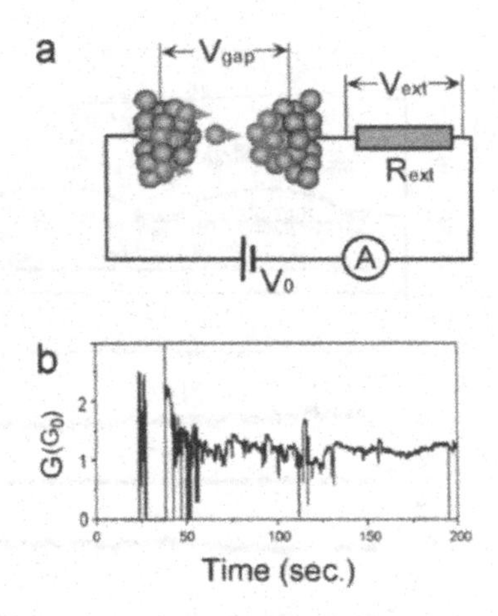

Fig. 5. (a). Metal atoms etched off left electrode are guided and deposited onto the right electrode by applying a voltage between the electrodes with an external resistor in series. As the gap between the two electrodes shrinks, the gap resistance decreases, which results in a drop in etching/deposition voltage and eventually terminates the etching/deposition processes. (b). Choosing a $R_{ext} < 12.7$ kΩ, an atomic-scale contact with quantized conductance can be fabricated. The conductance change during the formation of a contact ($R_{ext} = 3$ kΩ and $V_0 = 1.2$ V). The large conductance fluctuations corresponding to constant breakdown and reformation of the contact are due to electromigration.

insulation layer, the unwanted leakage current can be reduced below 1 pA. When the gap is decreased to a few nm or less, R_{gap} begins to decrease because of electron tunneling across the gap. Finally, once a contact is formed between the two electrodes, R_{gap} depends on the size of the contact.

We can easily understand how Eq. 4 leads to self-termination. Initially $R_{gap} >> R_{ext}$, $V_{gap} \sim V_0$ according to Eq. 1. So the etching and deposition take place at full speed. As the gap narrows, the tunneling probability across the gap rises and R_{gap} decreases. This results in a decreasing V_{gap}, or a slowdown in the etching and deposition rates. Eventually, when $R_{gap} << R_{ext}$, $V_{gap} \sim 0$, which terminates the etching and deposition and leaves us with a small gap or contact between two facing electrodes. The self-termination effect is further enhanced by the exponential dependence of the etching and deposition current density, J, on V_{gap}, according to [29]

$$J(V_{gap}) \propto \exp(\alpha e V_{gap}/k_B T) \tag{5}$$

where, α is usually around 0.5, e is the electron charge, k_B is the Boltzman constant and T is temperature. For example, if V_{gap} drops from an initial 1.5 V to 0.75 V, the etching and deposition current density will drop by ~3,000,000 times!

The final gap width or contact size is determined by R_{ext}. If R_{ext} is greater than $h/2e^2$, the resistance for a single atom contact between two electrodes, it terminates with a small tunneling gap between the electrodes. On the other hand, if R_{ext} is smaller than $h/2e^2$, it terminates with an atomic-scale contact between the electrodes.

Fig. 5b shows the conductance (normalized against G_0) between two Cu electrodes during electrochemical etching and deposition in water with R_{ext} preset at 3 kΩ (< 12.7 kΩ). The initial conductance due to ionic conduction is negligibly small comparing to G_0. A few minutes after applying a 1.2 V voltage, the conductance suddenly jumps to ~2 G_0 and the deposition terminates itself as a contact is formed between the electrodes. However, the initial contact breaks within seconds and the conductance drops back to zero. Once the contact is broken, the voltage across the electrodes goes back to the maximum value, according to Eq. 4, and the etching and deposition starts over again. Indeed, several seconds later, a new contact with conductance near 1 G_0 is reformed. So the method has a self repairing mechanism. While the conductance tends to stabilize, a large noise (sharp spikes) in the conductance is clearly visible. Zooming-in the noise reveals stepwise fluctuations in the conductance between ~1 G_0 and ~0 G_0, corresponding to a constant breakdown and reformation of the contact. The breakdown is likely due to electromigration, i.e., a high current in the atomic-scale wire tends to drag atoms along with it and cause mechanical instability. Since the effect of electromigration increases with the current, we expect a smaller electromigration effect at lower bias voltage. It has been, indeed, found that the breakdown becomes a lot less likely when one reduces the bias voltage. The reformation is, as discussed above, due to the re-establishment of a large electrodeposition and etching voltage once the conductance drops to zero.

A smaller R_{ext} results in a contact with conductance at a higher value. During the formation of the contact, the conductance varies in a stepwise fashion that can be recorded when the variation is slow enough. The measured I-V characteristic curves of the contacts show that the current is linearly proportional to the bias voltage for the contacts. This ohmic behavior was also found by Hansen et al. in the atomic-scale wires formed by STM [30].

3. Molecular Adsorption onto Atomic-Scale Metal Wires

In contrast to its semiconductor counterpart, the metal wires have several distinct features. First, they exhibit the conductance quantization phenomenon even at room temperature and in electrolytes. This allows us to study various electrochemical processes, such as double layer charging, ionic and molecular adsorption and electrochemical potential-induced surface stress, taking place on the wires. The second feature is that metal wires with quantized conductance are atomically thin, which means virtually every atom in the wire is a surface atom. Such a large surface to volume ratio suggests a very sensitive dependence of the wire conductance on the molecular adsorption. Finally, the thinnest portion of the metal wire is only a few atoms long, which can accommodate only a few molecules. Because the conductance of the entire system is dominated by the thinnest portion, the atomic-scale metal wires are particularly suitable for studying the adsorption of a few molecules or ions. Several studies on the molecular adsorption onto atomic-scale metal wires have been recently carried out [13, 14, 31], which show sensitive dependence of both the conductance and atomic configuration of the atomic wire on the molecular adsorption.

Before discussing experimental results, it is beneficial to consider how a conductor changes its conductance upon molecular adsorption onto its surface [32–34]. This phenomenon has actually been known for decades in the classical regime in which diffusive scattering of electrons determines the conductance. It has attracted renewed interests in recent years because of its direct relevance to chemical sensor applications and surface friction studies [35]. Zhang et al. have recently shown that the self-assembly of thiol compounds onto a Au film can be easily detected by measuring the conductance of the Au film [36]. Several groups have studied the conductance changes of thin film electrodes due to electrochemically controlled molecular and anion adsorptions [37–40].

It is widely accepted that the adsorbate-induced conductance change is due to the scattering of conduction electrons in the wire by the adsorbates as the electrons impinge on the surface of the wire [35, 41]. In some systems, a reduction in the conduction electron density by adsorbates is also believed to play a role [33, 38]. For metal films thicker than a few nm, the semiclassical models of Fuch [42] and Sondheimer [43] have been frequently used. Ishida has developed a microscopic theory that allows one to evaluate the conductance change in terms of adsorbate-substrate bonding length and adsorbate-adsorbate distance [41]. Recently Persson has extended the semiclassical models and developed a simple relation between the adsorbate-induced conductance change and the density of states of the adsorbed molecules, $\rho_a(\varepsilon_F)$, at the Fermi energy of the metal [35]. According to his theory, the conductance change per adsorbate molecule is

$$\frac{\Delta G}{G} \sim -\frac{\Gamma \rho_a(\varepsilon_F)}{d}, \qquad (6)$$

where d is the thickness of the metal film and Γ is the width of $\rho_a(\varepsilon)$ that has usually a bell shape. Since different molecules have different $\rho_a(\varepsilon_F)$, the conductance change should be specific for each adsorbate, which has been confirmed for classical conductors [33, 35]. Eq. 6 also shows that the adsorbate-induced conductance change is inversely proportional to the thickness of the film. So in terms of sensor applications, thinner films mean higher sensitivity.

Based on the semiclassical theories, an atomic-scale wire should give the highest sensitivity. However, as we have discussed in the introduction, the conductance of such wires is quantized and the semiclassical theories are not strictly applicable. A theory for the quantum ballistic regime has not yet been fully developed, but the adsorbate scattering is also expected to be an important mechanism. A number of theoretical attempts have been made to calculate the effects of defects on conductance quantization [44–46]. Brandbyge et al. have treated a defect by an effective potential superimposed on an otherwise smooth wall of the wire [46]. Their numerical calculations as well as tight binding method-based calculations [45] show that the scattering by defects lowers the wire conductance, as found in the classical theories. Consequently the conductance steps shift away from the integer multiples of G_0 to fractional values.

Since a metal wire with conductance at the lowest step consists of only a few atoms [12], a microscopic theory that treats the individual atoms and adsorbate molecules seems necessary. Landman et al. has simulated the adsorption of a methyl thiol onto a Au wire made of a chain of four Au atoms between two well-defined electrodes [47]. They found that the molecule binds strongly to the Au atom and becomes a part of the wire. The final conductance of molecule-Au wire can be even greater than the Au wire. Using a self-consistent scheme, Lang has calculated the conductance through a chain of three Al atoms connected to Jellium electrodes [48]. Substituting one of the Al atoms with a sulfur atom decreases the conductance. The theoretical works mentioned above predict an adsorbate-induced conductance change, but a satisfactory theory is not yet complete. Experimental data in this area should stimulate further theoretical efforts, which will be discussed below.

3.1. Effects on the conductance—molecular adsorption

Bogozi et al. have studied effects of molecular adsorption on the conductance of atomic-scale Cu wires experimentally using several molecules, including dopamine, 2,2′-bipyridine (22BPy), mercaptopropionic acid (MPA) [31]. These molecules have different binding strengths to Cu electrodes. Dopamine is an important neurotransmitter, which is expected to weakly adsorb onto Cu [49]. MPA adsorbs onto the electrodes via the strong sulfur-metal bond [49–51]. 22BPY binds to metal electrodes via N-metal interaction [52], which allows individual 22BPY molecules to stand vertically on Au [53, 54] and Cu electrodes [55], as shown directly by STM. The 22BPY binding strength to Cu electrodes is stronger than that of dopamine, but weaker than MPA.

They fabricated the atomic-scale Cu wires using the electrochemical method described in Section 2. They started with a Cu wire with conductance at a chosen quantum step in a blank electrolyte, then injected sample molecules into the electrolyte and

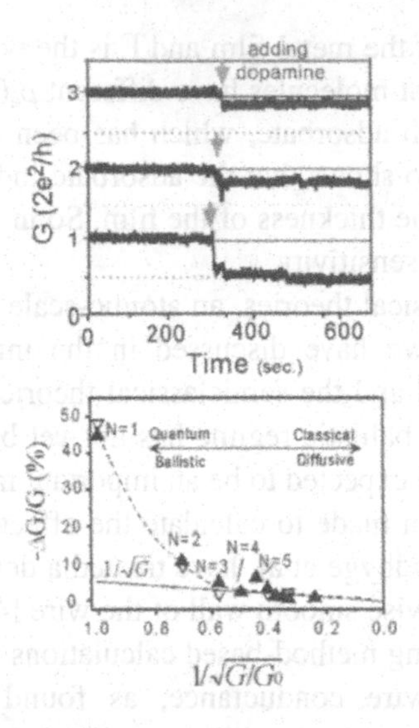

Fig. 6. Top: Conductance change of Cu wires with conductance quantized at 1 G_0, 2 G_0, 3 G_0, ..., 9 G_0 upon dopamine adsorption from the electrolyte. Bottom: The relative change, $-\Delta G/G$ versus $1/\sqrt{G}$ for various quantum steps (*N*). The solid line marks the limit of the semiclassical theories for diffusive scattering. The slope of the solid line is ~0.05 which reflects the binding strength between dopamine and the Cu wires.

monitored the subsequent conductance as a function of time. Fig. 6 shows the conductance of Cu wires with conductance quantized at various steps upon introduction of dopamine into the electrolyte. For a wire with conductance at the lowest quantum step (1 G_0), the conductance decreases drastically a few seconds after the introduction of a drop of dopamine (1 mM), and it then stabilizes at ~0.5 G_0. Adding another drop of dopamine causes no more changes in the conductance, which indicates no further adsorption once the wires are covered with the molecules. The observed conductance change is not due to mechanical disturbance during the injection of the sample solution into the electrolyte because the conductance does not change when adding blank electrolyte in the same way.

For a wire with conductance at the second step (2 G_0), dopamine adsorption caused also a sharp decrease in the conductance but the net change is about 0.25 G_0, much smaller than that of the first quantum step. For a wire at the third step (3 G_0), the conductance decrease is even smaller. At 7th conductance step, the decrease becomes too small to be accurately measured with the setup. The relative conductance change, $-\Delta G/G$, is plotted in Fig. 6. It shows a change as high as 50% for the first conductance step, but the change decreases rapidly for at higher steps as the quantum regime is eventually replaced by the classical diffusive scattering regime. So in terms of chemical sensor applications, a wire with conductance at the lowest quantum step gives the highest sensitivity.

The decrease in the conductance supports the scattering of the conduction electrons by the adsorbates as the dominant mechanism. The observation that the adsorbate-induced conductance change increases as the diameter (*d*) of the wire decreases is not surprising according to Eq. 6, from the semiclassical Persson's theory [35]. Because the number of the quantum modes is proportional to the square of the wire diameter, we have $G = NG_0 \sim d^2$ or $d \sim \sqrt{G}$. Substituting this relation into Eq. 6, we have

$$-\frac{\Delta G}{G} \sim \Gamma \rho(E_F) \frac{1}{\sqrt{G}} \qquad (7)$$

In order to examine the above relation, we plotted $-\Delta G/G$ versus $1/\sqrt{G}$ in Fig. 6. A solid straight line in the figure marks the simple dependence given by Eq. 7. The slope of the straight line is proportional to $\Gamma \rho_s(E_F)$, which depends on the electronic states of the adsorbate. For high conductance steps that correspond to thicker wires, the experimental data can be roughly described by the simple dependence with a slope of ~0.05. However, for the conductance at the lowest few steps, $-\Delta G/G$ increases much faster than $1/\sqrt{G}$ as G decreases. The failure of the semiclassical theories may be attributed to the following reasons. A wire with the conductance at the lowest quantum step is determined by only a few atoms, so virtually every atom is a surface atom and the wire cannot be simply treated as a cylinder with a smooth wall. As found by the molecular dynamics simulations, the adsorbate molecules may get incorporated into the wires [47].

Upon adsorption of MPA and 22BPY onto atomic-scale Cu wires, the conductance decreases also abruptly but with some important differences. First, the extrapolated slopes in the $-\Delta G/G$ versus $1/\sqrt{G}$ plot are 0.25 and 0.1 for MPA and 22BPY, respectively, which are in contrast to 0.05 for dopamine. Since the slopes reflects the adsorbate density of states at the Fermi level of the wire, different slopes are due to different densities of states of the molecules. Fig. 7 compares the relative conductance changes, $-\Delta G/G$, versus G for the three molecules. They all decrease as G increases which is fully expected. The data shows that MPA induces the greatest change, dopamine the least and 22BPY in between, which is consistent with the relative binding strengths of the three molecules.

Large fluctuations in the conductance are often observed after introducing molecules into the solution (Fig. 8). These fluctuations tend to be stepwise and may be

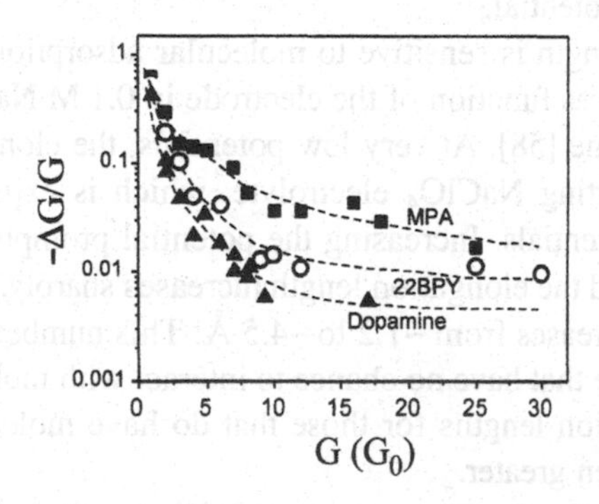

Fig. 7. Comparison of dopamine-, MPA- and 22BPY-induced conductance changes of Cu wires with conductance at various quantum steps. The changes are consistent with the relative binding strengths of the three molecules.

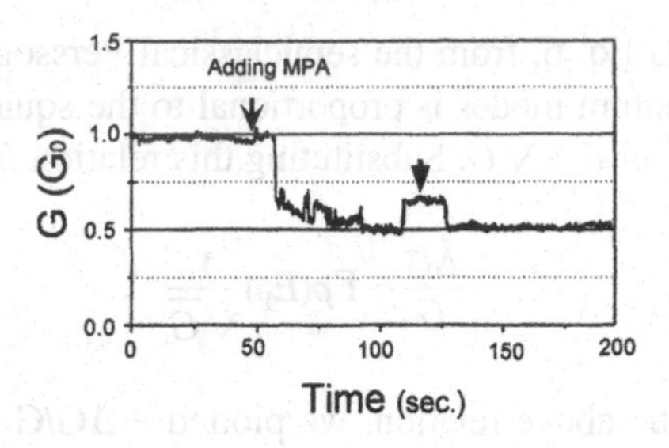

Fig. 8. Stepwise fluctuations (arrow) are frequently observed in the adsorbate-induced conductance change.

attributed to the rearrangement of Cu atoms in the wires. This is expected because the strong binding between the molecules and metal atoms weaken the bindings between metal atoms and thus leads to atomic rearrangement in the atomic-scale wires. This atomic rearrangement may affect mechanical properties of the wires that can be measured experimentally, which will be discussed next.

3.2. Effects on mechanical properties

In order to study molecular adsorption effect on the mechanical stability of the wires, He et al. [56] studied atomic-scale metal wires created by STM as a function of potential. The electrochemical potential allowed them to control the amount of adsorption in a flexible and reversible fashion. They measured the length that each wire can be elongated before losing stability, which provides important mechanical stability information about the wires. The elongation length was determined from the individual conductance traces and the elongation rate set by the z-piezoelectric transducer (see Fig. 3a). By repeatedly forming and breaking the wires, the elongation length averaged over a thousand individual traces for the wires quantized at the lowest quantum step was extracted at each potential. In the absence of molecular adsorbates, the elongation length is about 1.2 Å, in good agreement with previous studies in controlled gas environment and in vacuum [19, 57]. The length is largely independent of the electrochemical potential.

The elongation length is sensitive to molecular adsorption. Fig. 9 plots the average elongation length as function of the electrode in 0.1 M $NaClO_4$ containing different amounts of adenine [58]. At very low potentials, the elongation length is similar to that in the supporting $NaClO_4$ electrolyte, which is expected since no adenine adsorption at low potentials. Increasing the potential prompts adsorption of adenine onto the Au wires, and the elongation length increases sharply. In 3.3 mM adenine, the elongation length increases from ~1.2 to ~4.5 Å! This number is averaged over many wires, including those that have no chance to interact with molecules before breaking, so the actual elongation lengths for those that do have molecules adsorbed on their surfaces might be even greater.

Similar to that in adenine, the presence of MPA greatly increases the average elongation length. However, for a given concentration, the effect of MPA is greater than that of adenine, which may reflect a greater binding strength between MPA and Au.

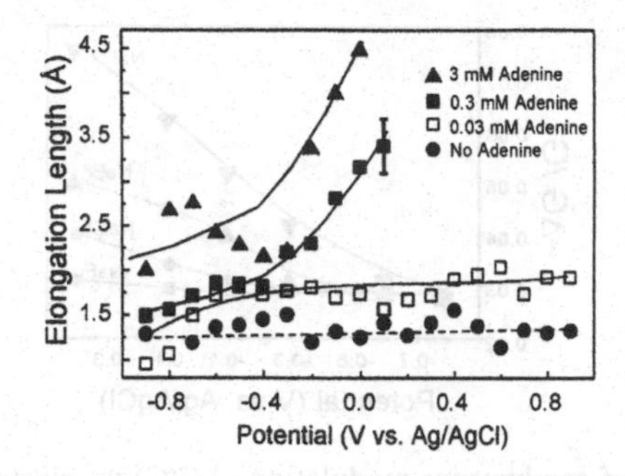

Fig. 9. Average elongation length of Au wires with conductance at the lowest quantum step (thinnest wire) as a function of potential in 0.1 M $NaClO_4$, 0. 03 mM adenine + 0.1 M $NaClO_4$, 0. 3 mM adenine + 0.1 M $NaClO_4$ and 3 mM adenine + 0.1 M $NaClO_4$.

For example, the wires can be repeatedly formed in 3 mM adenine, but they became too "sticky" to be repeatedly formed in 3 mM MPA. The presence of 22BPY also increases the elongation length, but it is considerably less dramatic than MPA and adenine, which is consistent with the relative binding strength of each molecule to Au electrode.

Because the elongation length measures how robust a wire can be pulled before losing stability and switching to a new atomic configuration, the large increase in the elongation length indicates an enhanced stability of the wires by molecular adsorbates. This effect may be used to improve the stability of metal wires with quantized conductance. So in addition to scattering conduction electrons discussed above, molecular adsorption may affect the stability and atomic configuration of the atomic-scale metal wires.

4. Anionic Adsorption

Adsorption of small anions, such as ClO_4^-, F^-, Cl^-, Br^- and I^-, onto an atomic-scale metal wire has also been studied [59]. Among these anions, ClO_4^- and F^- adsorb only weakly onto Au and Ag electrodes, but Cl^-, Br^- and I^- adsorb onto the electrodes much more strongly with a binding strength in the order of $Cl^- < Br^- < I^-$ [60]. Another important property of the anions is that their adsorption strengths changes with the electrochemical potential, so one can control the adsorption strengths via the electrochemical potential.

In order to accurately determine anionic adsorption-induced conductance change, the electrochemical potential is modulated and both the amplitude and phase shift of the induced conductance modulation are determined. This AC technique removes various noises including mechanical drift effects in a similar way to a lock-in amplifier. The phase shift between the induced conductance and the applied electrochemical potential is ~180°, which means an increase in the potential causes a decrease in the conductance. Since the adsorption strengths of the anions increase as the potential, the phase shift shows that anionic adsorption induces a decrease in the conductance.

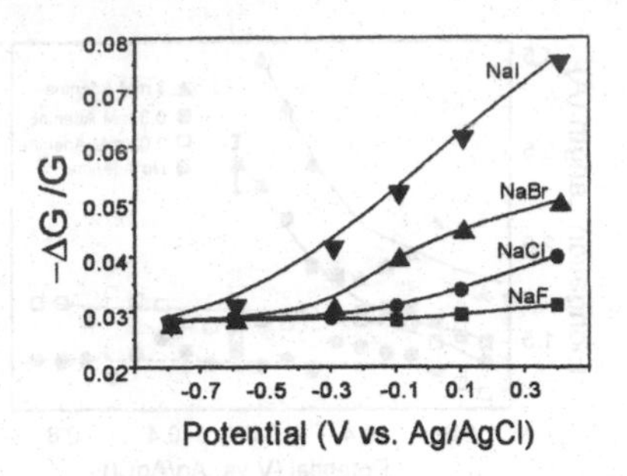

Fig. 10. The dependence of conductance modulation ($\Delta G/G$) vs. electrochemical potential for Au wires with conductance near 1 G_0 in the electrolytes containing F^-, Cl^-, Br^- and I^- anions. The amplitude of the applied electrochemical potential modulation is 0.05 V and the frequency is 1 kHz.

Fig. 10 plots $\Delta G/G$, the amplitude of the induced conductance normalized against the conductance vs. the electrochemical potential in electrolytes containing various anions. At low potentials, $\Delta G/G$ is about the same for all the anions, which is consistent with the fact that no anion adsorption takes place at very negative potentials. The potential-induced conductance modulation in the low potential regime is solely determined by the double layer charging effect [59]. Increasing the potential, $\Delta G/G$ increases for all the electrolytes but the amount of the increase differs dramatically from one anion to another. For example, in 0.5 M NaF, the change is only ~10% from −0.7 V to +0.8 V. In sharp contrast, the change in 0.01 M NaI + 0.5 M is as large as ~300%! The amount of change varies in the order of $F^- < Cl^- < Br^- < I^-$, in good agreement with adsorption strengths of these anions. A decrease in the conductance of classical metal thin films due to anion adsorption has been observed in electrolytes containing F^-, Cl^-, Br^- and I^- [37, 38, 40, 61–63].

The anion adsorption-induced conductance change could also be attributed to the scattering of conduction electrons in the wires by the adsorbed anions on the surfaces of the wires [13, 31]. The adsorption of these anions are known to result in a large amount of charge transfer between the anions and the electrodes, which may play an important role either contributing the scattering or changing the electron density in the wires [64]. Other mechanisms, such as shift in PZC (Potential of Zero Charge) upon specific anionic adsorption, may also have a contribution. Further studies will be needed for a complete understanding of the anionic adsorption onto atomic-scale wires.

5. Summary

Atomic-scale wires can be fabricated either mechanically by separating two electrodes from contact or electrochemically by etching a thicker wire down to the atomic scale. In contrast to nanowires fabricated with other techniques, these atomic-scale wires have several important features. First, they are naturally connected to two bulk electrodes, so one can easily measure and study the electrical properties of each of the wires during and after the fabrication. Second, these wires are shorter than the mean free path of conduction electrons, so the electron transport is ballistic along the wires. Finally, they are as thin as the electron wavelength so that electrons can form well defined quantum modes in the transverse direction of the wires, which gives rise to conductance quantization.

Using conductance quantization as a signature, one can fabricate atomic-scale metal wires for various applications. One important possible application is to study ionic and molecular adsorption onto the atomic-scale wires by monitoring the conductance changes. The high sensitivity is expected since virtually every atom in the thinnest portion of the wire is a surface atom that is available to interact with molecules or ions. Indeed, experiments have shown abrupt decreases in the conductance upon molecular and ionic adsorption. The decreases correlate well with the adsorption strengths of the ions and molecules. While both the ionic and molecular adsorption-induced conductance decreases can be partially attributed to the scattering of conduction electrons in the wires by the adsorbates, it is also evident that the strong interactions between adsorbates and the wires can cause atoms in the wires to rearrange into different configurations. Because the wires are often as short as a few atoms, enough to accommodate a single or a few ions or molecules on their surfaces, so one can study the adsorption of single ions and molecules with the atomic-scale wires. The atomic-scale wires can last for up to several days, but further improvement in the stability is desired for many practical applications. One possible way is to "coat" them with appropriate molecules. Recent experiment has shown that strong molecule-wire interactions can significantly enhance the wire mechanical stability. Higher stability can also be achieved using somewhat thicker wires at some cost of sensitivity. Practical applications demand us to understand how molecular adsorption affects the conductance and atomic configurations of the wires, which has just begun. Many important basic questions remain to be addressed, which demands further efforts from both experimentalists and theorists.

Acknowledgment

We thank Prof. Landman for helpful discussions and NSF (CHE-9818073), DOE (DE-FG03-01ER45943) and EPA (R829623) for financial support.

References

1. D. A. Wharam, T. J. Thornton, R. Newbury, M. Pepper, H. Ahmed, J. E. F. Frost, D. G. Hasko, D. C. Peacock, D. A. Ritchie and G. A. C. Jones, "One-dimensional transport and the quantization of the ballistic resistance", *J. Phys. C* **21** (1988) 209–214.
2. B. J. van Wees, H. van Houten, C. W. J. Beenakker, J. G. Williams, L. P. Kouwenhowen, D. van d. Marel and C. T. Foxon, "Quantized conductance of point contact in a two-dimensional electron gas", *Phys. Rev. Lett.* **60** (1988) 848–850.
3. U. Landman, W. D. Luedtke, N. A. Burnham and R. J. Colton, "Atomistic mechanisms and dynamics of adhesion, nanoindentation, and fracture", *Science* **248** (1990) 454–457.
4. J. K. Gimzewski and R. Moller, "Transition from the tunneling regime to point contact studied using scanning tunneling microscopy", *Phys. Rev. B* **36** (1987) 1284–1287.

5. J. M. Krans, C. J. Muller, I. K. Yanson, T. C. M. Govaert, R. Hesper and J. M. van Ruitenbeek, "One-atom point contact", *Phys. Rev. B* **48** (1993) 14721–14724.
6. R. Landauer, "Spatial variation of currents and fields due to localized scatters in metallic conduction", *IBM J. Res. Dev* **1** (1957) 223–231.
7. J. M. Krans, J. M. van Ruitenbeek, V. V. Fisun, I. K. Yanson and L. J. d. Jongh, "The signature of conductance quantization in metallic point contacts", *Nature* **375** (1995) 767–769.
8. J. I. Pascual, J. Mendez, J. Gomez-Herrero, A. M. Baro, N. Garcia and V. T. Binh, "Quantum contact in gold nanostructures by scanning tunneling microscopy", *Phys. Rev. Lett.* **71** (1993) 1852–1855.
9. L. Olesen, E. Lægsgaard, I. Stensgaad, F. Besenbacher, J. Schiotz, P. Stoltze, K. W. Jacobsen and J. K. Norskov, "Quantized conductance in an atom-sized point contact", *Phys. Rev. Lett.* **72** (1994) 2251–2254.
10. J. L. Costa-Kramer, N. Garcia, P. Garcia-Mochales and P. A. Serena, "Nanowire formation in macroscopic metallic contacts-quantum-mechanical conductance tapping a table top" *Surf. Sci.* **342** (1995) L1144–L1146.
11. H. Yasuda and A. Sakai, "Conductance of atomic-scale gold contacts under high-bias voltages", *Phys. Rev. B* **56** (1997) 1069–1072.
12. H. Ohnishi, Y. Kondo and K. Takayanagi, "Quantized conductance through individual rows of suspended gold atoms", *Nature* **395** (1998) 780–783.
13. C. Z. Li, H. Sha and N. J. Tao, "Adsorbate effect on conductance quantization of metallic nanowires", *Phys. Rev. B.* 58 (1998) 6775–6778.
14. C. Z. Li, H. X. He, A. Bogozi, J. S. Bunch and N. J. Tao, "Molecular detection based on conductance quantization of nanowires", *Appl. Phys. Lett.* **76** (2000) 1333–1336.
15. N. Agrait, J. G. Rodrigo and S. Vieira, "Conduture steps and quantization in atomic-size contacts", *Phys. Rev. B* **52** (1993) 12345–12348.
16. J. I. Pascual, J. Mendez, J. Gomez-Herrero, A. M. Baro, N. Garcia, U. Landman, W. D. Luedtke, E. N. Bogachek and H.-P. Cheng, "Properties of metallic nanowires: from conductance quantization to localization", *Science* **267** (1995) 1793–1796.
17. D. P. E. Smith, "Quantum Point Contact Switches", *Science* **269** (1995) 371–373.
18. Z. Cai, Y. He, H. Yu and W. S. Yang, "Observation of conductance quantization of ballistic metallic point contacts at room temperature", *Phys. Rev. B* **53** (1996) 1042–1045.
19. G. Rubio, N. Agrait, S. Vieira, "Atomic-sized metallic contacts: Mechanical properties and electronic transport", *Phys. Rev. Lett.* **76** (1996) 2302–2305.
20. C. J. Muller, J. M. van Ruitenbeek, L. J. de Jongh, "Conductance and supercurrent discontinuities in atomic-scale metallic constrictions of variable width" *Phys. Rev. Lett.* **69** (1992) 140–143.
21. U. Landman, W. D. Luedtke, B. E. Salisbury and R. L. Whetten, "Reversible manipulation of room temperature mechanical and quantum transport properties in nanowire junctions", *Phys. Rev. Lett.* **77** (1996) 1362–1365.
22. C. Z. Li and N. J. Tao, "Quantum transport in metallic nanowires fabricated by electrochemical deposition/dissolution", *Appl. Phys. Lett.* **72** (1998) 894–897.
23. C. Z. Li, A. Bogozi, W. Huang and N. J. Tao, "Fabrication of stable metallic nanowires with quantized conductance", *Nanotechnology* **10** (1999) 221–223.
24. A. F. Morpurgo, C. M. Marcus and D. B. Robinson, "Controlled Fabrication of Metallic Electrodes with Atomic Separation.", *Appl. Phys. Lett.* **14** (1999) 2082.
25. J. Z. Li, T. Kanzaki, K. Murakoshi and Y. Nakato, "Metal-dependent conductance quantization of nanocontacts in solution", *Appl. Phys. Lett.* **81** (2002) 123–125.
26. F. Elhoussine, S. Matefi-Tempfli, A. Encinas and L. Piraux, "Conductance quantization in magnetic nanowires electrodeposited in nanopores", *Appl. Phys. Lett.* **81** (2002) 1681–1683.
27. S. Nakabayashi, H. Sakaguchi, R. Baba and E. Fukushima, "Quantum contact by colliding 2D fractal", *Nano Lett.* **2** (2001) 507–510.

28. S. Boussaad and N. J. Tao, "Atom-Size contacts and gaps between electrodes fabricated with a self-terminated electrochemical method", *Appl. Phys. Lett.* **80** (2002) 2398–2400.
29. A. J. Bard and L. R. Faulkner, *Electrochemical Methods*, John Wiley and Sons, Inc., New York (1980).
30. K. Hansen, S. K. Nielsen, M. Brandbyge, E. Lægsgaard, I. Stensgaard and F. Besenbacher, "Current-voltage curves of gold quantum point contacts revisited", *Appl. Phys. Lett.* **77** (2001) 708–710.
31. A. Bogozi, O. Lam, H. X. He, C. Z. Li, N. J. Tao, L. A. Nagahara, I. Amlani and R. Tsui, "Molecular adsorption onto metallic quantum wires", *J. Am. Chem. Soc.* **123** (2001) 4585–4590.
32. D. Schumacher, in *Surface Scattering Experiments with Conduction Electrons* Vol. **128**, Edited by G. Hohler, Springer, New York (1993).
33. E. T. Krastev, D. E. Kuhl and R. G. Tobin, "Multiple mechanisms for adsorbate-induced resistivity oxygen and formate on Cu(100)", *Surf. Sci.* **387** (1997) L1051–L1056.
34. M. Rauh, B. Heping and P. Wissmann, "The effect of CO adsorption on the resistivity of thin Pd films", *Appl. Phys. A* **61** (1995) 587–590.
35. B. N. J. Persson, "Applications of surface resistivity to atomic scale friction, to the migration of hot adatoms and to electrochemistry", *J. Chem. Phys.* **98** (1993) 1659–1672.
36. Y. Zhang, R. H. Terrill and P. W. Bohn, "In-plane resistivity of ultrathin gold films: a high sensitivity, molecularly differentiated probe of mercaptan chemisorption at the liquid-metal interface", *J. Am Chem Soc.* **120** (1998) 9969–9970.
37. R. I. Tucceri and D. Posadas, "A surface conductance study of the anion adsorption on gold", *J. Electroanal. Chem.* **191** (1985) 387–399.
38. R. I. Tucceri and D. Posadas, "The effect of surface charge on the surface conductance of silver in surface inactive electrolytes", *J. Electroanal. Chem.* **283** (1990) 159–166.
39. D. Korwer, D. Schumacher and A. Otto, "Resistance changes of thin film electrodes of silver", *Ber. Bunsenges. Phys. Chem.* **95** (1991) 1484–1488.
40. C. Hanewinkel, A. Otto and T. Wandlowski, "Change in surface resistance of an Ag(100) electrode by adsorbed bromide", *Surf. Sci.* **429** (1999) 255–259.
41. H. Ishida, "Microscopic theory of surface resistivity" *Phys. Rev. B* **52** (1995) 10819–10822.
42. K. Fuchs, H. H. Wills, "The conductivity of thin metallic films according to the electron theory of metals" *Proc. Cambridge Philos. Soc.* **34** (1938) 100–108.
43. E. H. Sondheimer, "The mean free path of electrons in metals", *Adv. Phys.* **1** (1952) 1–42.
44. K. Nikolic and A. MacKinno, "Conductance and conductance fluctuations of narrow disordered quantum wires" *Phys. Rev. B* **50** (1994) 11008–11017.
45. P. Garcia-Mochales and P. A. Serena, "Disorder as origin of residual resistance in nanowires", *Phys. Rev. Lett.* **79** (1997) 2316–2319.
46. M. Brandbyge, K. W. Jacobsen and J. K. Norskov, "Scattering and conductance quantization in three-dimensional metal nanocontacts", *Phys. Rev. B* **55** (1997) 2637–2650.
47. H. Hakkinen, R. N. Barnett and U. Landman, "Gold nanowires and their chemical modifications", *J. Phys. Chem* **103** (1999) 8814–8817.
48. N. D. Lang, "Resistance of atomic wires", *Phys. Rev. B* **52** (1995) 5335–5342.
49. M. J. Giz, B. Duong and N. J. Tao, "*In situ* STM study of self-assembled mercaptopropionic acid monolayers for electrochemical detection of dopamine", *J. Electroana. Chem.* **465** (1999) 72–79.
50. L. H. Dubois and R. G. Nuzzo, "Synthesis, structure and properties of model organic surfaces", *Annu. Rev. Phys. Chem.* **43** (1992) 437–463.
51. C. A. Widrig, C. A. Alves and M. D. Porter, "Scanning tunneling microscopy of ethanethiolate and n-octadecanethiolate monolayers spontaneously adsorbed at gold surfaces", *J. Am. Chem. Soc.* **113** (1991) 2805–2809.

52. D. F. Yang, D. Bizzotto, J. Lipkowski, B. Pettinger and S. Mirwald, "Electrochemical and second harmonic generation studies of 2,2′-bipyridine adsorption at the Au(111) electrode surface", *J. Phys. Chem.* **98** (1994) 7083–7089.
53. F. Cunha and N. J. Tao, "Surface charge induced order–disorder transition in an organic monolayer", *Phys. Rev. Lett.* **75** (1995) 2376.
54. F. Cunha, N. J. Tao, X. W. Wang, Q. Jin, B. Duong, J. D'Agnese, "Potential induced phase transitions in 2,2′-bipyridine and 4,4′-bipyridine monolayers on Au(111) studied *by in situ* STM and AFM", *Langmuir* **12** (1996) 6410.
55. N. J. Tao and C. Z. Li, *unpublished.*
56. H. X. He, Shu, C. Z. Li and N. J. Tao, "Adsorbate effect on the mechanical stability of atomically thin metallic wires", *J. Electroanal. Chem.* **522** (2002) 26–32.
57. J. L. Costa-Krämer, N. García, P. García-Mochales, P. A. Serena, M. I. Marqués and A. Correia, "Conductance quantization in nanowires formed between micro and macroscopic metallic electrodes", *Phys. Rev. B* **55** (1997) 5416.
58. N. J. Tao, J. A. DeRose and S. M. Lindsay, "Self-Assembly of molecular superstructures studied by insitu scanning tunneling microscopy—DNA bases on Au(111)", *Journal of Physical Chemistry* **97** (1993) 910–919.
59. B. Xu, H. X. He and N. J. Tao, "Controlling the conductance of atomically thin metal wires with electrochemical potential", *J. Am. Chem. Soc.* **124** (2002) 13568–13575.
60. J. Lipkowski, Z. C. Shi, A. C. Chen, B. Pettinger and C. Bilger, "Ionic adsorption at the Au(111) electrode", *Electrochemica Acta* **43** (1998) 2875–2888.
61. G. A. Fried, Y. Zhang and P. W. Bohn, "Effect of molecular adsorption at the liquid-metal interface on electronic conductivity: the role of surface morphology", *Thin Solid films* **401** (2001) 171–178.
62. R. I. Tucceri and D. Posadas, "Surface structural changes of evaporated thin silver film electrodes during the adsorption of chloride ion", *J. Electroanal. Chem.* **270** (1989) 415–419.
63. H. Winks, D. Schumacher and A. Otto, "Surface resistance measurements at the metal/electrolyte inteface of Ag(100) and Ag(111) thin film electrodes", *Surf. Sci.* **400** (1998) 44–53.
64. J. W. Schultze and D. Rolle, "The partial discharge of electrosorbates and its influence in electrocatalysis", *Can. J. Chem.* **75** (1997) 1750–1758.

Chapter 6

Structural Study of Metal Nanowires

Varlei Rodrigues and Daniel Ugarte
Laboratório Nacional de Luz Síncrotron, Caixa Postal 6192—CEP 13084-971, Campinas, São Paulo, Brazil

1. Introduction

The structural and mechanical properties of nanometric metal wires (nanowires—NWs) represent a fundamental issue for the understanding of different phenomena such as friction, fracture, adhesion, etc. [1]. Nanometric systems are formed by a reduced number of atoms and the majority of them are located at the surface; this may lead to new and different attributes (structural, optical, etc.) when compared to their macroscopic counter part.

The plastic deformation of bulk materials is mainly associated to the motion of dislocations; in contrast, NWs display an enhanced strength whose value is consistent with a system free of dislocations [2–4]. At very low scale (<10 nm), dislocations should be completely suppressed because the involved stress becomes comparable to the intrinsic lattice strength; in consequence, the plastic deformation of nanostructures should be associated to collective slips of entire atomic planes or order-disorder transitions [2, 5–7]. In this sense, the atomic arrangement and morphology of the system play a central role [2, 5–10] and, a microscopic analysis is required to get a deeper understanding. From this point of view, High Resolution Transmission Electron Microscopy (HRTEM) is a powerful tool, because it allow us to image nanometric structures with atomic resolution (see Fig. 1), and even analyze dynamical atomistic processes by means of time-resolved observations.

Recently, new and strong motivation was raised in the field of NW stretching, because the mechanical elongation of contacts represents the basic experiment to analyze the electrical transport properties of nanometric conductors (see Fig. 2) [11, 12]. For example, Ohnishi et al. [13] have used a simplified Scanning Tunneling Microscope (STM) to generate Au NWs *in situ* in a HRTEM, while simultaneously recording its conductance; these studies revealed the existence of suspended chains of Au atoms [13].

In this chapter, we address the study of the atomic arrangement of NWs generated *in situ* in a HRTEM by mechanical stretching. Just before rupture, NWs are crystalline

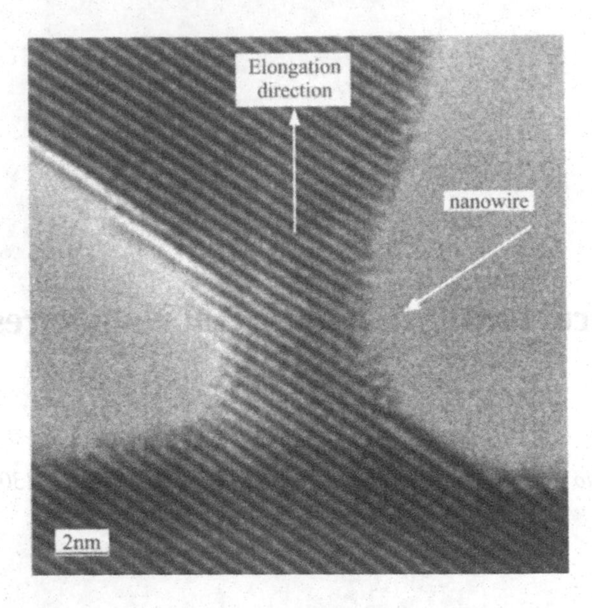

Fig. 1. Atomic resolution image of a Au NW generated by the elongation of a point contact. Dark lines are (1 1 1) planes.

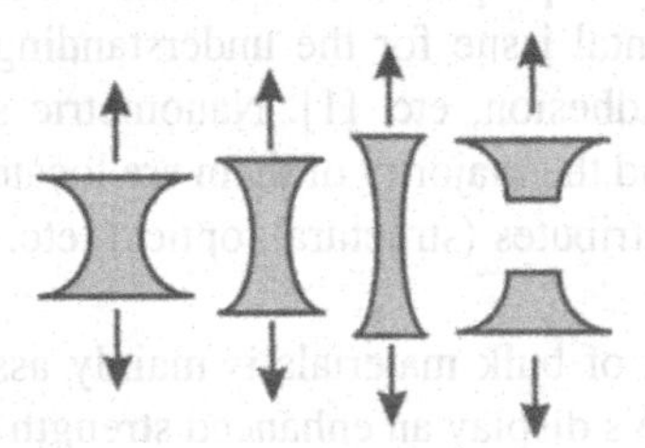

Fig. 2. Schema of NW generation by mechanically stretching a macroscopic contact; just before rupture, the wire is composed of only a few atoms.

and free of defects, adopting only a few number of atomic arrangements. The derived structural information has allowed us to understand some open questions on the conductance behavior of metal nanowires.

2. Morphology Imaging of Metal NWs

One fundamental characteristic of nanometric systems is the fact that the majority of their atoms are located at the surface, what can lead to new and interesting mechanical, electrical, optical, magnetic properties, as well as unexpected atomic arrangements and morphologies [14]. Several techniques has been employed to study the structure of atomic scale systems, however, High Resolution Transmission Electron Microscopy can be considered one of the most powerful tools, because it provides real space imaging of the atomic structure [15]. Here, we will not describe general details about the HRTEM basis or conventional applications, and we will focus on the image interpretation for a particular case, the face-centered-cubic (fcc) metals; in fact, the NW studies discussed in this chapter deal with fcc metals (Au, Ag, Pt, etc.).

When the HRTEM sample is a few atomic planes thick, the image contrast can be basically viewed as the projection of the atomic potential [16]. In this way, the image will show a characteristic pattern, that can be interpreted as the bidimensional projection of the crystals along the electron beam direction. This allows the measurement of the angles and distances between atomic planes [16] for crystals oriented along high symmetry axes, but in general the image reveals no useful information for an arbitrary direction.

Fig. 3(a) shows the cubic unit cell for a fcc crystal and Fig. 3(b–d) display the expected images when this structure is observed along the three main zone axes ([1 0 0], [1 1 0] and [1 1 1]). Analyzing the HRTEM image, we can determine the observation axis based on the projection pattern geometry, being squares for [1 0 0], flattened hexagons for [1 1 0] and perfect hexagons for [1 1 1] direction. As mentioned before, these geometric motifs are related to atomic planes parallel to the electron beam, as indicated in Fig. 3, where the interplanar distances should be the expected values for the studied material. From the images, it is also possible to determine the crystallography directions in the plane of the sample (a necessary information to determine the NW axis orientation and/or the elongation direction). As a practical example, Fig. 4 displays an atomic resolved micrograph of a Au NW; the flattened hexagon pattern indicates that the observation axis is the [1 1 0] direction and, the angles and distances between the atomic planes corroborate this interpretation. Also, it is possible to deduce that the NW axis is along the [−1 1 0] direction (compare with Fig. 3(c)).

Besides the atomic arrangement of the nanosystems, the shape is particularly important because it determines properties related to quantum confinement. To predict/interpret the morphology of atomic scale metal systems, one powerful tool is the geometrical Wulff construction [17, 18]. This approach yields the crystal shape by predicting the relative size of the lower energy facets of the crystal [18]. For a given facet, the surface free energy is given by the γ_{ijk}, where ijk are the Miller index of the corresponding atomic plane. The geometry of the nanosystem is obtained by taking the normal distance from the particle center to any facet as proportional to its surface free energy. This procedure is illustrated for two-dimensions in Fig. 5, where the $\gamma_{100}/\gamma_{111} = (3)^{-1/2}/2$ (~0.87, value frequently used to predict Au nanoparticles shape [18]). Fig. 6 shows some ideal morphologies for different $\gamma_{100}/\gamma_{111}$ values; it must be noted that the size of the {1 1 1} facets increase with respect to the {1 0 0} ones as a function of the increasing $\gamma_{100}/\gamma_{111}$ ratio.

Although the Wulff construction neglects several effects as edge/corner energies, it has been used successfully to predict metal nanoparticles shapes, even for complex structures as truncated decahedra [18] and it has been successfully applied to understand NW morphologies [8, 9].

In next section, we will describe the experimental procedure used to study NWs *in situ* in the HRTEM, explaining the basic experimental methodology for the NW generation.

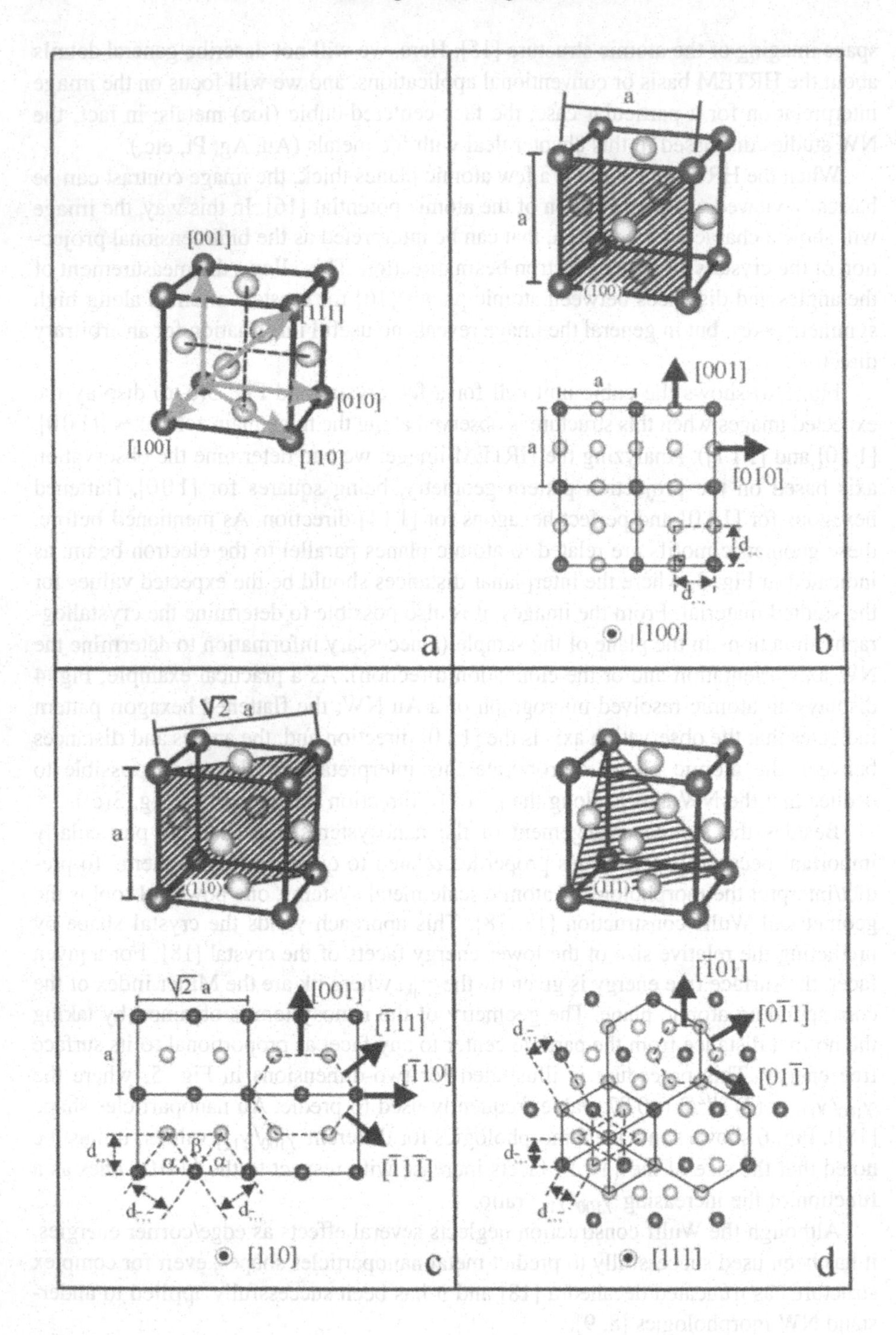

Fig. 3. (a) fcc crystal structure; (b–d) atomic projections when observed along the [1 0 0], [1 1 0] and [1 1 1] direction, respectively. In (c) $\alpha = 54°44'$ and $\beta = 70°32'$.

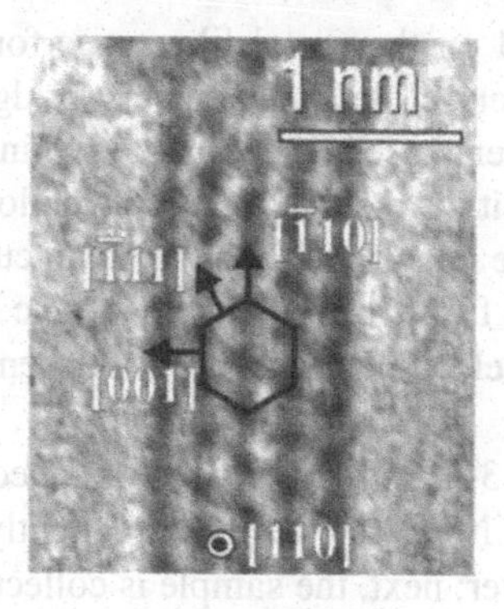

Fig. 4. Atomic resolution image of a Au NW. The flattened hexagon pattern indicates that the observation axis is the [1 1 0] direction. Also, from the flattened hexagon orientation is possible to determine crystallography directions on the plane of the NW. In this case, the NW is generated along the [−1 1 0] direction, and the lateral faces are [0 0 1] planes.

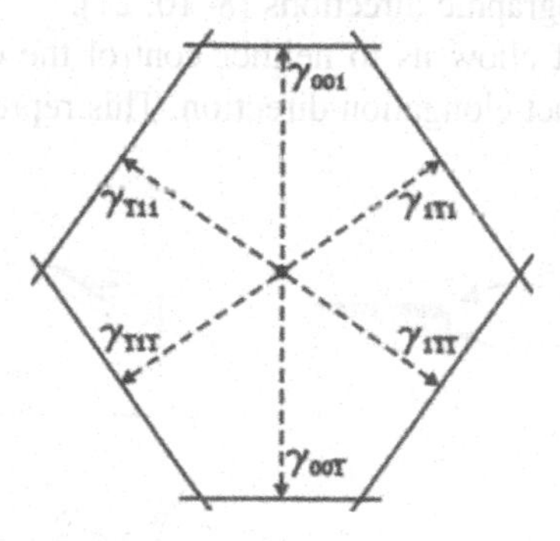

Fig. 5. Schematic draw of a bi-dimensional projection along the [1 1 0] direction of a fcc single crystal. The particle shape is delimited by {1 1 1} and {1 0 0} planes whose distances from the center of the particle are proportional to γ_{111} and γ_{100} respectively (see text for explanation).

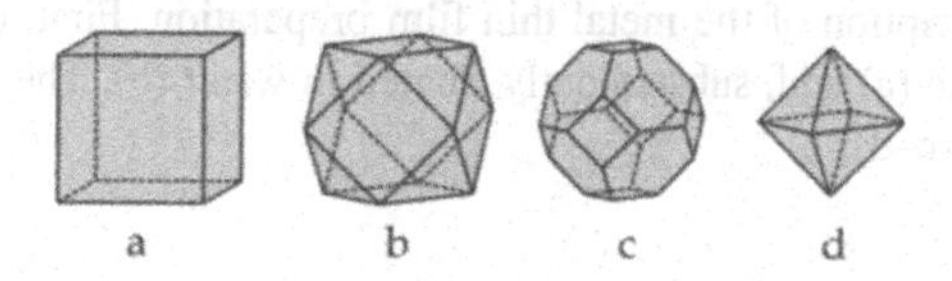

Fig. 6. Geometrical shapes predicted for nanoparticles using the Wulff approach, for different $\gamma_{100}/\gamma_{111}$ values: (a) 0.58 ($3^{-1/2}$), (b) 0.87 ($3^{1/2}/2$), (c) 1.15 ($2/3^{-1/2}$) and (d) 1.73 ($3^{1/2}$).

3. Experimental Methods

We have generated NWs *in situ* in a HRTEM using the method developed by Takayanagi's group [19]. Inside the microscope, the NWs are generated by the following procedure: the microscope electron beam is increased to a current density of ~120 A/cm^2 and focussed on the metal film to perforate and grow neighboring holes. When two holes are very close, a nanometric bridge is formed between them (schema in Figs 7 and 8). When these bridges are very thin (1–2 nm) and close to rupture, the electron beam intensity is reduced to its conventional value (10–30 A/cm^2) in order to perform the real-time imaging. When the beam current is too high (as during the hole generation step), the film vibrates and no atomic imaging can be performed; this forces us to reduce the electron beam (as in conventional operation for image acquisition).

The thin films (thickness 3–10 nm) have been obtained by thermal evaporation of metal on a substrate (usually NaCl crystals). Subsequently, it was detached from the substrate by floating it in water; next, the sample is collected on a TEM holey carbon grid, remaining self-supported on the regions hanging over the holes (Fig. 7) [20]. The TEM sample is mounted on a conventional single tilt TEM sample holder, operated at room temperature. In particular, our sample consisted of a self-supported polycrystalline metal thin film. The use of a polycrystalline sample is very useful because it enables us to generate NWs between apexes of different orientations and elongate them in different crystallographic directions [8–10, 21].

This method does not allow us to neither control the crystal orientation of the NW apexes, nor the contact elongation direction. This represents a serious difficulty

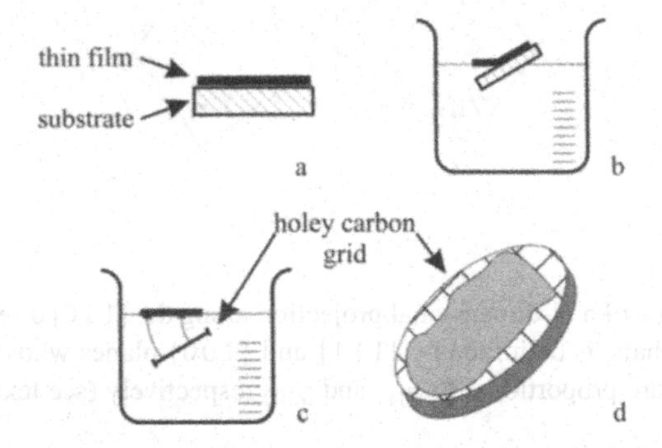

Fig. 7. Schematic description of the metal thin film preparation. First, the metal is thermally evaporated in a substrate (a) and, subsequently floated in water (b). Then, the film is collected on a holey carbon grid (c–d).

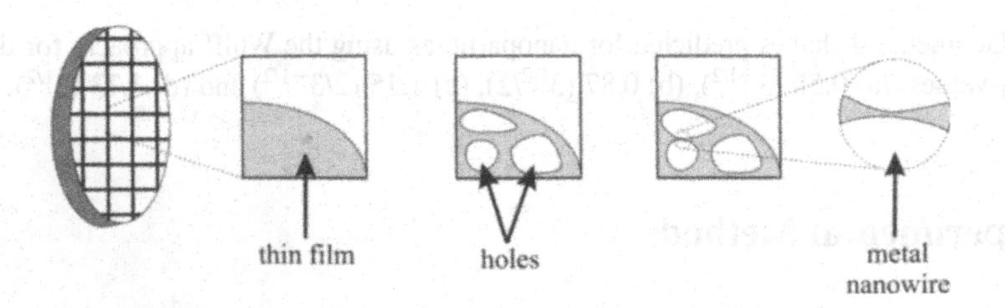

Fig. 8. Metal NWs are obtained by generating holes on a thin film by focusing the TEM electron beam (see text for explanation).

because the NW structure must be aligned with the TEM electron beam to be imaged with atomic resolution. On the other side, it is possible to study NWs generated in a large variety of conditions and understand the general role for the NW formation and the adopted atomic arrangement. This point should be taken into account when comparing the results obtained with different approaches. For example, Takayanagi's group has actually developed this procedure, but they used a monocrystalline Au (1 0 0) film [19]. In consequence, they have mainly observed NWs generated along the [1 1 0] and a few cases along the [1 0 0] direction, while we have imaged NWs formed for a wide range of possible directions (discussed below in Section 4.1.1). We must note that specially designed sample holders have been also employed to study metal NWs [13, 22–24]. For example, in a landmark paper in the field, Ohnishi et al. [13] have described the use of a simplified STM mounted in the TEM sample holder in order to generate the NW and simultaneously measure their electrical properties.

The procedure of making holes in a polycrystalline sample, described above, has allowed us to generate NWs with a remarkable stability. In fact, the NW, its apexes and the surrounding regions are all parts of a unique metal film and, in consequence, they form a monolithic block. NWs formed by a few atomic layers usually show a long lifetime in the range of 1–10 min. Although this stability, the generated NWs elongate spontaneously, get thinner and, then, break due to the relative slow movement of the NW apexes. This apex displacement is probably due to a film deformation induced by thermal gradients or just by low frequency vibration of the thin metallic film membrane, as usually observed in TEM thin film work. Fig. 9 shows a series of snapshots

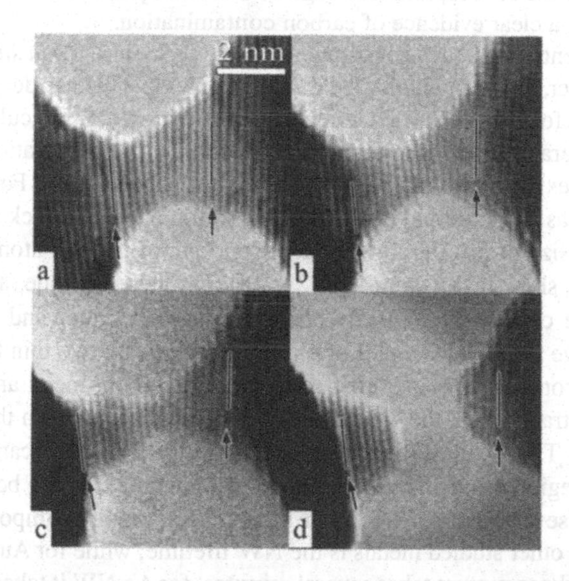

Fig. 9. (a–d) time evolution for a Au NW stretched approximately along the [1 1 1] direction. 0, 120, 193 and 215 s, respectively. The elongation can be followed by the inclusion of planes in the NW neck: 12, 16, 18 planes from (a) to (c) (counted between the two undeformed regions indicated by arrows). In (d), the two apexes reorganize and retract after the rupture. Atomic positions appear dark.

of a complete Au NW elongation process along the [1 1 1] direction. The stretching can be followed by the sequence of inclusions of (1 1 1) planes (dark lines) on the contact region [8]. Also, it is possible to measure directly the relative distance between two undeformed regions of the apexes; in this way, the apex retraction can be quantitatively determined with a precision of ~0.01–0.02 nm.

All images were acquired close to Scherzer defocus [16] and a digital camera (Gatan MSC794, acquisition time ~1 s) was used to acquire atomic resolved micrographs. However, the main method used to image the NWs has been a high sensitivity TV camera (Gatan 622SC, 30 frame/s) associated with a conventional video recorder; here, the images were obtained by digitizing the video film *a posteriori*. In order to enhance signal-to-noise ratio, several frames (3–5) are usually added; however, a great care must be taken to avoid drift between frames because it easily degrades the atomic resolution. In this case, the cross-correlation between frames are analyzed; if any shift is verified, the frames are realigned before adding. This procedure has shown to be very efficient and, has enabled us to register the NW real-time formation, evolution and rupture, even for low atomic number metals [25]. It is important to remark that no additional image treatment or filtering has been employed in our HRTEM images.

One important point to be considered when studying such tiny nanostructures is the presence of contaminants and, the most critical one in TEM work is amorphous carbon [26]. In order to overcome this problem, an intense electron irradiation was used to clean the metal surface, transforming carbon into bucky-onions [27, 28]. The NW cleanness is indicated by a clear faceting of the NW apexes and, a high mobility of surface atoms; another important test is to observe the reorganization/retraction of the apex tip after the NW rupture (see Fig. 9). For example, a long-live sharp tip after the NW break is a clear evidence of carbon contamination.

The steps mentioned above constitute the basic procedure for *in situ* HRTEM NW analysis. However, we must emphasize that the study of NWs has demonstrated to be a new challenge for each metal. Each material has required a particular setting, from the sample preparation (thin film thickness, etc.) to the NW generation and imaging. For this kind of experiments, Au seems to be a well-behaved case. For this metal, we have used NaCl as substrate and evaporated a Au film of ~5 nm thick in high-vacuum (average grains size of 50–100 nm). In addition, due to its high atomic number, the HRTEM images show high contrast [8, 21, 25]. On the other side, all other studied metals are more complicated, mainly because they are lighter and reactive. As an example, we have protected samples of oxidizing metals by two thin films of carbon. The procedure consists of evaporating sequentially: carbon, metal and again carbon layers on a substrate. Then, the "metal sandwich" is detached from the substrate and collected on the TEM grid, as described above. In the TEM, the carbon is removed from a sample region using an intense electron irradiation, this can be very time consuming, taking several hours (6–8 h) in some cases. Another important difference between Au and other studied metals is the NW lifetime; while for Au it is possible to generate NWs whose rupture takes several minutes, for Ag NW it takes a few seconds. This fact has lead to serious difficulties for imaging Ag wires because our time resolution is ~1/30 s.

This approach was successfully employed for several metals, revealing remarkable differences between macroscopic and nanometric systems, as well as the influence of the surface energies to determine the morphology of NWs. Also, mechanical properties and defect generation process in NWs could be analyzed. In the next sections, we will describe in more detail illustrative examples of atomic arrangement, morphologies and mechanical properties of metal NWs studied by time-resolved HRTEM.

4. Results and Discussion

In this study, we have analyzed NWs in the atomic size regime, addressing their structure, morphology and mechanical behavior. In the following, we will focus on a few illustrative examples of this kind of studies. In Section 4.1, the atomic arrangement and morphology of Au and Ag NWs will be discussed; we will show that NWs adopt only a few preferred atomic arrangement and that their mechanical properties are dependent of the NW elongation direction. Section 4.2 will focus on an important problem of nanometric systems: the formation of extended defects, which play important role on the mechanical properties [29]. Finally, the last Section 4.3 will concentrate on the properties of nanocontacts composed of a suspended chain of atoms [8, 9, 13, 29–31].

4.1. Atomic arrangement and morphology of NWs

4.1.1. Au NWs

The majority of studies of NWs properties have been focussed on Au due to its well-known ductility, which helps to generate the wires by mechanical stretching. Also, it is a low reactive material, what renders easy the sample preparation and manipulation. As its atomic number is high ($Z = 79$), the HRTEM images show high contrast. Finally, it is a fcc material, then the images can be easily understood (see Section 3).

Using the procedure described in the Section III, we have observed hundreds of Au NW elongation and break processes and, we have been able to make a reliable statistical analysis of their characteristics [8]. For the last stages of elongation and just before rupture we have found that Au NWs:

i. are crystalline and free of defects;
ii. adopt only three kinds of atomic configurations, characterized by the fact that one of the [1 1 1]/[1 0 0]/[1 1 0] zone axes lay parallel to the elongation direction (see Fig. 10).

The first observation is explained because defects are not energetically favored in nanostructures, as indicated by theoretical predictions [5, 6]. This question will be discussed more deeply in the Section 4.2.

As for the second observation, this phenomenon agrees with physical intuition of metal atom behavior, because it is well known that Au atoms try to maximize their packing, in other terms, they prefer to maximize the number of neighbors. Then, Au NWs adjust their atomic arrangement to have one of the [1 1 1]/[1 0 0]/[1 1 0] axes

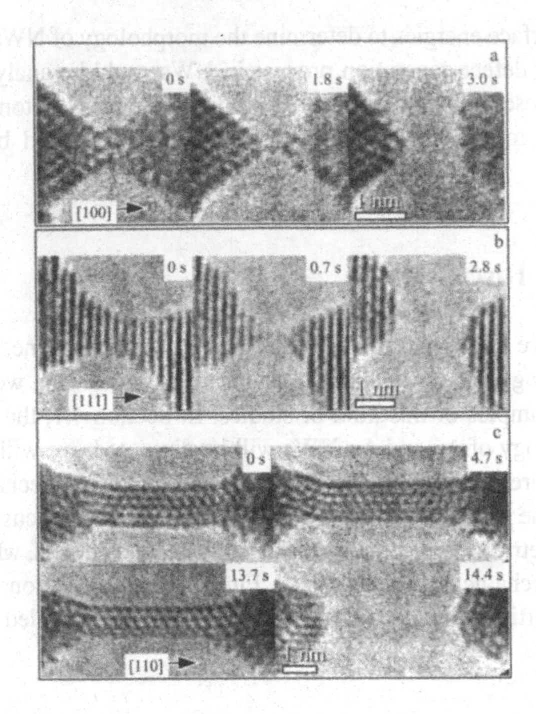

Fig. 10. High resolution images showing the time evolution of the three kinds of observed Au NWs: (a) [1 0 0], (b) [1 1 1] and (c) [1 1 0]. NWs generated along the [1 0 0] and [1 1 1] directions display bi-pyramidal morphology, while the ones formed along the [1 1 0] direction show rod-like shape.

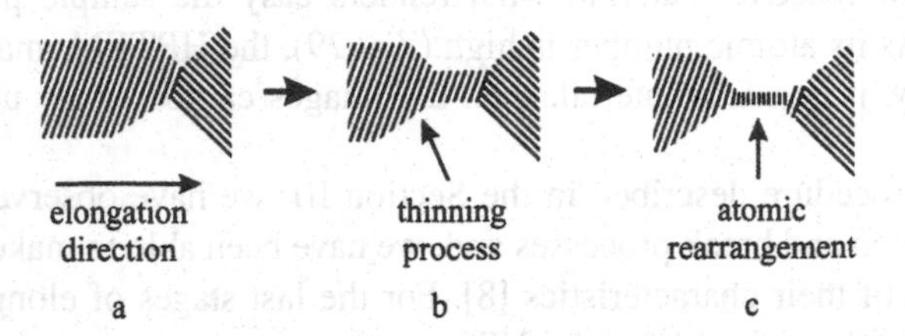

Fig. 11. Schema showing that the contact thinning process takes place mainly on the apex having a zone axis ([111]/[1 0 0]/[1 1 0]) closer to the elongation direction. When the NW is composed of a few atoms, its atomic arrangement adjusts to induce that the zone axis become parallel to the elongation direction.

aligned with the stretching direction, because they optimize the atom packing in the NW cross-section.

We must consider that the NWs were generated from a polycrystalline film. Then, the NW apexes may be two different crystal grains, which are not necessarily aligned; also the elongation does not follow any preferential direction (see Fig. 11(a)). In our experiments, it has been observed that the thinning process takes place preferentially at the apex where one of the [1 1 1]/[1 0 0]/[1 1 0] directions is the closest to the elongation direction (Fig. 11(b)). Also, when the nanoconstriction is composed of only a few atoms, the NW adjusts its atomic arrangement in the narrowest neck region so that the adopted orientation gets parallel to the stretching direction (Fig. 11(c)).

Besides the existence of three types of Au NWs, the constriction morphology and mechanical behavior can be divided in two groups:

i. NWs along [1 1 1] and [1 0 0] directions adopt bi-pyramidal shape constrictions [1, 32], evolving to form one-atom-thick contacts; usually they form a suspended atom chain structure (Fig. 10(a, b)) [13, 30, 31]. This evolution can be associated with a ductile behavior. ATCs usually were 2–4 atoms long and displayed bond distance ~0.36 nm. This topic will be discussed more deeply in Section 4.3.
ii. NWs along [1 1 0] direction display a rod-like morphology [19, 22, 23] with aspect ratio in the 3–6 range (Fig. 10(c)). They break abruptly when they are rather thick, ordinarily 3–4 atomic layers, as expected for a brittle material.

The morphologies adopted by the three kinds of Au NWs can be understood by using the geometrical Wulff construction [18]. As a basis for our analysis, we will use the expected shape of a Au nanoparticle, i.e. a truncated cubo-octahedra with regular hexagonal faces ($\gamma_{100}/\gamma_{111} = 2/(3)^{-1/2}$, see Fig. 6(c)). Looking at the cubo-octahedra along the [1 1 1] axis (Fig. 12(a)), we can deduce that a [1 1 1] NW pyramidal apex would be generated by alternating 3 (1 1 1) and 3 (1 0 0) facets. In these terms, the [1 1 1] NW can be visualized as the extension of the cubo-octahedra in the [1 1 1] direction, as shown in Fig. 12(a). When the NW is oriented along a [1 0 0] axis, the pyramid becomes composed by four (1 1 1) facets, as indicated in Fig. 12(b). On the other side, the rod-like shape of NWs along [1 1 0] axis can be explained by the existence of two families of low energy (1 1 1) planes that lie parallel to this direction (Fig. 12(c)). In consequence, pyramidal short constrictions become energetically unfavoured and an elongated shape is stabilized. Fig. 13 shows top and particular side views of proposed atomic arrangement for the three Au NW families. The rod marked 8/6 (Fig. 13(c)) corresponds to the [1 1 0] projection of an Au_{38} truncated cubo-ctahedron [18]; this pillar would be generated by the alternate stacking of atomic planes with 8 and 6 atoms, along the [1 1 0] axis (marked with different tones).

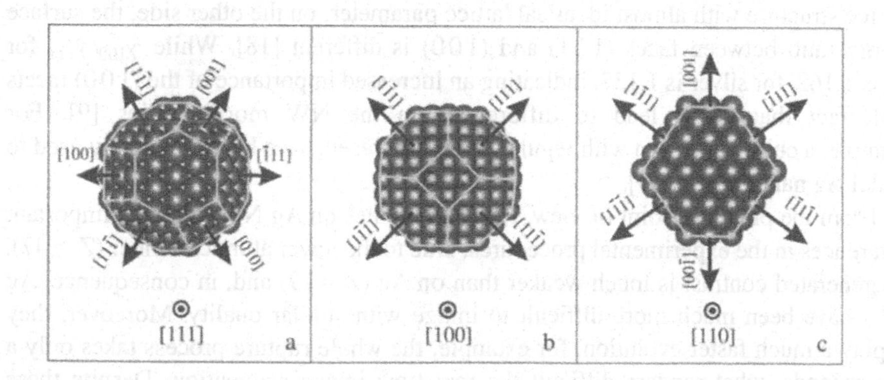

Fig. 12. Truncated cubo-octahedra with regular hexagonal faces viewed along the [111], [100] and [110] directions, respectively. The faceting determine the three different morphologies observed for Au NWs.

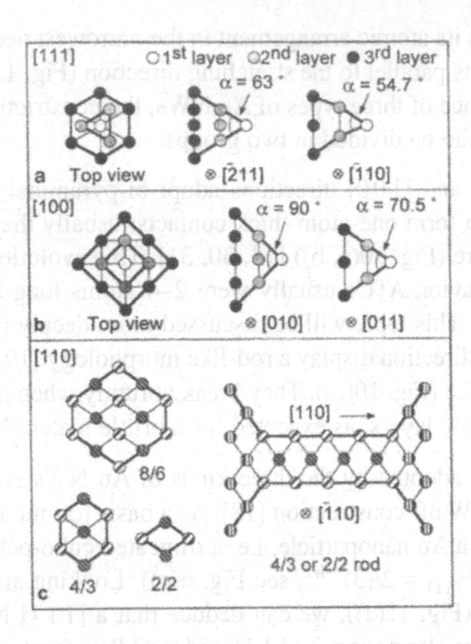

Fig. 13. Top and side views schemas of possible atomic arrangement of Au NWs. Observation axes and opening tip angles (α) are indicated. For the [1 1 0] direction (c), we have choosen a particular side view of a rod (correspond to 4/3 or 2/2), formed by 3 (1 1 1) planes (see text for explanation).

4.1.2. Ag NWs

The morphologies adopted by Au NWs were explained in simple terms of the surface energy of different facets [8]. In order to verify the validity of this model and get a deeper understanding of the mechanisms involved in the NW formation and rupture, it would be necessary to extended this kind of studies to other metals. For this purpose, Ag represents an excellent case in point, because it is quite similar to Au, adopting fcc structure with almost identical lattice parameter; on the other side, the surface energy ratio between faces (1 1 1) and (1 0 0) is different [18]. While $\gamma_{100}/\gamma_{111}$ for Au is 1.162, for silver is 1.137, indicating an increased importance of the (1 0 0) facets [18], fact that could lead to differences in the NW morphologies [9]. For example, a cubo-ctahedron with regular triangular facets have been frequently used to model Ag nanoparticles [9].

From the practical point of view, HRTEM studies on Ag NWs showed important differences in the experimental procedures. Due to the lower atomic number ($Z = 47$), the generated contrast is much weaker than on Au ($Z = 79$) and, in consequence, Ag NWs have been much more difficult to image with similar quality. Moreover, they display a much faster evolution, for example, the whole rupture process takes only a few seconds, what renders difficult the real time image acquisition. Despite these difficulties, we have been able to observe Ag NW elongation process; their analysis also revealed that they are crystalline and free of defects [2, 5, 8, 9]. Even so, they have shown some striking and unexpected structural behavior, which contrast with Au results discussed precedently [8, 13, 19, 21, 31]:

i. NWs generated along the [1 0 0] direction evolve to form one-atom-thick contacts (Fig. 14);
ii. [1 1 1] NWs breaks abruptly when they are rather thick.
iii. high aspect-ratio (5–10) rod-like NWs along the [1 1 0] direction are the most frequently observed morphology (Fig. 15);

In analogy to Au, suspended atom chains of Ag can be formed; nevertheless, they are much less frequently observed in our HRTEM experiments [8, 21, 31], being generated when one of the junction apexes is oriented along a [1 0 0] direction. The Ag chains are usually 2–4 atoms long with a bond lengths in the 0.33–0.36 nm range, quite similar to Au ATCs results [8, 13, 30, 31]. Here, an important difference should

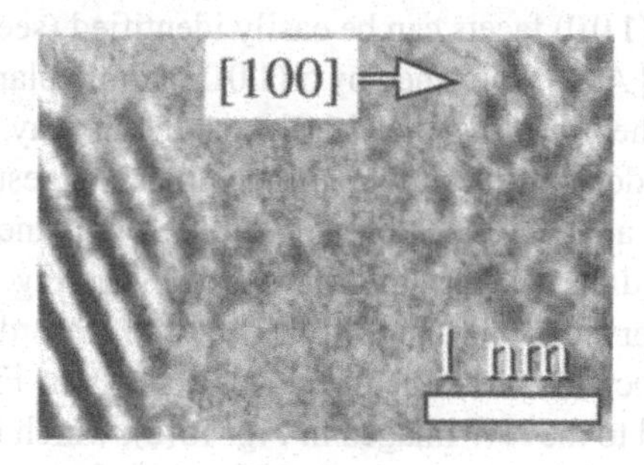

Fig. 14. Atomically resolved image of a Ag NW generated along the [1 0 0] direction. Note the formation of a one-atom-thick contact. Atoms appear dark.

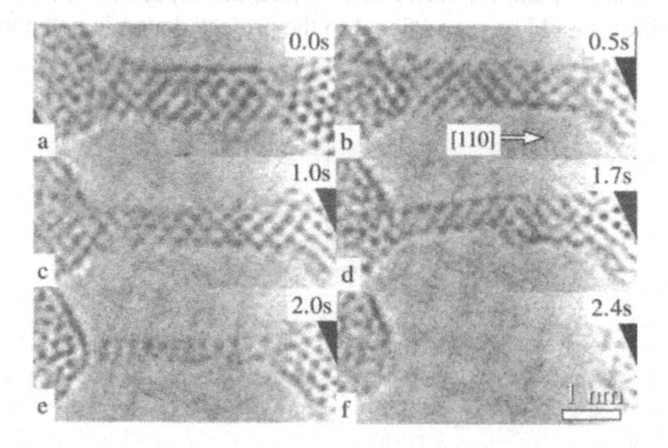

Fig. 15. Atomic resolution micrographs showing the elongation, thinning and rupture of a rod-like silver NW along the [1 1 0] direction (see text for explanation).

be noted: [1 1 1] Ag NWs display a fast and abrupt rupture, preventing the formation of ATCs (within our time resolution).

For Ag, the formation of NWs along the [1 1 0] direction is the predominant case, then it must be analyzed in more detail. Fig. 15 shows a series of snapshots of a complete elongation/thinning process of a pillar shaped NW. Atom lines along the NW are (200) planes (interplanar spacing is $d_{200} = 0.2$ nm). The first snapshot display a rod like contact composed of 5 atomic planes (Fig. 15(a)). Subsequently, it becomes thinner by losing one atomic plane at a time (Fig. 15(b), (c)). In Fig. 15(d), the thinnest region of the NW seems to have the same diameter as in the previous one (Fig. 15(c)), however, the contrast pattern is quite different. The image shows darker dots for the external planes (tube-like), whereas the central layer contrast is much weaker, indicating significant atomic rearrangement. Finally, in the last stage before breaking, the wire consists of two (200) atomic planes (Fig. 15(e)).

In order to use the Wulff method [8, 18] to model the morphology of Ag NWs, we must consider a truncated cubo-ctahedron with regular triangular (1 1 1) facets [33], where the relevance of (1 0 0) facets can be easily identified (see Fig. 16(a)) [33]. The cross section of a [1 1 0] Ag NW, formed by 5 (1 0 0) atomic planes, can be derived by looking at this cubo-ctahedron along the [1 1 0] axis. In this way, the rod-like NW seen in Fig. 15(b) can be modeled as the atomic arrangement suggested in Fig. 16(b). This NW is generated by the alternate stacking of two different planes containing 11 and 8 atoms (marked 11/8, and displayed with different colors in Fig. 16(b)). In the experiment, the rods are observed along the [1–10] axis and, in order to render easy the comparison, these projections are indicated as side views in Fig. 16. The same discussion can be extended to the NW imaged in Fig. 15(c), which is 3 (200) planes thick and composed by the stacking, along the [1 1 0] axis, of planes containing 4 and 3 atoms (Fig. 16(c)).

To model the next stage of the NW evolution (Fig. 15(d)), we should consider some basic concepts of HRTEM image formation for thin objects. In a first approximation, at the Scherzer defocus [16] and for such a thin object, the expected contrast at each atomic column position should be proportional to the projected atomic potential or, in

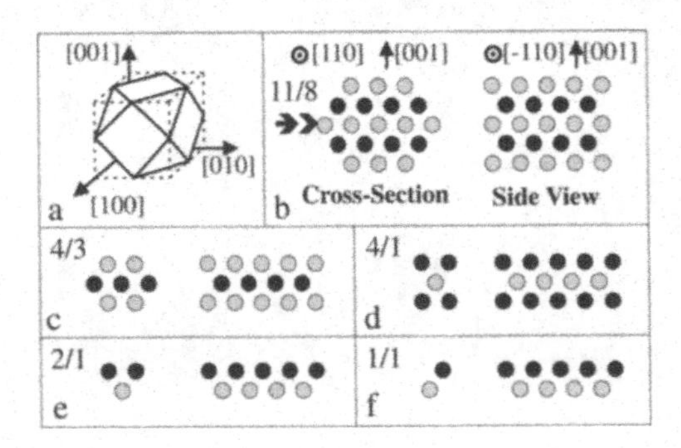

Fig. 16. (a) Truncated cubo-ctahedron with regular triangular (1 1 1) facets used to model the Ag NWs. (b–f) Scheme of the cross-sections and side views of proposed atomic arrangements for [1 1 0] rod-like Ag nanocontacts.

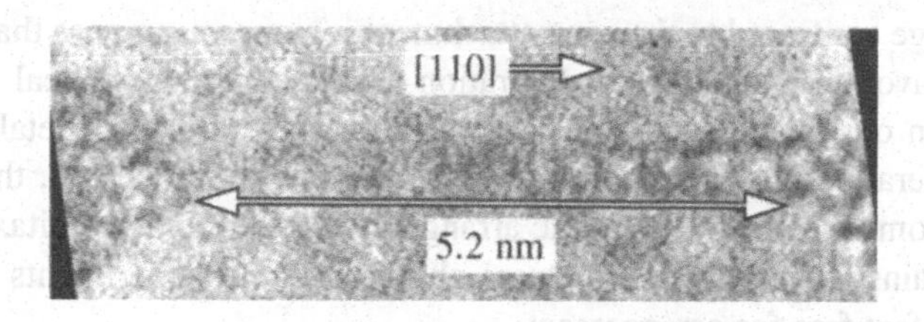

Fig. 17. HRTEM image of a high aspect-ratio (~12) pillar like [1 1 0] Ag NW (width ~0.4 nm and length ~5 nm).

other words, the number of atoms along the observation direction. In Fig. 15(d), the NW contrast is tube-like, with the external planes much darker than the central one, indicating that the central atomic columns should contain less atoms than the border. In this way, the contrast pattern in Fig. 15(d) could be obtained when the 4/3 NW loose the two most external lines of atoms, generating the 4/1 structure (Fig. 16(d)). Finally, the thinnest Ag [1 1 0] NW consists of two atomic planes (Fig. 15(e)), what can be generated by two different atomic arrangements: a 2/1 structure with a triangular cross-section (Fig. 16(e)) or the two parallel atom chains marked as 1/1 (Fig. 16(f)). In this case, the signal-to-noise ratio in the image does not allow us to identify which one is actually imaged. This question could be resolved using their electrical transport behavior, what indicates that the image in Fig. 15(e) should be associated to the 2/1 rod (Fig. 16(e)) [9]. This point will be discussed in more details in Section 5.

Also, we have been able to observe that the Ag rod-like contacts described as 4/1 can attain aspect ratio >10 (as exemplified in Fig. 17). Besides, in some nanocontact formation the apexes retracts, causing the 4/1 NW to elongate by a factor ~1.5–3, without thinning. This lengthening reflects the enhanced strength of this atomic configuration.

4.1.3. Summary

Dynamical HRTEM has allowed us to observe the elongation of metal NWs with atomic resolution. Just before rupture, NWs are crystalline and free of defects being properly described by a perfect fcc structure; in addition, they adopt merely a few kinds of atomic arrangement (3 for Au [8] and 3 for Ag [9]). The NW morphology and mechanical behavior depend strongly of the crystallographic direction of the NW stretching. The observed morphologies have successfully modeled using the geometrical Wulff construction; in these terms, the differences among Au and Ag NW structures are accounted for the different surface energy anisotropy for each of these metals.

It must be noted that some authors have reported the observation of helical Au NWs [34] using a similar experimental procedure. This apparent contradiction with the results described above can be explained by taking into account the crystallinity of the metal thin film used to produce the NWs. As mentioned before, we have used polycrystalline samples, while Takanayagi's group uses monocrystalline Au [1 0 0] thin films [19]. The very long range order of monocrystalline films allows the frequent formation of extremely long rod-like [1 1 0] wires formed by 3–4 atomic layers (aspect ratio >10, see Fig. 1 in [19]). As the number of atoms in the volume of long

NW is very large compared to the apex attachment region, we suppose that the energy balance will favor the helical reconfiguration predicted by theoretical calculations [35, 36, 37]. In our experiments the polycrystalline nature of the metal films, only allows the generation of much shorter [1 1 0] NWs (aspect ratio 3–6), then the apex crystal order dominates the NW atomic arrangement generating an epitaxial relation. This also explains why the NW structures observed in our experiments can be considered as a defect free fcc arrangement.

4.2. Defect generation in NWs

The experiments described in the precedent Sections deal with the tensile deformation of nanometric metal junctions. From many aspects (simple geometry, etc.), this case is very well suited to study the effect of finite size on mechanical deformation processes. In general terms, the plastic deformation of bulk materials is associated to the motion of defects, mainly dislocations. In contrast, force measurements have revealed that Au NWs display an enhanced strength (20–40 times stiffer than bulk), as expected for system free of dislocations [2–4]. Besides, theoretical and experimental studies have verified that the NW elongation occurs by elastic strain stages, followed by sharply defined yielding points, originated by structural reorganizations [1, 2, 5, 7, 32]. In fact, very low scale systems (~10 nm) should be completely free of dislocation because the required stress value is comparable to the intrinsic lattice strength [5, 6]; then, the plastic deformation of a NW should be basically associated to collective slips of entire atomic planes or to order-disorder transitions [5, 6]. From this point of view, the conventional macroscopic approaches to describe plastic behavior of materials can not be applied to NWs and, atomistic analyses are required. In this sense, dynamic HRTEM observations, as shown in previous sections, play a fundamental role to provide experimental insight into the microscopic mechanisms that are responsible for the NW deformation.

For example, Fig. 18 shows atomic resolution micrographs of the elongation of a Pt pyramidal tip along [1 1 1] direction. This sequence provides an instructive example of the formation of a planar defect. In Fig 18(b), the Pt NW was under mechanical tensile and, a minor shear stress (the upper apex is moving mainly up, but with a small lateral component to the right). Although the applied stress, the narrowest (upper) part of the lower apex (imaged with atomic resolution) stayed defect-free for

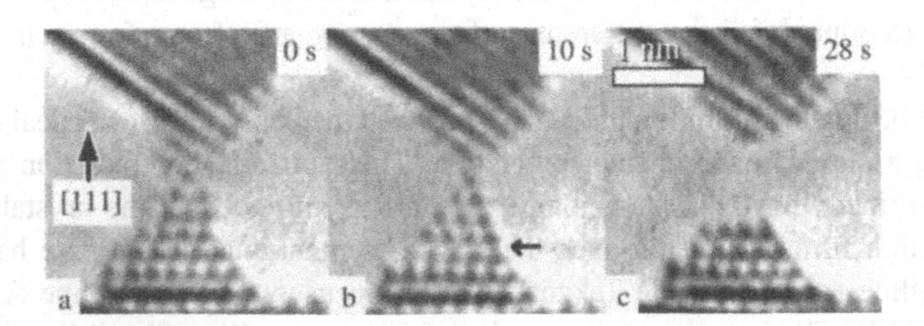

Fig. 18. (a–c) Serie of images showing a [1 1 1] Pt NW under tensile stress. An arrow indicates a twin defect generated at the 5th layer of the lower apex in (b).

the first layers counting from the tip; the generation of a planar defect was only observed at the 5th layer (indicated by an arrow in Fig. 18(b)). The defect can be easily identified as a twin by the angular change of the right apex side, which is due to transition from a (1 0 0) facet below the defect, to a (1 1 1) facet above (the opposite happens at the left apex side).

It would be very important to understand why the twin occurs at the 5th plane of the atomically sharp apexes and, no twin was observed to form closer to the tip. In order to do so, we have estimated the energy relaxed by the twin formation; the NW apex was modeled as formed by seven stacked layers along the [1 1 1] direction, with the twin defect located at different heights (see Figs. 19, 20). Basically, the tip shape was determined using the Wulff method and, for Pt only [1 1 1] and [1 0 0] facets should be considered [18, 33]. Besides surface energy considerations, additional geometrical constrains must be take into account to build the [1 1 1] Pt apex. Firstly, the tip of an atomically sharp [1 1 1] Pt apex should be identical to the Au one shown in Fig. 13(a) (Au and Pt surfaces tensions are quite similar [33]). In this configuration, the atom at the tip (1st layer) sits at the center of a regular triangle (2nd layer), that is itself lying on a hexagon (3rd layer). Secondly, the triangular shape of the second layer fixes the apex [1 0 0] facets to be two-atom-row wide (see Fig. 20(a)). When modeling a twinned apex, we must consider that the two crystalline regions (a top and a base) connected by the twin represent crystal domains with different orientation, but sharing the atomic positions at the twin plane (see Fig. 19). A twin can also be viewed as the rotation of one domain by 60° with respect to the other, the (1 1 1) (or the (1 0 0)) facet at the top domain become a (1 0 0) (or the (1 1 1)) facet at the base domain (see Fig. 19); in consequence the facet width at the twin (constrain determined by the atomically sharp tip, or top domain) fixes the base facet width (see Fig. 19).

The total energy of the tip was obtained using the Embedded Atom Method (EAM). The Pt potential derived following the approach of Foiles et al. [29, 38], which has been extensively applied to metal surfaces and clusters [39]. As a simplification, our calculation has not considered any structural relaxation, the position of the atoms follows a geometrically perfect/twinned fcc lattice [29]. These calculations aim to catch the fundamental role of tip morphology (or surface energy) on the energy balance of a twinned apex. The EAM energy estimation is summarized in Table 1; note

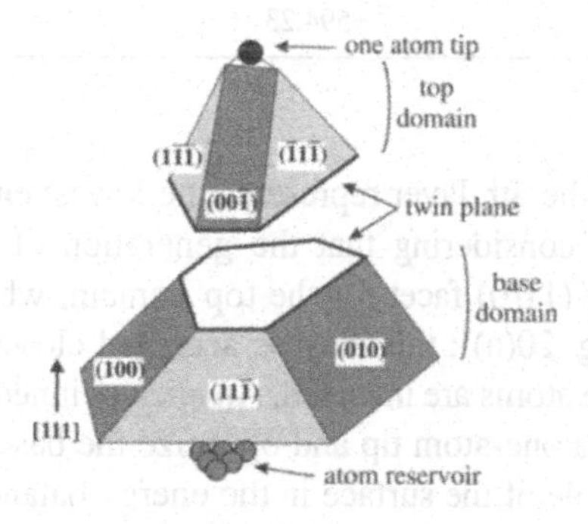

Fig. 19. Schema of the geometrical constrains used to build a twinned [1 1 1] Pt tip. See text for details.

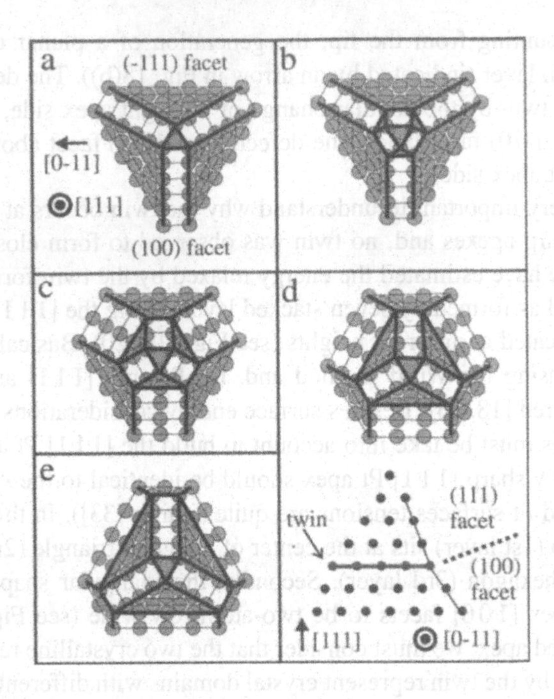

Fig. 20. Schemas of the proposed atomic arrangements for twinned Pt [1 1 1] tips. From (a) to (e) are shown tips with the twin at the: 2nd, 3rd, 4th, 5th and 6th layer, respectively. Thin lines indicate the atomic layer and thick lines mark the facet borders. In (d) is shown a top and side view for the apex twinned at the 5th layer, in order to perform a direct comparison with the experimental image in Fig. 18(b).

Table 1. EAM calculation of total cohesive energy for twinned Pt tips

Twin position (n)	Total energy (E_n) [eV]	$\Delta E = E_n - E_4$ [eV]
2	−592.17	+3.54
3	−592.17	+3.54
4	−595.71	
5	−595.49	+0.22
6	−594.23	+1.48

that the apex twinned at the 4th layer represents the lowest energy configuration. This can be easily explained considering that the generation of an atomically sharp tip requires a two-atom-row (1 0 0) facet for the top domain, what, in fact, represents an unfavorable faceting (Fig. 20(a)); this may be accepted closer to the tip, but becomes unsustainable when more atoms are involved. An apex twinned at the 4th layer can both fill the need to generate a one-atom tip and optimize the base surface energy; emphasizing the fundamental role of the surface in the energy balance of the NW structure.

To analyze the elastic energy released by the twin generation, we have considered that the initial stage of the tip is the lowest energy apex (the apex twinned at the 4th layer). The third column of Table 1 shows the total energy difference between this and the other configurations. From this analysis, the first available configuration to absorb accumulated stress is the tip with a twin at the 5th layer ($\Delta E = +0.22$ eV), in excellent agreement with our experimental observation (Fig. 18(b)).

In summary, these results show that the elongation of Pt NWs would induce the formation of twins located a few atomic planes away from the narrowest constriction, what implies that NWs would have the narrowest region free of extended defects (first 4 atomic planes for a [1 1 1] Pt NW). This behavior could indicate a transition size between the macroscopic plasticity mechanism and the nanometric scale where extended defects can not be sustained [1, 6]. This also explains why Au and Ag nanowires were observed to be free of defects in the atom size range (discussed in the Section 4.1).

4.3. Metal atom chain contacts

One of the most amazing observations associated to the study of NWs is the frequent formation of the thinnest possible wire, made of a linear chain of suspended metal atoms. The properties of such tiny system have attracted a great interest since their discovery [13, 30]. For monovalent metals, the ATCs show a conductance of 1 G_0 [8, 13, 30]. Then, in a conventional electrical experiment applying a voltage of 100 mV, a current of ~7.75 μA travels through the one-atom-thick conductor; this implies an impressive current density of the order of 10^{13} A/m^2. Another striking property of ATCs is the existence of very long interatomic bonds in the 0.29–0.36 nm range [8, 13, 22, 30, 31, 40] what is much longer than the nearest-neighbor distance, for example ~0.289 nm in macroscopic Au.

The experimental procedure described in Section III allows an easy generation and HRTEM observation of Au ATCs with atomic resolution [31]. In fact, suspended chains can also be observed for other metal such as Ag and Pt (see Fig. 21); the chains are formed by 2–4 suspended atoms and, they also show the long interatomic distances (0.3–0.36 nm) for all studied metals.

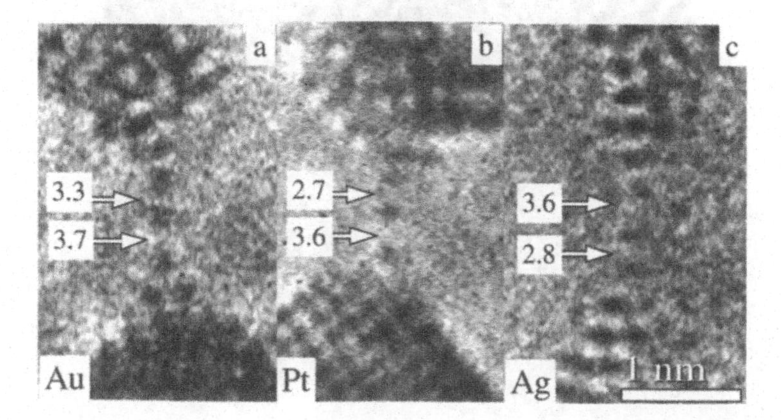

Fig. 21. HRTEM images of Au, Pt and Ag ATCs, respectively. Bond lengths are indicated in Å and error bar is ~0.2 Å. Note that in all cases, we observe interatomic bonds with lengths above 3 Å.

The observation of the individual atoms forming an ATC is a very instructive example of the nowadays capabilities of the HRTEM technique. However, it is quite intriguing that such a tiny structure could be stable inside the harsh environment of the 300 kV HRTEM, used in this study. In fact, ATCs show a remarkable stability and, their life-time is in tens of seconds range, even under electron bombardment of the microscope beam. The remarkable stability of ATC has also been evidenced by the experimental study of Rubio et al. [2], who have used a modified Atomic Force Microscope (AFM) to show that one-atom-thick metal contacts are much stronger (elastic modulus is 20–40 times) than macroscopic Au. The enhanced strength of ATCs can be indirectly deduced from the images displayed in Fig. 22. This sequence shows the formation, elongation and rupture of a Au atom-chain; just after rupture, the right apex makes a sudden displacement of ~0.7 nm (Fig. 22(d)). Although we can not measure the force [2], this quick apex movement indicates that the ATC was actually under tension and, that the atom chain was robust and strong enough to keep both sides of the contact together.

The dynamical HRTEM observation of atom chains has also revealed another amazing behavior: once the ATC is generated, it can move easily on the tip surface while holding the tensile effort. The sticking position on the apex changes frequently during the elongation; this behavior is easily identified in the images as a change of the angle between the ATC and the apex surface. Image sequences, in Fig. 23, show this behavior for Au (Fig. 23(a)) and Pt (Fig. 23(b)) ATCs. In both evolutions, the chain elongation and the apex movements are easily accommodated by a flexible chain attachment to the surface and, also, through the chain diffusion on the apex surface.

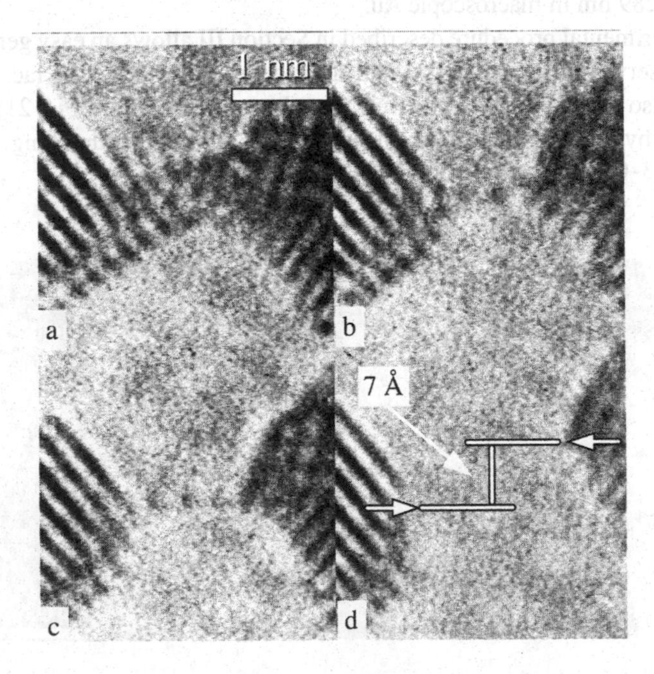

Fig. 22. Image sequence showing a sudden apex displacement just after the atom chain rupture. This event indicates that the ATC is robust and strongly bounded to the tips, holding the tips together, despite the stretching movement. (a–d) 0 s, 3.0 s, 5.4 s and 6.1 s, respectively.

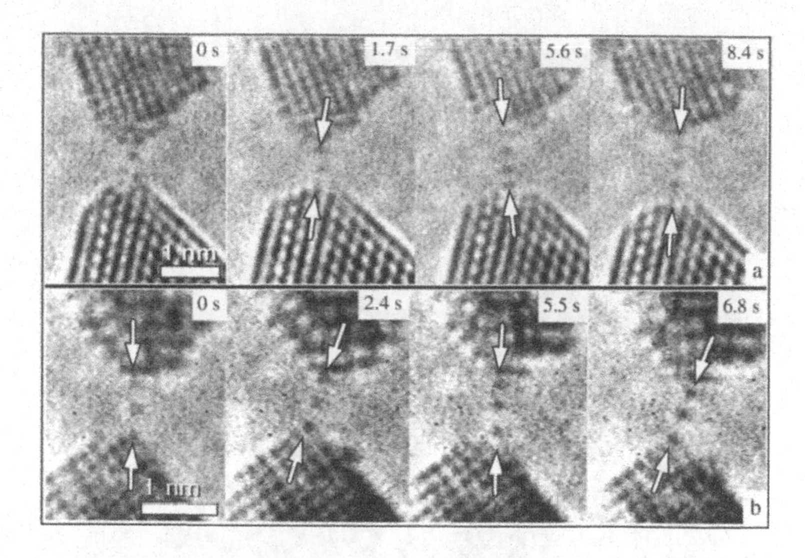

Fig. 23. Temporal evolution of the ATC position on the apex surface for (a) Au and, (b) Pt. Note the movement of the ATCs; arrows indicate the ATC sticking sites.

Although, the ATC extreme seems to move rather easily on the apex surface, we must remark that the chain is strongly bounded to the apexes, transmitting the tensile stress and inducing the formation of sharp and well-faceted tips. This apparently contradictory behavior can be easily understood by comparing it with the analogous situation of Au atoms on compact surfaces. In this case, atoms show a rather low diffusion barrier (surface potential corrugation of ~0.1 eV [41]), while in order to evaporate, the cohesion energy is much higher, being ~3.8 eV for bulk atoms and ~1 eV for surfaces atoms [2].

In order to understand the ATC generation process, it is very important to get atomistic information on the NW thinning process. Fig. 24 is an illustrative example of the time evolution of the Au ATC formation that reveals some fundamental characteristics of the process, in particular the relation between apex separation and the constriction structure. From Fig. 24(a) to (b), the apex retracted ~0.12 nm ($\pm$0.02 nm) and the NW rearranged to form a 2-hanging-atoms ATC with ~0.78 nm in length. Subsequently (Fig. 24(c)), the neck region rearranged (apex retraction below our experimental sensitivity) and, the ATC evolved to show 3 hanging atoms and became ~1.45 nm in length. Finally, an additional apex separation of ~0.10 nm induces the ATC rupture (Fig. 24(d)). These observations indicate that the ATC structure is due to a combination of atomic rearrangement in the apexes and the direct stretching of the chain. In this way, ATC length and interatomic distances can not be directly correlated or derived from the apex separation measurement, as some simple structural pictures predict [30, 42]. In fact, the complex interplay between apex rearrangement and ATC actual elongation has also been observed in molecular dynamic simulations [1, 4, 43].

Another fact revealing the complex relation between ATC structure and apex movement arises from an uneven bond length distribution along the suspended chain;

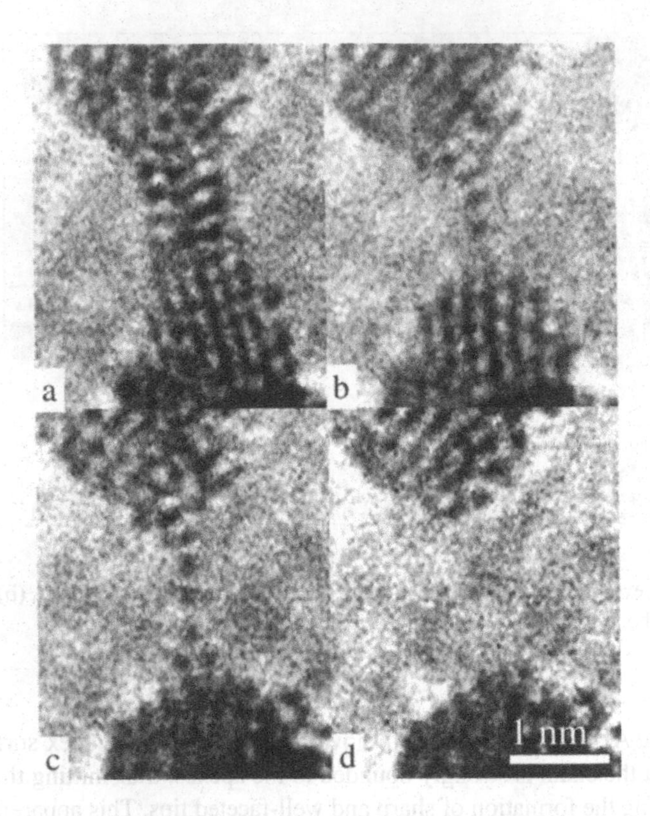

Fig. 24. HRTEM images of the evolution of a thick NW into an atom chain formed by 3 suspended Au atoms. From (a) to (d): 0 s, 0.64 s, 1.12 s and 3.72 s, respectively.

this contrasts with the conventional assumption for metal systems which predict a regular atom separation inside the stretched ATC. This property can be observed in Fig. 25, where it can be noted that the ATC sticking bonds (distance from apex to first ATC atom) are ~0.24 (± 0.02) nm, while bonds between suspended atoms are 0.34 and 0.36 nm. This phenomenon is observed in the majority of ATC images and, it is in agreement to recent theoretical calculations [43, 44]. Additional insights can be derived by looking carefully at the evolution of bond lengths during the tensile ATC deformation; Fig. 26 shows the effect of an inter-apex distance increase on an ATC formed by two suspended atoms. The elongation of ~0.15 nm induces the bond between the suspended atoms to increase from 0.28 to 0.35 ± 0.2 nm (absorbing about one half of the deformation), while the sticking bonds remain always shorter [44].

Among the intriguing questions raised by the discovery of the suspended atom chain structure, the observation of metal bond lengths above 0.31 nm has raised an intense controversy [8, 9, 13, 30, 31, 42, 44–50]. In fact, the physical origin of these long bonds has been a challenge for theoretical explanation. Many numerical techniques, with different level of complexity, have been applied [1, 32, 43, 44, 47, 51–58], but, the rupture of a pure metal chain is always predicted to occur for distances

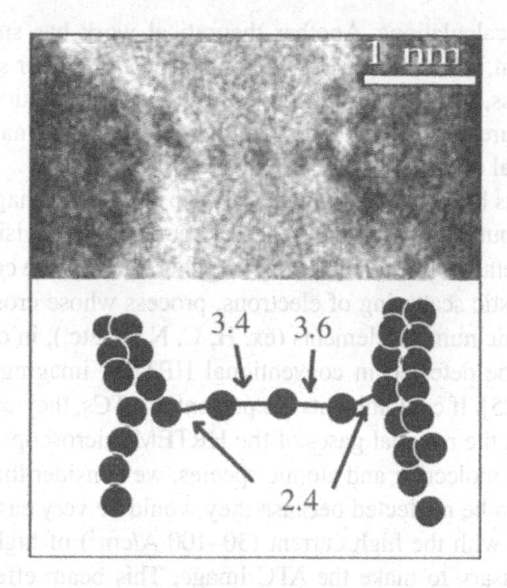

Fig. 25. HRTEM image of an atom chain formed by 3 Au atoms (upper part) and its schema showing the suspended atom chain bond distances in Å (lower part).

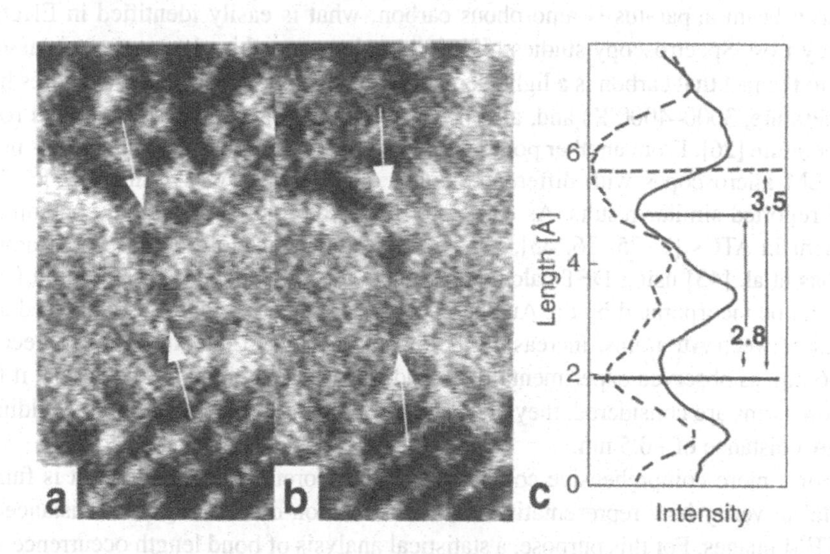

Fig. 26. Time-sequence of the elongation of an ATC formed by two atoms: (a) 0 s and (b) 3.2 s. At the right, we show the intensity profiles along the ATC in (a) and (b). Profiles directions are indicated with arrows and bond lengths between the suspended atoms (intensity minima) are indicated in Å.

(0.30–0.31 nm) below the values derived from HRTEM observations (0.34–0.36 nm). Several possible explanations of this discrepancy have been proposed; for example, in the experiments the ATCs are formed by a tensile deformation, what means that the atomic chain is under stress and not at equilibrium, fact that is not always easily incorporated in the calculations. Another theoretical work has suggested a ziz-zag geometry metal chain, where a spinning movement would render some atoms invisible [47]. Nevertheless, this model is disproved by image simulations [25] what indicate that this structure would be easily identified with the signal-to-noise ratio of reported experimental images [8, 13, 31].

As the long bonds have been mainly measured by HRTEM imaging, some authors have conjectured about the possible presence of spurious and invisible atoms intercalated in the linear metal atom chain [43, 49]. In general terms, the contrast of HRTEM images is due to elastic scattering of electrons, process whose cross-section is much weaker for low atomic number elements (ex. H, C, N, O, etc.); in consequence, these light atoms cannot be detected in conventional HRTEM imaging, as confirmed by image simulations [25]. If contaminants are present in ATCs, they could be most probably originated from the residual gases of the HRTEM microscope vacuum chamber. Among the different molecular and atomic species, we consider that very light atoms such as hydrogen can be neglected because they would be very easily swept off from the observed region with the high current (30–100 A/cm^2) of high energy electrons (200–300 kV) necessary to make the ATC image. This beam effect, actually based on moment transfer ("knock on") is a well-known phenomenon in the electron microscopy observation of low atomic number systems such as biological or polymeric samples [26]. The most frequent contamination product inside a high energy electron beam apparatus is amorphous carbon, what is easily identified in Electron Energy Loss Spectroscopy studies [59]. In simple terms, this phenomenon is associated to the fact that carbon is a light element which is not easily volatile (requires high temperature, 3000–4000 K) and, also because the pure carbon phase is solid at room temperature [26]. From another point of view, it must be emphasized that groups using HRTEM microscopes with different vacuum qualities (UHV [13] and HV [8, 31]) have reported similar results. As carbon seems to be the best potential contaminant element in ATCs [8, 25, 26, 43], it has been studied in more detail. For example, Legoas et al. [43] using DFT calculations, have recently shown that an atom of carbon can be incorporated by the Au chain. The ATC linear structure is conserved and, C sits between Au atoms, increasing the distance between the Au atoms to become ~0.36 nm, as observed experimentally. In addition, these authors predicted that if two carbon atoms are considered, they are also incorporated in the linear chain, yielding a Au–Au distance of ~0.5 nm.

For a more comprehensive comparison with theoretical calculations, it is fundamental to verify how representative is the observation of the long bond distances in HRTEM images. For this purpose, a statistical analysis of bond length occurrence was derived by analyzing a series of Au ATC video recording; the resulting histogram is displayed in Fig. 27. A total of 41 bond lengths between suspended atoms was included; this may seem rather low, however these represent a huge amount of data if we consider the difficulty of getting atomic resolved images from single atoms observed in a dynamical process. If we assume that all ATC bonds are due to contamination, for example carbon generating a Au-Au distance of 0.36 nm, the histogram should show a peak around 0.36 nm with a width associated to the estimated error bar (0.02 nm) (see Fig. 28(a)). In contrast, despite the rather low count histogram (Fig. 27), it indicates the existence of a continuous population of distances from

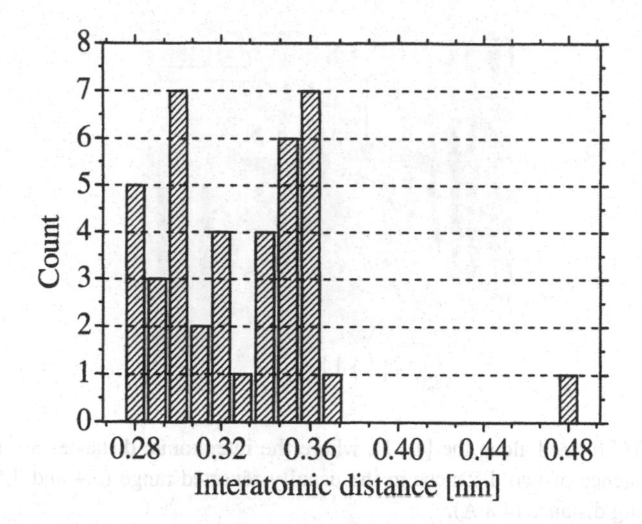

Fig. 27. Histogram of bond lengths measured between hanging atoms in Au ATCs. A total of 41 different distances are included. The analyzed ATCs were generated from wires elongated along different crystallographic directions.

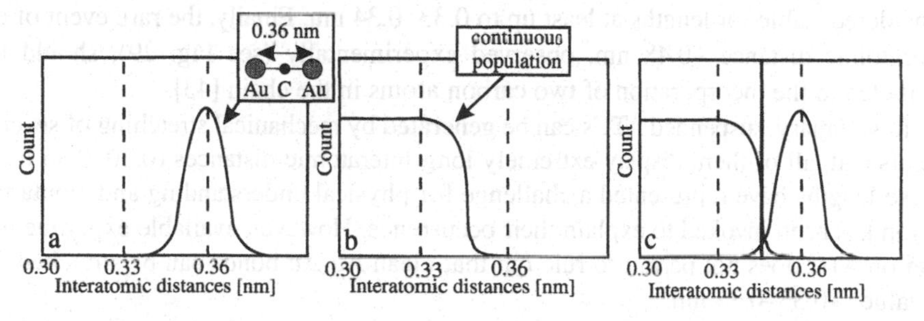

Fig. 28. Schematic hypothetical histograms of bond lengths between hanging atom in metals ATCs. (a) Expected curve profile when long bond lengths are exclusively due to carbon contamination. (b) Expected histogram for the case of clean metal bonds being elongated. (c) Expected profile assuming the coexistence of (a) and (b) regimes: superposition of the (a) and (b) histograms. In (c) we suggest a possible threshold distance between the two regimes (~0.33–0.34 nm).

0.29 to 0.36 nm (range much larger than the measurement error bar) as should be expected for elongated clean bonds (Fig. 28(b)). In consequence, the experimental bond length distribution do not corroborate the hypothesis of purely carbon contaminated ATCs during HRTEM work. However, it must be remarked that on the basis of the available experimental data and, in a rigorous sense, we can not fully eliminate the hypothesis that carbon is responsible for long interatomic distances around 0.36 nm within the measurement error bar (0.02 nm). The most logical scenario is to accept the hypothesis that both clean and contaminated bond could coexist in the experiments, yielding a histogram as shown in Fig. 28(c). Firstly, the physical origin of very long

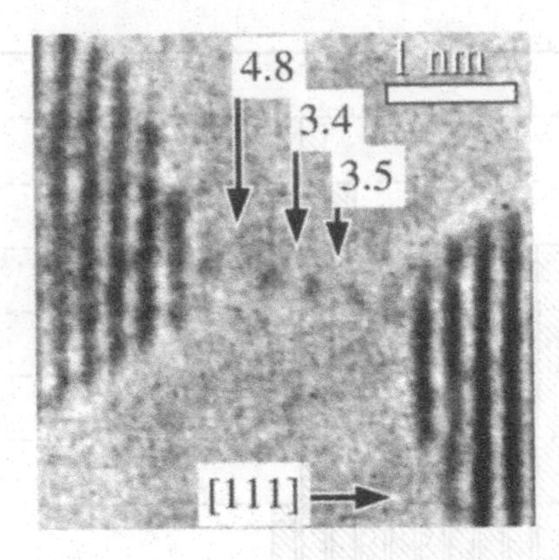

Fig. 29. Au ATC formed along the [1 1 1], where the interatomic distances are marked in Å. Note the coexistence of two distances in the usually reported range (3.4 and 3.5 Å) and one anomalously long distance (4.8 Å).

bonds around 0.36 nm is associated to carbon contamination. Secondly, considering the continuous bond length population and the error bar value for length determination, the experimental histogram indicates that clean metal–metal bond should be considered stable for lengths at least up to 0.33–0.34 nm. Finally, the rare event of an interatomic distance ~0.48 nm, observed experimentally (see Fig. 29), should be attributed to the incorporation of two carbon atoms in the chain [43].

In summary, suspended ATCs can be generated by mechanical stretching of several metals and, all of them display extremely long interatomic distances (0.30–0.36 nm). These lengths have represented a challenge for physical understanding and, contamination has been invoked to explain their occurrence. However, available experimental data on ATC does not permit to rule out that clean Au-Au bonds can be stretched up to values ~0.33–0.34 nm.

5. NW Structure vs. Quantized Conductance Behavior

The interpretation of the quantized conductance behavior of metal NWs [8, 9, 11, 12, 21, 46, 52, 60] is one of the main contributions derived from the study the atomic arrangement of nanometric metal wires generated by mechanical elongation. The usual experimental procedure to probe the electrical transport of NWs is performed by putting in contact two clean metal surfaces and subsequently separating them. During the contact stretching and just before rupture, atomic size NWs are generated. The conductance measured during this process displays flat plateaus connected by abrupt jumps whose value is approximately a conductance quantum $G_0 = 2e^2/h$ (where e is the electron charge and h is Planck's constant) [11, 12].

It must be emphasized that the study of such atomic size structure requires attention with the experimental conditions and, much care with vacuum cleanness conditions is extremely important to generate reliable experimental measurements on NW properties [8, 56]. Several approaches have been used to study metal NW electrical

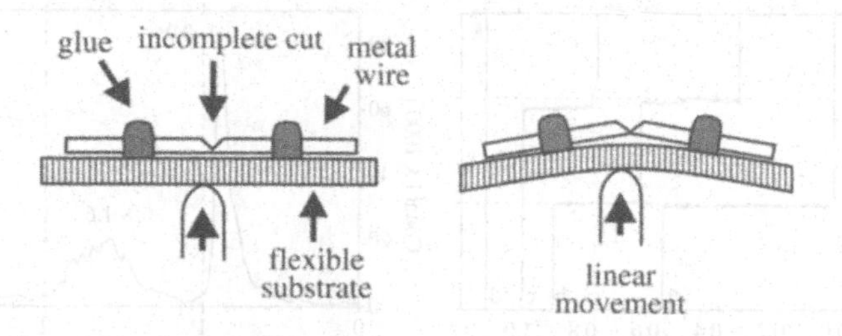

Fig. 30. Schema of the Mechanically Controllable Break Junction instrument. A substrate is bent to break the wire in the pre-fragilized position. Then, the separated parts are put together again and retracted, in a controlled way, in order to generate NWs.

transport behavior, such as Scanning Tunneling Microscope [52, 53] and Mechanically Controllable Break Junction setup (MCBJ) [46, 61–64]. In particular, we have used a MCBJ operated in ultra-high-vacuum (UHV-MCBJ) to study the conductance of metal NWs [63, 64]. In this approach, a macroscopic wire is glued on a flexible substrate in two points; then it is fragilized between the two fixing parts by an incomplete cut. By bending the substrate *in situ* in the UHV, we break the wire and produce two clean metal surfaces (see Fig. 30). Using the same bending movement, the fresh tips are put together and separated repeatedly in order to generate and elongate NWs. Its important to remark that in this configuration the NWs are generated from surfaces obtained in UHV ($<10^{-8}$ Pa), so it is expected to have a clean sample for few hours [63].

The electrical transport of the metal NWs has been measured using the two-points configuration. This implies that the conductance measurement probes the NW itself (the narrowest region of the contact) and the two leads (NW apexes). Although, the NW should show conductance quantization ($G = \alpha G_0$, where the factor α is an integer), the influence of the leads should act as additional resistance, diminishing the conductance and leading to non-integer α values. Experimentally, α is frequently closer to integer values but slightly lower.

The left side of Fig. 31 shows examples of conductance measurements obtained for Au using the UHV-MCBJ [63, 64]. The curves display abrupt jumps separating horizontal plateaus, whose values are very close to the expected integer multiples of G_0, indicating the quantized character of the electrical transport. However, all them display a different profile [2, 46, 52–54, 56, 61, 65]. In particular, this graphic displays four types of profiles that are obtained in experiments of conductance in the [0, 2.7] G_0 range: a) merely a plateau at 1 G_0; (b) plateaus at 1 and 2 G_0; (c) only a plateau at 2 G_0; (d) abrupt rupture, no plateaus below 2.7 G_0.

The variety of the conductance profiles occurs because in this kind of experiment we can not control the structure of the NW; then, each new measurement should correspond to a new NW with a different atomic arrangement [8, 9, 21]. To overcome this difficulty, most NW studies rely on the analysis of average behaviors of many curves. The most frequently used procedure consist in building histograms from each individual electrical transport measurement, where it is represented the number of times that one value of conductance occurs. Then, the so-called global histogram is

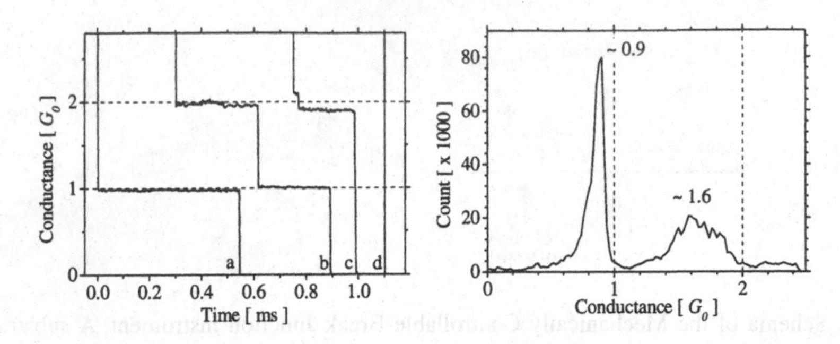

Fig. 31. Left side: typical example of conductance behavior of Au NWs obtained using an UHV-MCBJ. (a) only one plateau at ~1 G_0. (b) plateaus at ~1 and ~2 G_0. (c) single plateau at ~2 G_0. (d) abrupt NW rupture. Right side: global histogram of a series of 500 Au NW conductance measurements. Note that the histogram peaks are below the expected quantized values.

constructed [54, 56, 66], what is formed by the simple linear addition of individual histograms from a series of measurements (see Fig. 31 right side for one example for Au NWs). The presence of peaks close to integer multiples of the conductance quantum has been considered for many authors as the proof of the quantized electron transport in metal NWs [54, 56, 66]. Even so, this statistical procedure hinders a detailed study of the NW transport properties, mainly the influence of the atomic structure in the electrical transport [8, 9, 21, 46].

In order to try to correlate the NW conductance with its atomic structure, we should consider that the HRTEM studies have showed that, before breaking, Au NWs adopt merely three kinds of atomic arrangements with two different mechanical properties. These observations limit to a very few the numbers of possibilities to analyze, and, provide us with a very good basis for modeling the electrical transport of Au NWs from an atomistic point of view.

The first point to be considered is that the conductance behavior is mainly determined by the shape of the narrowest region of the NW. It has been shown that one-atom-thick Au contacts should have a conductance close to a single quantum [11, 12, 13, 30, 44, 51, 67–71]. HRTEM studies have revealed that NWs generated along [1 1 1] and [1 0 0] axes form one-atom-thick contacts just before rupture and, then, we should expect their lowest conductance as a plateau close to 1 G_0 in the conductance curve. On the other hand, rod-like [1 1 0] contacts should not display the 1 G_0 plateau, because they break abruptly rather thick, when formed by 3–4 atomic layers. This criterion allows us to discriminate between the formation of pyramidal cases and rod like wires. It would be very interesting to also recognize if 1 G_0 plateaus are generated by either the [1 1 1] or the [1 0 0] NW. One possible attempt can be done by considering that the NW cross-section evolution is somewhat related to the conductance profile [32, 52, 54, 55]. Fig. 13 shows that [1 1 1] NWs have lower opening angles than the [1 0 0] NWs, or in a rather simple view, before the formation of the one-atom-thick contact, [1 1 1] NWs would have a contact composed by 3 Au atoms while [1 0 0] NWs would have 4 atoms. In this way, we should expect that [1 1 1] NWs should be the better candidates to pass through both 1 and 2 G_0 conductance states.

Table 2. Statistical analysis of the occurrence of each kind of Au NW conductance evolution

	[100] NWs only 1 G_0	[1 1 1] NWs 1 G_0 and 2 G_0	[1 1 0] NWs no 1 G_0
Expected values	23% (3/13)	31% (4/13)	46% (6/13)
Series 1	(18 ± 2)%	(34 ± 2)%	(48 ± 2)%
Series 2	(23 ± 2)%	(33 ± 2)%	(44 ± 2)%
Series 3	(19 ± 2)%	(31 ±2)%	(50 ± 2)%

To verify the proposed model we can perform a statistical analysis of the occurrence of each conductance pattern. In the MCBJ experiment, we do not control the crystal orientation of the apexes forming the NW, nor the direction of elongation. Also, we have observed in HRTEM that NWs elongated along an arbitrary direction adopt the atomic arrangement (NW type) of the closest zone axis (see Section 4.1); so, we can consider that there is no preferential configuration. The occurrence rate for each NW type would be proportional to the multiplicity of their zone axis: 4 for the [1 1 1] NWs, 3 for [1 0 0] and 6 for [1 1 0]. In summary, 4/13 of the generated NWs will be oriented along [1 1 1] direction and will show both 1 and 2 G_0 plateaus (Fig. 31(b)); 3/13 will be [1 0 0] NWs and will display only the 1 G_0 plateau (Fig. 31(a)); and 6/13 will be rod-like contacts and will not show the 1 G_0 plateau (Fig. 31(c), 31(d)).

We have counted how many conductance curves showed each of the conductance behavior patterns in several series of 500 consecutive NW generations. Table 2 displays the result of the analysis and, a remarkable agreement with our statistical prediction can be verified, corroborating the proposed correlation between atomic arrangement and conductance behavior [8, 10]. This result emphasizes the importance of taking in account the atomic arrangement of the Au NWs to understand their conductance behavior.

Recently, Rego et al. [10] have reported an thorough study of Au NWs properties combining conductance measurements from the UHV-MCBJ, real-time HRTEM imaging of the NW atomic rearrangements and, also, calculations of the electronic transport properties for the observed NW structures. For this purpose, they have used an approach introduced by Emberly and Kirczenow [72, 73] that is based on the Extended-Hückel-Theory (EHT), the latter being employed to obtain the molecular orbitals (MO) of the Au NWs [74]. In this way, they derived a consistent correlation between the structural evolution of the three kinds of Au NW's and the conductance behavior in the 0–3 G_0 conductance range.

Another important example to be considered is Ag, because the generated NWs are quite different from Au ones; it should be expected that these differences would reflect in the conductance behavior. Fig. 32 shows some typical examples of conductance measurements for Ag NWs and, also, a global histogram obtained from 500 consecutive conductance measurements. This histogram shows large peaks at 1 G_0, ~2.4 G_0 and ~4 G_0, while a similar result for Au shows peaks at ~0.9 and ~1.6 G_0 and no one is observed at ~2.4 G_0 (see Fig. 31).

To correlate the experimental results of the conductance and structural behaviors, we have performed theoretical conductance calculations based on the EHT method for

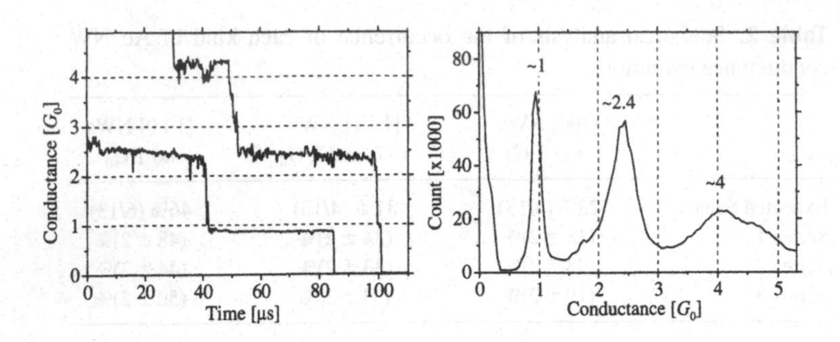

Fig. 32. Right side: Typical conductance curves obtained for Ag NWs. Left side: Global histogram of conductance where the peaks are located at 1, 2.4 and 4 G_0.

the proposed structures derived in Section 4.1.2 [9]. The calculations show that [1 0 0] Ag ATCs display the expected conductance of ~1 G_0 [8, 13]. The 1/1 [1 1 0] rod (Fig. 17(f)), composed of two parallel atom chains, shows a conductance close to 1.8 G_0. As the Ag global histogram (Fig. 32) does not show a peak associated with this value, it can be deduced that this atomic arrangement may not be stable and then it will not occur in our experiments (see Section 4.1.2). As for the 2/1 [1 1 0] NW, the model predicts a conductance of ~2.5 G_0, what is in remarkable agreement with the main peak of the conductance histogram. Finally, the conductance calculation for the 4/1 rod yields G ~3.8 G_0, which should be associated with the observed conductance peak at ~4 G_0. In this way, we have been able to determine the atomic structures yielding the observed conductance peaks.

In summary, when a conductor diameter is of the order of the electron wavelength, the wire can be viewed as an electron wave guide, with well defined transmission modes, or conduction channels. The number of occupied channels determines the conductance value and this number is mainly determined by the narrowest wire region; the conductance behavior is associated to the NW structural evolution. We have illustrated this fact for Au and Ag, whose monovalent nature could lead to expect a similar electronic structure. However, the conductance behavior is quite different, because the electrical transport properties of NWs is determined by their preferred atomic structure, emphasizing the importance of considering the atomic arrangement in conductance studies of nanometric systems.

6. Conclusion

Dynamical observations using HRTEM has allowed us to perform a thorough analysis of the atomic arrangement evolution of Au, Ag, and Pt NWs generated by mechanical stretching. Atomically resolved images have revealed that just before rupture, NWs are crystalline and free of defects, adopting only a reduced number of structures whose formation is determined by surface energy (faceting pattern) considerations. In particular, HRTEM has shown to be a fundamental tool to identify and analyze atom-size structures such as suspended chains.

The structural information derived from HRTEM observations was essential to understand the variable conductance behavior of NWs generated by mechanical stretching [8, 9]. These results emphasize the need of considering an atomistic description when analyzing the transport properties of nanoscopic systems.

Acknowledgement

The authors acknowledge the invaluable help of the LNLS staff. Also, the authors are grateful to L. G. C. Rego, A. R. Rocha, S. B. Legoas and D. S. Galvão for the theoretical support. VR and DU are indebted to LNLS, FAPESP (contracts 96/12546-0, 96/08353-2, 98/13501-6, 01/13025-4 and 96/04241-5) and CNPq for funding.

References

1. U. Landman, W. D. Luedtke, N. A. Burnham and R. J. Colton, *Science* **248** (1990) 454.
2. G. Rubio, N. Agraït and S. Vieira, *Phys. Rev. Lett.* **76** (1996) 2302.
3. S. P. Jarvis, M. A. Lantz, H. Ogiso, H. Tokumoto and U. Dürig, *Appl. Phys. Lett.* **75** (1999) 3132.
4. G. Rubio-Bollinger, S. R. Bahn, N. Agraït, K. W. Jacobsen and S. Vieira, *Phys. Rev. Lett.* **87** (2001) 26101.
5. U. Landman, W. D. Luedtke and R. N. Barnett in *Nanowires*, Edited by P. A. Serena and N. García, Kluwer Academic Publishers, Dordrecht (1997) pp. 109–132.
6. U. Durig in *Nanowires*, Edited by P. A. Serena and N. García, Kluwer Academic Publishers, Dordrecht (1997) pp. 275–300.
7. J. A. Torres, E. Tosatti, A. Dal Corso, F. Ercolessi, J. J. Kohanoff, F. D. Di Tolla and J. M. Soler, *Surf. Sci.* **426** (1999) L441.
8. V. Rodrigues, T. Fuhrer and D. Ugarte, *Phys. Rev. Lett.* **85** (2000) 4124.
9. V. Rodrigues, J. Bettini, A. R. Rocha, L. G. C. Rego and D. Ugarte, *Phys. Rev. B* **65** (2002) 153402.
10. L. G. C. Rego, A. R. Rocha, V. Rodrigues and D. Ugarte, *Phys. Rev. B* **67** (2003) 045412–045421.
11. J. M. van Ruitenbeek in *Metal Clusters at Surfaces*, Edited by K.-H. Meiwes-Broer, Springer-Verlag, Berlin Heidelberg, New York (2000) pp. 175–210.
12. N. Agraït, A. Levy Yeyati and J. M. van Ruitenbeek, *Physics Reports* **377** (2003) 81–279.
13. H. Ohnishi, Y. Kondo and K. Takayanagi, *Nature* **395** (1998) 780.
14. S. Ino, *J. Phys. Soc. Jpn.* **21** (1966) 346–351.
15. Z. L. Wang, in *Characterization of Nanophase Materials*, Edited by Z. L. Wang, Wiley-VCH, Weinheim (2000) pp. 37–80.
16. D. B. Williams and C. B. Carter, *Transmission Electron Microscopy*, Plenum Press, New York (1996).
17. G. Wulff, *Z. Kristallog.* **34** (1901) 449.
18. L. D. Marks, *Rep. Progr. Phys.* **57** (1994) 603.
19. Y. Kondo and K. Takayanagi, *Phys. Rev. Lett.* **79** (1997) 3455.
20. P. J. Goodhew, in *Thin Foil Preparation for Electron Microscopy*, Edited by A. M. Glauert, Elsevier Amsterdam (1985).
21. V. Rodrigues and D. Ugarte, *Europ. J. Phys. D* **16** (2001) 395.
22. T. Kizuka, *Phys. Rev. Lett.* **81** (1998) 4448.

23. T. Kizuka, *Phys. Rev. B* **57** (1998) 11158.
24. R. Lohmus, D. Erts, A. Lohmus, K. Svensson, Y. Jompol and H. Olin, *Phys. Low-Dimens. Struct.* **3** (2001) 81.
25. H. Koizumi, Y. Oshima, Y. Kondo and K. Takayanagi, *Ultramicroscopy* **88** (2001) 17.
26. L. Reimer, *Transmission Electron Microscopy*, Springer Verlag, Berlin (1997).
27. D. Ugarte, *Nature* **359** (1992) 707.
28. D. Ugarte, *Chem. Phys. Lett.* **209** (1993) 99.
29. V. Rodrigues and D. Ugarte (submitted).
30. A. I. Yanson, G. R. Bollinger, H. E. van den Brom, N. Agraït and J. M. van Ruitenbeek, *Nature* **395** (1998) 783.
31. V. Rodrigues and D. Ugarte, *Phys. Rev. B* **63** (2001) 073405.
32. M. R. Sørensen, M. Brandbyge and K. W. Jacobsen, *Phys. Rev. B* **57** (1998) 3283.
33. B. D. Hall, M. Flüeli, R. Monot and J.-P. Borel, *Phys. Rev. B* **43** (1991) 3906.
34. Y. Kondo and K. Takayanagi, *Science* **289** (2000) 606.
35. E. Tosatti and S. Prestipino, *Science* **289** (2000) 561.
36. E. Tosatti, S. Prestipino, S. Kostlmeier, A. Dal Corso and F. D. Di Tolla, *Science* **291** (2001) 288.
37. E. A. Jagla and E. Tosatti, *Phys. Rev. B* **64** (2001) 205412.
38. S. M. Foiles, M. I. Baskes, M. S. Daw, *Phys. Rev. B* **33** (1986) 7983.
39. P. R. Schwoebel, S. M. Foiles, C. L. Bisson and G. L. Kellogg, *Phys. Rev. B* **40** (1989) 10639.
40. Y. Takai, T. Kawasaki, Y. Kimura, T. Ikuta and R. Shimizu, *Phys. Rev. Lett.* **87** (2001) 106105.
41. A. Pimpinelli, J. Villain and C. Godreche, *Physics of Crystal Growth*, Cambridge University Press, Cambridge (1998) p. 121.
42. R. H. M. Smit, C. Untiedt, A. I. Yanson and J. M. van Ruitenbeek, *Phys. Rev. Lett.* **87** (2001) 266102.
43. E. Z. da Silva, A. J. R. da Silva and A. Fazzio, *Phys. Rev. Lett.* **87** (2001) 256102.
44. H. Häkkinen, R. N. Barnett and U. Landman, *J. Phys. Chem. B* **103** (1999) 8814.
45. S. B. Legoas, D. S. Glavão, V. Rodrigues and D. Ugarte, *Phys. Rev. Lett.* **88** (2002) 076105.
46. J. M. Krans, J. M. van Ruitenbeek, V. V. Fisun, I. K. Yanson and L. J de Jongh, *Nature* **375** (1995) 767.
47. D. Sánchez-Portal, E. Artacho, J. Junquera, P. Ordejon, A. Garcia and J. M. Soler, *Phys. Rev. Lett.* **83** (1999) 3884.
48. H. Häkkinen, R. N. Barnett, A. G. Scherbakov and U. Landman, *J. Phys. Chem. B* **10** (2000) 9063.
49. N. V. Skorodumova and S. I. Simak, *Condens. Matter* (2002) 0203162.
50. C. Untiedt, A. I. Yanson, R. Grande, G. Rubio-Bollinger, N. Agraït, S. Vieira and J. M. van Ruitenbeek, *Phys. Rev. B* **66** (2002) 085418.
51. T. N. Todorov and A. P. Sutton, *Phys. Rev. Lett.* **70** (1993) 2138.
52. L. Olesen, E. Lægsgaard, I. Stensgaard, F. Besenbacher, J. Schiøtz, P. Stoltze, K. W. Jacobsen and J. K. Nørskov, *Phys. Rev. Lett.* **72** (1994) 2251.
53. J. I. Pascual, J. Méndez, J. Gómez-Herrero, A. M. Baró, N. García N, U. Landman, W. D. Luedtke, E. N. Bogachek and H. P. Cheng, *Science* **267** (1995) 1793.
54. M. Brandbyge, J. Schiøtz, M. R. Sørensen, P. Stoltze, K. W. Jacobsen, J. K. Nørskov, L. Olesen, E. Lægsgaard, I. Stensgaard and F. Besenbacher, *Phys. Rev. B* **52** (1995) 8499.
55. A. M. Bratkovsky, A. P. Sutton and T. N. Todorov, *Phys. Rev. B* **52** (1995) 5036.
56. K. Hansen, E. Lægsgaard, I. Stensgaard and F. Besenbacher, *Phys. Rev. B* **56** (1997) 2208.
57. M. Okamoto and K. Takayanagi, *Phys. Rev. B* **60** (1999) 7808.
58. S. R. Bahn and K. W. Jacobsen, *Phys. Rev. Lett.* **87** (2001) 266101.
59. R. F. Egerton, *Electron Energy-Loss Spectroscopy in the Electron Microscope*, Plenum Press, New York (1996).

60. J. M. Krans, C. J. Muller, N. van der Post, F. R. Postma, A. P. Sutton, T. N. Todorov and J. M. van Ruitenbeek, *Phys. Rev. Lett.* **74** (1995) 2146.
61. C. J. Muller, J. M. van Ruitenbeek and L. J. de Jongh, *Phys. Rev. Lett.* **69** (1992) 140.
62. C. J. Muller, J. M. van Ruitenbeek and L. J. de Jongh, *Physica C* **191** (1992) 485.
63. V. Rodrigues, Master Thesis, Universidade Estadual de Campinas (1999).
64. V. Rodrigues and D. Ugarte, *Rev. Sci. Instrum.* (submitted).
65. J. L. Costa-Krämer, N. Garcia, P. Garcia-Mochales and P. A. Serena, *Surf. Sci.* **342** (1995) L1144.
66. J. L. Costa-Krämer, P. Garcia-Mochales, M. I. Marques and P. A. Serena in *Nanowires*, Edited by P. A. Serena and N. García, Kluwer Academic Publishers, Dordrecht (1997) pp. 171–190.
67. H. van Houten and C. Beenakker, *Phys. Today* **49** (July 1996) 22.
68. C. J. Muller, J. M. Krans, T. N. Todorov and M. A. Reed, *Phys. Rev. B* **53** (1996) 1022.
69. C. Sirvent, J. G. Rodrigo, S. Vieira, L. Jurczyszyn, N. Mingo and F. Flores, *Phys. Rev. B* **53** (1996) 16086.
70. A. Levy Yeyati, A. Martín-Rodero and F. Flores, *Phys. Rev. B* **56** (1997) 10369.
71. E. Scheer, N. Agraït, J. C. Cuevas, A. Levi Yeyati, B. Ludoph, A. Martín-Rodero, G. R. Bollinger, J. M. van Ruitenbeek and C. Urbina, *Nature* **394** (1998) 154.
72. E. G. Emberly and G. Kirczenow, *Phys. Rev. B* **58** (1998) 10911.
73. E. G. Emberly and G. Kirczenow, *Phys. Rev. B* **60** (1999) 6028.
74. S. P. McGlynn, L. G. Vanquickenborne, M. Kinoshita and D. G. Carroll, *Introduction to Applied Quantum Chemistry*, Holt, Rinechart and Winston Inc., New York (1972).w

Chapter 7

Metal Nanowires Synthesized by Solution-Phase Methods

Yugang Sun and Younan Xia
Department of Chemistry, University of Washington, Seattle, WA 98195-1700, USA

1. Introduction

One-dimensional (1D) nanostructures (such as wires, rods, belts and tubes) have been a subject of intensive research due to their unique applications in fabricating nanoscale electronic, photonic, electrochemical, electromechanical, and sensing devices [1–8]. Nanowires with uniform diameters also represent an ideal model system to investigate the dependence of transport and optical properties on size confinement or reduction [9]. In addition, there are many other areas where nanowires could be exploited to significantly enhance the functionality or performance of a material; typical examples include catalysis, formation of superstrong and tough composites, and fabrication of specialized scanning probes [10]. As a result, a tremendous amount of effort has recently been devoted to the synthesis, characterization, and utilization of nanowires with well-controlled dimensions and properties [11–13].

This chapter concentrates on metal nanowires, a class of 1D nanostructures that will serve as the necessary functional components (e.g., interconnects) in building nanoelectronic circuitry [14]. Recent studies by several groups have also demonstrated the potential use of metal nanowires as active components in fabricating nanoscale sensors. For instance, sensors for hydrogen gas have been fabricated by Penner et al. from arrays of mesoscopic Pd wires, which were prepared by electrodeposition on graphite and subsequently transferred onto the surface of a cyanoacrylate film [15, 16]. Exposure to diluted hydrogen gas caused a rapid reversible decrease in the resistance of these wires, and this change correlated well with the concentration of hydrogen gas in the range from 2 to 10%. Tao et al. demonstrated another kind of molecular sensor that involved the use of two thin wires of gold separated by a nanoscale gap [17, 18]. When molecules were adsorbed onto this gap, the quantized conductance across this gap decreased to different fractional values. This design was sensitive enough to detect the adsorption event of an individual molecule because the quantized conductance was mainly determined by a few atoms located at the ends of these metal nanowires. These applications will continue to provide compelling motivation for research into new routes to the synthesis of metal nanowires.

In addition to their superior performance as electrical or thermal conductors, nanostructures made of noble metals (e.g., Ag and Au) exhibit strong absorption in the visible and near infrared regions, an optical feature substantially different from those of planar metal surfaces or bulk materials [19–21]. This colorimetric phenomenon has also been known as surface plasmon resonance (SPR), which mainly arises from the collective oscillation of conduction electrons in response to optical excitation [22]. Different from nanoparticles (or 0D nanostructures), there exist two SPR bands for nanowires (or nanorods): one caused by the coherent electronic oscillation along the short axis (the transverse mode); and the other (the longitudinal mode, with a longer wavelength and stronger intensity) resulted from the coherent electronic oscillation along the long axis [19]. By tuning the aspect ratio of a gold or silver nanorod, it was possible to continuously change the position of its longitudinal SPR band to cover the spectral region from ~400 nm to near-infrared. Because of their distinctive absorption spectra, 1D nanostructures of noble metals can serve as biological labels and light absorbers to be used in the red and near infrared regimes [23]. Recent work by Keating et al. also demonstrated that cylindrical rods made of alternating segments of various metals could be potentially utilized as nanoscale barcodes [24, 25]. In this case, the coding was based on the difference in optical reflectivity displayed by various metals, and readout could be readily achieved using an optical microscope. In a more recent study, Schultz et al. have synthesized striped nanorods consisting of silver (or gold) and a magnetically active metal such as Ni [26]. These nanorods could be manipulated and aligned under a magnetic field to allow for the convenient resolution of coded information over large areas. Since the position and intensity of SPR band of each metal segment are sensitive to the dielectric constant of environment, such 1D nanostructures also provide a useful platform to fabricate combinatorial sensors capable of detecting adsorption events of mixed molecules on their surfaces. Metal nanowires containing striped patterns could, in principle, be used to identify which molecule(s) had been immobilized onto the surface, in much the same way as conventional microarrays where the information is encoded by the spatial location.

Nanowires made of noble metals have also been demonstrated as waveguiding structures to conduct photons. For example, Lyon et al. have observed unidirectional propagation of plasmon over a distance as long as 10 μm [27]. In their experiments, the photons were selectively coupled into the plasmon mode of a silver nanowire (~20 nm in diameter) by controlling the incident excitation wavelength and wire composition. The excited mode could propagate in a nonemissive fashion along the longitudinal axis of this nanowire before it was emitted as an elastically scattered photon at the other end. In related studies, this metal-dependent plasmon propagation has also been explored by Atwater et al. to fabricate nanoscale optical waveguides [28]. These demonstrations, together with a fundamental interest in understanding the mechanism, will make significant contributions to the general field of nanophotonics.

2. Scope and Objectives

This chapter provides a brief overview of several solution-phase methods that have been demonstrated for synthesizing nanowires from metals. It is organized into four major sections. The first section explicitly discusses chemical methods that involve the use of a capping reagent to kinetically control the growth rates of various surfaces. We take the *modified polyol process* developed in our group as an example to illustrate the concept. This method has allowed for the synthesis of silver nanowires with uniform diameters as thin as ~30 nm. Our discussion include: (*i*) a plausible mechanism proposed for the growth of silver nanowires in an isotropic medium; (*ii*) microscopic evidence for the morphological evolution of silver nanostructures; (*iii*) structural characterization of the silver nanowires; (*iv*) dimensional control over these silver nanowires; and (*v*) correlation between the capping reagent and silver surfaces. The second section provides a brief survey of several methods that rely on the use of various templates to dictate the process of anisotropic 1D growth. We cover the following examples: deposition along step edges presented on the surfaces of a solid substrate (e.g., silicon and graphite); filling the channels contained in a porous solid (e.g., thin membranes with cylindrical pores, mesoporous materials, and carbon nanotubes); templating against the surfaces of currently existing nanostructures that have wire-like morphologies (e.g., carbon nanotubes, DNA and other biological macromolecules, viruses, and nanowires made of various materials). We also discuss soft templates self-assembled from organic molecules, with focus being placed on rod-shaped micelles, hexagonal phases of liquid crystals, and block copolymers. The third section assesses the potential of assembly in generating nanowires. This strategy represents a relatively new approach, and building blocks that have been fully explored so far are mainly based on nanoparticles. The last section is concluded with personal perspectives on the trends towards which future work regarding the synthesis of metal nanowires might be directed. The emphasis of this chapter is limited to synthetic methods, with only a brief discussion on some of the intriguing applications associated with metal nanowires. It is also worth noting that most of the synthetic approaches discussed in this chapter can also be (or have been) extended to materials other than metals [29].

The objectives of this chapter are the following: (*i*) to address experimental issues associated with the facile synthesis of metal nanowires *via* chemical methods; (*ii*) to evaluate the potential of current methods that have broad flexibility in generating metal nanowires with well-controlled sizes and properties; and (*iii*) to examine the capability and feasibility of new methods potentially useful in the synthesis of metal nanowires.

3. Solution-Phase Methods that Involve No Templates

The growth of nanowires from an isotropic medium is relatively simple and straightforward if the solid material has a highly anisotropic crystal structure. As recently demonstrated with trigonal phase selenium (*t*-Se) and tellurium (*t*-Te), uniform nanowires could be easily grown from their solutions with lengths up to ~50 μm [30–34].

For metals that are often characterized by isotropic crystal structures (almost all metals crystallize in the cubic lattices), symmetry-breaking is required in the nucleation step if one wants to obtain wire-like morphologies. In these cases, anisotropic confinements have to be applied to induce and maintain anisotropic growth. Confinements that have been extensively explored include physical templates such as channels in porous materials and structures self-assembled from various building blocks. Capping reagents have also been examined to kinetically control the growth rates of metal surfaces and thus to achieve anisotropic growth. This section focuses on the latter approach that involves the use of a capping reagent to generate silver nanowires through a *modified polyol process*. The next section deals with synthetic routes whose success relies on the use of various templates.

3.1. Silver nanowires synthesized through the polyol process

Xia and Sun recently demonstrated a solution-phase method that generated silver nanowires by reducing silver nitrate with ethylene glycol in the presence of poly(vinyl pyrrolidone) (PVP) [35–37]. A plausible mechanism for the formation of silver nanowires is schematically illustrated in Fig. 1. In this so-called *polyol process*, which has been extensively exploited by many research groups to produce nanoparticles from various metals [38, 39], ethylene glycol serves as both solvent and reducing agent. The key to the evolution of a wire-like morphology is the use of a polymeric capping reagent (PVP) and the introduction of seeds (e.g., Pt nanoparticles) to the reaction mixture [37]. Only when silver nitrate is reduced in the presence of seeds, silver nanoparticles characterized by a bimodal size distribution will be formed *via* heterogeneous and homogeneous nucleation processes, respectively. In the case of heterogeneous nucleation, the pre-formed Pt nanoparticles serve as nuclei for the growth of silver because their crystal structures and lattice parameters match well. This nucleation process results in the formation of silver nanoparticles with diameters of 20–30 nm in the reaction system. At the same time, most silver atoms form nanoparticles with diameters <5 nm through a homogenous nucleation process. In the following steps, silver nanoparticles with larger sizes will grow at the expense of smaller ones through the Ostwald ripening process [40]. With the assistance of PVP (a capping polymer whose function will be discussed at the end of this section), most silver particles can be confined and directed to grow into nanowires with uniform diameters. In this synthesis, the cubic symmetry associated with seeds was reduced on a lower one when two single crystalline seeds fused together to form a twined crystallite.

The potential of this polyol process has been demonstrated by generating twinned crystalline nanowires of silver with diameters in the range of 30–60 nm, and lengths up to ~50 μm. We believe this synthesis could be scaled up to produce silver nanowires on the gram-scale. The availability of silver nanowires in large quantities should have a great impact on their use in electronic industry. For instance, the loading of silver in polymeric conductive composites could be greatly reduced if nanoparticles are replaced by nanowires having higher aspect ratios [41]. The reduction in silver loading could decrease the consumption of coinage metals, as well as the weight of electronic devices.

3.2. Microscopic studies on the evolution of silver nanowires

The morphologies of silver nanostructures obtained at various stages of the growth process were characterized using both scanning electron microscopy (SEM) and transmission electron microscopy (TEM). Fig. 2 gives a set of images that were sampled from the reaction mixture after $AgNO_3$ had been added for 0, 10, 20, 40, 50, and 60 min, respectively. This synthesis was carried out under a set of most commonly used conditions, with the solution being heated at 160°C and the concentrations of $PtCl_2$, $AgNO_3$, and PVP ($M_w \approx 55{,}000$, calculated by the repeating unit) being at 6.8×10^{-6}, 0.027, and 0.16 mol · dm^{-3} (final concentrations in the reaction mixture). This set of images correlate well with the proposed mechanism outlined in Fig. 1, clearly showing the evolution of silver nanostructures from zero- to one-dimensional morphology as the reaction proceeded. Fig. 2A gives a high-resolution TEM (HRTEM) image of some Pt seeds formed in situ in the solution mixture. These Pt seeds had a spherical shape with an average diameter of ~5 nm, and their lattice fringes were spaced 0.23 nm apart, which is in good agreement with the *d* value calculated for {1 1 1} planes of face-center-cubic (*fcc*) platinum (0.226 nm). As soon

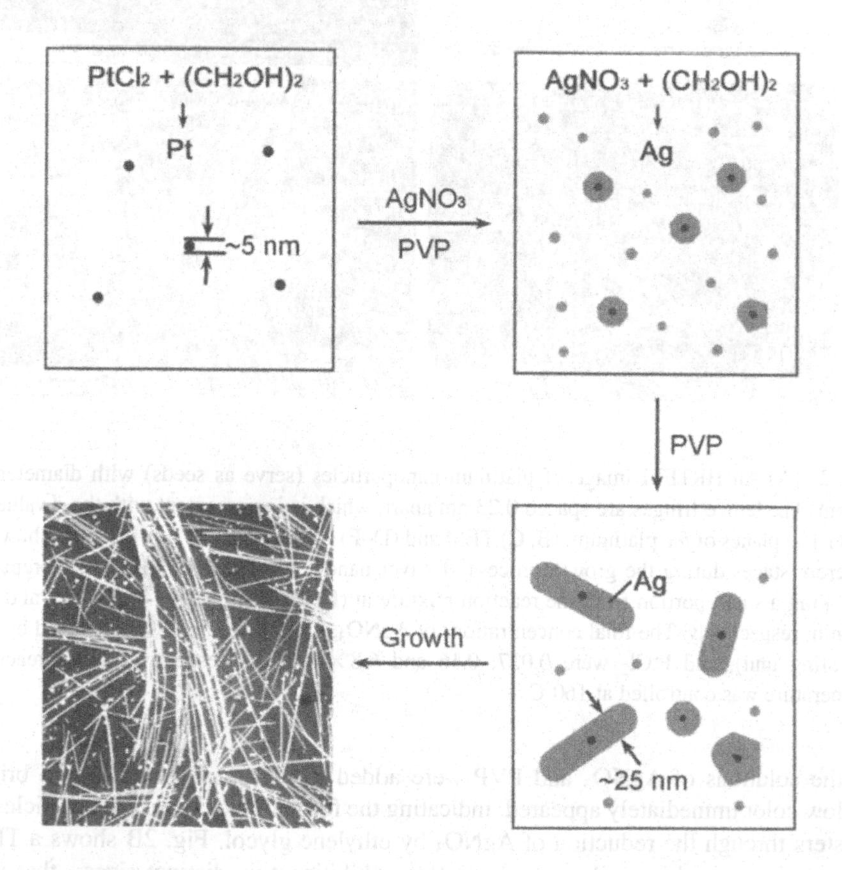

Fig. 1. A schematic illustration of the experimental procedure that generates silver nanowires through a Pt-seeded, PVP-mediated polyol process.

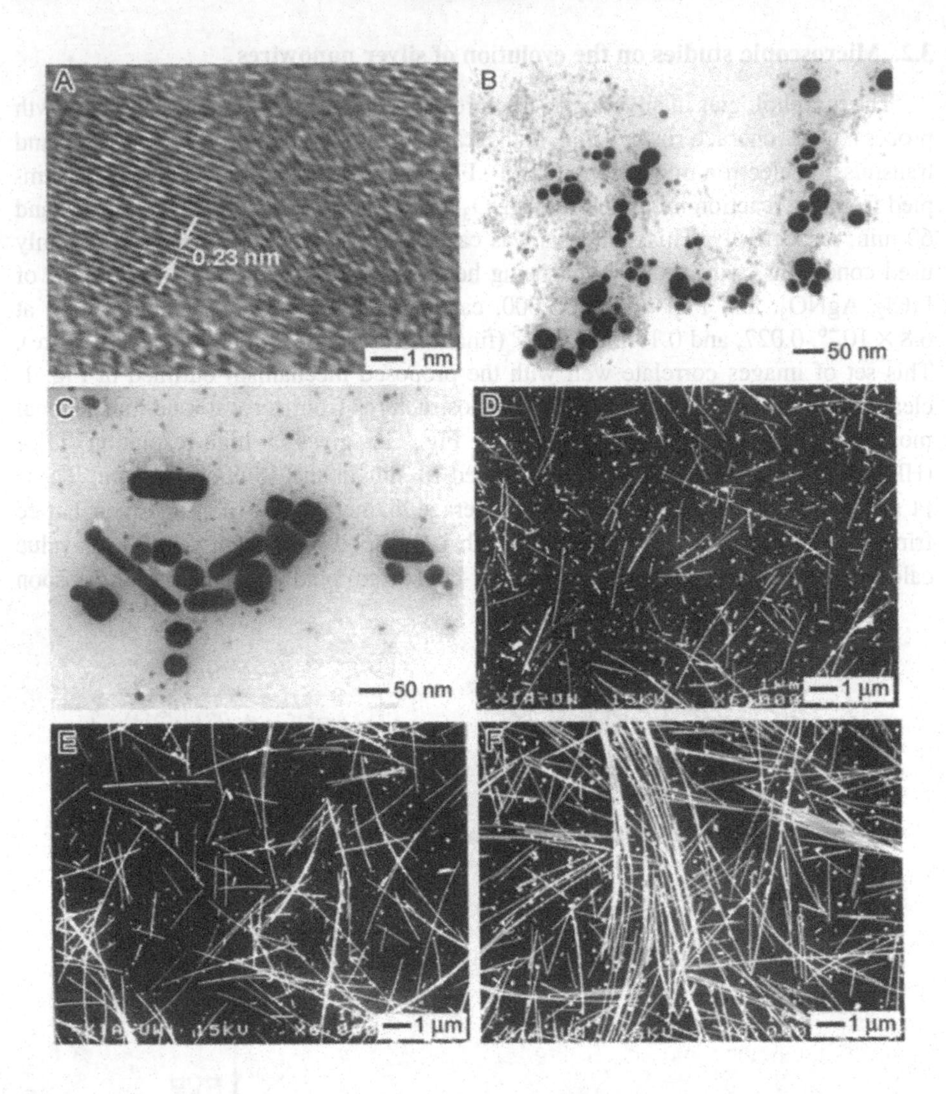

Fig. 2. (A) An HRTEM image of platinum nanoparticles (serve as seeds) with diameter of ~5 nm. The lattice fringes are spaced 0.23 nm apart, which is in agreement with the *d* value of the (1 1 1) planes of *fcc* platinum. (B, C) TEM and (D–F) SEM images of five samples, showing different stages during the growth process of silver nanowires. These samples were prepared by taking a small portion from the reaction mixture at (B) 10, (C) 20, (D) 40, (E) 50, and (F) 60 min, respectively. The final concentrations of $AgNO_3$, PVP ($M_w \approx 55{,}000$, calculated by the repeating unit), and $PtCl_2$ were 0.027, 0.16 and 6.8×10^{-6} M, respectively. The reaction temperature was controlled at 160°C.

as the solutions of $AgNO_3$ and PVP were added to the reaction mixture, a bright yellow color immediately appeared, indicating the formation of silver nanoparticles or clusters through the reduction of $AgNO_3$ by ethylene glycol. Fig. 2B shows a TEM image of the product—silver nanoparticles exhibiting two distinct sizes—that was obtained after the solution of $AgNO_3$ had been added for 10 min. The majority of these silver nanoparticles had a size <5 nm, and they were presumably formed through homogeneous nucleation. Some silver nanoparticles with diameters as large as 20–30 nm were also present, and they were mainly grown on the Pt seeds through heterogeneous nucleation. As the reaction mixture was continued with heating at ~160 °C, small silver nanoparticles slowly dissolved into the solution and recrystallized on the larger ones through a process commonly known as Ostwald ripening [40]. After the reaction had proceeded for ~20 min, the mixture became opaque, and rod-like nanostructures could be observed in the sample under an optical microscope (in the dark-field mode). As shown in Fig. 2C, some of the large silver nanoparticles had, indeed, grown into highly anisotropic 1D nanostructures with the assistance of PVP, while others continued to grow into larger particles with more isotropic shapes. In principle, this growth process should continue until all silver nanoparticles with diameters < 5 nm had been completely consumed (Fig. 2D–F). Both the length and number of silver nanowires greatly increased as the reaction time was elongated. In a typical synthesis, nanowires obtained at 40 min were relatively short in length (<3 μm, Fig. 2D), and they could grow up to 20–50 μm in length after the reaction had proceeded for 60 min (Fig. 2F).

3.3. Structural characterization of the silver nanowires

The product of a typical synthesis contained both nanowires and nanoparticles of silver (see, for example, Fig. 2F), albeit such a mixture could be readily separated into pure components using centrifugation (with acetone added as the co-solvent). The separation efficiency could be evaluated from the UV-visible extinction spectra since the SPR peaks of silver nanowires and nanoparticles are often located at well-separated wavelengths. Fig. 3A shows the UV-visible extinction spectra of an as-synthesized sample before and after it had gone through three rounds of centrifugation. In this case, silver nanowires were captured in the precipitate that exhibited two strong peaks centered at ~380 and ~350 nm [42]. Silver nanoparticles, on the other hand, were retained in the supernatant that displayed a broad peak (at ~436 nm) characteristics of silver colloids. Fig. 3B shows an SEM image of the purified silver nanowires, clearly indicating the removal of silver particles *via* centrifugation. Statistical evaluation suggested that these uniform nanowires had a mean diameter around 40 nm, with a standard deviation of ~5 nm. The XRD pattern shown in Fig. 3C indicates that these silver nanowires were crystallized purely in the *fcc* phase. It is worth noting that the ratio (~3) between the intensities of (1 1 1) and (200) peaks is approximately 0.5 higher than that taken from conventional isotropic silver powders. This result means that silver nanowires synthesized using this polyol method were significantly enriched in {1 1 1} crystallographic planes. Fig. 3D gives a TEM image of several such nanowires, indicating the level of uniformity and perfection that could be routinely achieved from this solution-phase route. The crystallinity and structure of these nanowires were further studied using HRTEM and electron diffraction. Previous studies have suggested a low threshold for twining along {1 1 1} planes of *fcc* metals such as silver and gold, especially when they are processed into nanowires with relatively high aspect ratios [43, 44]. Such 1D nanostructures tend to nucleate and grow as crystals twined by {1 1 1} planes along their longitudinal axis. Fig. 3E shows the HRTEM image of

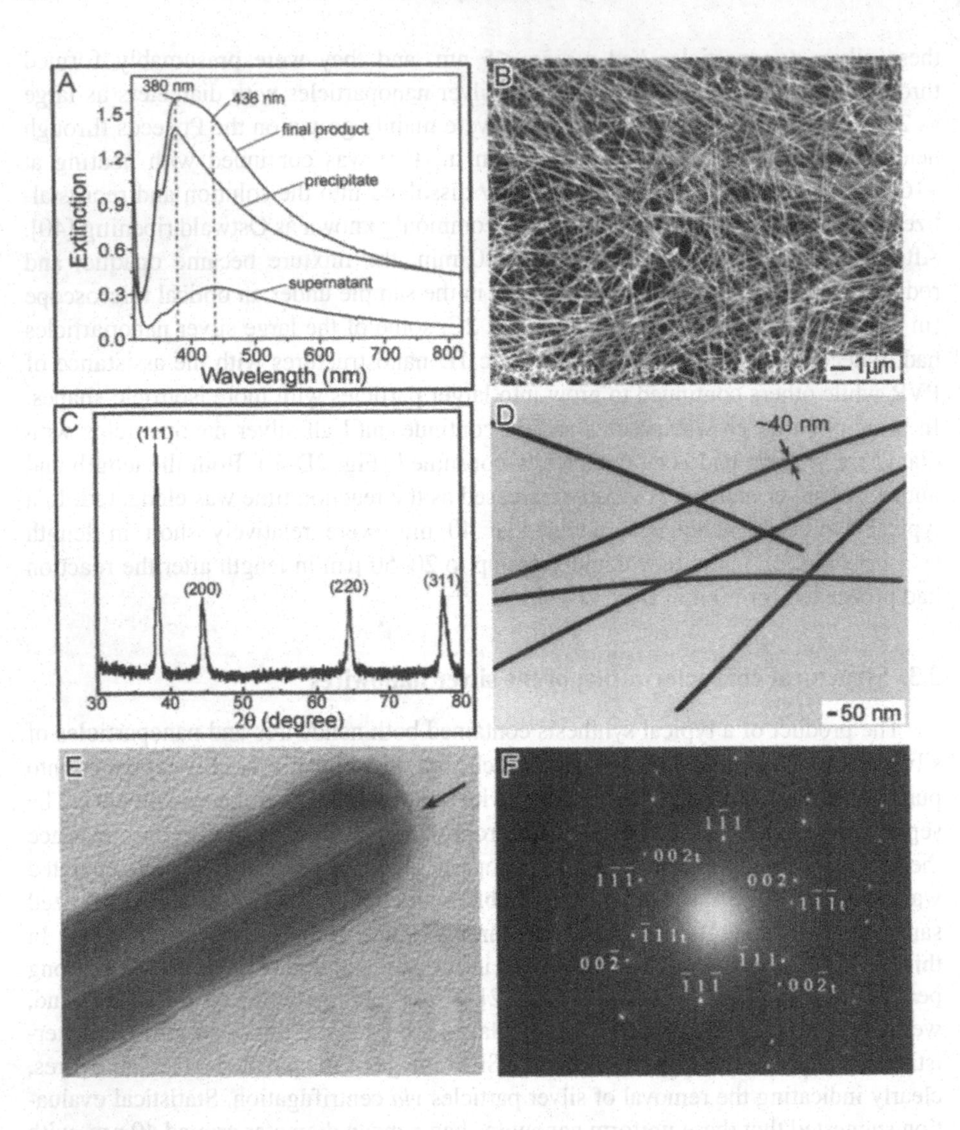

Fig. 3. (A) UV-visible extinction spectra of the final product before and after purification. (B) SEM and (D) TEM images of a purified sample of silver nanowires. (C) The XRD pattern of as-synthesized silver nanowires. All the peaks could be indexed to *fcc* silver. (E) HRTEM image of the end of an individual silver nanowire, indicating a twin plane parallel along its longitudinal axis (shown by an arrow). (F) The corresponding convergent beam electron diffraction pattern. Indices without subscript refer to one side of this nanowire, and indices with the subscript "t" refer to another side. These two patterns have the reflection symmetry about the {1 1 1}-type planes. Reaction conditions are the same as those indicated in Fig. 2.

an individual silver nanowire that contains a twin plane parallel to its longitudinal axis (as indicated by an arrow). Fig. 3F shows the selected-area electron diffraction (SAED) pattern obtained by directing the e-beam parallel to [1 1 0] zone axis. Such a pattern is characteristics of reflection twin for an *fcc* metal, with the mirror being positioned parallel to {1 1 1} planes. The two sets of diffraction spots could be assigned to each side of the nanowire.

3.4. Dimensional control over the silver nanowires

The reduction power of ethylene glycol, the Ostwald ripening process, and the rate of seed-mediated growth were all found to strongly depend on the reaction conditions (e.g., temperature, concentrations of reactants, molar ratio between the repeating unit of PVP and $AgNO_3$, degree of polymerization of PVP, and concentration of seeds). Such a dependence provides a simple and convenient means to control the dimensions of silver nanowires. For example, Fig. 4A, B show the SEM images of products synthesized at two different temperatures. The sample prepared at 100 °C (Fig. 4A) only yielded nanoparticles (rather than nanowires) even after the reaction mixture had been

Fig. 4. (A) An SEM image of the product obtained after the solution was heated at 100 °C for 20 h. (B) An SEM image of the product obtained when the solution was heated at 185 °C for 1 h. (C) A TEM image of the product obtained when the concentration of $PtCl_2$ (seeding source) was increased by 10 times (6.8×10^{-5} M). The average diameter of these nanowires decreased from ~40 to ~30 nm. (D) An SEM image of the silver nanowires obtained with a self-seeding process. In this case, the solutions of $AgNO_3$ and PVP (with lower concentration of 0.04 M) were injected into the reaction system over a period of 10 min. The average diameter of these nanowires became thicker to ~60 nm. Other conditions are the same as those indicated in Fig. 2.

heated for more than 20 h. The size and shape of these silver nanoparticles were both characterized by a moderately high polydispersity. We believe that the dissolution of small silver nanoparticles and the diffusion of silver atoms across the surfaces of seeds both require a sufficiently high temperature. The relatively low temperature involved in this synthesis could not provide enough energy required for the activation of certain specific facets critical to the anisotropic growth of nanowires. In comparison, much higher temperatures tended to favor the formation of silver nanowires with lower aspect ratios. Fig. 4B shows an SEM image of silver nanowires that were synthesized at 185 °C (with other conditions unchanged). These nanowires had a mean diameter of 39 ± 3 nm, and an average length of 1.9 ± 0.4 μm. Note that the average length of these nanowires had been greatly reduced (by as much as 90%) in comparison with those grown at 160 °C (Fig. 3B). In other demonstrations, we found that the diameter of silver nanowires could be slightly varied by changing the concentration of $PtCl_2$ that served as the seeding precursor. For instance, the mean diameter of silver nanowires reduced from 40 to 30 nm (Fig. 4C) when the concentration of $PtCl_2$ was increased by ~10 times. Recently, we discovered that silver nanowires with similar quality could also be synthesized using a *self-seeding* process, in which solutions of $AgNO_3$ and PVP (both in ethylene glycol) were simultaneously injected into a hot bath of ethylene glycol (at 160 °C) through a two-channel syringe pump [36]. By controlling the injection rate, the silver nanoparticles formed at the initial stage of the reduction reaction could serve as seeds for the subsequent growth of silver nanowires. The exclusion of seeds made of an exotic material is favorable for the synthesis of silver nanowires with lower levels of impurity. Fig. 4D shows an SEM image of silver nanowires synthesized under such a seeding condition. The mean diameter of these nanowires had increased by ~50% (to ~60 nm) as compared with those synthesized through the commonly used procedure (Fig. 3B), probably due to a relatively loose control over the size of silver seeds by PVP. Nevertheless, no difference was observed for yield, uniformity, and crystallinity no matter which metal (Pt or Ag) was used as the seeding material.

3.5. The role of poly(vinyl pyrrolidone)

The exact function of PVP in this polyol process is yet to be completely understood. It is believed that PVP kinetically controlled the growth rates of various faces of silver through adsorption and desorption on these surfaces (Fig. 5). Chemical interaction (e.g., the formation of coordination bonds between PVP and silver surfaces) might also be involved because there seemed to exist a selectivity for the functional group on the capping reagent. Similar to the "poisoning" mechanism proposed to account for the anisotropic growth of other materials [45], PVP macromolecules could selectively interact with different faces of silver nanostructures through Ag-O and Ag-N coordination bonds (Fig. 5A). As a result, the growth rates of some surfaces (covered by PVP) would be greatly reduced, leading to a highly anisotropic growth for the silver nanoparticle. If the concentration of PVP was too high, all surfaces of the silver nanoparticle would likely be covered by PVP (Fig. 5B). In this case, selectivity would be lost and silver nanoparticles would be obtained as the major products [46, 47]. Several other polymers with oxygen atoms have also been evaluated for the

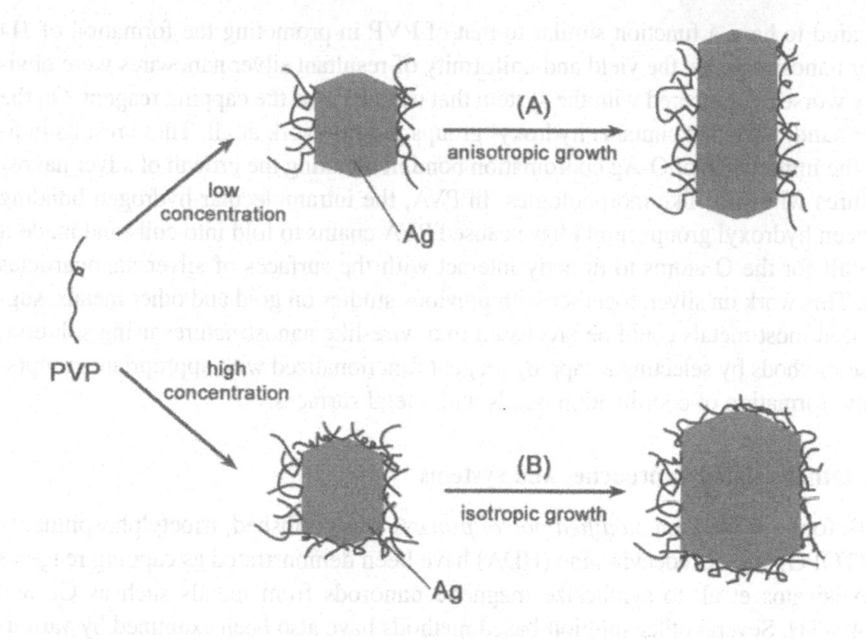

Fig. 5. Illustration of the possible function of PVP in controlling the growth of silver nanostructures. Some surfaces of silver nanoparticles are selectively covered with PVP when the concentration of PVP is not high, leading to an anisotropic growth (process A). All the surfaces of silver nanoparticles are fully covered with PVP when the concentration of PVP is high enough, leading to an isotropic growth (process B).

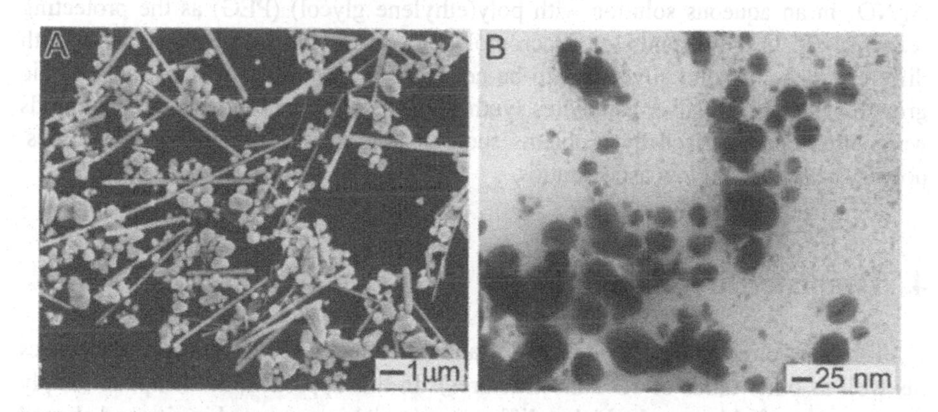

Fig. 6. (A) SEM and (B) TEM images of the as-synthesized silver products when PVP was replaced with other capping polymers: (A) poly(ethylene oxide) (PEO, $M_w \approx 100{,}000$) and (B) poly(vinyl alcohol) (PVA, $M_w \approx 50{,}000$–$85{,}000$).

synthesis of silver nanowires, but none of them worked as effectively as PVP did. Fig. 6 shows SEM and TEM images of two as-synthesized samples that were prepared with PVP being replaced by the same amount of poly(ethylene oxide) (PEO, $M_w \approx 100{,}000$) and poly(vinyl alcohol) (PVA, $M_w \approx 50{,}000$–$85{,}000$), respectively. Although PEO appeared to have a function similar to that of PVP in promoting the formation of 1D silver nanostructures, the yield and uniformity of resultant silver nanowires were obviously worse as compared with the system that used PVP as the capping reagent. On the other hand, PVA that contains hydroxyl groups did not work at all. These results indicate the importance of O-Ag coordination bond in dictating the growth of silver nanostructures with wire-like morphologies. In PVA, the intramolecular hydrogen bonding between hydroxyl groups might have caused PVA chains to fold into coils and made it difficult for the O atoms to directly interact with the surfaces of silver nanoparticles [48]. This work on silver, together with previous studies on gold and other metals, suggest that most metals could be processed into wire-like nanostructures using solution-phase methods by selecting a capping reagent functionalized with appropriate group(s) for the formation of coordination bonds with metal surfaces.

3.6. Other related approaches and systems

Before our work on *modified polyol process* was published, trioctylphosphineoxide (TOPO) and hexadecylamine (HDA) have been demonstrated as capping reagents by Alisivatos et al. to synthesize magnetic nanorods from metals such as Co and Ni [49–51]. Several other solution-based methods have also been examined by various groups for generating silver nanowires without the use of any templates. For instances, silver nanowires were successfully prepared by reducing $AgNO_3$ with a developer that contained AgBr nanocrystallites [52], or by inducing arc discharge between two silver electrodes immersed in an aqueous $NaNO_3$ solution in the presence of PVA [53]. Silver nanorods have also been produced from an aqueous $AgNO_3$ solution by irradiating with ultraviolet light in the presence of PVA [54], or by electroreduction of $AgNO_3$ in an aqueous solution with poly(ethylene glycol) (PEG) as the protecting reagent [42]. In these syntheses, macromolecules that could selectively interact with different faces of solid silver had to be added to induce and direct the anisotropic growth of silver. The silver nanowires synthesized using these solution-based methods were often characterized by problems such as low yields, irregular morphologies, polycrystallinity, and low aspect ratios.

4. Template-Directed Synthesis

Template-directed synthesis provides another effective route to metal nanowires (as well as other kinds of 1D nanostructures). In this approach, the template simply serves as a scaffold around which a different material is generated in situ and shaped into wire-like nanostructures. A wealth of templates have been successfully demonstrated by various research groups, and typical examples include step edges on the surfaces of solid substrates, channels within porous materials, mesoscale structures self-assembled from surfactants or block copolymers, biological macromolecules such as DNA strains or rod-shaped viruses, and 1D nanostructures synthesized using other methods. One could have a physically or chemically templating process, depending on whether the templates are chemically involved in the reaction. For physical templating, the templates have to be selectively removed (by chemical etching or calcination) in a post-synthesis process in order to harvest the nanowires. Different from solution-phase methods discussed in the previous section, metal nanowires synthesized using template-directed methods are usually characterized by polycrystalline structures or low yield. Limited by space, this section only provides a brief discussion on these methods, with a focus on their capability and feasibility.

4.1. Step edges on solid substrates

Decoration of step edges on the surface of a solid substrate with metal has already been demonstrated as a general and convenient route to metal nanowires (Fig. 7). The metal can be applied using procedures such as vapor deposition, ion or electron beam epitaxial growth, and electrochemical deposition. Metal nanowires with lengths up to hundreds of micrometers have been obtained as arrays on solid supports that could be subsequently released into the free-standing form or transferred onto other substrates. Vicinal single-crystal surfaces (e.g., V-shaped grooves on silicon and InP substrates that could be conveniently patterned using lithography and anisotropic etching) have been explored by various groups as a class of templates to demonstrate the potential of step edge decoration as a means for preparing nanowires [55–59]. Large arrays of parallel nanowires (20–120 nm in width) have been fabricated from Pd and Ni by templating against V-grooves on the surfaces of InP(001) wafers [57]. Thin nanowires of Ag, Al, Ga, and Ge have also been prepared by templating against V-grooves on silicon wafers [58, 59]. It is worth noting that Sugawara et al. have reported the fabrication of three-dimensional (3D) arrays of iron nanowires by templating against relief structures patterned on the surfaces of NaCl(1 1 0) crystals [60].

In contrast to single-crystal substrates, Penner et al. have recently grown metal nanowires on the stepped surface of a highly oriented pyrolytic graphite *via* electrodeposition [15, 16]. Two types of materials have been employed in their studies, including noble metals (e.g., Pd, Cu, Au and Ag) and electronically conductive metal oxides (MnO_2, Cu_2O, and Fe_2O_3) that could be reduced to corresponding metals (Mn, Cu, and Fe) by hydrogen gas at elevated temperature. Palladium nanowires, which represent the most intensively studied example, were electrodeposited from aqueous solutions of Pd^{2+} onto step edges present on a graphite surface. These wire arrays were then transferred onto a cyanoacrylate film supported on a glass slide, and tested as hydrogen gas sensors and hydrogen-actuated switches. Such individual wires were characterized to be disconnected by 15–50 nanogaps. In the sensing process, these nanogaps closed in the presence of hydrogen gas and reopened in its absence because the lattice constants of Pd could reversibly expand by several percent under exposure to hydrogen gas. The change in resistance for these Pd wire arrays

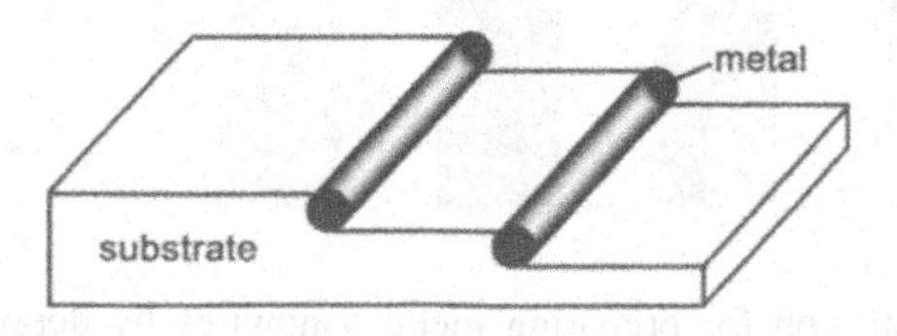

Fig. 7. Preparation of metal nanowires through decorating step edges on the surfaces of solid substrates.

(that attributed to the close/open status of nanogaps) could be related to hydrogen concentration over a range from 2 to 10%. Formation of Mo nanowires is a typical example for the latter case, in which two separated steps were involved [61]. In the first step, "precursor" wires composed of molybdenum oxides (MoO_x) with diameters from 20 nm to 1.3 μm were selectively electrodeposited at step edges on a graphite surface from an alkaline solution of MoO_4^{2-}. These wires of MoO_x were then reduced in H_2 atmosphere at 500 °C to produce nanowires of Mo that retained the dimensional uniformity and hemicylindrical shape of the parent oxide wires but were smaller in diameter with a reduction of 30 to 35%. The conductivity of individual embedded wires was similar to that of bulk Mo. The arrays of metal nanowires prepared using this method could be directly employed to fabricate sensors and other functional devices.

4.2. Channels in porous materials

Channels in porous materials provide another class of templates for use in the synthesis of metal nanowires (Fig. 8). Metals that have been tested for use in this method include noble metals, magnetic metals, and their mixtures or alloys. The only requirement seems to be that the metals could be loaded into the pores through vapor-phase evaporation, liquid-phase injection, or solution-phase chemical or electrochemical deposition. The resultant nanowires could exist as arrays within the pores, or be released from the templates and collected as an ensemble of free-standing nanostructures. Two types of porous membranes are commonly used as the template in such syntheses: polymer films containing track-etched channels and alumina films containing anodically etched pores. Both of them can be directly obtained from commercial sources (e.g., Nuclepore, Poretics, and Whatman). For track-etching, polymer foils (6–20 μm thick) made of the desired material are irradiated with heavy ions from nuclear fissions to create damaged tracks in the material. These tracks are then amplified through chemical etching to form uniform, cylindrical pores across the membranes [62]. The pores formed using this method are often randomly scattered across

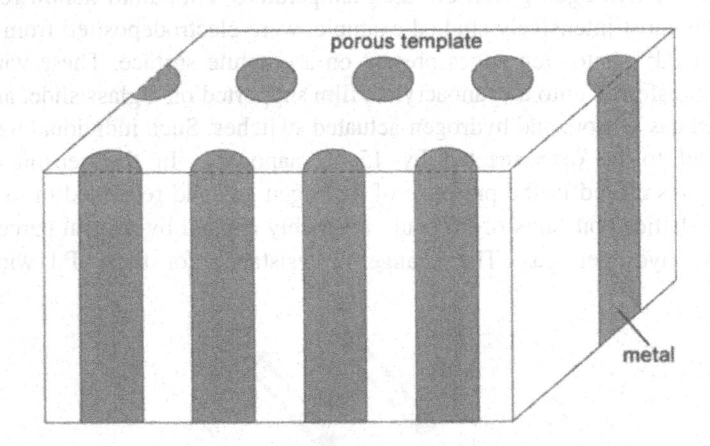

Fig. 8. Schematic illustration for preparing metal nanowires by depositing or filling metals within the pores of nanoporous materials (e.g., membranes containing nanochannels, mesoporous silica and carbon nanotubes).

the membrane surface; and their orientation can be tilted by as large as 34 degrees with respect to the surface normal [63]. Porous alumina membranes are usually prepared via anodization of aluminum foils in an acidic medium [64], and are characterized by a hexagonally packed array of cylindrical pores with relatively uniform size. Unlike the polymer membranes fabricated by track-etching, the pores in alumina membranes have little or no tilting with respect to the surface normal and the pore density is also much higher.

Membrane-templated synthesis has been extensively explored by Martin [65–68], Moskovits [69–71], Searson [72–77], and many other research groups [78–80]. The nanowires synthesized using this method often exhibited unique properties different from the bulk materials. For instance, cyclic voltammetric detection limits for electroactive species at an array of cylindrical gold nanoelectrodes (10 nm in diameter) could be three orders of magnitude lower than those observed at a conventional disk microelectrode made of gold [67]. Arrays of ferromagnetic nanowires of metals (such as Ni, Co, and Fe) with distinctive magnetic behaviors have also been prepared through this template-directed process [81–83]. Much enhanced coercivities as high as 680 Oersteds (only tens of Oersteds for bulk Ni) and remnant magnetization up to 90% have been observed for arrayed Ni nanowires of 30 nm in diameter [81]. Metal nanowires with controllable crystallinity and microstructures have also been prepared by a number of groups to investigate their structure–property relationship. For instance, the superconducting transition temperature for Pb nanowires with diameter of ~50 nm exhibited different values for polycrystalline and single crystalline samples, and their transition temperatures were found to be 5.93 and 7.30 K, respectively [84]. Although electroless deposition was able to generate single crystalline nanowires (such as Ag) in the channels of polycarbonate membranes via a self-catalyzed process [85], pulse electrodeposition seems to be the most powerful method to selectively grow single crystalline or polycrystalline nanowires [84, 86]. It has also been shown that the formation of single crystalline wires of Pb required a greater departure from the equilibrium conditions (e.g., greater overpotential) than the growth of polycrystalline ones. Complex nanostructures and nanoscale devices (such as rod-shaped rectifiers and nanobarcodes) have also been fabricated through layer-by-layer self-assembly strategy [24, 87–89]. In addition to vapor evaporation and solution deposition, metals with relatively low melting points (such as Bi) could be directly injected as liquids into the nanochannels of an anodic alumina membrane [90, 91]. The major advantage associated with membrane-based templates is that the dimensions and compositions of metal nanowires can be easily controlled by varying experimental conditions.

In addition to porous membranes, mesoporous materials with much smaller pore diameters (1.5–30 nm) have also been explored as physical templates for the preparation of ultrafine metal nanowires. Two types of mesoporous silica have been tested: the MCM series (e.g., MCM-41 and MCM-48) characterized by a cubic lattice of pores, and the SBA family (such as SBA-15) containing an array of hexagonal pores. Preparation of metal nanowires in mesoporous materials involves at least two steps: filling the pores with a metal precursor, and the conversion of this precursor into a metal. The metal nanowires obtained from these templates usually have a polycrystalline structure, and they may also exist as discrete nanoparticles. The materials that have been tested for this process include noble metals, such as Ag, Au, Pt, Pd, and bimetallic AuPt alloy [92–97]. Similar to mesoporous materials, single-wall-carbon-nanotubes (SWNTs) with open ends can also be readily filled with liquid metals or metal precursors through capillary action [98–102]. Due to the small diameter of carbon nanotubes, it has been difficult to achieve filling with high yields. As reported by Sloan et al., the yield of filling SWNTs with Ru by wet chemistry was only 2% [98]. More recently, they also put forward a more efficient liquid phase method that could increase the yield of filling SWNTs with silver to 50% by employing eutectic melting system composed of KCl and UCl_4 or solid solution system formed from AgCl and AgBr [99].

4.3. Self-assembled molecular structures

Mesophase structures self-assembled from surfactants provide another class of useful and versatile "soft" templates to synthesize metal nanowires at relatively large scales. As shown in Fig. 9, surfactant molecules can self-assemble into rod-like micelles when their concentration is appropriate (Fig. 9A). These micelles can serve as templates to promote the formation of metal nanorods when coupled with chemical or electrochemical reduction of metal precursors (Fig. 9B). After the surfactants have been removed, nanorods can be collected as relatively pure samples (Fig. 9C). Based on this principle, Murphy et al. have reported a seed-mediated route to the synthesis of metal nanorods with controllable aspect ratios [103–107]. In this case, gold or silver nanoparticles with diameters of 3–5 nm were added as seeds to a solution that

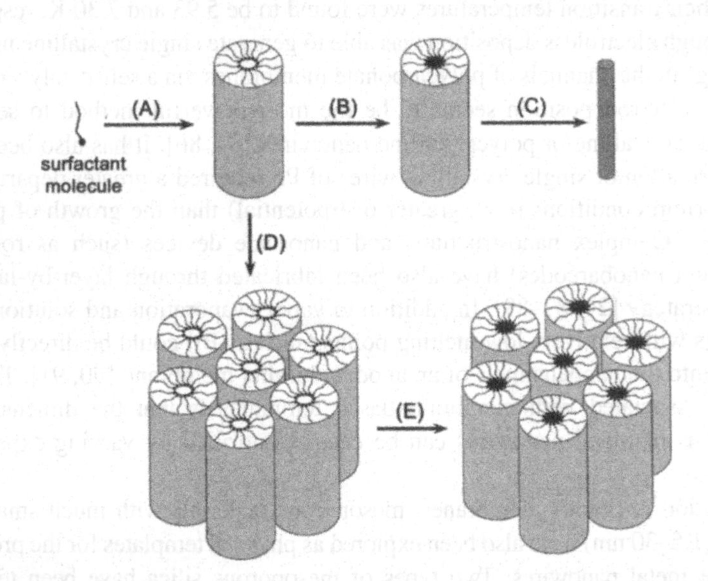

Fig. 9. Schematic illustration of the formation of metal nanowires by templating against surfactant mesophases. (A) formation of micellar cylinder; (B) reduction of metal salts in aqueous phase which are encapsuled within the micellar cylinder; (C) removal of surfactant with appropriate solvents to obtain individual nanowires; (D) formation of liquid crystalline phase with high concentration of surfactant; (E) reduction of metal salts resulting in the formation of arrays of metal nanowires.

contained rod-like micelles (formed from cetyltrimethylammonium bromide, or CTAB) and metal salt such as $HAuCl_4$ or $AgNO_3$. When a weak reducing agent (e.g., ascorbic acid) was added, the seeds served as nucleation sites for the growth of nanorods within the confinement of micellar phase. The lateral dimensions and aspect ratios of these 1D nanostructures can be controlled by varying the concentration of seeds. Another useful route to gold nanorods was demonstrated by Wang et al. that involved the use of CTAB and another much more hydrophobic cationic surfactant (such as tetraoctylammonium bromide, TOAB) to form rod-like micelles [108]. Gold nanorods could be generated via electrochemical reaction within a cell that contained a gold anode and a platinum cathode. In a third demonstration, crystalline Ag nanowire arrays with high aspect ratios have been fabricated by Yan et al. when electrodeposition was performed in a reverse hexagonal liquid-crystalline phase composed of sodium bis(2-ethylhexyl) sulfosuccinate (AOT), *p*-xylene, and water [109]. In this case, rod-like micelles further assembled into a homogeneous normal or reverse hexagonal liquid-crystalline phase when the concentrations of surfactants were sufficiently high (Fig. 9D). Regular arrays of metal nanowires were obtained when templating against such a liquid-crystalline phase (Fig. 9E). Metal nanowires could be synthesized in large quantities by using this kind of templates, albeit the preparation and removal of micellar phase are difficult and tedious.

Diblock copolymers—polymers formed by joining two chemically distinct segments of polymers (or blocks) end-to-end with a covalent bond—have also been explored as templates to synthesize metal nanowires. Because chemically distinct polymers tend to be immiscible, a large collection of such polymers will separate into different phases. These polymers can only phase separate on the length scale because of the covalent bonds. As a result, regular arrays of cylinders can be formed in the self-assembled structures of diblock copolymers [1 1 0, 1 1 1]. These arrayed patterns can be selectively decorated both physically and chemically (e.g., adsorption and coordination) with precursors to metals, making diblock copolymers a powerful templating system for the synthesis of 1D nanostructures of metals [112–114]. Various block copolymers have been tested for the synthesis of silver nanowires, with notable examples including the double-hydrophilic block copolymer of poly(ethylene oxide)-*block*-poly(methacrylic acid) [115], and block copolymers consisting of carbosilane dendrimers and polyisocyanopeptides [116]. Lopes et al. have systematically studied the nonequilibrium self-assembly process of metal nanoparticles on the surfaces of thin films formed from diblock copolymers [117], showing that metal nanoparticles tended to aggregate into chains inside the polystyrene block (one of the blocks of polystyrene-*b*-polymethylmethacrylate block copolymer) with a selectivity approaching 100%. Composition of metal nanowires synthesized based on diblock copolymers could be expected to vary in a broad range because the functional groups of blocks could be selected to interact with interested materials. However, these nanowires are often polycrystalline and aggregate in the form of bundles.

4.4. Currently existing 1D nanostructures

Currently existing 1D nanostructures provide another class of useful templates to synthesize metal nanowires. When templates are relatively inert, metals could be directly deposited on their surfaces (*via* physical evaporation or chemical deposition) to form wire-like nanostructures. For example, Dai et al. reported the formation of metal nanowires on the surfaces of SWNTs through direct evaporation [118, 119]. The pre-formation of a thin layer of Ti was critical to form a continuous layer of other metals (such as Au, Pd, Fe, Al, Pb) because Ti could improve the wetability of SWNTs' surfaces, as well as increase the interaction between metals and SWNTs. Otherwise, direct deposition of metals only led to the formation of discrete particles as a result of dewetting. Since carbon nanotubes as long as half a centimeter can now be synthesized, this approach has the potential to generate relatively long metal nanowires. In other demonstrations, chain-like biomolecules were used to assemble metal ions into strains through interaction between their side active groups and the metal ions. These metal ions could then be reduced into nanoparticles along the linear skeleton of a biomolecule. When a reservoir of the metal ions is also present, these nanoparticles could be further connected into continuous wires. For instance, DNA has been explored as templates by Braun et al. to generate metallic nanowires from Ag and Pt, which could be subsequently used as interconnects to fabricate simple circuits [120–122].

Some nanowires could also serve as chemical templates to generate 1D nanostructures that might be difficult (or impossible) to directly synthesize or fabricate. As demonstrated by several groups, some nanowires can be transformed into new substances without changing their 1D morphology when they react with proper reagents under carefully controlled conditions. For example, Yang et al. demonstrated free-standing nanowires of noble metals (e.g., Au, Ag, Pd and Pt) could be prepared via a redox reaction that involved $LiMo_3Se_3$ nanowires (as the reducing agent) and aqueous metal ions (e.g., $AuCl_4^-$, Ag^+, $PdCl_4^{2-}$, $PtCl_4^{2-}$) [123]. Xia et al. have recently showed that Ag nanowires could serve as chemical templates to form Au (or Pd, Pt) nanotubes by reacting with appropriate chemicals [124]. One of the major problems associated with this approach is the difficulty to control the composition and crystallinity of the final products. The nanowires synthesized using this approach are often polycrystalline and have kinks. When the product has a larger molar volume than the starting material, the reaction might stop at a certain point due to the stress accumulated around the template.

5. Assembly of Metal Nanoparticles into 1D Nanostructures

One family of radically different approaches to fabrication of 1D nanostructures is based on self- or mechanically manipulated assembly, with nanoparticles as the building blocks. A variety of strategies have been demonstrated. For example, Schmid et al. illustrated the fabrication of supported and insulated nanostructures by filling the pores in an alumina membrane with gold nanoparticles or clusters [125]. Rudimentary nanowires were obtained by sintering the nanoparticles confined within membrane pores at 300 °C for several hours. Korgel and Fitzmaurice showed that ellipsoidal silver nanoparticles could assemble themselves into chain-like structures (~7 nm thick) from the solution phase and subsequently fuse into bundles that could extend unbroken up to 200 nm in length [126]. More recently, Schmid et al. further demonstrated that gold clusters such as $Au_{55}(PPh_3)_{12}Cl_6$ could be assembled into 1D architectures at water/dichloromethane interface with the templating of isooctyl-substituted poly-(*para*-phenylenethinylen) [127]. When PVP was present in the solution phase, 2D networks of cluster-loaded polymer chains were also formed at the liquid/air interface using Langmuir-Blodgett technique [128]. The gold clusters seemed to serve as joints between the PVP polymer chains. Velev et al. demonstrated that metallic nanoparticles suspended in a liquid medium could be assembled into microwires up to 5 mm in length via dielectrophoresis [129]. In this approach, the mobility of and attractive interaction between particles were caused by an alternating electric field. In addition to these self-assembly methods, the improved spatial resolution of scanning probe microscopes has also made them ideal tools for organizing individual nanoparticles into complex nanostructures one by one. For example, Koel et al. have demonstrated the construction of interconnected, multiparticle nanostructures (e.g., linear chains, circular rings, and pyramidal lattices) by manipulating the positions of individual gold nanoparticles with the tip of a scanning force microscope [130–132]. They have also fabricated a straight line composed of equally spaced gold nanoparticles (30 nm in diameter), which could be further explored as nanoscale plasmon waveguides [28]. Similar linear chains of gold nanoparticles (15 nm in diameter) have also been used as templates to generate continuous nanowires of gold supported on the surface of a SiO_2 substrate via the hydroxylamine-seeding process [132]. Although the capabilities of these assembly methods have been demonstrated to certain extents, their development into practical routes to fabrication of nanowires still requires great ingenuity. For example, the manipulation method based on scanning probes is a serial process and its slow speed will limit its use only to very small scale production. The self-assembly approaches do not have this limitation, but, at the current stage of development, they all lack a good control over the dimensions and morphology of resultant nanostructures.

6. Concluding Remarks

We have briefly discussed several chemical methods that have been demonstrated for the synthesis of metal nanowires: solution-phase methods that generate 1D structures by kinetically controlling the growth rates of various faces through the use of capping reagents; methods that achieve highly anisotropic growth by templating against various types of nanostructures with 1D morphologies; and assembly of particulate building blocks. Each method has its specific merits and inevitable weaknesses. For example, the templateless methods usually yield nanowires with highly crystalline structures and well-controlled compositions, albeit it is nontrivial to select the appropriate capping reagent for each metal or to achieve a precise control over the dimension and monodispersity of resultant nanowires. The template-directed methods, on the other hand, provide a better control over the uniformity and dimensions. The use of templates, however, greatly limits the scale of nanowires that could be produced in each round of synthesis. Removal of templates in the post-synthesis process may also cause damages to the nanowires. In addition, nanowires synthesized via template-directed methods were usually polycrystalline in structure, an unwanted feature that may limit their use in device fabrication and/or fundamental studies.

As the use of metal nanowires grows in importance in a broad range of applications, the demand for new synthetic methods will certainly increase. Preliminary work from many research groups suggest that the ultimate use of nanowires strongly depends on our ability to precisely control or fine tune their dimensions, chemical compositions, surface properties, and crystal structures. Judged against this metrics, all the methods described in this chapter need to be greatly improved before they will find widespread use in various applications.

Another challenge faced by chemically synthesized metal nanowires is their assembly into complex structures or device architectures. Recent work by Lieber et al. has started to address this issue, and several promising methods have been demonstrated for assembling nanowires into simple test patterns such as arrays of junctions [133]. In their approach, microfluidic channels were used as templates to guide the motion and assembly of nanowires. Another potentially useful method for assembling nanowires is based on the so-called *Langmuir-Blodgett* (LB) technique. In this case, metal nanowires can be coated with uniform dielectric sheaths to form nanocables that will be subsequently used as building blocks for LB assembly (Fig. 10A). Using this approach, it should be possible to fabricate large arrays of metal/dielectric/metal junctions with relatively high densities. With silver nanowires synthesized by the polyol process as an example, we have recently made some progress along this direction. For example, we have been able to coat the surfaces of these silver nanowires with uniform sheaths of amorphous silica with well-controlled thickness to form $Ag@SiO_2$ coaxial nanocables (Fig. 10B) [134]. At the moment, we are seeking effective ways to assemble these 1D building blocks into the test patterns illustrated in Fig. 10A. We believe that the ability to form such functional architectures represents one of the

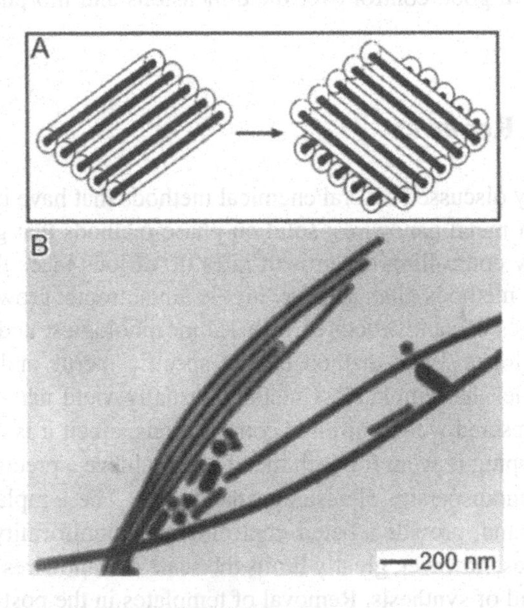

Fig. 10. (A) Schematic illustration for the formation of 2D arrays and junctions by assembling the coaxial $Ag@SiO_2$ nanocables. (B) TEM image of $Ag@SiO_2$ coaxial nanocables synthesized by directly coating silver nanowires with amorphous silica.

critical steps towards the fabrication of nanoelectronic devices. Only at that point, nanowires (made of metals and other materials) will become an important addition to the field of nanoscale science and technology.

Acknowledgments

This research has been supported in part by an AFOSR-DURINT subcontract from SUNY-Buffalo, a Career Award from the NSF (DMR-9983893), a Fellowship from the David and Lucile Packard Foundation, and a Research Fellowship from the Alfred P. Sloan Foundation. Y.X. is a Camille Dreyfus Teacher Scholar. We would like to thank Dr. B. Gates, Y. Yin, B. Mayers, T. Herricks, and Yu Lu for their help with X-ray diffraction, electron diffraction, and electron microscopic studies.

References

1. M. S. Gudiksen, L. J. Lauhon, J. Wang, D. C. Smith and C. M. Lieber, *Nature* **415** (2002) 617.
2. X. Duan, Y. Huang, J. Wang and C. M. Lieber, *Nature* **409** (2001) 66.
3. M. H. Huang, S. Mao, H. Feick, H. Yan, Y. Wu, H. Kind, E. Weber, R. Russo and P. Yang, *Science* **292** (2001) 1897.
4. Z. W. Pan, Z. R. Dai and Z. L. Wang, *Science* **291** (2001) 1947.
5. K. B. Jirage, J. C. Hulteen and C. R. Martin, *Science* **278** (1997) 655.
6. Y. Cui and C. M. Lieber, *Science* **291** (2001) 851.
7. S. J. Tans, M. H. Devoret, H. Dai, A. Thess, R. E. Smalley, L. J. Geerligs and C. Dekker, *Science* **386** (1997) 474.
8. A. I. Yanson, G. R. Bollinger, H. E. van den Brom, N. Agrait and J. M. van Ruitenbeek, *Nature* **395** (1998) 783.
9. M. R. Black, Y.-M. Lin, S. B. Cronin, O. Rabin and M. S. Dresselhaus, *Phys. Rev. B* **65** (2002) 195417.
10. J. Hu, T. W. Odom and C. M. Lieber, *Acc. Chem. Res.* **32** (1999) 435.
11. S. M. Prokes and K. L. Wang, *MRS Bulletin* **24**(8) (1999) 13.
12. C. R. Martin, *Science* **266** (1994) 1961.
13. R. M. Penner, *J. Phys. Chem. B* **106** (2002) 3339.
14. C. M. Lieber, *Sci. Am.* **285** (2001) 58.
15. F. Favier, E. C. Walter, M. P. Zach, T. Benter and R. M. Penner, *Science* **293** (2001) 2227.
16. E. C. Walter, F. Faview and R. M. Penner, *Anal. Chem.* **74** (2002) 1546.
17. C. Z. Li, H. X. He, A. Bogozi, J. S. Bunch and N. J. Tao, *Appl. Phys. Lett.* **76** (2000) 1333.
18. H. X. He, S. Boussaad, B. Q. Xu, C. Z. Li and N. J. Tao, *J. Electroanal. Chem.* **522** (2002) 167.
19. M. A. El-Sayed, *Acc. Chem. Res.* **34** (2001) 257.
20. P. B. Johnson and R. W. Christy, *Phys. Rew. B* **6** (1972) 4370.
21. S. L. Westcott, S. J. Oldenburg, T. R. Lee and N. J. Halas, *Chem. Phys. Lett.* **300** (1999) 651.
22. C. F. Bohren and D. R. Huffman, *Absorption and Scattering of Light by Small Particles*, Wiley, New York (1983).
23. R. Elghanian, J. J. Storhoff, R. C. Mucic, R. L. Letsinger and C. A. Mirkin, *Science* **277** (1997) 1078.
24. S. R. Nicewarner-Peña, R. G. Freeman, B. D. Reiss, L. He, D. J. Peña, I. D. Walton, R. Cromer, C. D. Keating and M. J. Natan, *Science* **294** (2001) 137.

25. I. D. Walton, S. M. Norton, A. Balasingham, L. He, S. F. Oviso, Jr., D. Gupta, P. A. Raju, M. J. Natan and R. G. Freeman, *Anal. Chem.* **74** (2002) 2240.
26. J. J. Mock, S. J. Oldenburg, D. R. Smith, D. A. Schultz and S. Schultz, *Nano Lett.* **2** (2002) 465.
27. R. M. Dickson and L. A. Lyon, *J. Phys. Chem. B* **104** (2000) 6095.
28. S. A. Maier, M. L. Brongersma, P. G. Kik, S. Meltzer, A. A. G. Requicha and H. A. Atwater, *Adv. Mater.* **13** (2001) 1501.
29. C. R. Martin, *Acc. Chem. Res.* **28** (1995) 61.
30. B. Gates, Y. Yin and Y. Xia, *J. Am. Chem. Soc.* **122** (2000) 12582.
31. B. Mayers, B. Gates, Y. Yin and Y. Xia, *Adv. Mater.* **13** (2001) 1380.
32. B. Gates, B. Mayers, B. Cattle and Y. Xia, *Adv. Func. Mater.* **12** (2002) 219.
33. B. Mayers and Y. Xia, *Adv. Mater.* **14** (2002) 279.
34. B. Mayers and Y. Xia, *J. Mater. Chem.* **12** (2002) 1875.
35. Y. Sun, B. Gates, B. Mayers and Y. Xia, *Nano Lett.* **2** (2002) 165.
36. Y. Sun and Y. Xia, *Adv. Mater.* **14** (2002) 833.
37. Y. Sun, Y. Yin, B. Mayers, T. Herricks and Y. Xia, *Chem. Mater.* (in press).
38. P. Toneguzzo, G. Viau, O. Acher, F. Fiévet-Vincent and F. Fiévet, *Adv. Mater.* **10** (1998) 1032.
39. G. Carotenuto, G. P. Pepe and L. Nicolais, *Eur. Phys. J. B* **16** (2000) 11.
40. A. R. Roosen and W. C. Carter, *Physica A* **261** (1998) 232.
41. F. Carmona, F. Barreau, P. Delhaes and R. Canet, *J. Phys. Lett.* **41** (1980) L-531.
42. J.-J. Zhu, X.-H. Liao, X.-N. Zhao and H.-Y. Chen, *Mater. Lett.* **49** (2001) 91.
43. G. Bögels, H. Meekes, P. Bennema and D. Bollen, *J. Phys. Chem. B* **103** (1999) 7577.
44. Z. L. Wang, M. B. Mohamed, S. Link and M. A. El-Sayed, *Surf. Sci.* **440** (1999) L809.
45. M. Almeida and L. Alcacer, *J. Cryst. Growth* **62** (1983) 183.
46. P.-Y. Silvert, R. Herrera-Urbina, N. Duvauchelle, V. Vijayakrishnan and K. T. Elhsissen, *J. Mater. Chem.* **6** (1996) 573.
47. P.-Y. Silvert, R. Herrera-Urbina and K. Tekaia-Elhsissen, *J. Mater. Chem.* **7** (1997) 293.
48. K. Masuda, H. Kaji and F. Horii, *J. Polym. Sci. Pol. Phys.* **38** (2000) 1.
49. V. F. Puntes, K. M. Krishnan and A. P. Alivisatos, *Science* **291** (2001) 2115.
50. V. F. Puntes, K. M. Krishnan and A. P. Alivisatos, *Top. Catalysis* **19** (2002) 145.
51. N. Cordente, M. Respaud, F. Senocq, M.-J. Casanove, C. Amiens and B. Chaudret, *Nano Lett.* **1** (2001) 565.
52. S. Liu, Y. Yue and A. Gedanken, *Adv. Mater.* **13** (2001) 656.
53. Y. Zhou, S. H. Yu, X. P. Cui, C. Y. Wang and Z. Y. Chen, *Chem. Mater.* **11** (1999) 545.
54. Y. Zhou, S. H. Yu, C. Y. Wang, X. G. Li, Y. R. Zhu and Z. Y. Chen, *Adv. Mater.* **11** (1999) 850.
55. F. J. Himpsel, T. Jung and J. E. Ortega, *Surf. Rev. Lett.* **4** (1997) 371.
56. D. Y. Petrovykh, F. J. Himpsel and T. Jung, *Surf. Sci.* **407** (1998) 189.
57. J. Jorritsma, M. A. M. Gijs, J. M. Kerkhof and J. G. H. Stienen, *Nanotechnology* **7** (1996) 263.
58. T. Muller, K.-H. Heinig and B. Schmidt, *Nucl. Instrum. Methods Phys. Res. B* **175–177** (2001) 468.
59. H. H. Song, K. M. Jones and A. A. Baski, *J. Vac. Sci. Technol. A* **17** (1999) 1696.
60. A. Sugawara, T. Coyle, G. G. Hembree and M. R. Scheinfein, *Appl. Phys. Lett.* **70** (1997) 1043.
61. M. P. Zach, K. H. Ng and R. M. Penner, *Science* **290** (2000) 2120.
62. R. L. Fleisher, P. B. Price and R. M. Walker, *Nuclear Tracks in Solids*, University of California Press, Berkeley, CA (1975).
63. J. C. Hulteen and C. R. Martin, *J. Mater. Chem.* **7** (1997) 1075.
64. H. Masuda and K. Fukuda, *Science* **268** (1995) 1466.
65. V. M. Cepak and C. R. Martin, *J. Phys. Chem. B* **102** (1998) 9985.
66. M. Nishizawa, V. P. Menon and C. R. Martin, *Science* **268** (1995) 700.

67. V. P. Menon and C. R. Martin, *Anal. Chem.* **67** (1995) 1920.
68. C. J. Brumlik, V. P. Menon and C. R. Martin, *J. Mater. Res.* **9** (1994) 1174.
69. D. Almawlawi, C. Z. Liu and M. Moskovits, *J. Mater. Res.* **9** (1994) 1014.
70. C. K. Preston and M. Moskovits, *J. Phys. Chem.* **97** (1993) 8495.
71. A. A. Tager, J. M. Xu and M. Moskovits, *Phys. Rev. B* **55** (1997) 4530.
72. K. M. Hong, F. Y. Yang, K. Liu, D. H. Reich, P. C. Searson, C. L. Chien, F. F. Balakirev and G. S. Boebinger, *J. Appl. Phys.* **85** (1999) 6184.
73. K. Liu, K. Nagodawithana, P. C. Searson and C. L. Chien, *Phys. Rev. B* **51** (1995) 7381.
74. K. Liu, C. L. Chien, P. C. Searson and K. Yu-Zhang, *Appl. Phys. Lett.* **73** (1998) 1436.
75. K. Liu, C. L. Chien and P. C. Searson, *Phys. Rev. B* **58** (1998) R14681.
76. G. Oskam, J. G. Long, A. Natarajan and P. C. Searson, *J. Phys. E: Appl. Phys.* **31** (1998) 1927.
77. L. Sun, P. C. Searson and C. L. Chien, *Appl. Phys. Lett.* **74** (1999) 2803.
78. Y. W. Wang, L. D. Zhang, G. W. Meng, X. S. Peng, Y. X. Jin and J. Zhang, *J. Phys. Chem. B* **106** (2002) 2502.
79. D. S. Xu, D. P. Chen, Y. J. Xu, X. S. Shi, G. L. Guo, L. L. Gui and Y. Q. Tang, *Pure Appl. Chem.* **72** (2000) 127.
80. J. K. N. Mbindyo, T. E. Mallouk, J. B. Mattzela, I. Kratochvilova, B. Razavi, T. N. Jackson and T. S. Mayer, *J. Am. Chem. Soc.* **124** (2002) 4020.
81. T. M. Whitney, J. S. Jiang, P. C. Searson and C. L. Chien, *Science* **261** (1993) 1316.
82. C. Schonenberger, B. M. I. van der Zande, L. G. J. Fokkink, M. Henny, C. Schmid, M. Kruger, A. Bachtold, R. Huber, H. Birk and U. Staufer, *J. Phys. Chem. B* **101** (1997) 5497.
83. G. Tourillon, L. Pontonnier, J. P. Levy and V. Langlais, *Electrochem. Solid-State Lett.* **3** (2000) 20.
84. G. Yi and W. Schwarzacher, *Appl. Phys. Lett.* **74** (1999) 1747.
85. M. Barbic, J. J. Mock, D. R. Smith and S. Schultz, *J. Appl. Phys.* **91** (2002) 9341.
86. M. E. T. Molares, V. Buschmann, D. Dobrev, R. Neumann, R. Scholz, I. U. Schuchert and J. Vetter, *Adv. Mater* **13** (2001) 62.
87. N. I. Kovtyukhova, B. R. Martin, J. K. N. Mbindyo, T. E. Mallouk, M. Cabassi and T. S. Mayer, *Mater. Sci. Eng. C* **19** (2002) 255.
88. N. I. Kovtyukhova, B. R. Martin, J. K. N. Mbindyo, P. A. Smith, B. Razavi, T. S. Mayer and T. E. Mallouk, *J. Phys. Chem. B* **105** (2001) 8762.
89. J. Haruyama, I. Takesue, S. Kato, K. Takazawa and Y. Sato, *Appl. Surf. Sci.* **175/176** (2001) 597.
90. Z. Zhang, J. Y. Ying and M. S. Dresselhaus, *J. Mater. Res.* **13** (1998) 1745.
91. Z. Zhang, D. Gekhtman, M. S. Dresselhaus and J. Y. Ying, *Chem. Mater.* **11** (1999) 1659.
92. M. H. Huang, A. Choudrey and P. Yang, *Chem. Commun.* (2000) 1063.
93. Y.-J. Han, J. M. Kim and G. D. Stucky, *Chem. Mater.* **12** (2000) 2068.
94. K.-B. Lee, S.-M. Lee and J. Cheon, *Adv. Mater.* **13** (2001) 517.
95. C.-M. Yang, H.-S. Sheu and K.-J. Chao, *Adv. Funct. Mater.* **12** (2002) 143.
96. S. Bhattacharyya, S. K. Saha and D. Chakravorty, *Appl. Phys. Lett.* **77** (2000) 3770.
97. H. Kang, Y.-W. Jun, J.-I. Park, K.-B. Lee and J. Cheon, *Chem. Mater.* **12** (2000) 3530.
98. J. Sloan, J. Hammer, M. Zwiefka-Sibley and M. L. H. Green, *Chem. Commun.* (1998) 347.
99. J. Sloan, D. M. Wright, H.-G. Woo, S. Bailey, G. Brown, A. P. E. York, K. S. Coleman, J. L. Hutchison and M. L. H. Green, *Chem. Commun.* (1999) 699.
100. A. Govindaraj, B. C. Satishkumar, M. Nath and C. N. R. Rao, *Chem. Mater.* **12** (2000) 202.
101. Z. L. Zhang, B. Li, Z. J. Shi, Z. N. Gu, Z. Q. Xue and L.-M. Peng, *J. Mater. Res.* **15** (2000) 2658.

102. W. K. Hsu, J. Li, H. Terrones, M. Terrones, N. Grobert, Y. Q. Zhu, S. Trasobares, J. P. Hare, C. J. Pickett, H. W. Kroto and D. R. M. Walton, *Chem. Phys. Lett.* **301** (1999) 159.
103. C. J. Murphy and N. R. Jana, *Adv. Mater.* **14** (2002) 80.
104. C. J. Johnson, E. Dujardin, S. A. Davis, C. J. Murphy and S. Mann, *J. Mater. Chem.* **12** (2002) 1765.
105. N. R. Jana, L. Gearheart and C. J. Murphy, *Adv. Mater* **13** (2001) 1389.
106. N. R. Jana, L. Gearheart and C. J. Murphy, *Chem. Commun.* (2001) 617.
107. N. R. Jana, L. Gearheart and C. J. Murphy, *J. Phys. Chem. B* **105** (2001) 4065.
108. Y.-Y. Yu, S. S. Chang, C.-L. Lee and C. R. C. Wang, *J. Phys. Chem. B* **101** (1997) 6661.
109. L. Huang, H. Wang, Z. Wang, A. Mitra, K. N. Bozhilov and Y. Yan, *Adv. Mater.* **14** (2002) 61.
110. F. S. Bates and G. H. Fredrickson, *Ann. Rev. Phys. Chem.* **41** (1990) 525.
111. F. S. Bates and G. H. Fredrickson, *Phys. Today* **52**(2) (1999) 32.
112. R. W. Zehner and L. R. Sita, *Langmuir* **15** (1999) 6139.
113. T. L. Morkved, P. Wiltzius, H. M. Jaeger, D. G. Grier and T. A. Witten, *Appl. Phys. Lett.* **64** (1994) 422.
114. T. Thurn-Albrecht, J. Schotter, C. A. Kastle, N. Emley, T. Shibauchi, L. Krusin-Elbaum, K. Guarini, C. T. Black, M. T. Tuominen and T. P. Russell, *Science* **290** (2000) 2126.
115. D. Zhang, L. Qi, J. Ma and H. Cheng, *Chem. Mater.* **13** (2001) 2753.
116. J. J. L. M. Cornelissen, R. van Heerbeek, P. C. J. Kamer, J. N. H. Reek, N. A. J. M. Sommerdijk and R. J. M. Nolte, *Adv. Mater.* **14** (2002) 489.
117. W. A. Lopes, *Phys. Rev. E* **65** (2002) 031606.
118. Y. Zhang and H. Dai, *Appl. Phys. Lett.* **77** (2000) 3015.
119. Y. Zhang, N. W. Franklin, R. J. Chen and H. Dai, *Chem. Phys. Lett.* **331** (2000) 35.
120. K. Keren, M. Krueger, R. Gilad, G. Ben-Yoseph, U. Sivan and E. Braun, *Science* **297** (2002) 72.
121. E. Braun, Y. Eichen, U. Sivan and G. Ben-Yoseph, *Nature* **391** (1998) 775.
122. W. E. Ford, O. Harnack, A. Yasuda and J. M. Wessels, *Adv. Mater.* **13** (2001) 1793.
123. J. H. Song, Y. Wu, B. Messer, H. Kind and P. Yang, *J. Am. Chem. Soc.* **123** (2001) 10397.
124. Y. Sun, B. T. Mayers and Y. Xia, *Nano Lett.* **2** (2002) 481.
125. G. Hornyak, M. Kröll, R. Pugin, T. Sawitowski, G. Schmid, J.-O. Bovin, G. Karsson, H. Hofmeister and S. Hopfe, *Chem. Eur. J.* **3** (1997) 1951.
126. B. A. Korgel and D. Fitzmaurice, *Adv. Mater.* **10** (1998) 661.
127. D. Wyrwa, N. Beyer and G. Schmid, *Nano Lett.* **2** (2002) 419.
128. T. Reuter, O. Vidoni, V. Torma and G. Schmid, *Nano Lett.* **2** (2002) 709.
129. K. D. Hermanson, S. O. Lumsdon, J. P. Williams, E. W. Kaler and O. D. Velev, *Science* **294** (2001) 1082.
130. R. Resch, C. Baur, A. Bugacov, B. E. Keol, A. Madhukar, A. A. G. Requicha and P. Will, *Langmuir* **14** (1998) 6613.
131. R. Resch, C. Baur, A. Bugacov, B. E. Keol, P. M. Echternach, A. Madhukar, N. Montoya, A. A. G. Requicha and P. Will, *J. Phys. Chem. B* **103** (1999) 3647.
132. S. Meltzer, R. Resch, B. E. Keol, M. E. Thompson, A. Madhukar, A. A. G. Requicha and P. Will, *Langmuir* **17** (2001) 1713.
133. Y. Huang, X. Duan, Q. Wei and C. M. Lieber, *Science* **291** (2001) 630.
134. Y. Yin, Y. Lu, Y. Sun and Y. Xia, *Nano Lett.* **2** (2002) 427.

Chapter 8

Chemical and Biomolecular Interactions in the Assembly of Nanowires

Achim Amma and Thomas E. Mallouk
Department of Chemistry, The Pennsylvania State University,
152 Davey Laboratory, University Park, PA 16802, USA

1. Introduction

Over the past decade, there has been remarkable progress in the synthesis of inorganic materials as high aspect ratio nanoparticles. In addition to nanotubes and nanowires derived from carbon and other group IV elements, numerous metals, oxides, and compound semiconductors have now been made as wires, tubes, scrolls, and ribbons. High quality single crystals of many of these nanomaterials can be obtained by solution or vapor phase syntheses. In fact, our understanding of these synthetic processes has now advanced to the point where it is possible to exert control over not only the physical dimensions (size and aspect ratio) of nanoparticles, but also to control their shape [1]. Thus, in addition to essentially one-dimensional nanostructures, numerous interesting shapes such as arrowheads, teardrops, tetrapods [2], prisms [3], and stars [4] have recently been synthesized.

The availability of these compositionally and morphologically varied nanowires and related high aspect ratio particles presents new opportunities that were not apparent in earlier work that focused on essentially zero-dimensional (quantum dot) nanoparticles. Chemically synthesized nanowires are now serious contenders as components of post-silicon, self-assembled nanocircuits [5, 6]. They are also being widely investigated as elements of sensor arrays [7, 8, 9], field emitter displays [10], optical circuits [11], and solar cells [12]. Beyond this, it is of course interesting to speculate about more advanced applications in micromachines and robotic devices.

In approaching each of these interesting applications, the barrier to progress is now increasingly becoming one of assembly rather than synthesis. We have at our disposal an unprecedented tool kit of nanoscale building blocks, but because of their small size they are inherently difficult to manipulate for measurement or for construction into functional superstructures. This chapter considers some of the physical and chemical principles of nanowire assembly, and describes the recent progress that has been made in using chemical and biomolecular interactions to assemble nanowires into interesting superstructures.

2. Phase Behavior of Nanowires

The problem of assembly of high aspect ratio particles was first considered theoretically by Onsager [13], who showed that an isotropic fluid containing them should spontaneously order into a nematic phase (in which the particles have local orientational order) above a certain critical density. This hard particle model does not consider attractive interactions between rods, and thus predicts a phase transition that is entirely entropy driven [14]. In order to understand how entropy can be maximized in a process that results in a more ordered phase, one needs to consider the entropy of the whole system. Each rod defines an excluded volume in the fluid that cannot be occupied by any part of another particle. When rods are packed together in parallel fashion, their excluded volumes overlap, and thus the free volume of the whole system (rods + free space) increases.

This kind of phase behavior has been observed on many length scales in experiments involving monodisperse collections of rigid rods. Depending on the conditions, nematic, smectic, or ordered crystalline phases are obtained. Fig. 1A shows an image of the smectic phase of β-FeOOH nanocrystals studied by Maeda and Maeda [15]. The ordering of these rods on the length scale of visible light explains the brilliant opalescence that had been observed decades earlier in colloidal β-FeOOH sols. In this particular image, one can see that the director of the smectic phase gradually rotates, and that there are numerous defects in the packing of the layers. These features are very typical of packings of different kinds of high aspect ratio nanoparticles that have been observed directly in fluid suspensions [16, 17], at the air-water interface [18, 19], or dried from suspensions onto electron microscope grids [20, 21, 22].

Adding other structure-directing components to a suspension of nanowires can give rise to new phases, although there is still a prevalent tendency for rods to align side by side in order to maximize the free volume. When a small quantity of colloidal spheres was added to suspensions of rodlike bacteriophage fd virus particles, several new phases were found that could be explained theoretically on the basis of hard particle interactions [23]. Mixed rod-sphere phases represent a balance between the entropy of mixing, which favors their formation at low sphere concentration, and phase separation, which occurs at higher sphere concentration because it minimizes the excluded volume. Among these new phases are those in which lamellae of rods and spheres alternate, as well as interesting filamentary phases in which columns of the lamellar structure permeate a nematic or smectic rod packing.

The phase behavior of rods becomes more complex in cases when the interaction is not purely a short-range repulsion. Real nanowires have both long range electrostatic (attractive and/or repulsive) as well as shorter range attractive van der Waals interactions, and these can lead to other kinds of organized aggregates or extended arrays. Fig. 1B shows ribbons of prismatic 16 nm long $BaCrO_4$ nanorods, which were stabilized in a surfactant microemulsion prior to drying for transmission electron microscopy (TEM) [24]. Association of hydrophobic surfactant tail groups is thought to promote

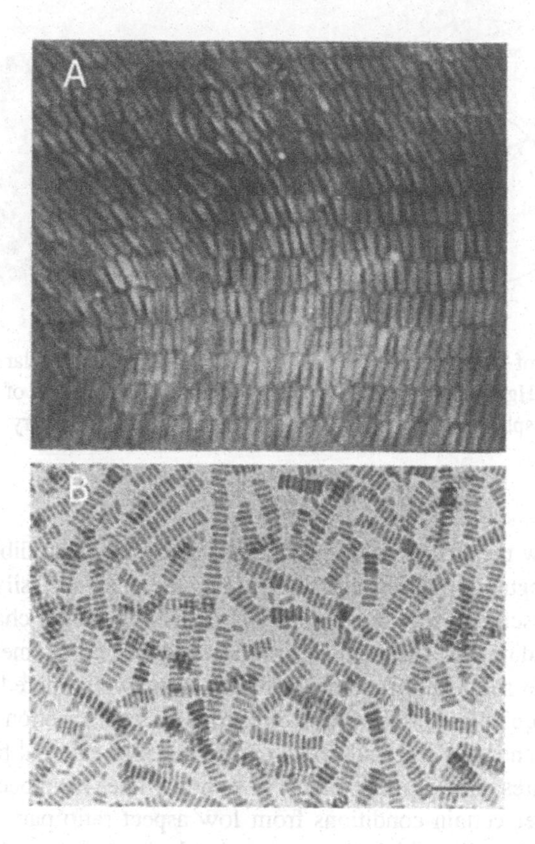

Fig. 1. Electron micrographs of (A) a smectic array of β-FeOOH nanocrystals [15], and (B) ribbons of $BaCrO_4$ nanocrystals in a surfactant microemulsion [24]. Copyright American Chemical Society and Macmillan Magazines, Ltd., reproduced with permission.

formation of the ribbon phase. Similar kinds of bundles have been observed at the air-water interface with surfactant-stabilized $BaCrO_4$ and Au rods [18, 19].

To date relatively little work has been done to tune the attractive and repulsive interactions between rods and spheres. In colloidal systems, if attractive interactions are too strong then disordered aggregates rather than ordered structures are likely to form. However, long range (e.g., like charge in low dielectric media) repulsions between nanowires may lead to interesting mixed phases similar to those that have been found in the "rod packings" of crystalline inorganic solids [25]. For example, garnet structures contain interwoven straight chains of corner-sharing silicate tetrahedra. An interesting crossbar structure is formed in Hg_3AsF_6 [26], the topology of which is shown in Fig. 2. In this structure, metallic chains of disordered Hg atoms run through the crystal in two dimensions, and are interleaved by roughly spherical AsF_6^- ions.

Another strategy for making ordered arrays of nanowires is to begin with a structure directing medium and then synthesize the wires within it. For example, diblock copolymer thin films which micro-phase separate into lamellar structures have been decorated with Au, Ag, Sn, In, Pb, and Bi metal nanoparticles by vapor deposition [27].

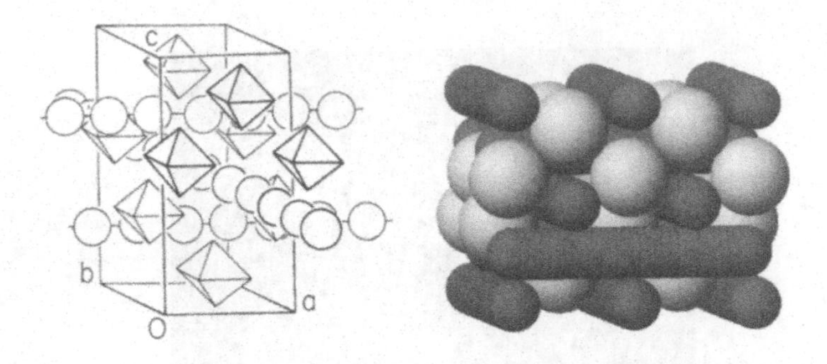

Fig. 2. The structure of "Alchemist's gold", $Hg_{2.86}AsF_6$ [26]. Left, molecular structure showing disordered chains of Hg atoms and AsF_6 octahedra; right, representation of the structure as a lattice of rods and spheres. Copyright Canadian Society of Chemistry, reproduced with permission.

The particles follow the stripes of one of the components of the diblock copolymer, aggregating into regularly spaced chains. Among these metals, silver seems to be unique in that it assembles into continuous filaments rather than chains of spherical particles at high loading [28]. Simulations of related diblock copolymer melts have predicted that rigid rods should under some conditions form well ordered stripe structures [29]. This result is quite interesting because it provides a prescription for making long straight chains of nanowires, particularly in the presence of external fields.

Finally, nanowires and wire-like chains of nanoparticles have been found to form spontaneously under certain conditions from low aspect ratio particles. Korgel and Fitzmaurice found that ellipsoidal silver nanocrystals can spontaneously fuse to form ordered arrays of high aspect ratio nanowires [30]. A similar accretion mechanism has been proposed by Kotov and coworkers for the formation of CdTe nanowires from smaller particles [31]. Chung et al. [32] observed regularly spaced wire-like chains of dodecanethiol-capped spherical silver nanoparticles in Langmuir films. In this case the mechanism of wire formation is not understood, but appears to involve long range repulsive interactions between bands of nanoparticles at the air–water interface.

3. Synthesis of Compositionally Varied Nanowires

In the preceding section it was noted that nanowires and spherical particle in structure-directing media (such as microemulsions, diblock copolymers, and Langmuir films) can organize into assemblies that are different from those predicted from simple hard particle repulsion. The ability to vary the composition along the length of a nanowire provides an additional handle on controlling these more complex interactions, because it allows one to change the surface chemistry on certain parts of the particle. In addition, it becomes possible to build in new electronic, magnetic, or optical functionality that could not easily be achieved in isotropic nanowires.

"Striped" nanowires are generally grown from the vapor or solution phase, using one end of the wire as a seed and changing the composition of the feedstock vapor or solution during growth. Recently, several papers have been published on semiconductor nanowires grown from reactive vapors and solutions [33, 34, 35, 36], and these structures are described in more detail elsewhere in this book. High aspect ratio segmented nanowires can also be grown electrochemically using porous membranes as templates. These kinds of structures follow from the earlier studies of Moskovits [37], Martin [38], and coworkers on the replication of anodic aluminum oxide (AAO) and track-etched polymer membranes. Commercially available (or easily synthesized) membranes containing 10^8–10^{11} pores/cm^2 are filled by electroplating, and the membranes are then dissolved to give suspensions of free-standing nanowires. Monodisperse collections of wires with diameters in the range of 20–300 nm and lengths of several microns are easily obtained by this method. By interrupting the plating process, changing the plating solution, and/or introducing self-assembly reactions between plating steps, it is possible to make many different kinds of structures in which composition is controlled along the length or around the circumference of the nanowire.

Fig. 3 illustrates some of the segmented and concentric nanostructures that have been made by membrane replication. Low-melting metals such as Au, Sn, Pb, Bi, Cu, and Sn can be grown as single crystal segments under appropriate plating conditions [39]. Saturated organic spacers [40] and electronically interesting conjugated molecules [41] can be inserted into the wires as monolayer junctions by interrupting the plating process with self-assembly steps. Similarly, photoconductive and rectifying semiconductor segments can be incorporated by combining electroplating and layer-by-layer assembly steps

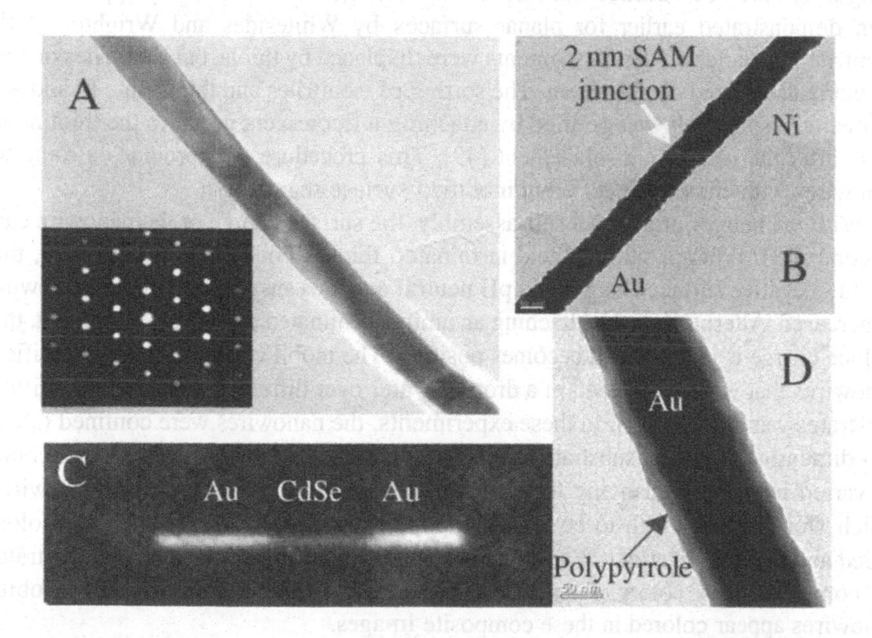

Fig. 3. In-wire and concentric nanostructures made by membrane replication methods. (A) 40 nm diameter single crystal Au nanowire, (B) 70 nm diameter wire containing a 2 nm thick monolayer tunnel junction, (C) 200 nm diameter gold wire containing a photoconductive CdSe segment, (D) 70 nm diameter gold wire with a 7-nm thick poly(pyrrole) coating.

within the pores [42, 43]. Finally, concentric structures can be synthesized either by coating the pore walls prior to electrodeposition of metal, or by removing the template and then coating the wires before or after release from the backing electrode [44].

4. Molecular and Biomolecular Interactions in Nanowire Assembly

The assembly of nanowires suspended from a fluid onto a substrate depends strongly on the interactions between the substrate and the nanowires. These interactions represent the sum of the electrostatic, van der Waals, capillary, and gravitational forces. The ability to modify the chemistry of the substrate and nanowire surfaces with charged groups, hydrophobic and hydrophilic monolayers, and with biomolecules that have specific molecular recognition capability makes it possible to control both the substrate-nanowire and nanowire-nanowire interactions.

4.1. Chemical interactions

The surface modification of substrates such as silicon and gold is well understood [see for example 45, 46, 47, 48]. Similar techniques can be used to functionalize the surface of metal nanowires. Martin et al. [49] showed that different regions of nanowires with alternating Au and Pt segments could be selectively functionalized by using a mixture of isonitriles and thiols. This orthogonal self-assembly approach had been demonstrated earlier for planar surfaces by Whitesides and Wrighton [50]. Isonitrile molecules on the Au segments were displaced by thiols, but isonitriles on the Pt segments resisted displacement. The sorting of isonitriles and thiols onto Pt and Au segments, respectively was verified by attaching a fluorescent group to the thiol molecule after the isonitrile displacement [49]. This procedure thus provides a route to nanowires with chemically different modified surface segments.

With the help of orthogonal self-assembly, the surface charge of the nanowire can be controlled. When a sulfonic acid terminated thiol is bound to a Au segment, the already negative surface charge (in a pH neutral aqueous suspension) on the nanowire is increased. Alternatively, by attaching an amine terminated thiol to a Au segment, the surface charge is inverted and becomes positive. The mobility of chemically modified nanowires that were suspended in a drop of water over different chemically modified substrates was studied [51]. In these experiments, the nanowires were confined to the two-dimensional droplet/substrate interface by gravity, and the in-plane density could be varied by simply changing the nanowire concentration. For micron-sized wires which were large enough to be imaged optically, successive snapshots were color-coded and digitally overlaid. When a nanowire becomes immobilized on the substrate, the complementary colors of that nanowire in each frame add up to black. Mobile nanowires appear colored in these composite images.

Using this technique, surface diffusion constants were measured for Au nanowires as a function of their length, diameter, and surface chemistry [52]. Nanowires with positive surface charge (derivatized with 2-mercaptoethylamine) were immobile on negatively charged substrates derivatized with 2-mercaptoethanesulfonic acid

(MESA). For negatively charged wires, an interesting deviation from the Stokes-Einstein theory was observed. The theory predicts that the diffusion coefficient should scale as $l^{1/3}d^{-2/3}$, where l is the nanowire length and d as the diameter. However, it was found that on both negatively charged and uncharged substrates, there was a stronger length dependence. It was concluded that the nanowires must have a frictional interaction with the substrate, which is expected to be vary linearly with the length of the nanowire. Experiments on neutral surfaces with partially negative nanowires of the same length but different diameters a;sp showed an interesting result. Larger diameter nanowires moved freely, but the smaller diameter nanowires adhered to the substrate. For the bigger diameter nanowires, the repulsive electrostatic interactions between the nanowires and the surface were dominant, but for smaller nanowires the shorter range attractive van der Waals force was more important.

This result shows that the different scaling behavior of opposing forces acting on the nanowires can lead to unusual changes in their mobility. Gravitational forces in particular are more important for larger nanowires, and for those of higher density. With relatively large (>200 nm diameter) Au wires, it was possible to use gravity to drive self-organization into small, oriented smectic arrays on surfaces [51]. Fig. 4 illustrates this idea. Negatively charged (MESA-covered) nanowires were suspended over a negatively charged Au/MESA surface. At low density, the wires are mobile on the surface, as indicated by the color coding of sequential images. However, as the wires diffuse into shallow pits in the surface, they become trapped, and eventually their density is high enough to drive the transition from the isotropic to the bundled ordered phase. Although the ordering is not perfect—because the relatively large wires aggregate irreversibly—the lithographically aligned pits enforce orientational order. This kind of patterned array is potentially useful because it represents one half

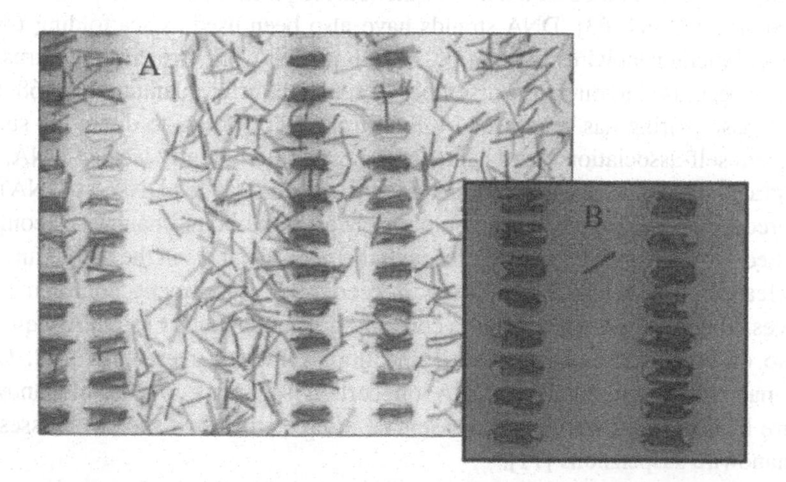

Fig. 4. MESA derivatized Au nanowires suspended in water on a topographically patterned hydrophilic surface. (A) Composite optical micrographs taken at 1 minute intervals; wires in successive frames are color-coded cyan, magenta, and yellow. The wires are mobile on all surfaces but are trapped by the wells. (B) Wire bundles are trapped in the wells after the mobile wires are washed away [51]. Copyright Wiley-VCH, reproduced with permission. See color plate 7.

of a crossbar array, which is an interesting structure for defect-tolerant computing architectures [53, 54, 55].

Template grown carbon nanowires have been prepared by decomposing polymer rods grown in AAO. In this case, somewhat different techniques were used to alter the surface chemistry [56]. Base-etching the AAO membranes gives hydrophilic wires, presumably because it ionizes polar phenolic and carboxylic acid groups on the carbon surface. These wires can be given a permanent negative surface charge by coupling the diazonium derivative of sulfanilic acid to the carbon surface. Alternatively, the surface could be rendered hydrophobic by pyrolysis in an inert atmosphere. Consistent with their lower density, relative to Au, these nanowires were only weakly retained by gravity in the kinds of pits shown in Fig. 4.

Orthogonal self-assembly has also been used to direct the alignment of single carbon nanotubes onto chemically defined areas of a substrate or onto chemically modified electrodes [57, 58]. Carbon nanotube devices have been fabricated using "pick and place" mechanical manipulation, for example by using an AFM tip [59, 60]. The chemical alignment technique offers a substantial advantage in terms of parallel fabrication of these kinds of devices.

4.2. Biomolecular interactions

Biochemical interactions such as DNA base pairing and protein-substrate binding have the advantage of very high specificity over simple anion-cation or van der Waals interactions. This means that in principle they offer the potential to program the assembly of much more complex structures from inorganic components. This idea was pioneered by Seeman, who showed that topologically complex structures such as knots and closed polyhedra could be made to assemble from appropriate DNA sequences [61]. Later, the programmed assembly of spherical Au particles using DNA interactions was demonstrated [62, 63]. DNA strands have also been used as scaffolding for the synthesis of metal nanowires [64, 65, 66, 67]. In related work, proteins and virus particles have been used to direct the assembly of semiconductor quantum dots [68, 69].

DNA base-pairing has been used by Keating and coworkers to direct the surface binding and self-association of Au nanowires. Using fluorescently labeled DNA, they demonstrated that Au nanowires derivatized with single-stranded DNA (ss-DNA) was able to recognize its complement in solution. Membrane-bound nanowires could be derivatized on their tips with ss-DNA through the open channels in the membrane, and then released. These tips were able to recognize fluorescent complementary sequences. Specific binding to surfaces containing complementary ss-DNA sequences was also observed for both segmented and non-segmented nanowires [70]. Using striped nanowires that bound ss-DNA on certain segments to control nanowire-nanowire binding, they were able to assemble simple structures such as crosses and T's in nanowire suspensions [71].

Another interesting application of these DNA-derivatized striped nanowires was to act as optical barcodes for complementary DNA sequences in solution [72]. Different metal stripes (for example, Au and Ag) have different reflectivity and can be easily distinguished by optical microscopy. Thus, a reflectance map of wires (which gives a map of the ss-DNA sequences on their surfaces, as read from the stripe pattern) can be

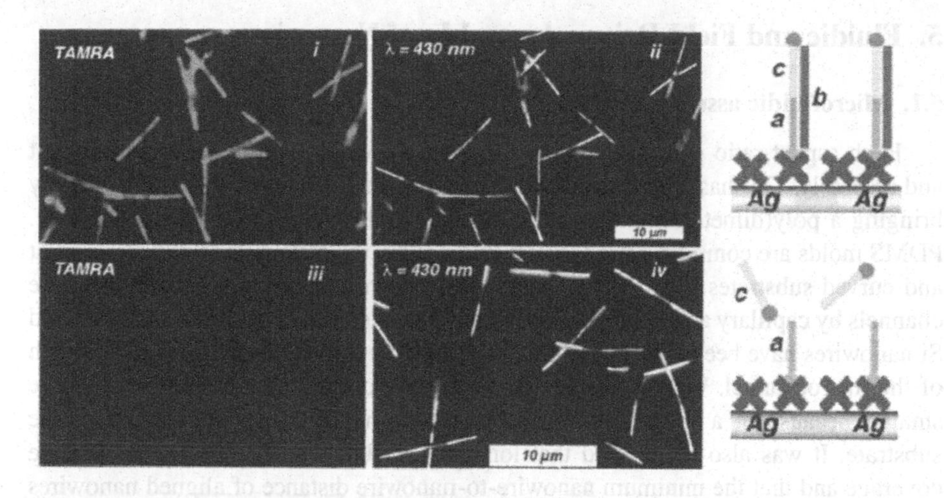

Fig. 5. Nanowire "bar codes" used for ss-DNA detection. The fluorescence image at the upper left shows binding of complementary DNA in a sandwich assay, which captures a Rhodamine-labeled complementary ss-DNA strand. The image at the upper right shows that the bar code on the same wires can be read by optical microscopy. The lower image shows a control experiment conducted without the target ss-DNA strand [72]. Copyright American Association for the Advancement of Science, reproduced with permission.

correlated with a fluorescence map of complementary DNA. Fig. 5 demonstrates this idea with a "sandwich" DNA assay. The target ss-DNA analyte sequence, which is complementary to both the ss-DNA on the nanowire and to fluorescently labeled ss-DNA in solution, gives a fluorescent signal only on complementary wires. No fluorescence is observed in the absence of the analyte sequence. If two or more kinds of nanowires are imaged together, only those with the complementary sequence are fluorescent. Effectively, this method allows one to make a gene chip in a drop of buffer solution using a collection of appropriately barcoded and DNA-derivatized nanowires.

Protein-substrate interactions have also been used to create or derivatize nanowire structures. Sapp et al. showed that latex and Au spheres could be bound to the ends of Au nanowires that were coated with conducting polymer films. The end of the conducting polymer sheath was covalently linked to a biotin derivative. The non-covalent interaction of biotin with avidin on the surface of the spheres was then used to link the spheres to the tips of the wires.

Proteins can form filamentary tertiary and quaternary structures, and these appear to be quite interesting for controlling the placement of wires on complementary surfaces. So far, there appear to be no examples in which these biomolecular nanostructures have been coverted into metallic wires. Matsui and coworkers [73] demonstrated that 20 nm diameter high aspect ratio protein tubules could bridge between lithographically defined Au pads on surfaces. Padilla et al. [74] assembled ca. 50 nm diameter filaments, as well as various polyhedral structures, from fusion proteins. It is clear that although this approach is at an early stage of development, it offers significant promise for organizing nanowires and other functional inorganic materials into deterministic superstructures.

5. Fluidic and Field-Driven Assembly of Nanowires

5.1. Microfluidic assembly

High aspect ratio particles lend themselves especially well to fluidic alignment and assembly. This has been demonstrated by using microchannels that are formed by bringing a poly(dimethylsiloxane) (PDMS) mold into contact with a flat substrate. PDMS molds are commonly used in soft lithography to create microstructures on flat and curved substrates [75, 76, 77]. Nanowire suspensions are introduced into these channels by capillary action [75, 78, 79] or at controlled flow rates [80]. GaP, InP, and Si nanowires have been found to align along the flow direction over the whole length of the microchannel. With increasing flow rate, the deviations in alignment become smaller because of a shear force that acts on the nanowires at the surface of the substrate. It was also established that longer flow duration leads to higher surface coverage and that the minimum nanowire-to-nanowire distance of aligned nanowires depends on the lengths of the nanowires.

Nanowire—substrate interactions can be modified chemically, as described above, and this has interesting and useful consequences for fluidic alignment. Nanowires with a negative surface charge were more rapidly deposited on amine-derivatized surfaces, which were positively charged under the same conditions [80]. By using patterned substrates, this complementary chemistry could be further employed to direct the position of nanowire deposition in the microchannels. Nanowires were preferentially found on the complementary modified pattern in the microchannel. Crossed nanowire junctions could also be created by a layer-by-layer approach. After each deposition, the flow direction could be changed leading to crossed and more complex nanowire superstructures. The nanowire-substrate interaction was found to be strong enough to enable sequential flow-steps without destroying the nanowire alignments of preceding flow steps.

When PDMS microchannels were filled by the capillary method, nanowire bundles (formed from $[Mo_3Se_3^-]_\infty$ molecular wires, conducting polymer nanowires, and carbon nanotubes [75]),—often with double helix formation between nanowires—were found at the corners of the microchannels. The bundles were continuous and extended up to several millimeters. The diameter of these bundles depended on the concentration of the nanowire suspension and microchannel volume. At low $[Mo_3Se_3^-]_\infty$ concentrations, nanowire bundles as small as 1–3 nm in diameter could be found [75]. The formation of bundles at the corners of the microchannels was explained in terms of a concentration effect that arises during solvent evaporation. Because the solvent (dimethylsulfoxide) wetted the PDMS walls of the microchannels, the meniscus migrated to the corners of the channels leading to a high concentration of wires in these areas.

5.2. Electric-field assisted assembly of nanowires

The alignment process of anisotropic nanoparticles suspended in a dielectric solvent was first studied by electro-optical techniques such as electric birefringence, electric light scattering, dichroism, and absorbance measurements [81, 82, 83]. These methods were used to probe the geometrical as well as the electrical properties of anisotropic nanoparticle suspensions. When an electric field is applied to a nanowire suspension, the electric field polarizes the nanowires, which forces them to align with their length oriented parallel to the direction of the electric field. The alignment quality is dependent on the field strength and, since the alignment process is in competition with the Brownian motion of the nanowires, on the thermal energy. Perfect alignment of suspended anisotropic nanoparticles is not obtainable since the required electric fields strengths cannot be reached in experimental conditions. In addition, there is a dependence on the anisotropy in electric polarizability of the nanowires, which has been found to increase with increasing nanowire length in gold nanowires and carbon nanotubes [81, 84].

For field-aligned metallic nanowires, the alignment quality can be determined by means absorbance measurements using polarized light. The absorption spectrum of a nanowire suspension in the absence of an electric field shows two maxima corresponding to the transverse and longitudinal resonances. With the application of an electric field, the nanowires align and in the ideal case of complete nanowire alignment, the transverse resonance can be extinguished by using light that is polarized parallel to the electric field. This leads to just one maximum in the absorbance spectrum. Alternatively, the absorbance maximum of the longitudinal resonance disappears when the light is polarized perpendicular to the electric field [81].

For most nanowire alignment purposes, alternating electric fields have been used. At high frequencies, alternating fields can prevent shielding of the polarization on the nanowire caused by solvents or ions surrounding the nanowires [85, 86]. Nanowires possess surface charges and are surrounded by ions of opposite charge. At low frequencies, these ions and polar solvent molecules move with the voltage variations of the alternating field and the induced polarization on the nanowire. At high frequencies, the surrounding ions and solvent molecules cannot follow these variations, and this results in greater net polarization (electrical dipole) of the nanowire ends. The dipole in these nanowires is attracted to the electrodes and moves towards the nearest electrode since the Coulomb force attracting the dipole is stronger for the nearest electrode. The Coulomb force increases with increasing dipole moment. Thus, longer nanowires experience a higher attractive force from the electrodes than do shorter wires, and this effect has been used to separate carbon nanotubes from smaller nanoparticles [84].

Alternating electric fields have also been used to align nanowires between two lithographically defined metal pads for electrical characterization [86, 87, 88]. In this case, isolated electrodes with an interdigitated finger pattern on a silicon dioxide substrate were used (Fig. 6a). The substrate and the electrodes were covered with 500 nm of an insulating layer of silicon nitride. This protective layer was thinned to 100 nm above the interdigitated finger electrodes. When an electric field is applied to the electrode fingers, the metal nanowires polarize. This causes them to move in the direction of increasing field strength. The highest field strengths on the electrodes were found where the protective silicon nitride layer was thinnest, i.e., above the finger electrodes. Field simulations showed that when the nanowires are in close range of these electrodes, the electric field strength increases proportionally to the inverse of the distance from the electrode [86]. This strong force results in the nanowire alignment between the two electrode fingers.

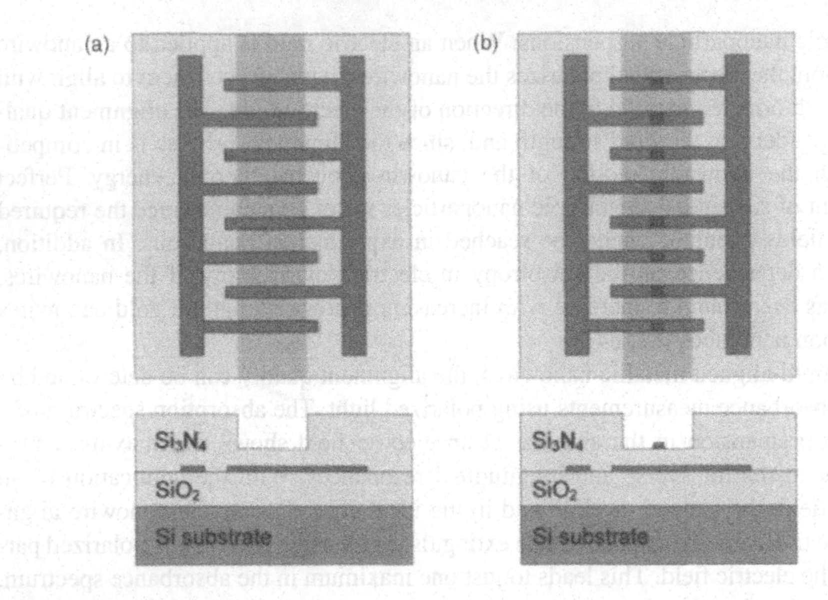

Fig. 6. Top- and cross-sectional views of the electrode structures used in the field assisted assembly experiments [86]. (a) The 3-μm-wide Ti/Au electrode fingers are spaced 5 μm apart. The Si_3N_4 is thinned over the fingers to enhance the electric field in that region. (b) The same electrode structure with $4 \times 4\ \mu m^2$ top electrodes added to improve the alignment precision. Copyright American Institute of Physics, reproduced with permission.

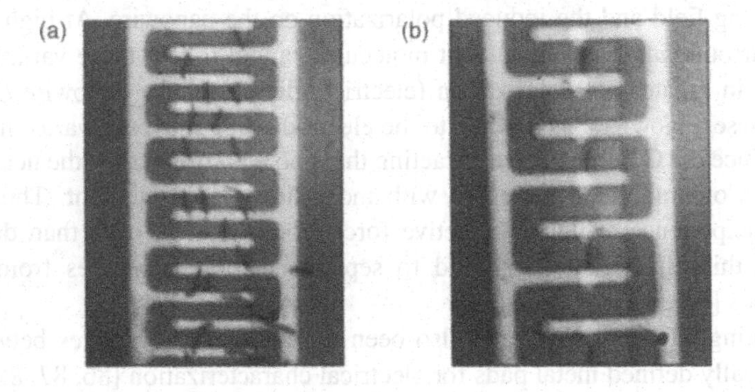

Fig. 7. Optical microscope image of 5 μm long, 200-nm-diameter Au nanowires aligned by applying a (a) 30 V, 1 kHz voltage to the structure without upper field electrodes, and (b) 20 V, 1 kHz voltage to the structure with upper field electrodes [86]. Copyright American Institute of Physics, reproduced with permission.

This design resulting in the alignment of several nanowires to each pair electrodes as shown in the optical micrograph in Fig. 7a. In order to align only a single nanowire to each pair of electrodes, it was necessary to concentrate the field strength into a small area on the electrodes fingers. This was done by placing small metal electrodes on top of the thin silicon nitride layer at the centers of the electrode fingers.

These isolated metal pads develop higher fields strengths than their surrounding areas because of capacitive coupling with the buried electrodes. In an AC field, nanowires are predominantly aligned to these pads. Nanowire alignment on top of these areas of adjacent electrodes extinguished the electric field strength between these areas and reduced it by factor two towards neighboring electrodes (Figs. 6b and 7b). As with carbon nanotube alignment, higher frequencies cause better alignment.

This electrofluidic alignment method has been further developed and crossed nanowire junctions have been obtained by applying a layer-by-layer technique [89]. By changing the electric field direction for each layer, nanowires could be assembled in different directions and crossed nanowire junctions were created.

5.3. Magnetic-field assisted assembly of nanowires

Nanowires can be made from ferromagnetic metals by solution organometallic techniques, or by electrochemical growth in membranes. These magnetic wires are interesting from the point of view of spintronics, because they can in principle be made with alternating stripes of hard and soft magnets. As a consequence of their small cross-sectional areas, relatively high current densities can be achieved at low absolute currents. To date, however, there have been only a few examples of magnetic-field assisted nanowire assembly.

Ni nanowires with diameters of ca. 200 nm and lengths of a few μm were assembled into long chains using magnetic interactions. Additionally, single nanowires were confined onto lithographically defined magnetic features on a substrate using external magnetic fields [90, 91]. These nanowires were synthesized in the cylindrical pores of AAO membranes. After their release from the membrane, the Ni nanowires were permanently magnetized along their long axes with an external magnetic field of 1 kG. These suspended magnetized nanowires showed a tendency to aggregate in a random manner in the absence of an external magnetic field. An external field smaller than 10 G can restrain the random agglomeration and align the suspended nanowires, resulting in head-to-tail nanowire chains, which can be several hundreds of μm long. These nanowire chain agglomerates are stable enough to withstand the capillary forces that arise from slow evaporation of the solvent.

The dynamics of nanowire chain formation on glass substrates were studied using video microscopy. The joining of two nanowires proceeds first with a slow motion of the nanowires towards each other which is progressively accelerated as the nanowires join. These events depend strongly on the viscosity of the solvent, as experiments in different solvents revealed, and on the force between the magnetic dipoles on the wires. Using these two factors, the dynamics of joining two nanowires were modeled and found to be in quantitative agreement with experiment [91]. The dipolar forces between two nanowires in close proximity were found to be stronger than the applied external magnetic field, enabling two nanowires to rotate slightly out of alignment with the external magnetic field in order to connect. Once the nanowires joined, the nanowire chains returned to their original orientation parallel to the external magnetic field.

The influence of magnetic fields on magnetized suspended nanowires was further extended to confine nanowires to the poles of magnetic electrodes. These Ni electrodes had an elliptical shape with a length of 80 μm and were magnetized along their long axes. The distance between the electrodes was 6 μm, short enough for a single nanowire to bridge the gap. Using a small external magnetic field in the direction of the long axis of the Ni electrodes, Ni nanowires were confined to the gap between two electrodes. The small external magnetic field helped to restrain the nanowires from agglomerating and aligned the suspended Ni nanowires in the direction of the long axis of the electrodes, positioning them ideally for bridging the gap between two adjacent electrodes. This setup could be used to measure the resistance of Ni nanowires. However, because of electrical contact problems attributed to the oxide layer on Ni, the experiments were repeated with Pt-Ni-Pt segmented nanowires that were confined to the electrodes in the same manner.

Co nanoparticles synthesized by the injection of an organometallic precursor into a hot surfactant mixture assemble into interesting superstructures upon evaporation of the solvent of the colloidal suspension [92, 93]. Depending on the reaction conditions, spherical Co nanoparticles or disks are obtained. The assembly of superstructures was attributed to a combination of surface tension, van der Waals interactions, and magnetic interaction between the nanoparticles. It was found that small spherical nanoparticles (10 nm and smaller in diameter) form defined two- and three-dimensional superstructures. Bigger spherical nanoparticles (16 nm in diameter) were found to produce chains and closed loops while nanowires formed ribbons of wires. The reason for the different assembly behavior of the different kind of nanoparticles was explained with superparamagnetic and ferromagnetic behavior of the nanoparticles. The small spherical nanoparticles behave as superparamagnetic particles, which function as single magnetic dipoles. Since this dipole can move freely in superparamagnetic particles, these particles do not show any preference on their orientation towards a neighboring particle and their superstructures were similar to the colloidal crystal structures of silica or polystyrene spheres. Since chains and closed loops were observed with bigger nanoparticles, these particles do show a preference on where the neighboring nanoparticle can be located. These particles must therefore have a permanent anisotropy in their magnetic dipole that controls the assembly of superstructures.

Nickel and iron nanowires have also been synthesized by solution-phase methods in the absence of templates [94, 95]. The Ni nanowire synthesis was based on the decomposition of a neutral cycloocta-1,5-diene complex, $Ni(COD)_2$. In order to control the shape of the nanoparticles, an amine (hexadecylamine) was added. It was postulated that this amine selectively coordinated certain crystal faces of the nanowire, promoting anisotropic growth. By using this method, nanowires with dimensions of 15 nm in length and 4 nm in diameter could be obtained. Iron nanowires were prepared similarly, using $Fe(CO)_5$ as a precursor and didodecyldimethylammonium bromide as the surfactant to control particle morphology. The Ni were shown to be ferromagnetic. This ferromagnetism is also believed to be responsible for the observed assembly of these Ni nanowires. The ferromagnetism creates a permanent magnetic dipole in the wires. In order to minimize the magnetostatic energy of the system, it is most favorable for the wires to align in a ribbon fashion, with their long axes parallel to each other.

6. Conclusions

The past decade has seen remarkable progress in the synthesis of nanowire and other shape-controlled particles of different compositions. One can now have confidence that it will ultimately be possible to make almost any inorganic material on the nanoscale with some measure of shape control, and that these nanoparticles will be sufficiently manufacturable for "real world" applications. The stage is now set for these particles to be used in nanoscale electronic devices and numerous other areas. The assembly of particles into predictable organized arrays is the key to realizing these applications. The basic physical chemistry of nanowire phase behavior is now well understood, and preliminary work has been done to understand the physical forces and scaling laws that can be used to organize different kinds of nanowire structures. Biomolecular interactions appear to be especially promising for creating complex organized assemblies of nanowires and other kinds of inorganic particles, but this field is now truly just beginning. The next decade should prove to be an exciting one for research on the assembly of nanowires and other particles into functional systems.

Acknowledgments

This work was supported by the Office of Naval Research and the Defense Advanced Research Projects Agency for support of this work on grant N00014-01-10659. We thank our colleagues Theresa Mayer, Thomas Jackson, Christine Keating, Seth Goldstein, and Michael Natan, as well as the students, postdocs, and visitors (Ben Martin, Sarah St. Angelo, Jeremiah Mbindyo, Brian Reiss, David Peña, Nina Kovtyukhova, Jong-Sung Yu, Sheila Nicewarner-Peña, Baharak Razavi, Jim Mattzela, and Peter Smith) who contributed to the work described in this review.

Literature Cited

1. X. Peng, L. Manna, W. Yang, J. Wickham, E. Scher, A. Kadavanich and A. P. Alivisatos, "Shape control of CdSe nanocrystals", *Nature* **404** (2000) 59.
2. L. Manna, E. C. Scher and A. P. Alivisatos, "Synthesis of soluble and processable rod-, arrow-, teardrop-, and tetrapod-shaped CdSe nanocrystals", *J. Am. Chem. Soc.* **122** (2000) 12700.
3. R. Jin, Y. Cao, C. A. Mirkin, K. L. Kelly, G. C. Schatz and J. G. Zheng, "Photoinduced conversion of silver nanospheres to nanoprisms", *Science* **294** (2001) 1901.
4. S.-M. Lee, Y. Jun, S.-N. Cho and J. Cheon, "Single-crystalline star-shaped nanocrystals and their evolution: programming the geometry of nano-building blocks", *J. Am. Chem. Soc.* **124** (2002) 11244.
5. C. M. Lieber, "The incredible shrinking circuit", *Scien. Amer.* **285** (2001) 58.
6. X. Duan, Y. Huang and C. M. Lieber, "Nonvolatile memory and programmable logic from molecule-gated nanowires", *Nano Lett.* **2**(5) (2002) 487.
7. J. Kong, M. G. Chapline and H. Dai, "Functionalized carbon nanotubes for molecular hydrogen sensors", *Adv. Mater.* **13** (2001) 1384.

8. F. Favier, E. C. Walter, M. P. Zach, T. Benter and R. M. Penner, "Hydrogen sensors and switches from electrodeposited palladium mesowire arrays", *Science* **293** (2001) 2227.
9. Y. Cui, Q. Wei, H. Park and C. M. Lieber, "Nanowire nanosensors for highly sensitive and selective detection of biological and chemical species", *Science* **293** (2001) 1289.
10. S. Fan, M. G. Chapline, N. R. Franklin, T. W. Tombler, A. M. Cassell and H. Dai, "Self-oriented regular arrays of carbon nanotubes and their field emission properties", *Science* **283** (1999) 512.
11. H. Kind, H. Yan, B. Messer, M. Law and P. Yang, "Nanowire ultraviolet photodetectors and optical switches", *Adv. Mater.* **14** (2002) 158.
12. W. U. Huynh, J. J. Dittmer, A. P. Alivisatos, "Hybrid nanorod-polymer solar cells", *Science* **295** (2002) 2425.
13. L. Onsager, "The effects of shapes on the interaction of colloidal particles", *Ann. N.Y. Acad. Sci.* **51** (1949) 627.
14. An interesting consequence of the hard particle model is that in the absence of attractive interactions, the transition to an ordered phase should depend only on density and not on temperature.
15. H. Maeda and Y. Maeda, "Atomic force microscopy studies for investigating the smectic structures of colloidal crystals of β-FeOOH", *Langmuir* **12** (1996) 12 1446.
16. B. J. Lemaire, P. Davidson, J. Ferré, J. P. Jamet, P. Panine, I. Dozov and J. P. Jolivet, "Outstanding magnetic properties of nematic suspensions of goethite (a-FeOOH) nanorods", *Phys. Rev Lett.* **88**(12) (2002) 125507.
17. L.-S. Li, J. Walda, L. Manna and A. P. Alivisatos, "Semiconductor nanorod liquid crystals", *Nano Lett.* **2**(6) (2002) 557.
18. F. Kim, S. Kwan, J. Akana and P. Yang, "Langmuir-Blodgett nanorod assembly", *J. Am. Chem. Soc.* **123** (2001) 4360.
19. P. Yang and F. Kim, "Langmuir-Blodgett assembly of one-dimensional nanostructures", *Chem. Phys. Chem.* **3** (2002) 503.
20. S.-S. Chang, C.-W. Shih, C.-D. Chen, W.-C. Lai and C. R. C. Wang, "The shape transition of gold nanorods", *Langmuir* **15** (1999) 701.
21. B. Nikoobakht, Z. L. Wang and M. A. El-Sayed, "Self-assembly of gold nanorods", *J. Phys. Chem. B* **104** (2000) 8635.
22. N. R. Jana, L. Gearheart and C. J. Murphy, "Wet chemical synthesis of silver nanorods and nanowires of controllable aspect ratio", *Chem. Commun.* (2001) 617.
23. M. Adams, Z. Dogic, S. L. Keller and S. Fraden, "Entropically driven microphase transitions in mixtures of colloidal rods and spheres", *Nature* **393** (1998) 349.
24. M. Li, H. Schnablegger and S. Mann, "Coupled synthesis and self-assembly of nanoparticles to give structures with controlled organization", *Nature* **402** (1999) 393.
25. S. Lidin, M. Jacob and S. Andersson, "A mathematical analysis of rod packings", *J. Solid State Chem.* **114** (1995) 36.
26. I. D. Brown, B. D. Cutforth, C. G. Davies, R. J. Gillespie, P. R. Ireland and J. E. Vekris, "Alchemists' gold, $Hg_{2.86}AsF_6$; an X-ray crystallographic study of a novel disordered mercury compound containing metallically bonded infinite cations", *Can. J. Chem.* **52** (1974) 791.
27. W. A. Lopes and H. M. Jaeger, "Heirarchical self-assembly of metal nanostructures on diblock copolymer scaffolds", *Nature* **414** (2001) 735.
28. W. A. Lopes, "Nonequilibrium self-assembly of metals in diblock copolymer templates", *Phys. Rev. E* **65**(031606) (2002) 1.
29. K. Chen and Y.-Q. Ma, "Ordering stripe structures of nanoscale rods in diblock copolymer scaffolds", *J. Chem. Phys.* **116**(18) (2002) 7783.
30. B. A. Korgel and D. Fitzmaurice, "Self-assembly of silver nanocrystals into two-dimensional nanowire arrays", *Adv. Mater.* **10**(9) (1998) 661.

31. Z. Tang, N. A. Kotov and M. Giersig, "Spontaneous organization of single CdTe nanoparticles into luminescent nanowires", *Science* **297** (2002) 237.
32. S.-W. Chung, G. Markovich and J. R. Heath, "Fabrication and alignment of wires in two dimensions", *J. Phys. Chem. B* **102** (1998) 6685.
33. P. D. Markowitz, M. P. Zach, P. C. Gibbons, R. M. Penner and W. E. Buhro, *J. Am. Chem. Soc.* **123** (2001) 4502.
34. M. T. Bjork, B. J. Ohlosson, T. Sass, A. I. Persson, C. Thelander, M. H. Magnusson, K. Deppert, L. R. Wallenberg and L. Samuelson, *Nano Lett.* **2** (2002) 87.
35. Y. Wu, R. Fan and P. Yang, "Block-by-block growth of single-crystalline Si/Si-Ge superlattice nanowires", *Nano Lett.* **2**(2) (2002) 83.
36. M. S. Gudiksen, U. J. Lauhon, J. Wang, D. C. Smith and C. M. Lieber, "Growth of nanowire superlattice structures for nanoscale photonics and electronics", *Nature* **415** (2002) 617.
37. D. Al-Mawlawi, C. Z. Liu and M. Moskovits, "Nanowires formed in anodic oxide templates", *J. Mater. Res.* **9** (1994) 1014; D. Routkevitch, T. Bigioni, M. Moskovits and J. M. Xu, "Electrochemical fabrication of CdS nanowire arrays in porous anodic aluminum oxide templates", *J. Phys. Chem.* **100** (1996) 14037.
38. C. R. Martin, "Nanomaterials: a membrane-based synthetic approach", *Science* **266** (1994) 1961; C. R. Martin, "Template synthesis of electronically conductive polymer nanostructures", *Acc. Chem. Res.* **28** (1995) 61; C. R. Martin, "Membrane-based synthesis of nanomaterials", *Chem. Mater.* **8** (1996) 1739.
39. M. Tian and M. H. W. Chan (unpublished results).
40. J. K. N. Mbindyo, T. E. Mallouk, I. Kratochvilova, B. Razavi, T. S. Mayer and T. N. Jackson, "Template synthesis of metal nanowires containing monolayer molecular junctions", *J. Am. Chem. Soc.* **124** (2002) 4020.
41. I. Kratochvilova, M. Kocirik, A. Zambova, J. Mbindyo, T. E. Mallouk and T. S. Mayer, "Room temperature negative differential resistance in molecular nanowires", *J. Mater. Chem.* **12** (2002) 2927.
42. N. I. Kovtyukhova, B. R. Martin, J. K. N. Mbindyo, P. A. Smith, B. Razavi, T. S. Mayer and T. E. Malllouk, "Layer by layer assembly of rectifying junctions in and on metal nanowires", *J. Phys. Chem. B* **105** (2001) 8762.
43. D. J. Peña, J. K. N. Mbindyo, A. J. Carado, T. E. Mallouk, C. D. Keating, B. Razavi and T. S. Mayer, "Template growth of photoconductive metal-CdSe-metal nanowires", *J. Phys. Chem. B* **106**(30) (2002) 7458.
44. J. S. Yu, J. Y. Kim, S. Lee, J. K. N. Mbindyo, B. R. Martin and T. E. Mallouk, "Template synthesis of polymer-insulated colloidal gold nanowires with reactive ends", *J. Chem. Soc., Chem. Comm.* (2000) 2445.
45. D. L. Allara, "Critical issues in applications of self-assembled monolayers", *Biosens. Bioelectron.* **10**(9–10) (1995) 771.
46. P. E. Laibinis, G. M. Whitesides, D. L. Allara, Y. T. Tao, A. N. Parikh and R. G. Nuzzo, "Comparison of the structures and wetting properties of self-assembled monolayers of normal alkanethiols on the coinage metal surfaces, Cu, Ag, Au", *J. Am. Chem. Soc.* **113**(19) (1991) 7152.
47. S. R. Wasserman, G. M. Whitesides, I. M. Tidswell, B. M. Ocko, P. S. Pershan and J. D. Axe, "The structure of self-assembled monolayers on silicon—a comparison of results from ellipsometry and low-angle x-ray reflectivity", *J. Am. Chem. Soc.* **111**(15) (1989) 5852.
48. D. L. Angst and G. W. Simmons, "Moisture absorption characteristics of organosiloxane self-assembled monolayers", *Langmuir* **7**(10) (1991) 2236.
49. B. R. Martin, D. J. Dermody, B. D. Reiss, M. Fang, L. A. Lyon, M. J. Natan and T. E. Mallouk, "Orthogonal self assembly on colloidal gold-platinum nanorods", *Adv. Mater.* **11** (1999) 1021.

50. J. J. Hickman, P. E. Laibinis, D. I. Auerbach, C. Zou, T. J. Gardner, G. M. Whitesides and M. S. Wrighton, "Toward orthogonal self-assembly of redox active molecules on Pt and Au—selective reaction of disulfide with Au and isocyanide with Pt", *Langmuir* **8** (1992) 357.
51. B. R. Martin, S. K. St. Angelo and T. E. Mallouk, "Interactions between suspended nanowires and patterned surfaces", *Adv. Funct. Mater.* (in press).
52. S. K. St. Angelo, C. C. Waraska and T. E. Mallouk, "Diffusion of gold nanorods on chemically functionalized surfaces", *Adv. Mater.* (submitted).
53. C. P. Collier, G. Mattersteig, E. W. Wong, Y. Luo, K. Beverly, J. Sampaio, F. Raymo, J. Stoddart and J. R. Heath, "A [2]catenane-based solid state electronically reconfigurable switch", *Science* **289** (2000) 1172.
54. T. Rueckes, K. Kim, E. Joselevich, G. Tseng, C.-L. Cheung and C. M. Lieber, "Carbon nanotube-based nonvolatile random access memory for molecular computing", *Science* **289** (2000) 94.
55. S. C. Goldstein, M. Budiu, "Nanofabrics: spatial computing using molecular electronics", Proceedings of 28th Annual International Symposium. *Computer Architecture* (2001) (in press).
56. A. Amma, B. Razavi, S. K. St. Angelo, T. S. Mayer and T. E. Mallouk, "Synthesis, chemical modification, and surface alignment of carbon nanowires", *Adv. Funct. Mater.* (submitted).
57. J. Liu, J. Casavant, M. Cox, D. A. Walters, P. Boul, W. Lu, A. J. Rimberg, K. A. Smith, D. T. Colbert and R. E. Smalley, "Controlled deposition of individual single-walled carbon nanotubes on chemically functionalized templates", *Chem. Phys. Lett.* **303** (1999) 125.
58. M. Burhard, G. Duesberg, G. Philipp, J. Muster and S. Roth, "Controlled adsorption of carbon nanotubes on chemically modified electrode arrays", *Adv. Mat.* **10**(8) (1998) 584.
59. M. S. Fuhrer, J. Nygard, L. Shih, M. Forero, Y.-G. Yoon, M. S. C. Mazzoni, H. J. Choi, J. Ihm, S. G. Louie, A. Zetti and P. L. McEuen, "Crossed nanotube junctions", *Science* **288** (2000) 494.
60. T. Rueckes, K. Kim, E. Joselevich, G. Y. Tseng, C.-L. Cheung and C. M. Lieber, "Carbon nanotube-based nonvolatile random access memory for molecular computing", *Science* **289** (2000) 94.
61. N. C. Seeman, "DNA nanotechnology: novel DNA constructions", *Annu. Rev. Biophys. Biomol. Struct.* **27** (1998) 2245.
62. C. A. Mirkin, R. L. Letsinger, R. C. Mucic and J. J. Storhoff, "A DNA-based method for rationally assembling nanoparticles into macroscopic materials", *Nature* **382** (1996) 607.
63. C. J. Loweth, W. B. Caldwell, X. Peng, A. P. Alivisatos and P. G. Schultz, "DNA-based assembly of gold nanocrystals", *Angew. Chem. Int. Ed.* **38** (1999) 1808.
64. E. Braun, Y. Eichen, U. Sivan and G. Ben-Yoseph, "DNA-templated assembly and electrode attachment of a conducting silver wire", *Nature* **391** (1998) 775.
65. O. Harnack, W. E. Ford, A. Yasuda and J. M. Wessels, "Tris(hydroxymethyl)phosphine-capped gold particles templated by DNA as nanowire precursors", *Nano Lett.* **2**(9) (2002) 919.
66. J. Richter, M. Mertig, W. Pompe, I. Monch and H. K. Schackert, "Construction of highly conductive nanowires on a DNA template", *Appl. Phys. Lett.* **78** (2001) 536.
67. F. Patolsky, Y. Weizmann, O. Lioubashevski and I. Willner, "Au nanoparticle nanowires based on DNA and polylysine templates", *Angew. Chem. Int. Ed.* **41** (2002) 2323.
68. T. Douglas and M. Young, "Host-guest encapsulation of materials by assembled virus protein cages", *Nature* **393** (1998) 152.
69. S.-W. Lee, C. Mao, C. E. Flynn and A. M. Belcher, "Ordering of quantum dots using genetically engineered viruses", *Science* **296** (2002) 892.

70. J. K. N. Mbindyo, B. D. Reiss, B. R. Martin, C. D. Keating, M. J. Natan and T. E. Mallouk, "DNA-directed assembly of gold nanowires on complementary surfaces", *Adv. Mater.* **13** (2001) 249.
71. B. D. Reiss, J. K. N. Mbindyo, B. R. Martin, S. R. Nicewarner, T. E. Mallouk, M. J. Natan and C. D. Keating, "DNA-directed assembly of anisotropic nanoparticles on lithographically defined surfaces and in solution", *MRS Symp. Proc.* **635** (2001) C6.2.1–6.2.6.
72. S. R. Nicewarner-Peña, R. G. Freeman, B. D. Reiss, L. He, D. J. Peña, I. D. Walton, R. Cromer, C. D. Keating and M. J. Natan, "Submicrometer metallic barcodes", *Science* **294** (2001) 137.
73. H. Matsui, P. Porrata and G. E. Douberly, Jr., "Protein tuule immobilization on self-assembled monolayers on Au surfaces", *Nano Lett.* **1** (2001) 461.
74. J. E. Padilla, C. Colovos and T. O. Yeates, "Nanohedra: using symmetry to design self assembling protein cages, layers, crystals, and filaments", *Proc. Nat. Acad. USA* **98** (2001) 2217.
75. B. Messer, J. H. Song and P. Yang, "Microchannel networks for nanowire patterning", *J. Am. Chem. Soc.* **122** (2000) 10232.
76. Y. Xia and G. M. Whitesides, "Soft lithography", *Angew. Chem. Int. Ed.* **37** (1998) 550.
77. D. C. Duffy, J. Cooper McDonald, J. A. Schueller and G. M. Whitesides, "Rapid prototyping of microfluidic systems in poly(dimethylsiloxane)", *Anal. Chem.* **70**(23) (1998) 4974.
78. Y. Wu, H. Yan, M. Huang, B. Messer, J. H. Song and P. Yang, "Inorganic semiconductor nanowires: rational growth, assembly, and novel properties", *Chem. Eur. J.* **8**(6) (2002) 1261.
79. H.-J. Muhr, F. Krumeich, U. P. Schönholzer, F. Bieri, M. Niederberger, L. J. Gauckler and R. Nesper, "Vanadium oxide nanotubes—a new flexible vanadate nanophase", *Adv. Mat.* **12**(3) (2000) 231.
80. Y. Huang, X. Duan, Q. Wei and C. M. Lieber, "Directed assembly of one-dimensional nanostructures into functional networks", *Science* **291** (2001) 630.
81. B. M. I. van der Zande, G. J. M. Koper and H. N. W. Lekkerkerker, "Alignment of rod-shaped gold particles by electric fields", *J. Phys. Chem. B* **103** (1999) 5754.
82. S. P. Stoylov, *Colloid Electro-opticals, Theory, Techniques, Applications*, Academic Press, New York (1973); E. Fredericq and C. Houssier, *Electric Dichroism and Electric Birefringence*, Edited by W. Harrington and A. R. Peacocke, Oxford University Press, Oxford (1973).
83. N. A. F. Al-Rawashdeh, M. L. Sandrock, C. J. Seugling and C. A. Foss, Jr., "Visible region polarization spectroscopic studies of template-synthesized gold nanoparticles oriented in polyethylene", *J. Phys. Chem. B* **102** (1998) 361.
84. K. Yamamoto, S. Akita and Y. Nakayama, "Orientation and purification of carbon nanotubes using ac electrophoresis", *J. Phys. D: Appl. Phys.* **31** (1998) L34.
85. J. O'M. Brockris and A. K. N. Reddy, *Modern Electrochemistry*, Plenum, New York (1970).
86. P. A. Smith, C. D. Nordquist, T. N. Jackson, T. S. Mayer, B. R. Martin, J. Mbindyo and T. E. Mallouk, "Electric-field assisted assembly and alignment of metallic nanowires", *Appl. Phys. Lett.* **77**(9) (2000) 1399.
87. N. I. Kovtyukhova, B. R. Martin, J. K. N. Mbindyo, T. E. Mallouk, M. Cabassi and T. S. Mayer, "Layer-by-layer self-assembly strategy for template synthesis of nanoscale devices", *Mat. Sci. Eng. C* **19** (2002) 255.
88. I. Kratochvilova, A. Zambova, J. Mbindyo, B. Razavi and J. Holakovsky, "Current-voltage characterization of alkanethiol self-assembled monolayers in metal nanowires", *Mod. Phys. Lett. B* **16**(5&6) (2002) 161.
89. X. Duan, Y. Huang, Y. Cul, J. Wang and C. M. Lieber, "Indium phosphide nanowires as building blocks for nanoscale electronic and optoelectronic devices", *Nature* **409** (2001) 66.

90. M. Tanase, D. M. Silevitch, A. Hultgren, L. A. Bauer, P. C. Searson, G. J. Meyer and D. H. Reich, "Magnetic trapping and self-assembly of multicomponent nanowires", *J. Appl. Phys.* **91**(10) (2002) 8549.
91. M. Tanase, L. A. Bauer, A. Hultgren, D. M. Silevitch, L. Sun, D. H. Reich, P. C. Searson and G. J. Meyer, "Magnetic alignment of fluorescent nanowires", *Nano Lett.* **1**(3) (2001) 155.
92. V. F. Puntes, K. M. Krishnan and A. P. Alivisatos, "Colloidal nanocrystal shape and size control: the case of cobalt", *Science* **291** (2001) 2115.
93. V. F. Puntes, D. Zanchet, C. K. Erdonmez and A. P. Alivisatos, "Synthesis of hcp Co nanodisks", *J. Am. Chem. Soc.* **124** (2002) 12874.
94. N. Cordente, M. Respaud, F. Senocq, M.-J. Casanove, C. Amiens and B. Chaudret, "Synthesis and magnetic properties of nickel nanorods", *Nano Lett.* **1**(10) (2001) 565.
95. S.-J. Park, S. Kim, S. Lee, Z. G. Khim, K. Char and T. Hyeon, "Synthesis and magnetic studies of uniform iron nanorods and nanospheres", *J. Am. Chem. Soc.* (2000) 1228581.

Part IV

Semiconductor and Nitrides Nanowires

Chapter 9

Group III- and Group IV-Nitride Nanorods and Nanowires

L. C. Chen
Center for Condensed Matter Sciences,
National Taiwan University, Taipei, Taiwan

K. H. Chen
Institute of Atomic and Molecular Sciences,
Academia Sinica, and Center for Condensed Matter Sciences,
National Taiwan University, Taipei, Taiwan

C.-C. Chen
Department of Chemistry, National Taiwan Normal University, and
Institute of Atomic and Molecular Sciences,
Academia Sinica, Taipei, Taiwan

1. Introduction

The objective of this chapter is to provide a concise review of the growth, characterization and properties of the nanorods and nanowires in two nitrides systems: the group III-nitrides (specifically, AlN, GaN, InN and ternary Ga-In-N) and group IV-nitrides (Si_3N_4 and ternary Si-C-N). During the last three decades, tremendous efforts have been devoted to the physics, chemistry and synthesis of GaN and related materials. The wurtzite polytypes of AlN, GaN and InN form a nearly continuous alloy system with direct band gaps between 1.9 and 6.2 eV, therefore, suitable for optoelectronic devices that are active at wavelengths ranging from the green well into the ultraviolet. In fact, all primary and mixed colors can be obtained by using nitride emitters as pumps. Progress in III-nitride based devices has been made at an astonishing rate in the last decade [1]. Besides the applications of III-nitrides in light emitting diodes and laser diodes, high-speed field-effect transistors, ultraviolet photodetectors and high-temperature electronic devices have been demonstrated [2–4]. Furthermore, GaN is also chemically inert and radiation resistant and has been considered as a stable photocatalyst in photo-electrochemical fuel cells [5], while

InN/Si tandem cells have been proposed for high efficiency solar cells [6, 7]. Of equal importance to just being an active component of the nitride optoelectronic devices, AlN has also attracted extensive interest for applications as electrical packaging material due to its high thermal conductivity and low coefficient of thermal expansion that closely matches that of silicon, and as components in structural composites owing to its excellent mechanical strength and chemical stability [8].

It should be noted that most of the developments in the last few decades on III-nitrides have been heading towards thin film applications. Fewer efforts were devoted for other morphological forms, such as quantum dots, nanorods and nanowires. The latter ones, in particular, have received increasing attention just recently. In terms of geometry scaling, nanorods and nanowires could roughly be defined as whiskers with diameter in the range of 1–100 nm. Therefore, previous research on the growth of whiskers is very useful for synthesizing nanorods and nanowires. Indeed, a variety of synthetic techniques, with the majority being related to vapor deposition methods, have been reported for the growth of III-nitrides nanorods and nanowires. Especially, when some catalysts are involved in the process, the nanowires and nanorods are grown by the well-known vapor-liquid-solid (VLS) mechanism [9]. Although this technique has been known for about four decades, the reported size of the VLS grown whiskers and fibers was usually of microns or submicrons in the past. The advancement in forming catalyst of nanometers is really the key to make nanorods and nanowires formation feasible. While precise control in their orientation, size and site remains to be a challenge, high-purity and -quality single crystal GaN, the most popular one in the family of group III-nitrides, nanowires and nanorods have been obtained by many research groups worldwide.

Among the IV-nitrides, the market for Si_3N_4 is most rapidly growing, particularly in chemical resistance applications, such as a passive thin film component in various electronic devices (i.e., electrical insulator and diffusion barrier) [10]. The whisker form of Si_3N_4, specifically, enjoys a broad range of applications in ceramic industries due to its high strength, lightweight, good resistant to thermal shock and oxidation. Obtaining uniform Si_3N_4 nanorods, which are expected to enhance ductility or superplasticity of ceramic reinforcing materials, remains a challenge. In contrast to the III-nitrides, the IV-nitrides have rarely been explored for optoelectronic applications. Only very recently, the ternary Si-C-N compound has just been identified to be a new phase, distinct from those of crystallized Si_3N_4-SiC or Si_3N_4-C mixtures [11–15], which might be a promising wide band gap material for next generation optoelectronic applications active in the range of wavelengths comparable to that of ternary Al-Ga-N.

While the chemical bonding and the atomic local order can be quite complex in the ternary Si-C-N system [16], a simple class would be the one exhibiting crystal structure similar or isomorphic to that of Si_3N_4. For instance, a ternary solid solution of $(Si; C)_3N_4$ is a compound that falls on the tie line between the known Si_3N_4 and the hypothetical C_3N_4. Assuming both phases are miscible, the $(Si; C)_3N_4$ compound has a three-dimensional network wherein all the C and Si atoms are bonded to four N atoms and all the N atoms are bonded to three C and/or Si atoms. Theoretical studies for the series of $Si_{3-n}C_nN_4$ ($n = 0, 1, 2, 3$) indicated that they are potentially superhard materials [17–19]. Indeed, polycrystalline hexagonal Si-C-N film exhibits a hardness value rivaling cubic BN [20]. Depending on the composition, the crystalline Si-C-N film shows direct band gap ranging from 3.8 to 4.6 eV, therefore, a great potential for blue and ultraviolet optoelectronic applications [21, 22]. Good thermal conductivity [23] and ultra-high oxidation resistance up to 1600 °C [24] have been reported in the amorphous Si-C-N phase. More recently, Si-C-N nanorods have been successfully grown by a catalyst-free approach and excellent field emission properties in this system have also been demonstrated [25, 26].

In this chapter, we will begin our journey through the wonders of synthetic techniques for making the III- and IV-nitrides nanorods and nanowires. This is followed by structural and bonding characterizations. Then, some electronic and optoelectronic properties will also be presented. Even though the development of the III- and IV-nitrides possessing one-dimensional nanostructures is still in its infancy, the advancement in fabrication technique as well as characterization technique enables optical, electrical and electronic property measurements in the nano-scale. Rapid progress in design and synthesis of semiconductor nanowires and potential applications in electronics and optoelectronics are highly anticipated.

2. Synthesis of Binary Group-III Nitride Nanorods and Nanowires

Although the first work of GaN nanorods, prepared using a carbon-nanotube-confined reaction [27], goes back for only about 5 years, numerous groups have successfully produced GaN and related nanorods and nanowires, by employing a variety of techniques. Despite the initial success of carbon-nanotube-confined method, the lack of mass production and control of the carbon nanotubes has undermined this method for wide application afterwards. Quite a large portion of this chapter is in fact devoted to the catalyst-assisted growth, e.g., VLS growth, as it has been developed for many decades. However, since the catalyst usually remains in the nanorods and nanowires, which may affect their intrinsic properties, removal of catalyst may be required, adding complexity to the process. Therefore, it is important to develop catalyst-free methods for forming nanorods and nanowires. Various catalyst-free approaches, including vapor solid growth, oxide- and chloride-assisted methods, sublimation, combustion and etching techniques will also be addressed in this chapter.

Among the binary group-III nitrides, both AlN and GaN have been researched extensively. While InN possesses very attractive properties, overall it has not received as much attention as that given to GaN and AlN. This is probably due to difficulties in growing high quality crystalline InN samples. InN is not different from GaN and AlN in the sense that it suffers from the same lack of a suitable substrate for epitaxial film growth and, in particular, a high native defect concentration. In addition, due to its poor thermal stability, InN cannot be grown at temperatures higher than 600 °C without extraordinary high nitrogen overpressure, which is practically difficult. All these have hindered its progress. There are only very few reports on InN nanowires wherein either a catalyst [28] or single molecular precursors exhibiting low decomposition temperature [29] is required to form InN nanowires at temperatures around 500 °C, which is below its dissociation temperature.

Depending on a specific material, satisfactory fabrication of nanowires and nanorods can be achieved by controlling the reaction chemistry, the activation method or the deposition variables. Several approaches for the production of GaN, InN and AlN nanowires and nanorods are described below that can be divided into four major categories. They are: (1) the confined chemical reaction, (2) catalyst-assisted growth, (3) vapor-solid and related growth methods, and (4) reactive ion etching. Whenever applicable, we would also emphasize the way of forming these one-dimensional nanostructures with size, orientation, and position control.

2.1 Confined chemical reactions: template-based methods

For template-based methods, materials that provide confined spaces in a nanometer scale such as carbon nanotubes [27], GaAs nanocolumns [30], and anodic alumina membrane [31] have been used. The template-based techniques have been applied to produce GaN, AlN and Si_3N_4 nanorods. To date, there is still no report on using template-based methods for fabricating InN nanorods or nanowires.

2.1.1. Carbon nanotube (CNT)

The use of CNT as a template to synthesize nanostructure was first demonstrated by Dai et al. in forming carbide nanorods through CNT reaction with volatile oxide species [32]. This method was later exploited to prepare nitride nanorods. By reacting Ga_2O with NH_3, Han et al. grew GaN nanowires with diameters between 4 and 50 nm and a length of up to 25 μm, similar to those of the original carbon nanotubes. The starting material for Ga_2O was a 4 : 1 molar mixture of Ga and Ga_2O_3 powder, over which the Ga_2O vapor pressure is generated through the following reaction upon heating [33]:

$$Ga_2O_3(s) + 4Ga(l) \rightarrow 3Ga_2O(g) \quad (1)$$

The pressure of Ga_2O vapor generated from this reaction is quite high, e.g., about 1 Torr at 900 °C and 7.2 Torr at 1000 °C [34]. In Han's original experiment, the Ga-Ga_2O_3 powder mixture was first placed in an alumina crucible and covered with a porous alumina plate with 3 to 5 μm diameter channels. The CNTs were then placed on top of the alumina plate. It is expected that the Ga_2O vapor generated from the Ga-Ga_2O_3 powder mixture would flow up toward the region of CNTs through the porous alumina plate and react with the CNTs and NH_3 gas. The main reaction involved in the formation of GaN nanorods can be expressed as:

$$2Ga_2O(g) + C(s) + 4NH_3(g) \rightarrow 4GaN(s) + H_2O(g) + CO(g) + 5H_2(g) \quad (2)$$

It is important to note that, in the absence of Ga-Ga_2O_3 powder mixture, no change of the CNTs was observed under the identical process condition. Furthermore, without the presence of CNTs, GaN powders were produced by the reaction between gaseous Ga_2O and ammonia [35]:

$$Ga_2O(g) + 2NH_3(g) \rightarrow 2GaN(s) + H_2O(g) + 2H_2(g) \quad (3)$$

These experimental results provide further support for the CNT-confined reaction for the formation of GaN nanorods through reaction (2).

In analogous efforts to grow GaN nanowires through the above reactions, carbon nanotubes confined reaction using Al and Al_2O_3 powders as reactants has also been employed to synthesize AlN nanowires in bulk at 950–1200 °C [36, 37]. However, unlike the case for Ga_2O, Al_2O is difficult to prepare at low reaction temperature (e.g., below 1000 °C) since it is hard to reduce Al_2O_3 by Al or carbon due to the high positive change of Gibbs free energy. Meanwhile, Al and NH_3 may react with each other at 900 °C through the following reaction [38]:

$$2Al(s) + 2NH_3(g) \rightarrow 2AlN(s) + 3H_2(g) \quad (4)$$

This reaction is highly exothermic, which may result in substantial local heating, therefore promoting the following reactions which reduce the Al_2O_3 in the mixtures of CNTs and Al:

$$Al_2O_3(s) + 2C(s) \rightarrow Al_2O(g) + 2CO(g) \quad (5)$$

$$Al_2O_3(s) + 4Al(s) \rightarrow 3Al_2O(g) \quad (6)$$

Finally, the AlN nanowires can be produced according to the following reaction:

$$2Al_2O(g) + C(s) + 4NH_3(g) \rightarrow 4AlN(s) + CO(g) + H_2O(g) + 5H_2(g) \quad (7)$$

Although the last reaction is quite similar to reaction (2), the production of gaseous Al_2O does involve pathways quite different from those of Ga_2O.

The presence of a small amount of oxygen or the addition of Al_2O_3 or Ga_2O_3 powders as reactants in the CNTs-confined reaction is very crucial. For instance, in an ideal CNTs-Al-NH_3-AlN system without oxygen, the CNTs would probably keep their original forms and no AlN nanowires could be formed. It should also be noted that the starting temperature (~950 °C) to grow AlN nanowires using CNTs template is significantly lower than that of the conventional carbothermal reduction and nitridation of α-Al_2O_3, wherein a reaction temperature of higher than 1700 °C is required [39, 40].

2.1.2. Anodic alumina membrane

The membrane with a nanosized pore is an important class of material. Besides being useful for selective ion-transport, molecular filtration and drug separation, it has also been applied as a template for synthesizing nanomaterials [41–44]. A commonly used membrane is anodic alumina membrane, which is a self-ordered nanochannel material formed by anodization of Al in an appropriate acid solution [45, 46]. Being relatively established, the anodic alumina membrane hence becomes a popular choice of template. The anodic alumina membrane possesses hexagonal ordered porous structure with nanochannel diameters ranging from 10 to 200 nm, typical channel length of tens of μm, channel density in the range 10^{10}–10^{12} cm^{-2}. The extremely high aspect ratio of these nanochannels is indeed difficult to achieve with conventional lithographic techniques. The capillary effect of the anodic nanochannels has been suggested to facilitate the formation of nanowires. Owing to these desirable geometric features and chemical stability, the nanochannel membrane-based method has been employed for fabricating various kinds of ordered nanowire arrays in the past [47–50], and more recently for GaN nanowires as well.

Ordered crystalline GaN nanowires embedded in the nano-channels of anodic alumina membrane were achieved by a direct reaction of Ga and NH_3 [51]. This simple reaction can be expressed as [35]:

$$2Ga(g) + 2NH_3(g) \rightarrow 2GaN(s) + 3H_2(g) \quad (8)$$

In this reaction, the metal Ga inside a crucible was placed in the high temperature zone and the temperature of crucible, under an ammonia flow, was increased to or above 900 °C, since below which hardly any GaN nanowires were formed. To increase the yield of GaN nanowires, powder Ga_2O_3 can also be added to Ga metal so that vapor Ga_2O is generated following reaction (1). In this case, the gaseous Ga_2O and NH_3 can react directly, leading to the formation of GaN *via* reaction (3) as described above. If the confined reaction is preformed with the addition of catalysts, then the gaseous Ga_2O and NH_3 can continuously be dissolved into the catalysts, leading to the formation of GaN *via* vapor-liquid-solid mechanism that will be described in section 2.2.

It should be noted that the nanochannel-confined formation of the GaN nanowires reported in the literature was not solely achieved by direct reactions but was usually assisted with some catalysts. In fact, Cheng et al. has employed metallic indium, *via* evaporation in vacuum on one side face of the anodic alumina membrane to form indium nanoparticles on one end of the uniform nanochannels, prior to the reaction processes [31, 52]. In the presence of indium nanoparticles, it is believed that the vapor-liquid-solid growth mechanism is responsible for the formation of the GaN nanowires. The as-synthesized products were shown to exhibit phase pure hexagonal wurtzite GaN structure, indicating that the indium nanoparticles only act as a catalyst in this case. The GaN nanowires so produced were also highly ordered. However, the vapor deposition method employing indium on the side face of the anodic alumina membrane as a catalyst cannot assure the GaN nanowires growth exclusively inside the nanochannels of the membrane. Overgrowth of the nanowires out of the uniform nanochannels of the membrane and aggregation from thereon were observed. In contrast, electrochemical deposition method was shown to be more advantageous for introducing In nanoparticles into the nanochannels of the membrane [53]. The anodic alumina membrane, as is also the case for carbon nanotubes, does indeed act as a template and plays a key role in controlling the size of the nanowires. If these templates are absent, only powders are produced by the abovementioned reactions. Despite the slight difference in the diameter distribution, a direct proportional relationship between the diameters of the nitrides nanowires and the outer diameters of the carbon nanotubes (or the channel sizes of the membrane) is usually observed for samples fabricated in a rather wide range of process temperatures and reaction times ranging from 30 to 200 min [36]. However, the template effect will eventually be lost if the reaction temperature is raised too high. For the case of AlN, the carbon nanotubes did not work well at temperature above 1500 °C, at which Al and NH_3 would grow directly into the AlN fibers possessing much larger diameters [54–56]. While the template-based technique is effective in producing nanowires with controllable size, the inability to isolate the product and then maintain the orientation alignment after removing membrane does undermine its wide application.

2.2. Catalyst-assisted growth: The vapor-liquid-solid (VLS) and the solution-liquid-solid (SLS) methods

The catalyst-assisted growth in general is referred to a process wherein some catalysts are involved during growth but do not incorporate in the final product. One important example is the so-called vapor-liquid-solid (VLS) growth, which has been shown over several decades to be responsible for whisker growth [9]. Of late, the catalytic growth of GaN nanowires has also been achieved by laser-ablation [57], a relatively new approach in comparison to the conventional chemical vapor deposition (CVD). Furthermore, catalytic growth may also be conducted with source materials in a solution form. This approach is thus termed as solution-liquid-solid (SLS) growth.

2.2.1. The vapor-liquid-solid (VLS) method

Up until now, the CVD-based catalyst-assisted VLS growth is still the most widely exploited approach for fabricating a variety of one-dimensional nanostructures. Experimental CVD setup for catalytic growth of nitride nanowires and nanorods is shown schematically in Fig. 1 [28, 58]. Typical scanning electron microscopy (SEM) images of the catalytic-grown nitride nanowires and nanorods are presented in Fig. 2. The group III elements such as Ga or In in the forms of foils or powders are placed in the bottom of an alumina boat that serves the purpose to provide vapor sources upon heating. The same boat is covered with a substrate and then loaded into the center of a conventional quartz tube furnace. Alternatively, the substrate can also be placed in a separate boat. For convenience, we would refer these two different arrangements to

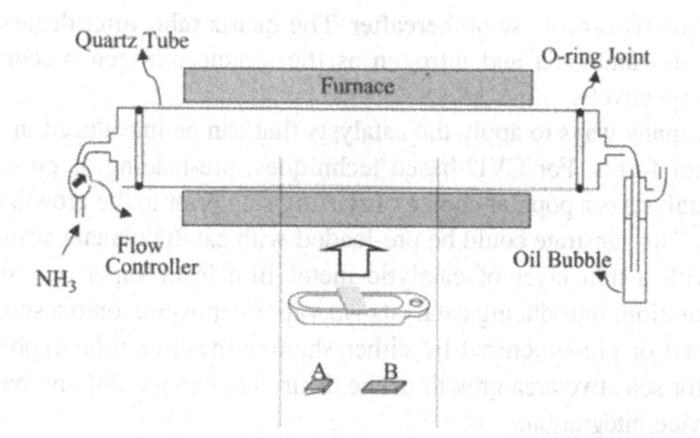

Fig. 1. Schematic experimental setup for the growth of GaN and InN nanowires and nanorods. Either "single-boat" (sample collected on a substrate placed on top of the crucible containing reactants) or "two-boat" (sample collected on a separate substrate, A and B, placed downstream of the reactants) configuration is used. Typical experimental condition: heating rate, 50 °C/min; reaction temperatures, 910 °C (GaN) or 500 °C (InN); and total reaction time, 12 h under an ammonium/nitrogen flow.

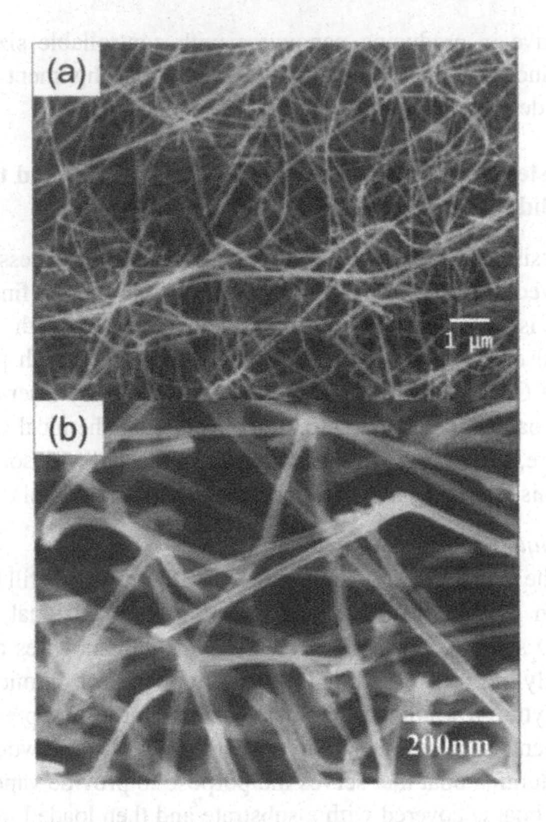

Fig. 2. Typical SEM images of (a) vermicelli-like GaN nanowires and (b) straight InN nanorods. [(a) With permission from C.-C. Chen et al., *J. Am. Chem. Soc.* **123** (2001) 2791]

"single-boat" or "two-boat" setup hereafter. The quartz tube, after degassing, is usually purged with ammonia and nitrogen as the atomic nitrogen precursor and the carrier gas, respectively.

There are many ways to apply the catalysts that can be introduced in either solid, liquid or vapor forms. For CVD-based techniques, pre-loading or pre-coating substrate with catalysts is a popular choice of starting step prior to the growth of nanorods or nanowires. The substrate could be pre-loaded with catalytic metal strip/powders or pre-coated with a thin layer of catalytic metal film from vapor, *via* sputtering or e-beam evaporation. Introducing catalysts *via* vapor deposition onto a substrate, especially patterned or pre-structured by either shadow mask or lithography, offers the opportunity for selective area growth of the desired materials [28] and paves the way for future device integration.

While transition metals are known to be efficient catalysts for growing many one-dimensional nanostructures, including CNT and nitride nanowires discussed here, transition metal oxides such as Ni and Fe oxides have a similar effect on nanowire growth [59]. These catalysts could also be dispersed onto the substrate using solutions of catalytic metal nitrates, $M(NO_3)_x$, metal complexes of ferrocene and related $C_{10}H_{10}M$, or iron phthalocyanine and related $C_{32}H_{16}N_8M$, where M stands for the

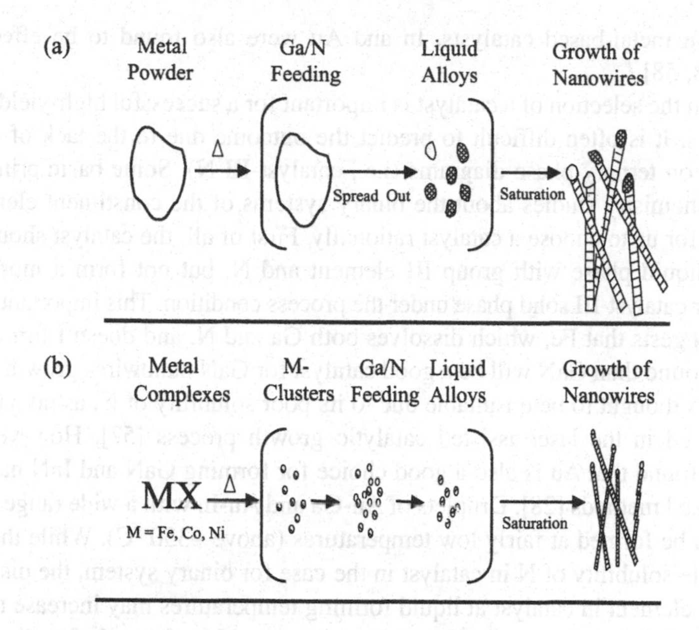

Fig. 3. Schematic illustration of two possible pathways of the VLS growth of GaN nanowires. (a) At the beginning, the catalysts of metal powder gradually form catalyst-gallium-nitrogen alloys after gallium vapor transport to the catalysts. Each alloy turns into a miscible liquid droplet and then spreads out, becoming many smaller droplets. Hence, after the concentration of gallium nitride reaches saturation in the droplet, the droplet can act as a nucleation site, and a GaN nanowire begins to grow in one direction. (b) At the initial heating stage, metal complexes are decomposed rapidly in the gas phase or on the substrate and sequentially generate many small metal clusters. These clusters form small catalyst-gallium-nitrogen liquid droplets and serve as nucleation sites after gallium and ammonium vapors are fed in. In comparison to path (a), the nanowires with smaller diameters could be obtained when the reaction goes through pathway (b). [With permission from C.-C. Chen et al., *J. Am. Chem. Soc.* **123** (2001) 2791]

catalytic metal, such as Fe, Ni and Co, etc. Schematic illustration of the two pathways wherein catalysts are introduced as metal powders or metal complexes of the VLS growth of nitrides nanowires is shown in Fig. 3. Using metal complexes as catalysts is thought to be helpful for producing nanowires with small diameters since the metal complexes would decompose rapidly into gaseous species upon heating and form metal clusters. A minimum diameter of 6 nm has been achieved in particular when Co phthalocyanine or Ni phthalocyanine was used as a catalyst [60]. Meanwhile, in the pulsed laser assisted catalytic growth [57], solid GaN/Fe composite was used as the target, which, upon pulsed laser irradiation, leads to the formation of ternary Ga-N-Fe liquid nanoclusters. The diameters of nanowires so produced were in the orders of 10 nm. Alternatively, GaN nanorods can also be produced by a dc arc discharge between a graphite cathode and a graphite anode drilled with a deep hole and then filled with a mixture of GaN, graphite and Ni powders [61]. However, mixed phases including carbon nanotubes, filled fullerene-like nanoparticles, GaN nanorods, together with carbon nanotubes encapsulated GaN nanorods, were produced in this method. Besides the transition metal-based catalysts, In and Au were also found to be effective as catalysts [28, 58].

Although the selection of a catalyst is important for a successful high-yield growth of nanowires, it is often difficult to predict the outcome due to the lack of detailed information on ternary phase diagrams (i.e., catalyst-III-N). Some basic principle in solid state chemistry studies about the binary systems of the constituent elements is still helpful for us to choose a catalyst rationally. First of all, the catalyst should form a miscible liquid phase with group III element and N, but not form a more stable catalyst-N or catalyst-III solid phase under the process condition. This important guiding principle suggests that Fe, which dissolves both Ga and N, and doesn't form a more stable compound than GaN will be a good catalyst for GaN nanowires growth. In contrast, Au was thought to be unsuitable due to its poor solubility of N, as no nanowires were observed in the laser-assisted catalytic growth process [57]. However, other researchers found that Au is also a good choice for forming GaN and InN nanowires by CVD-based methods [28]. Droplets of Au-Ga and Au-In with a wide range of composition can be formed at fairly low temperatures (above ~350 °C). While there may be no or little solubility of N in catalyst in the case for binary system, the dissolution of group III element in catalyst at liquid forming temperatures may increase the solubility of N in the ternary system. Hence nanowires could be formed after the concentration of both N and group III element dissolving in catalysts reached saturation. Fig. 4 shows a transmission electron microscopy (TEM) image of a single InN nanowire capped with a nearly equi-axed Au nanoparticle. Energy dispersive x-ray spectrometry (EDX) performed on the nanoparticle and the nanowire indicates that

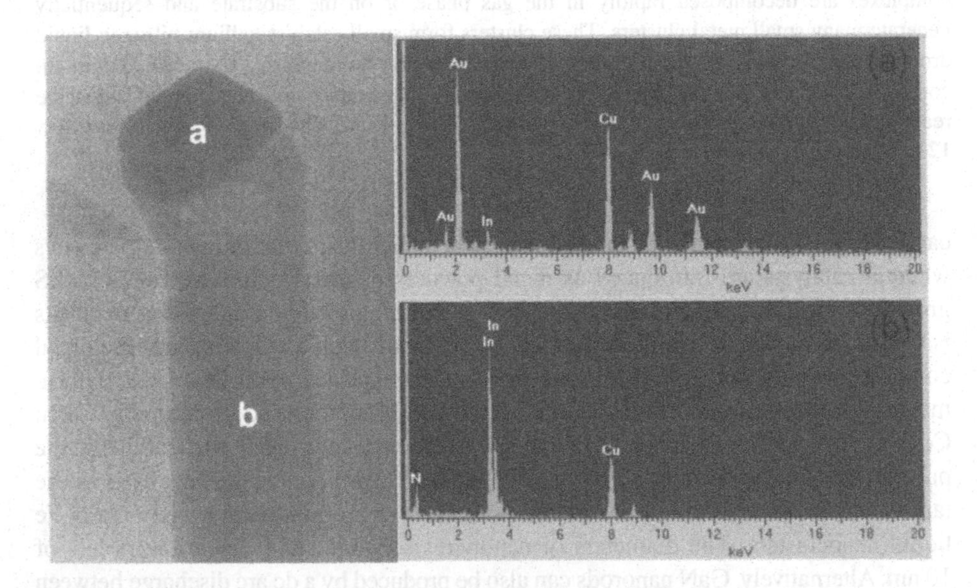

Fig. 4. TEM image of a single InN nanorod terminated with an Au nanoparticle. The corresponding EDX spectra of the nanoparticle and the rod stem are shown in (a) and (b), respectively. Cu signals are generated from microgrid mesh that supports the nanowires. [With permission from C. H. Liang et al., *Appl. Phys. Lett.* **81** (2002) 22]

the ends of the nanowires are composed primarily of Au and the rest of the nanowires is InN with no detectable Au incorporation. The presence of Au at the ends of the InN nanowires provides a strong evidence for a VLS growth mechanism.

2.2.2. Size and site control

One of the technical challenges for employing catalysts is to produce a uniform dispersion and yet maintain the small size of the catalyst. In general, correlation in the size of the catalytic nanoparticles and resulting diameters of the nanowires or nanorods is very high (Fig. 5). However, the average diameter of the nanowires or nanorods is usually larger than that of the catalyst initially applied. Poorly dispersed catalytic nanoparticles and thin film coating often lead to significant agglomeration of catalysts during heating. Since the process temperature for nanowires is relatively high, considerable self- and surface-diffusion occur. Therefore, mixing "diffusion barrier" material with catalyst is a viable technique for preventing agglomeration of the catalyst. Indeed a high-yield method for synthesizing single-crystal GaN nanowires has been reported using more complex reaction of Ga and SiO_2 mixtures with NH_3 in the presence of the Fe_2O_3 catalyst supported by Al_2O_3 [62]. The addition of SiO_2 is thought to be helpful for reducing the melting temperature of the catalyst by forming $FeSi_2$ as well as enhancing the production of high pressure Ga_2O gas owing to an eutectic in the Ga-SiO_2 system or through the following reaction:

$$4Ga(s, l) + SiO_2(s) \rightarrow 2Ga_2O(g) + Si(s) \tag{9}$$

The catalysts continuously dissolve the gaseous Ga_2O and NH_3, leading to the formation of GaN via reaction (3). In this process, while Al_2O_3 does not participate directly in the catalytic reaction for forming GaN nanowires, the addition of Al_2O_3 to Fe_2O_3 appears to be helpful for preventing the agglomeration of the Fe-containing

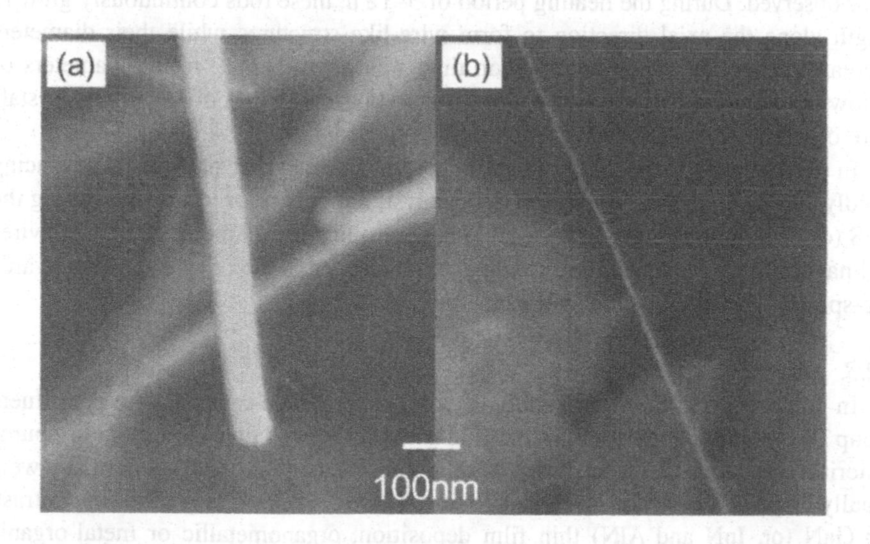

Fig. 5. FESEM images of InN nanowires prepared by VLS growth with average initial catalytic particle size of (a) 70 nm and (b) 10 nm.

catalytic droplets, therefore, leading to the production of GaN nanowires with a much smaller diameter (10–50 nm) than those produced with Fe_2O_3 only (80–200 nm).

In addition to the agglomeration problem of catalyst, the non-uniform distribution of the constituent group III elements and/or N in the catalytic reaction can also cause a broad diameter distribution of the nanowires. For instance, in a two-boat setup, as was used for synthesizing GaN nanowires by Chen et al. [60], metal Al, Ga and In, employed as the source materials in the CVD-based methods, are usually placed in the front boat from the gas inlet while the substrates loaded with catalytic materials are placed in the second boat situated downstream. At the process temperature, the group III elements become molten droplets, from which the corresponding vapor transports to the catalyst placed downstream of the vapor. It was found that the diameter distribution of the nanowires collected downstream was strongly dependent on the distance between the source material of group III element and the catalyst [60]. In general, the diameter of the nanowires decreases with increasing boat-to-boat distance, presumably due to the depletion of the reactant. The diameters of the nanowires were mostly in the range of 50–150 nm when the catalysts were placed at a distance of about 1 cm from the group III elements. In contrast, the diameters of the nanowires were significantly reduced to 20–50 nm when the distance between the two was increased to 10 cm. In a single-boat setup, the vapor of the constituent element transports to the catalyst placed directly above, resulting in nanowires with uniform diameter since the source-to-substrate distance is same.

Once the experimental setup for the catalytic growth of nanowires is chosen, the total reaction time is yet another critical factor for determining the diameters and lengths of the resulting nanowires. Chen et al. have investigated the structure and morphology of the GaN nanowires produced with a reaction time varying from 3 to 48 h, keeping the other process conditions the same [60]. Under a reaction time of 3 h, a large number of short rod-like structures with diameters of several hundred nm were observed. During the heating period of 3–12 h, these rods continuously grew in length along the axial direction to form wire-like structures while their diameters decreased. However, when the reaction time was more than 12 h, the diameters of nanowires were dramatically increased. In fact, unusual shapes of GaN bulk crystals were observed when the total reaction time exceeded 48 h.

In summary, the technology of obtaining ultra-fine catalytic particles is advancing rapidly. However, the size control of the catalyst, not only prior to but also during the VLS (or SLS) growth is the first key step in controlling the diameter of the nanowires and nanorods. The precision positioning of the catalyst is the other key step towards site-specific growth of the desired nanostructures.

2.2.3. Source gas considerations

In comparison to the wide variety of ways to introduce catalysts, the constituent group III elements and N are introduced in a less elaborated manner. Inorganic source materials of group III elements, such as solid metals and their oxide counterparts were usually employed to be the metal source for III-nitride nanowires growth. In contrast, for GaN (or, InN and AlN) thin film deposition, organometallic or metal-organic (MO) precursors such as trimethylgallium (or, its counterparts for In and Al) for metal and ammonia for nitrogen are the most commonly used ingredients. Recently,

Han et al. reported a pyrolysis route to prepare GaN nanorods where both Ga and the catalyst come from organic sources [63]. Specifically, ferrocene ($C_{10}H_{10}Fe$) and gallium dimethylamide ($Ga_2[N(CH_3)_2]_6$) were used under an ammonia atmosphere. The sizes of the GaN nanorods fabricated by this technique were typically 15–70 nm in diameter and 3–30 μm in length. Due to the presence of carbon in the source materials, carbon nanotubes were also formed at process temperatures of 900 and 1000 °C. However, the use of proper organic sources may still be advantageous for growing nanowires with uniform diameters and at low temperatures, as will be shown below.

For most cases of catalyst-assisted and CVD-based growth of GaN nanowires using Ga metal, a process temperature exceeding 850 °C is usually required. Furthermore, the nanowires are formed very near the metal source (~1–10 cm) and their average diameter decreases with increasing distance between the metal source and the catalyst-covered substrate [60]. This experimental evidence implies that the reactant depletion problem could be a concern in the growth of GaN nanowires using Ga metal due to its low vapor pressure (~10^{-4} Torr at 900 °C [64]). Thus, it is crucial to select a precursor material that can provide sufficient Ga source at low temperature. Chang and Wu have shown that, using gallium acetylacetonate ($(CH_3COCHCOCH_3)_3Ga$) and ammonia gas, catalytic growth of GaN nanowires can be conducted at a temperature as low as 620 °C [65]. In comparison to Ga metal, the low decomposition temperature (~196 °C) of gallium acetylacetonate can provide more sufficient Ga vapor than can the Ga metal. By employing a two-temperature-zone furnace, the concentration of Ga vapor can be controlled by the temperature of the first zone (e.g., typical operation temperature to be controlled at 130–185 °C for the case of Ga acetylacetonate), which is independent of the temperature for GaN nanowires growth in the second zone (e.g., typical operation temperature to be controlled at 600–1000 °C). The gallium vapor produced in the first zone was carried by an ammonia flow into the higher temperature zone for further catalytic growth. In this configuration, rather straight nanowires were formed with diameters in the range 15–60 nm and lengths of several tens of microns. It should be mentioned that no carbon nanotube is observed in the products although the gallium acetylacetonate contains carbon element.

2.2.4. Orientation control and substrate considerations

In line with the technology for homo-epi and hetero-epi film growth, one can expect to grow well-aligned nanorods under layer-by-layer growth conditions, provided that suitable substrates are used. However, due to the lack of lattice-matched substrate, growing epitaxial III-N film has long been a challenging issue. Up until now, sapphire substrate, despite its large lattice mismatch with the III-N crystal, is still the most commonly used substrate for epi-growth. Indeed, when a GaN (001) film epitaxially grown on a sapphire (001) substrate, hereafter expressed as GaN (001) substrate for convenience, was used as the substrate, highly oriented GaN nanorods were obtained by the catalyst-assisted growth method as well as other non-catalytic methods, of which some will be described in section 2.3.

Herein, we describe a catalyst-assisted oriented growth process of GaN nanorods in more detail. Prior to GaN nanorods growth, the GaN (001) substrate was first selectively deposited with nickel layer of 5 μm × 5 μm square arrays to form the mask

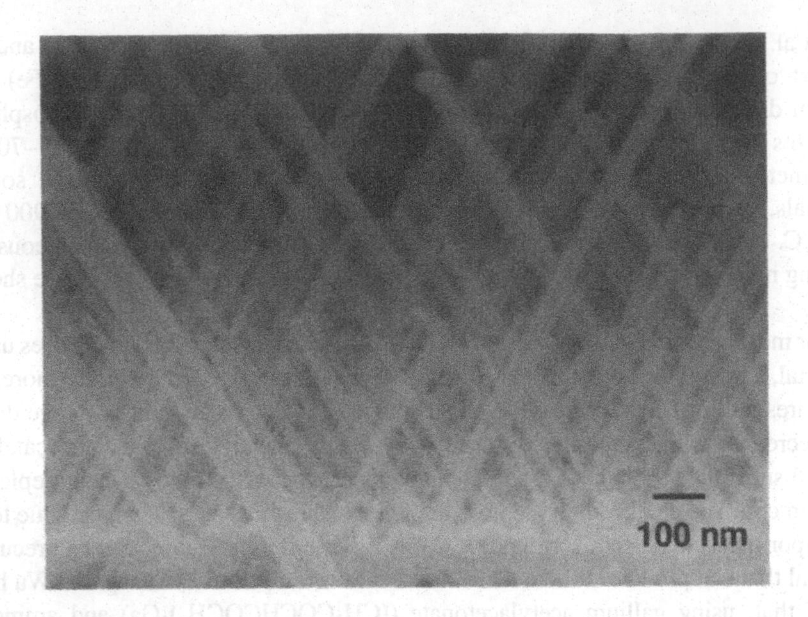

Fig. 6. Directionally aligned GaN nanorods formed in the side trench of the (001) oriented GaN substrate.

before being introduced to the inductively coupled plasma (ICP) system. The ICP operated at 200–250 Watt under chlorine and argon ambient was applied to etch away the unmasked GaN for 10 minutes. After the ICP process, the residual nickel was cleaned by nitric acid and hydrochloric acid of 1:3 volume ratio. GaN (001) substrate with square trenches thus formed was further coated with thin Au layer as the catalyst for the subsequent VLS growth of GaN nanorods. Following the standard VLS process for GaN nanorods growth at 850 °C for three hours, directionally aligned GaN nanorods were formed in the trench of the substrate (Fig. 6). As can be seen in this figure, the nanorods were grown along (100) and (1–10) surfaces (in hexagonal coordination), which are 60 degrees to each other and in agreement with the angle between the two growth directions. Strong epitaxy between the GaN substrate and the nanorods can be concluded from the directionality of the growth. In principle, other substrates that have been employed for epi-film growth may also be suitable for growing oriented nanorods, although there has been no report so far.

2.2.5. *Solution-liquid-solid (SLS) method*

The majority of the catalytic syntheses have been performed in conjunction with chemical vapor deposition techniques through the VLS growth mechanism. However, a related approach termed as solution-liquid-solid (SLS) growth has been demonstrated by Buhro and co-workers to prepare nanowires of several III-V materials in solutions [66–68]. Notably, nanowires and nanorods of InN, which is most challenging among all the III-nitrides, have also been produced by this SLS method [69]. Interestingly, the SLS growth of InN can be activated at 203 °C by a process wherein nanometer-sized metal droplets serve as catalytic sites for fiber formation.

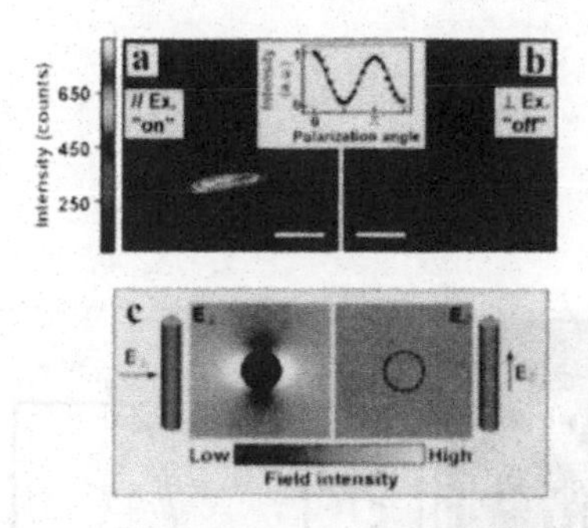

Plate 1. (a) PL image of a single 20-nm InP nanowire with the exciting laser polarized along the wire axis. Scale bar, 3 μm. (b) PL image of the same nanowire as in (a) under perpendicular excitation. Inset, variation of overall photoluminescence intensity as a function of excitation polarization angle with respect to the nanowire axis. The PL images were recorded at room temperature with integration times of 2 s. (c) Dielectric contrast model of polarization anisotropy. The nanowire is treated as an infinite dielectric cylinder in a vacuum while the laser polarizations are considered as electrostatic fields oriented as depicted. Field intensities ($|E|^2$) calculated from Maxwell's equations clearly show that the field is strongly attenuated inside the nanowire for the perpendicular polarization, $E_{\perp}$, whereas the field inside the nanowire is unaffected for the parallel polarization, $E_{//}$. Adapted from [75].

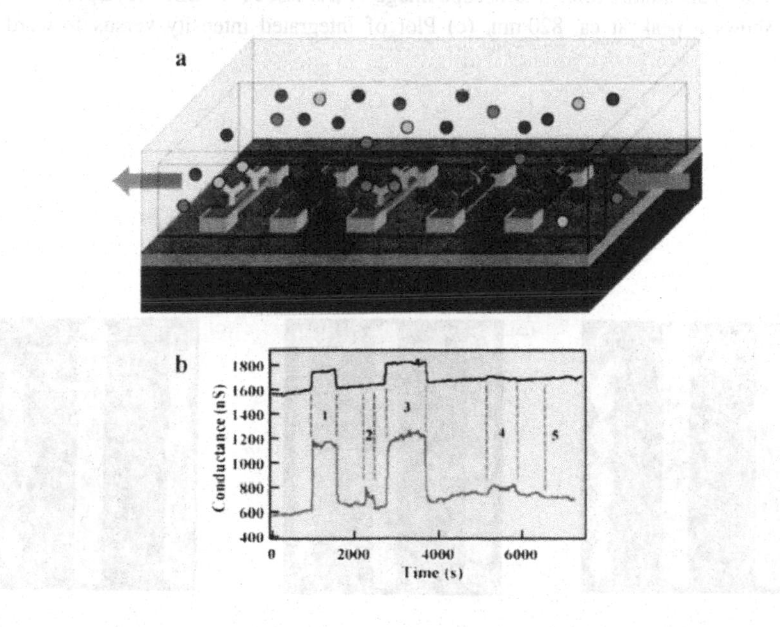

Plate 2. NW multiplexing sensor. (a) Schematic illustrating the concept of multiplexing. Multiple SiNWs can be differentially modified and the analyte solution can be introduced to all the NWs simultaneously with a common microfluidic channel. The transparent part stands for a microfluidic channel and the blue arrow points toward the flow direction. (b) Plot of conductance versus time for simultaneous real-time detection of fPSA and PSA-ACT. The fPSA antibody-modified SiNW (black) used an antibody specific for fPSA only whereas the PSA-ACT antibody-modified SiNW (red) used an antibody raised against fPSA, but has cross-reactivity with PSA-ACT. Region, 1, 2, 3, 4, and 5 correspond to 50 ng/ml fPSA, 50 ng/ml PSA-ACT, 50 ng/ml fPSA+50 ng/ml PSA-ACT, 4 ng/ml fPSA, and 5 μg/ml HAS, respectively. Unlabeled regions correspond to buffer. Arrows indicate the points when solutions were changed. Adapted from [90].

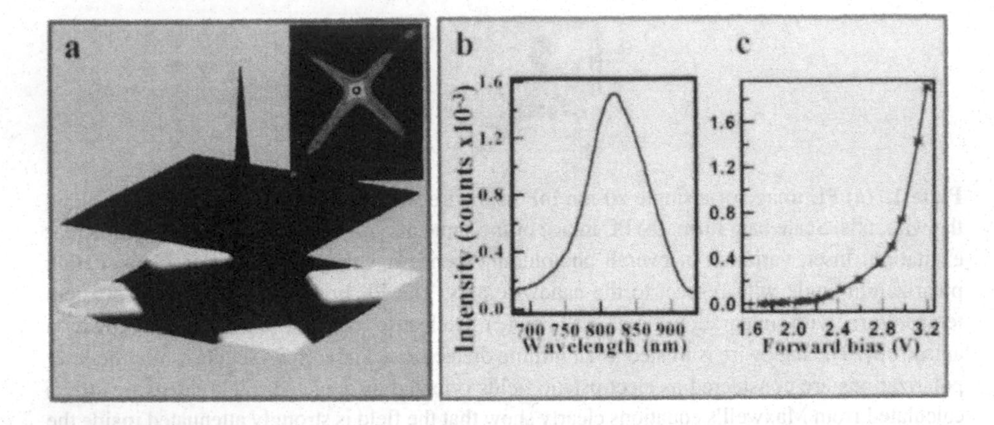

Plate 3. Crossed NW LED. (a) (top) Three-dimensional (3D) plot of light intensity of the electroluminescence from a crossed NW LED. Light is only observed around the crossing region. (bottom) 3D atomic force microscope image of a crossed NW LED. (b) Spectrum of the emission shows a peak at ca. 820 nm. (c) Plot of integrated intensity versus forward bias voltage.

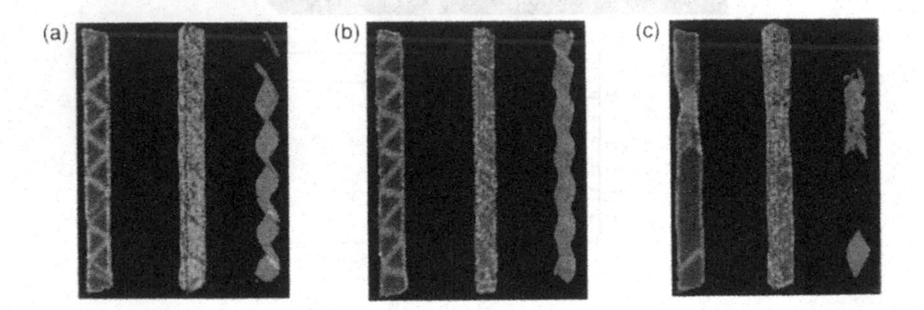

Plate 4. Deformed configurations of Cu nanowires at a strain of 10%; (a) $\dot{\varepsilon} = 1.67 \times 10^7\ s^{-1}$; (b) $\dot{\varepsilon} = 1.67 \times 10^8\ s^{-1}$ and (c) $\dot{\varepsilon} = 1.67 \times 10^9\ s^{-1}$. In each picture, the left image shows an internal cross-section, the center images shows a solid view of the wire with all atoms, and the right image shows only atoms involved in defects. Graphics generated using VMD1.7 [57].

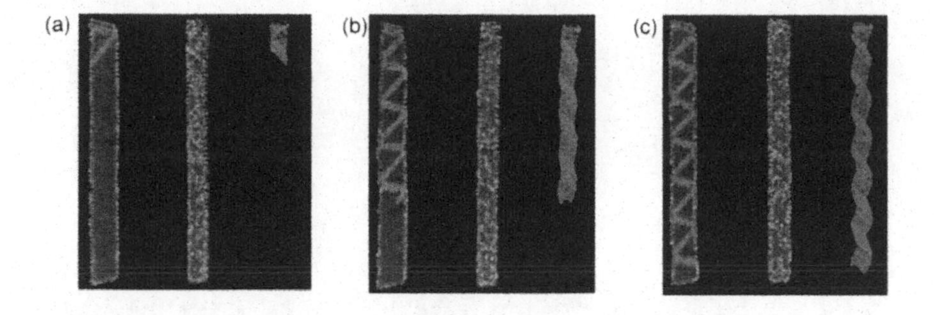

Plate 5. The propagation of slip bands at a strain rate of $1.67 \times 10^8\ s^{-1}$, (a) $\varepsilon = 4.67\%$, (b) $\varepsilon = 5.33\%$, and (c) $\varepsilon = 6.0\%$. The specimen size is $5 \times 5 \times 60$.

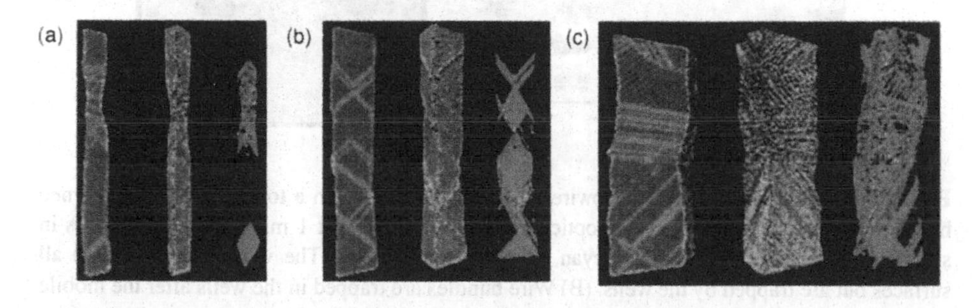

Plate 6. Deformed configurations of Cu nanowires of different sizes ($\dot{\varepsilon} = 1.67 \times 10^9\ s^{-1}$); (a) specimen with a cross-sectional size of 5 lattice spacings ($\varepsilon = 9.17\%$); (b) specimen with a cross-sectional size of 10 lattice spacings, ($\varepsilon = 18.3\%$); and (c) specimen with a cross-sectional size of 20 lattice spacings ($\varepsilon = 23.3\%$). In each picture, the left image shows an internal csection, the middle image shows an external view, and the right image shows only atoms involved in defects.

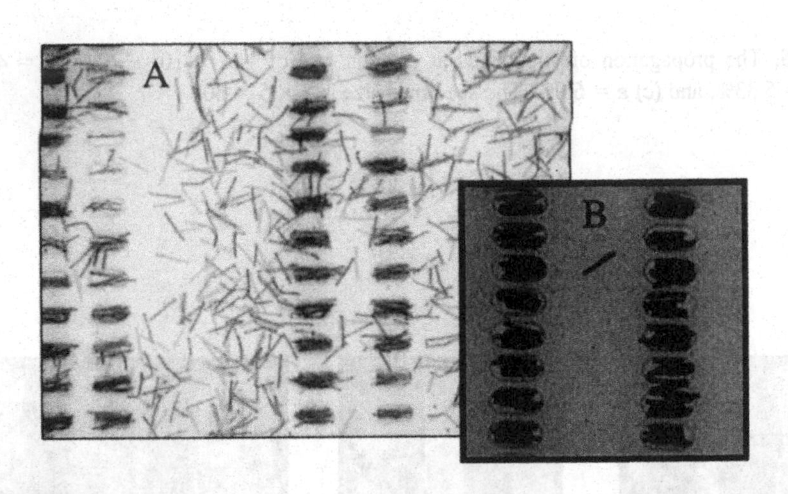

Plate 7. MESA derivatized Au nanowires suspended in water on a topographically patterned hydrophilic surface. (A) Composite optical micrographs taken at 1 minute intervals; wires in successive frames are color-coded cyan, magenta, and yellow. The wires are mobile on all surfaces but are trapped by the wells. (B) Wire bundles are trapped in the wells after the mobile wires are washed away [51]. Copyright Wiley-VCH, reproduced with permission.

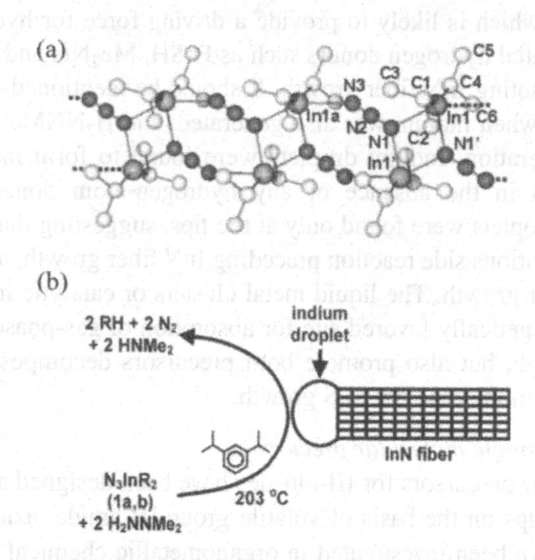

Fig. 7. (a) Ball-and-stick representation of the crystal structure of $[iPr_2InN_3]_n$ (**1a**). (b) Proposed solution-liquid-solid growth mechanism for InN nanowire synthesis from the diakyl(azido)indane precursors N_3InR_2 **1a** or **1b**. [With permission from S. D. Dingman et al., *Angew. Chem. Int. Ed.* **39** (2000) 1470].

The precursors employed in this relatively low temperature and solution-phase growth for In N nanowires were iPr_2InN_3 (**1a**) and tBu_2InN_3 (**1b**) that were prepared in one-pot procedures from the corresponding trialkylindanes *via* diakylmethoxyindane intermediates. In the solid state, **1a** and **1b** are iso-structural, ladder-like polymers, as illustrated in Fig. 7(a). The products are nearly insoluble in hydrocarbons at room temperature, but dissolve in Lewis-basic solvents such as pyridine and acetone. Reactions of the precursors **1a** or **1b** with the mild reductant 1,1-dimethylhydrazine, H_2NNMe_2, produced crystalline InN. Thermolysis of **1a** in refluxing diisopropylbenzene (203 °C) with H_2NNMe_2 gave crystallites with a coherent length of 18 nm, as determined by x-ray diffraction (XRD) data, whereas the same process without H_2NNMe_2 gave amorphous structure. Meanwhile, thermolysis of **1b** with and without H_2NNMe_2 also produced InN crystallites. However, the coherence length was slightly smaller (12 nm and 7 nm, with and without H_2NNMe_2). Monitoring of the H_2NNMe_2 reactions by nuclear magnetic resonance spectrometry (NMR) and gas chromatograph mass spectrometry (GC-MS), which detects the alkane and $HNMe_2$ byproducts, suggested the process shown in Fig. 7(b). TEM investigation of the materials produced by thermolysis of **1a** or **1b** in the presence of H_2NNMe_2 indicates polycrystalline InN fibers, of diameters around 20 nm and lengths in the range of 100–1000 nm. Moreover, each InN fiber exhibited a metallic indium droplet attached to its tip. In contrast, the dominant products obtained from thermolysis of **1b** without H_2NNMe_2 were very fine-grain equidimensional InN nanocrystallites ($<$10 nm); only a very small fraction ($<$2%) of stunted InN fibers was present.

A primary role of H_2NNMe_2 was to act as a stoichiometric hydrogen-atom donor to assist alkane elimination. Note that the N-N bond in H_2NNMe_2 was cleaved and

N_2 was generated, which is likely to provide a driving force for hydrogen donation, whereas other potential hydrogen donors such as PhSH, Me_2NH and $PhNH_2$ were not so effective in promoting InN fiber growth. It should be mentioned that fiber growth was observed only when indium was also generated, and H_2NNMe_2 was effective to induce indium generation. Indium droplets were found to form in the presence of H_2NNMe_2, but not in the absence of any hydrogen-atom donating co-reactant. Moreover, the In droplets were found only at the tips, suggesting that, the In droplets formed in an adventitious side reaction preceding InN fiber growth, and later on acted as a catalyst in fiber growth. The liquid metal clusters or catalytic metal droplets act not only as the energetically favored site for absorption of gas-phase reactants in the case for VLS growth, but also promote both precursors decomposition and crystal lattice construction in the case for SLS growth.

2.2.6. Pyrolysis of single molecular precursors

Single molecular precursors for III-nitrides have been designed and developed by a few research groups on the basis of volatile group-III amide, azide and hydrazide compounds and have been investigated in organometallic chemical vapor deposition (OMCVD) studies [70–72]. More recently, Parala et al. reported the growth of InN whiskers by CVD under very specific conditions using the single molecule precursor (*Azido*[bis(3-dimethylamino)-propyl]indium) (AZIN), $[N_3In[(CH_2)_3NMe_2]_2]$ [29]. The AZIN precursor, which was synthesized first by slightly modifying the synthetic procedure reported earlier by Fischer et al. [72], was taken in a reservoir and nitrogen was used as a carrier gas with ammonia as an additional reactive gas for growing InN at a typical substrate temperature of 500 °C. Interestingly, the growth of InN whiskers was achieved only when bare *c*-plane sapphire $[Al_2O_3(0001)]$ was used. When the deposition was conducted on nitridated sapphire substrate, e.g., forming an AlN buffer layer with thickness of the order of 20–50 nm, which is a common practice to have a close lattice match with the desired III-nitrides, dense and preferentially oriented thin film was formed instead.

It is noted that, in the pyrolysis technique using AZIN single molecular precursors, InN nanowires were observed with a growth period of about 15 min, which is much shorter than that normally reported in CVD-based catalytic growth and other vapor-solid methods described below. Another intriguing point to be noted here is that the InN fibers produced by the pyrolysis of AZIN do contain In droplets at the tips. Although conventional metal catalysts such as Fe, Ni, Au were not introduced deliberately, it is very likely that indium was generated at the very beginning stage during pyrolysis of AZIN and later on served the role as catalyst for further fiber growth *via* VLS mechanism. This "self-catalyst" behavior is quite similar to that observed in the formation of InN nanowires by solution-liquid-solid method in the presence of H_2NNMe_2 as described by Buhro et al. [67].

2.3. Vapor-solid (VS) growth and other methods

In sections 2.1 and 2.2, we have described several processes involving the use of either templates or catalysts to grow the nitrides nanowires. The preparation and removal of the templates or catalysts in most cases cannot be done by just a few simple steps and thus are not cost effective. Hence, growth techniques that do not require any catalyst and template are highly desirable. Vapor-solid growth has been proved to be an effective catalyst-free (and template-free) method for synthesizing whiskers [73]. Both VLS and VS techniques involve the vapor phase deposition. In the VS process, chemical species diffuse toward a substrate through an interfacial gas layer and are adsorbed on the surface of the substrate. Subsequently, nucleation and whisker growth occur, accompanied by the elimination of gas byproducts. In contrast, in VLS growth, a liquid-phase catalyst is involved for adsorbing vapor reactants and forming eutectics. Therefore, the diameters of the whiskers are controlled by the size of the catalyst used for the VLS growth, whereas the same are controlled by the diffusion rate and the surface migration of gas phase species for the VS growth. The key feature shared by the following VS and other catalyst-free methods is the existence of a large difference in growth rate along different crystal orientation under those process conditions. The occurrence of highly anisotropic growth may be inherent in the wurtzite III-N system, which offers an opportunity for forming the one-dimensional nanostructures in a wide range of process conditions. Nevertheless, cubic III-N nanowires were also observed by using this type of technique.

2.3.1. Direct reaction

He et al. has recently demonstrated growth of GaN nanowires by a direct reaction of Ga and NH_3 [74, 75]. In fact, the same research group has also reported the formation of thicker rods (500 nm to 10 μm) using this direct reaction technique. The key process parameters that control the diameter of the rods were the temperature and NH_3 flow rate. In general, the diameter of the nanowires increased with temperature and NH_3 flow rate, in the temperature range of 825–1000 °C and NH_3 flow rate range of 20–120 sccm. At temperatures above 925 °C, hexagonal crystal platelet growth was favored over that of the nanostructures. The growth process can be divided into three stages: first occurrence upon heating was the formation of a nearly amorphous Ga-rich GaN matrix, from which hillocks of thin platelets appeared with further heating, and finally growth of nanowires or rods from the platelets or of platelets themselves at the expense of the nanostructures, depending on the actual process temperatures and NH_3 flow rates. In the latter two stages, the morphological development of the GaN platelets and the nanostructures can be explained based on the influence of various growth conditions on the growth rates along different crystal axes.

Fig. 8 illustrates the various pathways that Ga and N atoms may arrive on the surface of one of the crystal platelets in the polycrystalline hillock. Ammonia molecules and Ga atoms arrive at the basal plane as indicated by **A**. The NH_3 can decompose stepwise to NH_2 and NH yielding N atoms, which react with the Ga to form GaN. At low temperatures, the number of sites for crystal plane growth is low and the average molecular diffusion length before desorption is short, therefore, formation of amorphous GaN is favorable (pathway **B**) or yet extremely fine crystal platelets on, but not incorporated into, the basal plane (pathway **C**). At the highest temperatures (~950 °C), the diffusion length of the molecules would be relatively long, allowing migration to favorable nucleation sites where they could undergo dissociation and reaction and then be incorporated into the existing crystal (pathway **D**). Another possibility is that, the molecules and atoms would desorb rapidly from sites where they are not strongly bound to the existing crystal (pathways **E-F**). Except at the highest

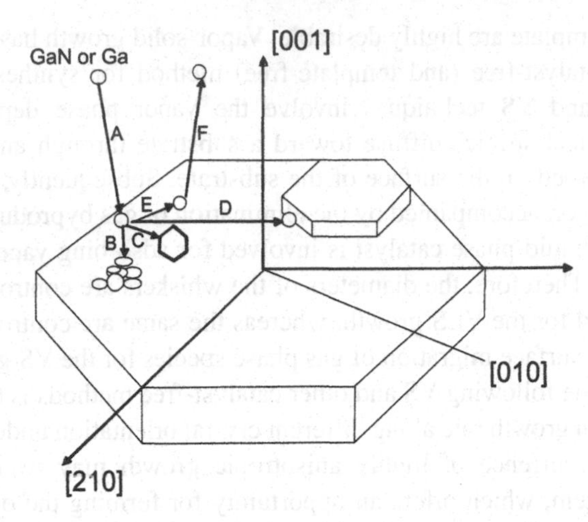

Fig. 8. A model of one of the crystal platelets found in the polycrystalline hillock which can be used to discuss the growth of the hexagonal platelets in the [001] direction. Ammonia molecules and Ga atoms arrive at the basal plane as indicated by **A**. The ammonia can decompose stepwise yielding N atoms, which react with the Ga atoms to form GaN. In any case, molecules on the basal plane have four pathways, as indicated as **B, C, D, E-F**. Depending on the process temperature and NH_3 flow rate, the generation of GaN molecules and the average molecular diffusion length before desorption or finding a nucleation site is different, resulting in different morphology. [With permission from M. He et al., *J. Cryst. Growth* **231** (2001) 357]

temperatures, growth of the crystal platelets in the [001] direction is very slow. Instead, edges provide more nucleation sites for further growth. SEM investigations did show that the nanowires were not formed as separate objects but were emerging from the side of the crystal platelets [75]. In this case, the nanowire diameters were the same as the thickness of the crystal plates from which they grew. Since at higher temperatures and NH_3 flow rates platelets are usually larger and thicker than their counterparts at lower temperatures and NH_3 flow rates, the average diameter of the nanowires increases with both temperatures and NH_3 flow rates.

2.3.2. Oxide-assisted growth

Another approach that does not involve metal catalysts and/or templates as described in sections 2.1 and 2.2 is the oxide-assisted growth approach. This approach was originally developed for bulk-quantity synthesis of Si and Ge nanowires [76, 77]. More recently, this technique was extended as a general synthetic route to III-V compounds, including GaN, nanowires [78]. More detailed descriptions of this technique are included in this book as a separate chapter. Briefly, GaN nanowires were obtained by laser ablation method using GaN targets mixed with 25 mol. % Ga_2O_3 and ablated at 1000 °C. The Ga_2O_3 mixed in the target is essential as no nanowire was obtained without Ga_2O_3 addition. A four-stage growth process is proposed to explain the formation of GaN nanowires as follows: (1) GaN in the target first decomposed into Ga and N upon laser ablation. (2) Ga atoms react with Ga_2O_3 in the target to produce Ga_2O. (3) As Ga_2O is much more volatile than Ga_2O_3 at the processing temperature, Ga_2O vapor can be transported to the deposition zone by the carrier gas. At the same time, the N is also transported to the deposition zone. (4) Ga_2O and N react at the deposition zone *via* the reaction:

$$3Ga_2O + 4N \rightarrow 4GaN + Ga_2O_3 \quad (10)$$

to form GaN nanowires finally. Interestingly, the GaN nanowires produced by this technique exhibited a cubic zinc-blende structure and grew along the [−111] direction, unlike the cases produced by VLS growth or direct reaction, wherein a hexagonal wurtzite structure is predominant. Moreover, a thin layer of amorphous GaO_x was found to wrap around the resultant GaN nanowires.

The key feature in the oxide–assisted growth is the production of the Ga_2O vapor. Peng et al. has reported a hot filament CVD approach under the presence of carbon for generating Ga_2O vapor [79]. A Ga-containing solid source, which was a mixture of Ga_2O_3 and C powders (with molecular ratio of 1 : 1) made under a hydraulic press of 3.2×10^8 Pa for 48 h, was mounted above the hot filament, while the substrate was placed below the filament. At a substrate temperature of 900 °C, reaction of NH_3 and the Ga-containing source is likely to proceed in two steps. The first step is

$$Ga_2O_3(s) + C(s) \rightarrow Ga_2O(g) + CO_2(g) \quad (11)$$

and the second step is the same as reaction (3), which has been described in section 2.1. The hot filament CVD method is also capable of growing bulk-quantity GaN nanowires. While the oxide-assisted growth *via* laser ablation of GaN-Ga_2O_3 mixed target was performed under hydrogen-free environment, GaN nanowires grown by the hot filament method were carried out in a reductive environment, although the same volatile oxide is involved in both cases. During cooling, the nanowires were also protected by the presence of ammonia flow in the hot filament method. Therefore, no amorphous oxide outer-layer was observed in the GaN nanowires, although Ga oxide was used. The absence of an oxide shell can be potentially very useful since it allows the electrical contact to be made onto the nanowires directly.

More recently, Zhang et al. have reported a large-scale production of InN nanowires through a gas reaction from mixtures of In metal and In_2O_3 powders (at a mole ratio of 4 : 1) with flowing ammonia in 700 °C using a conventional quartz tube furnace [80]. The materials resulting from such a reaction for two hours produced a high yield of wire-like structures having diameters of 10–100 nm and a maximum length of several hundred micrometers. The formation pathways of InN nanowires, although not been discussed in details in the original paper by Zhang et al., may involve the following two important steps. First, molten In reacts with the In_2O_3 powders to produce In_2O vapor. Second, gaseous In_2O and NH_3 react leading to solid products deposition on the substrate *via* the following pathway:

$$In_2O(g) + 2NH_3(g) \rightarrow 2InN(s) + H_2O(g) + 2H_2(g) \quad (12)$$

Since no metal particles were observed at the ends of nanowires, the formation of InN nanowires was not dominated by the VLS process, but was by the VS process wherein the conditions favored anisotropic growth along different crystal directions. Except that the generation of a volatile oxide is induced differently, this vapor-solid method is quite similar to the hot filament approach. One important feature of these two VS processes is that they were carried out in a reductive environment. Indeed, X-ray photoelectron spectroscopic (XPS) investigation revealed strong peaks for In and N bonding with no obvious peak for the In_2O_3 or $In(OH)_3$, confirming the absence of an oxide shell.

2.3.3. *Chloride-assisted growth*

Up until now, there is still no report on using oxide-assisted method for fabricating AlN nanorods or nanowires. Meanwhile, nitridation reactions between N_2 and Al powders have been commonly used for production of AlN materials. The morphologies of the AlN products depend on the reaction conditions and nature of the initial Al powders. Good selectivity for either equiaxed AlN nanoparticles or AlN nanowires has been achieved by nitridation of nanocrystalline Al metals at 1000–1100 °C [81, 82]. The selectivity was most strongly influenced by the presence or absence of $AlCl_3$, a promoter of nanowire growth, during nitridation. With no $AlCl_3$ addition, only a small nanowire fraction was obtained. When 9–12 wt% of $AlCl_3$ was added, over half of the product was nanowires, while addition of 93 wt% of $AlCl_3$ produced large yields (~90%) of long AlN nanowires (up to >30 μm) that possess among the smallest diameters (20–100 nm) known to date for this system.

It was proposed that the equiaxed nanoparticles grew from liquid Al, whereas the nanowires grew from gaseous Al species. The dramatic promotion in nanowire formation upon addition of $AlCl_3$ subjected to rapid heating implied a mechanism in which $AlCl_3$ acted as a vapor-transport agent. $AlCl_3(g)$ and Al(l) are known to form significant quantities of AlCl(g) and $AlCl_2(g)$ in equilibrium at the nitridation temperatures [83, 84]. Moreover, AlCl(g) and $AlCl_2(g)$ react with N_2 whereas $AlCl_3(g)$ does not [84]. Hence, the $AlCl_3$ functions primarily as a vapor-transport agent that generates volatile, reactive Al mono- and di-chlorides, which, in turn, are the nutrient species for nanowire growth through VS mechanism. However, generation of the AlCl and $AlCl_2$ competes with sublimation of $AlCl_3$ out of the furnace. Therefore, rapid heating was needed to increase the nanowire fraction in part by decreasing the amount of $AlCl_3$ sublimating out of the hot zone before nitridation temperatures were reached.

2.3.4. *Combustion technique*

Combustion synthesis is ignited by heating the reactant uniformly in a furnace to the ignition temperature, such that the reaction starts simultaneously throughout the sample. Recently, AlN nanowires with diameters in the range of 70–500 nm and a maximum length of about 2 mm have been fabricated by using combustion synthesis technique [85, 86]. The reactants used by Lee et al. [85] were Al powder (<5 μm) mixed with 50 wt % AlN diluent (1.5 μm) and a 3 wt % of $MgCl_2$ powder was used as a vapor-transport agent. Chen et al. [86] used fine Al powder with an average diameter of 0.5–1 μm, along with NH_4F and NH_4Cl (26% mol). Chlorine gas is known to fracture the intrinsic surface oxides on Al powder [87], hence chlorides were used in both cases. Addition of NH_4F and NH_4Cl mixtures was indispensable in the latter case, as they could act as solid nitrogen source and diluent agent.

Reactions occurring during combustion process could be quite complex. Multi-stage reactions were suggested. For instance, in the process reported by Lee et al., the first stage reaction was ignited near the melting temperature of Al, following which a temperature excursion was experienced. In the process reported by Chen et al., the reactants NH_4X, X = F or Cl, first decomposed into HX and NH_3 at temperatures of about 400 °C, then these components reacted further with Al to form the mediate products AlX_x(g) and $[Al_x(NH)_y]_n$(s), and finally to form AlN(s), HX and NH_3.

2.3.5. *Sublimation method*

Straight, smooth and long GaN nanorods can also be produced from a simple sublimation process [88]. In this technique, hexagonal GaN powders formed by a conventional gas reaction method were ball milled, using hardened agate balls with a diameter of 10 mm, for 18 h as the starting materials. The as-milled powders consisted of spherical particles with diameters ranging from 60 to 200 nm. After heating to 930 °C for 3 h under an ammonia flow, straight GaN nanorods with diameters of 10–45 nm and lengths of up to hundred microns were found on the $LaAlO_3$ substrate placed downstream of the ammonia flow in the tube furnace. The fact that a simple sublimation method would lead to the formation of nanorods is likely duc to a highly anisotropic growth in that the growth rate, which usually depends upon the process conditions, along the different crystallographic orientation is very different. In fact, more than twenty years ago, Ogino et al. reported that they had grown needle-like GaN crystals by sublimation at above 1500 °C [89]. Using the similar sublimation technique but with milled powders, Li et al. were able to form GaN nanorods at a relatively lower temperature more recently [88]. The milling process before annealing was crucial to the formation of GaN nanorods at low temperature as annealing of the non-milled samples with the same condition and duration of time did not lead to any nanorod formation. This is because the milled powders had very high specific surfaces and were unstable in thermal annealing process, which made it easy to sublimate and re-condense into nanorods.

2.3.6. *Oriented, epitaxial growth and substrate consideration*

Very recently, Kim et al. have demonstrated oriented GaN nanorods on a sapphire substrate by hydride vapor phase epitaxy (HVPE) [90]. The HVPE method was first employed by Maruska and Tietjen to grow single crystal epitaxial GaN thin film [91]. In the experiment by Kim et al., samples were grown on sapphire substrates and the substrate temperature was held at about 480 °C. In the growth process, the Ga precursor was synthesized in the lower region of the reactor at 750 °C, wherein the reaction of HCl gas in an N_2 diluent gas with Ga metal took place, resulting in GaCl. This precursor was then transported to the substrate area with temperature of 478 °C, where it was mixed with NH_3 to form GaN. The HCl flow rate was in the range of 30–90 sccm, the NH_3 flow rate was 2000 sccm, and the growth time was about 1–3 h. As a result, the straight and well-aligned GaN nanorods with high density were formed on the sapphire substrate without any catalyst or template layer. Typical diameter of the GaN nanorods produced by this HVPE method was about 80–120 nm.

In fact, formation of aligned nanorods (which were usually termed as "nanocolumns" in large volume of film-based reports) was commonly observed in epitaxial growth, e.g., MOCVD or molecular beam epitaxy (MBE), of GaN and related nitrides. Yoshizawa et al. have reported growth of self-organized GaN nanocolumns on (001) sapphire substrate by RF-radical source molecular beam epitaxy (RF-MBE) [92], in which Ga was evaporated from a conventional effusion cell and the reactive nitrogen beam was generated by a 13.56 MHz RF-radical plasma source. Self-organization is believed to occur through a complex interplay between the lattice strain, surface energy and surface migration. The nanorods were grown at high density, with the *c*-axis maintained perpendicular to the substrate surface. Under certain process conditions (RF input power at 450 W and substrate temperature at 800 °C), the average rod diameter appeared to be constant at around 80–90 nm independent of the nitrogen flow rate, and only the rod density changed with the nitrogen flow rate. At a constant nitrogen flow rate (2 sccm), the diameter of the GaN nanorods decreased with decreasing RF input power. A rod diameter of 53 nm was the smallest obtained. One explanation for relatively small rod diameter is that lowering of the RF input power may result in a decrease in the kinetic energy of the active nitrogen, and thus possibly shortening of the migration length on the substrate surface and then suppression of lateral growth.

2.4. Reactive ion etching method: An example of "top-down" method

All the techniques described in sections 2.1–2.3 can be categorized as "bottom up" technique, in that the formation of nanorods is indeed built stepwise from atoms or molecules to clusters, and nanorods. Recently, Yu et al. have reported a method of fabricating GaN nanorods of controllable diameter and density using inductively coupled plasma reactive ion etching (ICP-RIE) [93]. This technique can be considered as a "top-down" method since the constituent elements are removed from an original object. In principle, one can also produce nanorods using other etching processes.

Metal-organic CVD grown GaN epi-film of microns on a (001) sapphire substrate was used as the etching object. The ICP-RIE was conducted under a gas mixture condition of $Cl_2/Ar = 10/25$ sccm with the ICP source power and bias power at 200 W. Typical etching time was 2 min. GaN nanorods of various diameters and densities were obtained by controlling the chamber pressure. A uniform etched surface without any nanorods formation was observed if the chamber pressure was kept at and below 2.5 mTorr. As the chamber pressure was increased to 10 mTorr, the GaN nanorods began to form and the density of nanorods increased, as the chamber pressure was further increased from 20 to 30 mTorr. The GaN nanorods density increased from 10^8 to 3×10^{10} cm^{-2}, while the diameter of the nanorods decreased from about 100 nm to 20–50 nm, as the chamber pressure increased from 10 to 30 mTorr. The creation of nanorods and the variations in their diameter and density seemed to be related to the crystalline quality of the initial GaN film and the ability of the ICP process to dissociate GaN bonding. Typical MOCVD grown GaN films on sapphire substrates are known to exhibit high dislocations and defects density on the order of 10^8–10^9 cm^{-2} due to film-substrate lattice mismatch. These dislocations and defects tend to have weaker binding energy and could be easily etched by the ICP process. Thus, the GaN nanorods density was quite comparable to the defect density of the starting GaN film subjected to ICP etching. However, the plasma density is a strong function of the chamber pressure [94, 95]. As the chamber pressure increases, the mean-free path decreases and the collision frequency increases, which leads to a significant reduction of the ion energy, therefore, a degradation of ICP dissociation ability.

3. Synthesis of Ternary Group-III Nitride Nanorods and Nanowires

It is well known that, by alloying InN together with GaN or AlN, the band gap of the resulting alloy can be tuned to a desirable value for each specific optoelectronic application. However, up to now, only a little research has been performed on ternary In-Ga-N nanorods and nanowires [96]. The experimental setup was the same as that shown in Fig. 1, except that mixture of both Ga and In metals was used as the source material. Ternary Ga-In-N nanorods and nanowires were obtained, though "alloying" was not effective by this simple technique. The products were either In-rich (if grown at 500 °C) or Ga-rich (if grown at 800 °C) with very little solubility of the other element. Detailed structural analyses indicated that the resulting nanorods were not single-phase crystals, but contained some ultra-fine inclusions of the secondary phase (shown in section 5.2). It should also be noted that there was no formation of any crystalline nanorods if the growth temperature was held between 550 and 800 °C.

In general, incorporation of indium in the ternary or quaternary III-N alloys is not easy even for film growth, although the InN molar fraction in In-Ga-N film has been pushed above 30%. Calculation based on a modified valence-force-field model suggests the existence of a solid phase miscibility gap primarily due to the large difference in inter-atomic spacing between InN and GaN [97]. The presence of this miscibility gap is expected to cause a significant problem for the growth of alloying film. Moreover, while GaN crystal growth was usually conducted at high temperatures (above 800 °C), InN and In-Ga-N crystal growth had to be performed at low temperatures (no more than about 550 °C) to prevent InN dissociation during growth. Nevertheless, the use of a high nitrogen flux and a high indium source flow rate allowed the high temperature (800 °C) growth of relatively high-quality In-Ga-N film by MOCVD method [98, 99]. From the reports cited above, it may be concluded that the major challenge for obtaining high-quality In-Ga-N is to conduct a growth process with a compromise in the growth temperature, since InN is unstable at typical GaN deposition temperatures.

4. Synthesis of Binary and Ternary Group-IV Nitride Nanorods and Nanowires

4.1. Si_3N_3 nanorods and nanowires

Owing to the wide range of applications of Si_3N_3 whiskers in the ceramic industry, especially they are best suited as reinforcement in composites of different matrix materials (metallic, ceramic and polymeric), a variety of process techniques have been developed since early seventies. The main fabrication routes for Si_3N_3 whiskers include chemical vapor deposition [100, 101], nitridation of silicon powders [102, 103], carbothermic reduction of a silica-containing compound [104, 105], combustion synthesis [106, 107] and reactive evaporation [108]. However, the diameters of the Si_3N_3 whiskers given in those reports are usually over 100 nm. It is not until late nineties, that Si_3N_4 nanowires with diameters less than 40 nm have been reported, for the first time, by using the CNT-confined reaction method analogous to that for fabricating GaN nanowires [27]. Briefly, through the reaction of CNTs with SiO vapor, which in turn was generated by a reaction between carbon and silica, under nitrogen (or ammonia) atmosphere at 1400 °C for one hour, the Si_3N_4 nanowires could be produced. These nanowires were found to be 4 to 40 nm in diameter and up to 50 μm in length, quite similar to the original dimensions of the CNTs. The main reaction scheme in this technique can be expressed as:

$$C(s) + SiO_2(s) \rightarrow SiO(g) + CO(g) \tag{13}$$

$$3SiO(g) + 3C(s) + 2N_2(g) \rightarrow Si_3N_4(s) + 3CO(g) \tag{14}$$

In fact, even before the CNT template was available, production of Si_3N_4 whiskers from SiO_2/C powder mixtures in a nitrogen-containing atmosphere through reactions (12) and (13) has already been demonstrated [109, 110]. However, the yield for Si_3N_4 whiskers is usually very low, e.g., ~26% after 48 hour growth at 1400°C [111]. Moreover, SiO and CO can act as intermediates in the formation of both SiC and Si_3N_4, making their production a competitive process. The presence of hydrogen increased the SiO_2/C reaction yield and the introduction of transition metals in the SiO_2/C powder mixtures could promote SiC and Si_3N_4 whiskers formation by a VLS mechanism [112]. Typical diameters of the whiskers produced by this method were still in the range of a few hundreds nm, presumably due to the high working temperature (~ 1400 °C).

Meanwhile, Si_3N_4 nanowires and SiO_2 amorphous nanowires with diameters of 10–70 nm and 10–300 nm, respectively, have been synthesized by heating Si powders or Si/SiO_2 mixtures with or without metal catalyst at 1200 °C at ambient pressure [113]. The following reaction steps were proposed for the formation of Si_3N_4 nanowires.

$$Si(s) + SiO_2(s) \rightarrow 2SiO(g) \tag{15}$$

$$6SiO(g) + 4N_2(g) \rightarrow 2Si_3N_4(s) + 3O_2(g) \tag{16}$$

The flowing of the N_2 gas leads to a low SiO partial pressure. A low partial pressure of SiO may result in a low supersaturation of Si_3N_4, which favors the growth of the one-dimensional wires over that of the particles [114–116]. It is also noted that there is little effect of the metal catalyst on the growth of the Si_3N_4 nanowires derived from Si or Si/SiO_2 mixtures. Furthermore, the absence of the metal/Si alloy droplets in the ends of the nanowires also suggests that VLS mechanism might not work in the experiment reported by Zhang et al. [113]. The most possible route was the VS mechanism described earlier instead. In the meantime, SiO_2 could also be formed through vapor phase reactions either between Si and O_2 vapors or between gaseous SiO itself, since they were abundant in the gas phase. Indeed, amorphous SiO_2 nanowires were observed and the crystalline Si_3N_4 nanowires were also covered with amorphous SiO_2 shells in the products from the Si-SiO_2-N_2 reactions. It is noted that the Si_3N_4 nanowires reported by Zhang et al. [113] exhibited much smaller diameter than those by conventional carbothermal reduction or earlier reports that also employed a similar Si/SiO_2 nitridation technique. Two factors may be attributed to this difference. One is the lower temperature and shorter synthesis time; the other may be due to the presence of the SiO_2 shells, which retard the lateral growth of the Si_3N_4 nanowires. This type of core-shell and cable-like structure has also been found in other nanowires.

Besides using Si, SiO_2/Si and SiO_2/C mixtures as the starting materials, preparation of Si_3N_4 nanorods by nitriding borosilicate glass with ammonia has also been reported very recently [117]. Vapor-solid growth mechanism was also suggested to be responsible for the formation of the Si_3N_4 nanorods. Again, the key step was the generation of SiO, which might be formed by solid-diffusion reaction between boron and silica. The resultant Si_3N_4 nanorods were 20–80 nm in diameter and 10–100 μm in length. Some nanorods were wrapped by amorphous Si_2N_2O.

4.2. Ternary Si-C-N nanorods and nanowires

This section describes a non-VLS and catalyst-free approach of growing Si-C-N nanorods and nanowires directly on Si substrate, distinctly different from many other techniques, whereby nanorods and nanowires were formed primarily as free-standing samples. In our previous works, polycrystalline Si-C-N films were deposited by microwave plasma enhanced chemical vapor deposition (MW-PECVD), wherein the nucleation density was low, while the growth rate was high, leading to formation of micron-sized crystals [12]. On the other hand, deposition of Si-C-N by using electron cyclotron resonance (ECR) PECVD led to formation of amorphous or nanocrystalline film with high nucleation density [118], due to the higher ionization fraction and more abundant nucleation species in the ECR-PECVD process [119]. A two-stage VS approach that combines the advantages of ECR- and MW-PECVD has thus been developed to produce dense Si-C-N nanorods [25, 26]. Specifically, a thin Si-C-N buffer layer was first deposited on Si substrates using ECR-PECVD. Then, the substrates were transferred to the MW-PECVD reactor for directional growth with higher rate. This approach to the Si-C-N nanorod growth is catalyst-free and self-mediated in that the buffer layer contains the same chemical elements as the nanorods. In addition, the number density and the diameter of the nanorods are controlled by the buffer layer, as will be illustrated below.

During the ECR-PECVD process, the substrate temperature was varied between 270 and 900 °C. The buffer layer deposited at 270 °C was primarily amorphous with few clusters of nanocrystalline aggregates. The number density of the nanocrystals increased significantly with substrate temperature in the ECR-PECVD process while the average grain size of the nanocrystals showed insignificant change. Above 640 °C, nearly complete coverage of the nanocrystalline phase was observed. Detailed microstructural study revealed that the buffer layer exhibited a bi-layer structure wherein the nanocrystalline phase was formed on top of the amorphous phase in the buffer layer, presumably due to the strain induced in the amorphous film after a critical thickness (approximately 50 nm) in the ECR-PECVD process [120]. While Si-C-N nanorods can be synthesized by MW-PECVD with a rather wide range of process parameters on Si-C-N buffer layers prepared by ECR-PECVD, the most critical parameters are

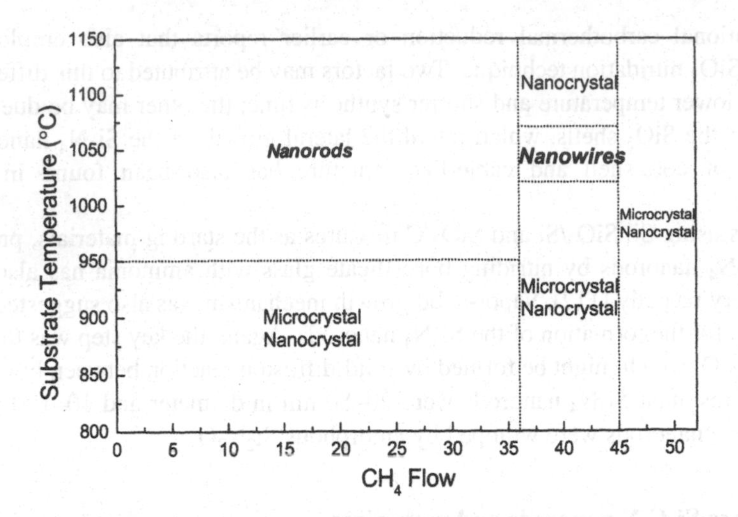

Fig. 9. "Phase fields" of the Si-C-N nanorods, nanowires and nanocrystals as well as microcrystals are indicated as a function of the substrate temperature and methane flow rate. [With permission from L. C. Chen et al., *J. Phys. Chem. Solids* **62** (2001) 1567]

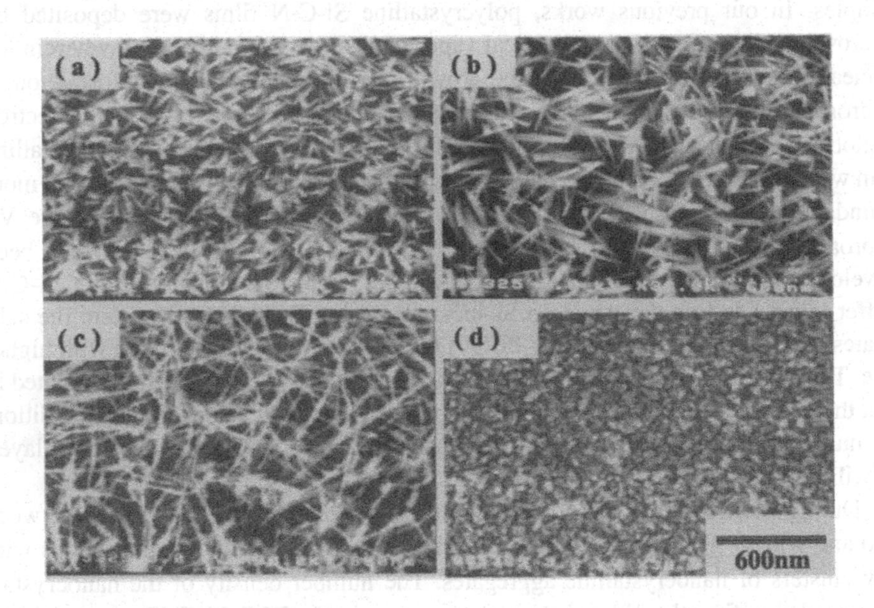

Fig. 10. SEM micrographs of the Si-C-N nanorods, nanowires and nanocrystals synthesized by MW-PECVD process using different methane flow rates but otherwise same growth parameters. For this specific set of experiments, the flow rates of H_2, N_2 and SiH_4 (10% SiH_4 in N_2 dilution) gases were fixed at 80, 80 and 4 sccm, respectively. The methane flow rate was varied: (a) 20, (b) 30, (c) 40 and (d) 50 sccm. During the MW-PECVD process, the chamber pressure was kept at 50 Torr with a microwave power of 2000 W. However, the substrate temperature gradually increased from 1000 °C, when 20 sccm methane flow rate was used, to about 1080 °C, when 50 sccm methane flow rate was used. [With permission from L. C. Chen et al., *J. Phys. Chem. Solids* **62** (2001) 1567]

the substrate temperature and the methane fraction in the source gas mixture during the MW-PECVD process (Fig. 9). Nanorods can be formed at a substrate temperature higher than 950 °C and up to a methane flow rate of 35 sccm. These process conditions appear to favor growth along a preferred orientation, leading to very high aspect ratios of the rods, even without any catalyst. The length and diameter of these nanorods were 0.7–2 μm and 10–60 nm, respectively. When the substrate temperature was below 950 °C or when the methane flow rate was higher than 45 sccm in the MW-PECVD process, microcrystalline and nanocrystalline phases were observed. Notably, wire-like materials were observed within a very limited range of the substrate temperature (~1050 °C) and methane flow rate (~40 sccm). SEM images of these phases are shown in Fig. 10. As shown in Fig. 11(a), cross-section transmission electron microscopy (TEM) image clearly indicates that the cluster of nanorods was grown on

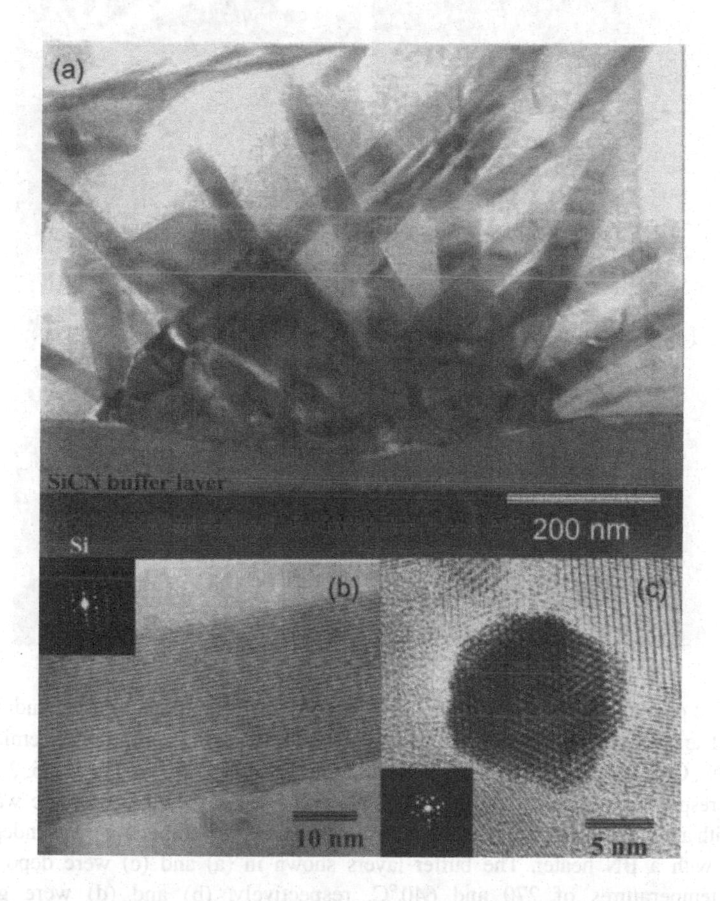

Fig. 11. (a) Cross-section TEM image of the Si-C-N nanorods grown on the Si substrate indicates that there exists an amorphous interlayer with nanocrystalline aggregates on top and that the nanorods only grow from the nanocrystals. High-resolution TEM images and the electron diffraction patterns of a Si-C-N nanorod viewed (b) in perpendicular and (c) in parallel to the long axis of the rod, indicating a two-fold and three-fold symmetry, respectively. [With permission from L. C. Chen et al., *J. Phys. Chem. Solids* **62** (2001) 1567]

top of a buffer layer and, apparently, the nanocrystals formed on top of the amorphous layer served as the "seeds" for subsequent growth of nanorods. High-resolution TEM images and their corresponding diffractions viewed in perpendicular and in parallel to the long axis of the rod are shown in Fig. 11(b) and (c), respectively. The axis of the nanorod shows hexagonal symmetry; whereas the perpendicular one shows twofold symmetry.

That the nanocrystals in the buffer layer serve as the seeds for nanorods growth can also be manifested in the correlation of their number densities and sizes (Fig. 12). The buffer layers shown in Fig. 12(a) and (c) were deposited by ECR-PECVD

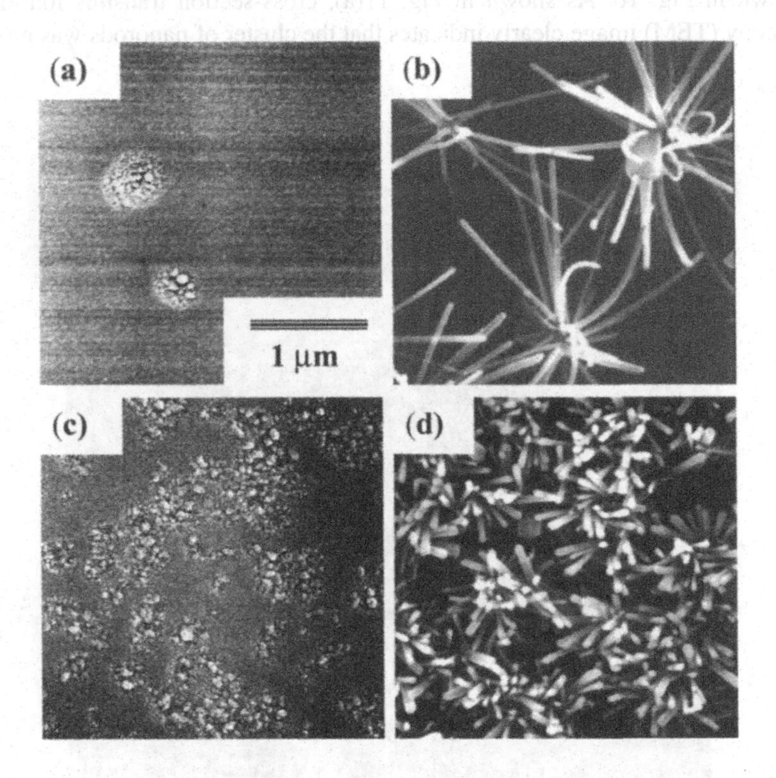

Fig. 12. Two pairs of SEM images of the Si-C-N buffer layers and their corresponding Si-C-N nanorods. The buffer layers were deposited by ECR-PECVD using a mixture of semiconductor grade H_2, N_2, CH_4 and SiH_4 (10% SiH_4 in N_2 dilution) gases with flow rates of 2.5, 2.5, 1.0 and 0.2 sccm, respectively. During the ECR-PECVD process, the chamber pressure was kept at 3 mTorr with a microwave power of 1200 W, while the substrate temperature was independently controlled with a BN heater. The buffer layers shown in (a) and (c) were deposited with substrate temperatures of 270 and 640 °C, respectively. (b) and (d) were grown by MW-PECVD using the buffer layers shown in (a) and (c), respectively. In MW-PECVD process, the same H_2, N_2, CH_4 and SiH_4 (10% SiH_4 in N_2 dilution) gases were used but with flow rates of 80, 80, 20 and 4 sccm, respectively. During the MW-PECVD process, the chamber pressure was kept at 50 Torr with a microwave power of 2000 W and a substrate temperature of 1000 °C. [With permission from L. C. Chen et al., *J. Phys. Chem. Solids* **62** (2001) 1567]

Fig. 13. SEM image of a dense (~10^{10} cm^{-2}), quasi-aligned Si-C-N nanorods sample grown by MW-PECVD under the same process conditions as described in Fig. 12. The buffer layer was prepared by ECR-PECVD also using similar process conditions as described in Fig. 12, except that the substrate temperature was higher. [With permission from F. G. Tarntair et al., *Appl. Phys. Lett.* **76** (2000) 2630]

process at two different temperatures. These two substrates with buffer layers were placed side by side in the same MW-PECVD run for subsequent growth of nanorods, and resulted in nanorods shown in Fig. 12(b) and (d). The SEM images clearly reveal that the number density and the size of the nanocrystals in the buffer layer are the key factors that control the number density and the diameter of the nanorods. A rod number density of ~10^{10} cm^{-2} has been achieved. Uniformly distributed quasi-aligned nanorods in which their long axis grows almost in perpendicular to the substrate have also been obtained (Fig. 13). It should be noted that the diameter of the nanorod remained quite constant, whereas the length of the nanorod progressively increased with time during the MW-PECVD process. A rough estimate of the average growth rate of the Si-C-N nanorods gave a value at about 0.25 μm/h. The slow growth rate makes Si-C-N nanorods preferable for length-controlled growth. It should be emphasized that direct growth of the nanorods on Si is advantageous for future integration into the existing microelectronics technology.

5. Structure and Bonding Characteristics

In sections 2–4, we have already illustrated some morphological and structural characteristics, especially on the process-related aspects. In this section, we would like to address some structure and bonding characteristics observed in the as-grown or ion-irradiated nanowires and nanorods, particularly if applicable, emphasizing the features that might be distinctive from their bulk or thin film counterparts.

5.1. Defect and/or strain in binary group III-nitride nanorods and nanowires

A typical x-ray diffraction pattern of the catalytic-grown GaN nanowires at 910 °C is shown in Fig. 14 [60]. All the sharp diffraction peaks in the pattern can be indexed to a hexagonal wurtzite structure. The lattice constants derived from the peak positions were $a = 0.3188$ and $c = 0.518$ nm, which agree well with the reported values of bulk GaN crystals [121, 122]. The fact that the diffraction peaks are identical to those of bulk GaN crystals and the strong diffraction peaks relative to the background signals suggest that the resulting GaN nanowires were not strained and had a high purity. Typical high-resolution TEM images of GaN nanowires also revealed nearly defect-free structure except some stacking faults (Fig. 15). Growth direction can also be determined from TEM along with electron diffraction analyses. Chen et al. [58] reported growth directions of both [110] and [100] in parallel to the long axis of the nanowires, whereas Han et al. [27] and Duan et al. [57] reported a common growth direction of [100].

The ability of obtaining high quality crystalline structure in GaN and other III-N nanowires is important for future opto-electronic applications. An improved crystal quality of InN nanowires over that of films was also achieved. Fig. 16 shows a comparison of the HRTEM images of the highly oriented MOCVD InN films and catalytic-grown InN nanowires prepared under very similar substrate temperature (500 °C). As the oriented InN film was grown on sapphire substrate, wherein a large lattice mismatch exists, high density of dislocation was still observed in the film, whereas there was hardly any in the "free-standing" nanowire.

5.1.1. Strained GaN nanowires

Seo et al. reported strained GaN nanowires synthesized by catalyst-assisted CVD, using the reaction of Ga metal and GaN powder mixtures with ammonia, on Si

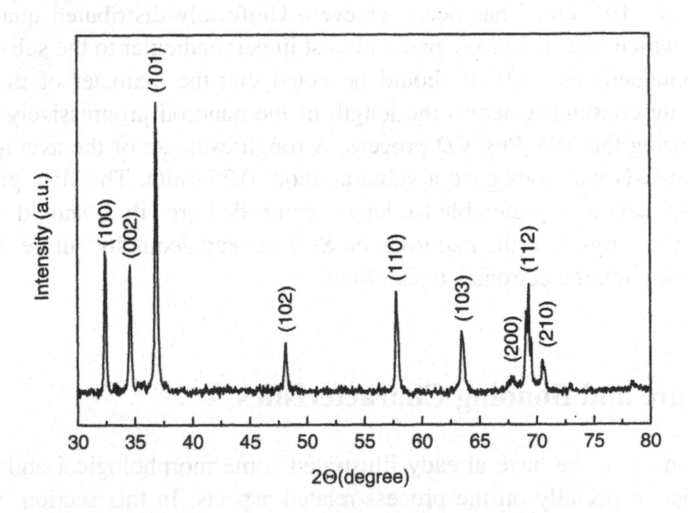

Fig. 14. X-ray powder diffraction of GaN nanowires, prepared by catalytic growth at 910 °C, as described in Fig. 1. All the peaks can be indexed to that of wurtzite GaN structure. [With permission from C.-C. Chen et al., *J. Am. Chem. Soc.* **123** (2001) 2791]

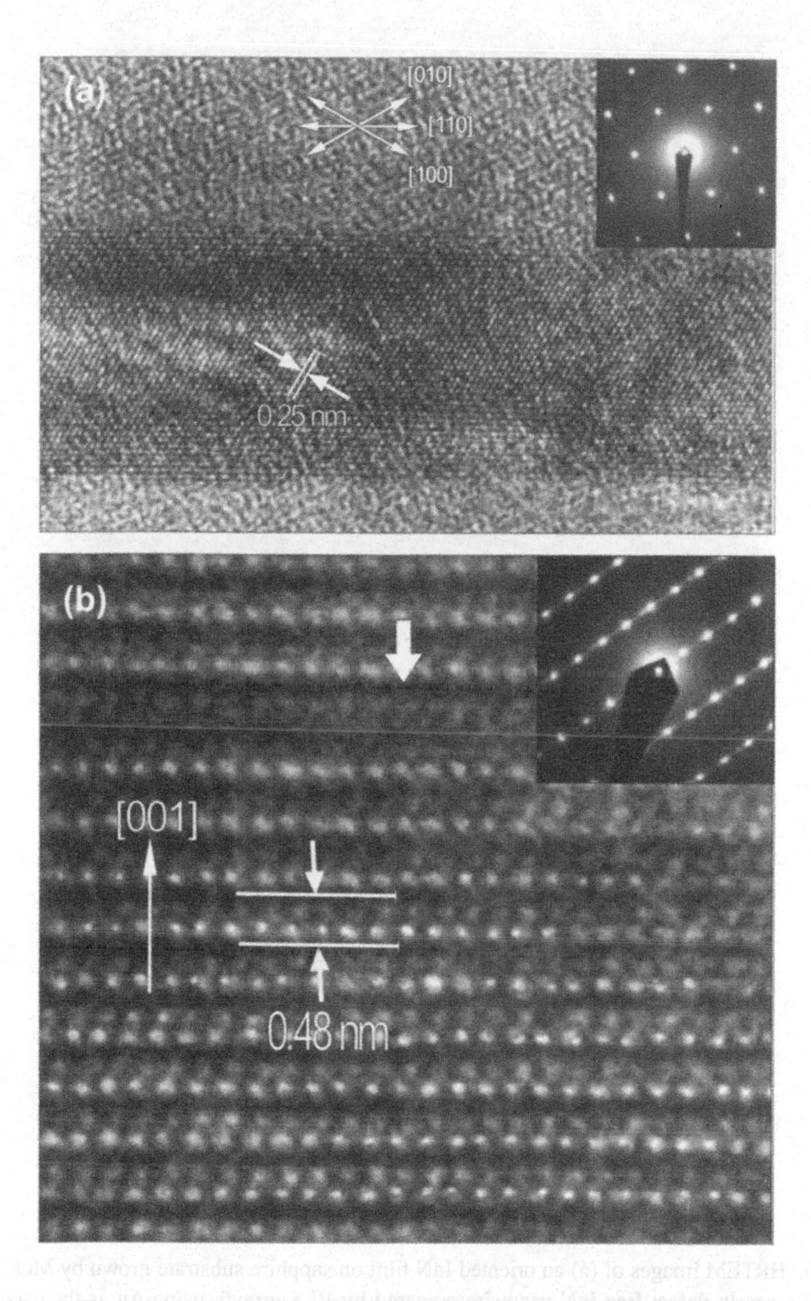

Fig. 15. (a) Typical high-resolution TEM image and its corresponding electron diffraction (inset) of a nearly defect-free GaN nanowire prepared using Co phthalocyanine as the catalyst. The direction of the long axis is [110], and the zone axis is [001]. The diameter of this single nanowire is ~8 nm. The space of ~0.25 nm between arrowheads corresponds to the distance between two (100) planes. (b) High-resolution TEM image and its corresponding electron diffraction (inset) of a faulted GaN nanowire prepared using iron powder as the catalyst. The direction of the long axis is [001], and the zone axis is [100]. The faulted structure is marked by an arrowhead. [With permission from C.-C. Chen et al., *J. Am. Chem. Soc.* **123** (2001) 2791]

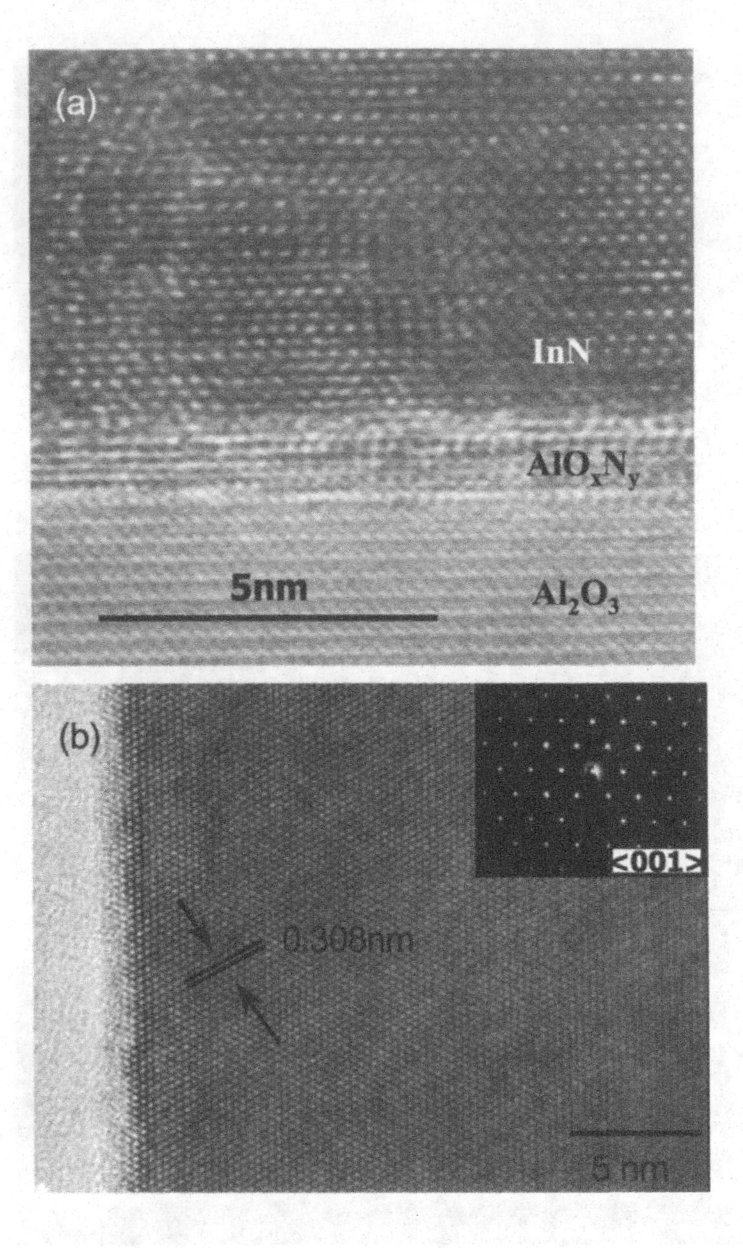

Fig. 16. HRTEM images of (a) an oriented InN film on sapphire substrate grown by MOCVD and (b) a nearly defect-free InN nanowire prepared by VLS growth using Au as the catalyst. Both InN film and InN nanowire were grown at a similar temperature (~520 °C). Prior to MOCVD growth, nitridation of the sapphire substrate was performed at 900 °C for half an hour to form a buffer layer so that the mismatch between the sapphire substrate and InN film can be reduced. However, dislocations of high density are clearly seen in the InN film. [(b) With permission from C. H. Liang et al., *Appl. Phys. Lett.* **81** (2002) 22]

substrates held at 900–1050 °C for 30 min to 2 h [123]. Prior to CVD process, catalytic iron nanoparticles were produced first by laser photo-dissociation of $Fe(CO)_5$ at 266 nm. The size of the Fe nanoparticles so produced was quite uniform at 8–10 nm and the resultant GaN nanowires had an average diameter of about 25 nm. Most of these GaN nanowires were single crystals and growth with their own specific growth direction, but the nanowires did not have an identical growth direction. A careful examination into x-ray diffraction and Raman scattering data revealed that the nanowires were strained. The full range XRD pattern of GaN nanowires together with those taken from GaN epilayer on sapphire and commercially available GaN powders (Aldrich) are shown in Fig. 17. The (002) peak of the GaN epilayer was at a position similar to that of the GaN powders and the lattice constants derived from these two XRD patterns agreed well with the reported values of bulk GaN crystals. In contrast, the peaks of GaN nanowires showed a shift to the higher diffraction angle. The magnitude of peak shift was $\Delta(2\theta) = 0.07$, 0.15 and 0.1 degree for (100), (002) and (101) peaks, respectively. Furthermore, these three peaks of the GaN nanowires and

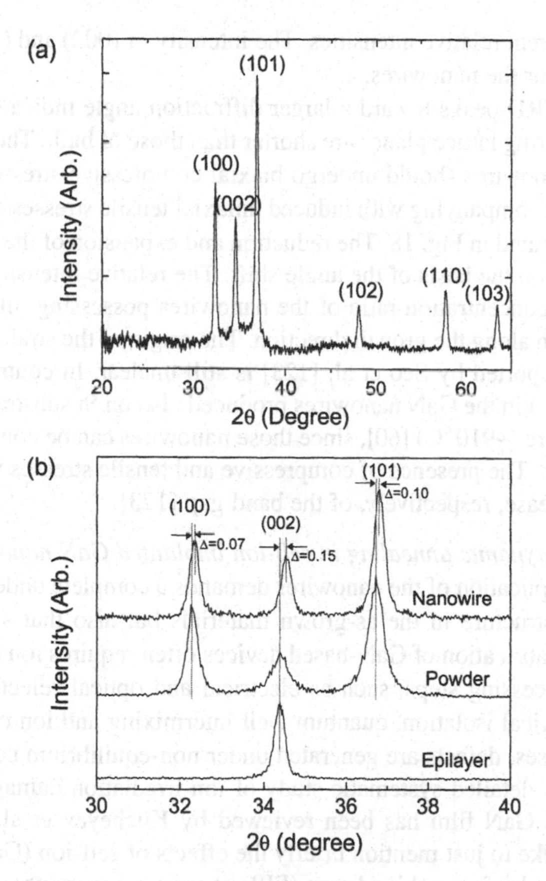

Fig. 17. (a) A full-range XRD pattern taken from the strained GaN nanowires. (b) The (100), (002) and (101) peaks for the strained GaN nanowires, epilayer and powders. [With permission from H. W. Seo et al., *J. Chem. Phys.* **116** (2002) 9492]

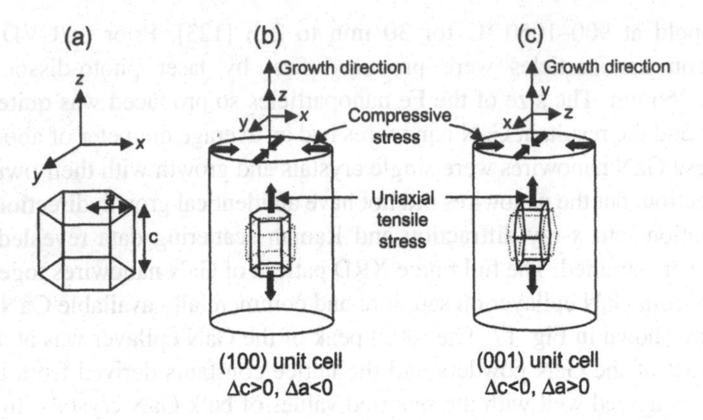

Fig. 18. Schematic diagrams for (a) a hexagonal unit cell with the lattice constants *a* and *c*, (b) the (100) unit cell that z-axis is aligned to the growth direction of a nanowire, and (c) the (001) unit cell that z-axis is perpendicular to the growth direction of a nanowire. [With permission from H. W. Seo et al., *J. Chem. Phys.* **116** (2002) 9492]

powders had different relative intensities. The intensity of (002) and (101) peaks were relatively higher for the nanowires.

The shift of XRD peaks toward a larger diffraction angle indicates that the separations of neighboring lattice planes are shorter than those of bulk. Therefore, it is suggested that the nanowires should undergo biaxial compressive stresses in the inward radial direction, accompanying with induced uniaxial tensile stresses along the growth direction, as illustrated in Fig. 18. The reduction and expansion of the lattice constants can be calculated on the basis of the angle shift. The relative intensity of XRD peaks would reveal the concentration ratio of the nanowires possessing different crystallographic orientation along the growth direction. The origin of the strain in the as-grown GaN nanowires reported by Seo et al. [123] is still unclear. In contrast, no apparent strain was observed in the GaN nanowires produced also on Si substrate and at a comparable temperature (~910 °C) [60], since those nanowires can be considered as "free-standing" samples. The presence of compressive and tensile stresses would lead to an increase and decrease, respectively, of the band gap [123].

5.1.2. Enhanced dynamic annealing in self-ion implanted GaN nanowires

The device application of the nanowires demands a complete understanding of not only the crystal structure in the as-grown materials but also that subjected to post growth process. Fabrication of GaN-based devices often requires ion bombardment as indispensable processing steps, such as electrical and optical selective-area doping, dry etching, electrical isolation, quantum well intermixing and ion cut. During these irradiation processes, defects are generated under non-equilibrium conditions at high quenching rate. A detailed systematic study of ion irradiation damage carried out in epitaxially grown GaN film has been reviewed by Kucheyev et al. recently [124]. Here, we would like to just mention briefly the effects of self-ion (Ga^+) implantation on GaN nanowires by focused ion beam (FIB) at room temperature.

In this preliminary study, catalytic-grown GaN nanowires were subjected to 50 keV Ga^+ focused ion beam with different fluence. The fluence dependent defect structure and the role of dynamic annealing (defect annihilation) were investigated by

SEM and TEM. A layer-by-layer defect accumulation was observed in the fluence range of 1×10^{14} to 1×10^{16} ions/cm^2. Unlike heavy ion irradiated epitaxial GaN film, large-scale amorphization appeared to be suppressed until a very high fluence of 2×10^{16} ions/cm^2. At a high fluence, point-defect clusters were identified as the major component in the irradiated nanowires. In contrast, the extended (planar) defects were reported as the major component in the disordered region of GaN film irradiated by heavy ion [125–127]. While both processes being nuclear energy loss dominated, enhanced dynamic annealing due to high diffusivity of mobile point defects in the confined geometry of nanowires is thought to be accountable for the observed differences.

5.2. Phase separation in ternary In-Ga-N nanorods and nanowires

As described in section 3, "alloying" was not effective in the ternary Ga-In-N systems. Fig. 19 shows the XRD patterns taken from typical In-rich and Ga-rich ternary Ga-In-N nanorods. The same for binary InN and GaN systems are also depicted. In general, the XRD patterns of Ga-rich or In-rich Ga-In-N nanorods were quite similar to those of binary GaN or InN nanorods, except that the In_2O_3 phase was also detected in the In-rich Ga-In-N system. Notably, the (100), (002) and (101) bands of Ga-rich $Ga_{1-x}In_xN$ ($x < 0.1$) nanowires either exhibited multiple-peak feature or had a side band slightly shift to the lower diffraction angle with respect to those of GaN nanowires. Literature report on the composition dependence of the lattice parameter

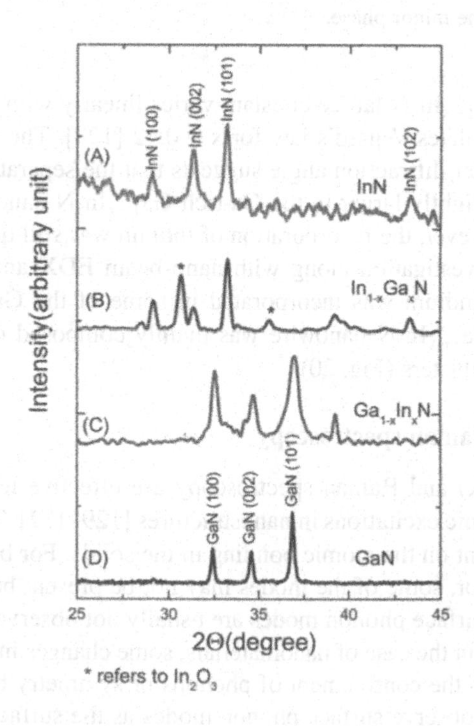

Fig. 19. X-ray powder diffraction patterns of Ga-rich and In-rich ternary Ga-In-N nanowires. For comparison, data taken from the binary GaN nanowires and InN nanowires are also shown.

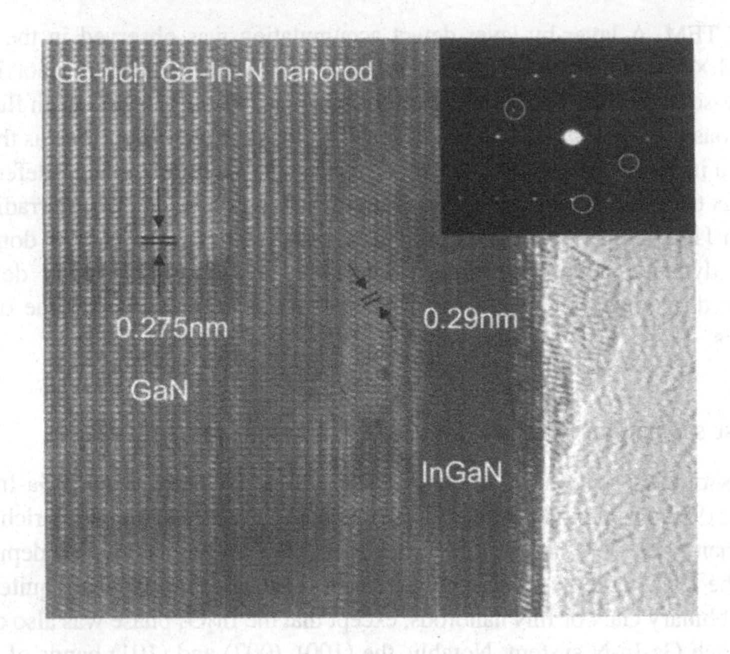

Fig. 20. Typical high-resolution TEM image and its corresponding electron diffraction (inset) of a Ga-rich ternary Ga-In-N nanowire. The matrix is primarily of GaN structure. An inclusion of ~3×10 nm size with a different structure from that of the matrix is clearly observed. The diffraction data indicates two superposed single crystal diffraction patterns, among which the circled spots are from the minor phase.

indicated that the $Ga_{1-x}In_xN$ lattice constant varies linearly with In mole fraction up to $x = 0.42$, but it violates Vegard's law for $x > 0.42$ [128]. The slight shift of XRD peaks toward a smaller diffraction angle suggests that the separations of neighboring lattice planes were slightly larger in the Ga-rich $Ga_{1-x}In_xN$ nanowires than those of GaN nanowires. However, the incorporation of indium was still quite small and inhomogeneous. TEM investigation along with nano-beam EDX analyses revealed that only a few at.% of indium was incorporated in some of the $Ga_{1-x}In_xN$ nanowires and the individual $Ga_{1-x}In_xN$ nanowire was mainly composed of GaN matrix with embedded InN nanoclusters (Fig. 20).

5.3. Infrared and Raman spectroscopy

Both infrared (IR) and Raman spectroscopy are effective tools for probing the phononic and electronic excitations in nanostructures [129–132]. The vibration modes are strongly dependent on the atomic bonding in the solids. For bulk crystalline insulator or semiconductor, some of the modes may not be present because of symmetry restrictions and the surface phonon modes are usually not observable due to their low intensities. However, in the case of nanomaterials, some changes in vibration properties are introduced due to the confinement of phonons or symmetry breaking; and it also becomes possible to observe surface phonon modes as the surface to volume ratio is large in the nanostructured materials. The position and the width of Raman peaks provide valuable information on the type of bonding, the structure perfection and the internal stress state of the nanostructured materials. Recently, the IR and Raman techniques have also been employed to investigate the vibration properties of GaN nanowires [31, 133].

5.3.1. Comparison of nanowires and epitaxial films in GaN

The space group of wurtzite GaN is $P6_3mc$. According to the factor group analysis at the zone-center, optical phonons belong to the following irreducible representations:

$$\Gamma_{opt} = A_1(z) + 2B_1 + E_1(x, y) + 2E_2 \tag{17}$$

where x, y, z in parentheses represent the direction of phonon polarization. For GaN, A_1 and E_1 modes are both Raman and IR active, two E_2 modes are only Raman active, and two B_1 modes are inactive in both Raman and IR. Since the GaN is noncentrosymmetric, the A_1 and E_1 modes are further split into longitudinal optical (LO) and transverse optical (TO) components. Fig. 21 shows a typical room temperature IR absorption spectrum of the GaN nanowires produced by catalytic VLS growth technique. The spectrum was calculated from a Kramers-Kronig analysis of the reflectance data recorded with a Bruker Equinox 55 FTIR spectrometer coupled with a Bruker IR Scope II [133]. Two phonon resonances at ~727 and 1117 cm^{-1} were observed. These two infrared-active phonon features should correspond to A_1(LO) and the two-phonon mode of E_1(TO) symmetries.

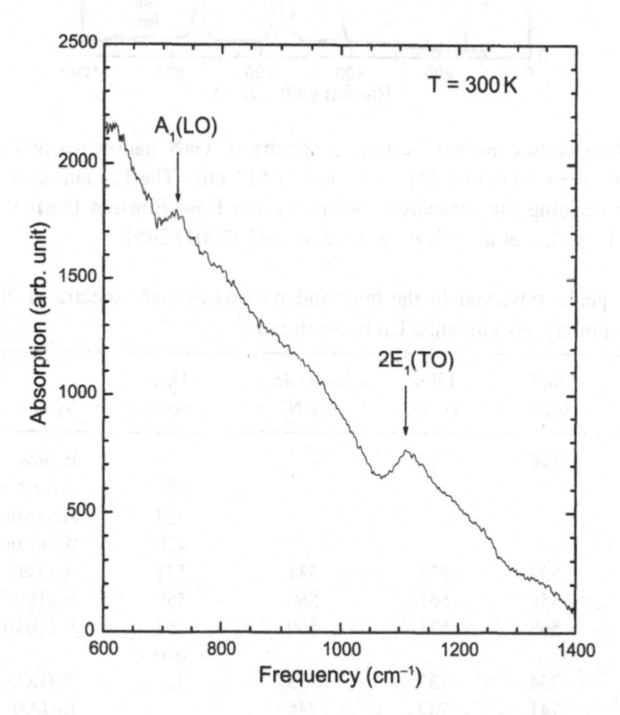

Fig. 21. Room temperature infrared absorption spectrum of catalytic-grown GaN nanowires at 910 °C by using Co powder as a catalyst. [With permission from H. L. Liu et al., *Chem. Phys. Lett.* **345** (2001) 245]

Fig. 22 shows the Raman spectra of GaN nanowires together with a 5 μm thick GaN epitaxial film for comparison. The spectrum of GaN nanowires is composed of several optical phonon modes and a broad photoluminescence background. The frequencies and assignments of these phonon modes determined by two different research groups are listed in Table 1. Results for the single crystal GaN are also included in this table for comparison.

The five Raman-active phonon bands observed by Liu et al. [133] at ~ 143, 531, 559, 567 and 728 cm^{-1} can be attributed to symmetries E_2(low), A_1(TO), E_1(TO),

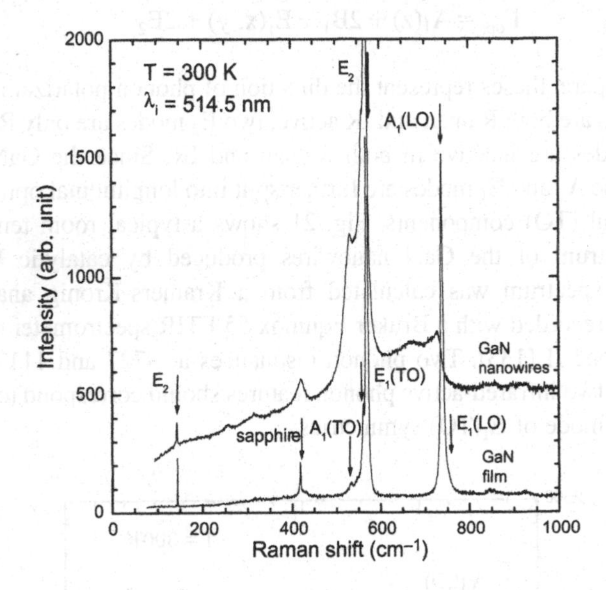

Fig. 22. Room temperature Raman scattering spectra of GaN nanowires and a 5 μm thick epilayer film with green excitation (Ar^+ ion laser at 514 nm). The Raman-scattering response was obtained by dividing the measured spectra by the Bose-Einstein thermal factor. [With permission from H. L. Liu et al., *Chem. Phys. Lett.* **345** (2001) 245]

Table 1. Raman peaks observed in the bulk and nanowires GaN spectra at 300 K and the corresponding symmetry assignments. Units are in cm^{-1}

Bulk GaN[a]	Bulk GaN[b]	Bulk GaN[c]	Film GaN[d]	Nanowires GaN[e]	This work	Symmetry assignment
144	145	144	145		143	E_2(low)
					254	Zone-boundary phonon
					314	Acoustic overtone
					420	Acoustic overtone
533	533	532	533	531	531	A_1(TO)
561	561	559	561	560	559	E_1(TO)
569	570	568	571	569	567	E_2(high)
					670	
735	735	734	737	733	728	A_1(LO)
743	742	741	742	745		E_1(LO)

With permission from H. L. Liu et al., *Chem. Phys. Lett.* **345** (2001) 245.
a, From Ref. [132]; b, From Ref. [133]; c, From Ref. [134]; d, From Ref. [130]; e, From Ref. [31].

E_2(high), and A_1(LO), respectively. The strong E_2(high) phonon line in the Raman spectra reflects the characteristics of wurtzite structure of GaN nanowires. The positions of the first four Raman modes shift very slightly to lower frequency (by a few cm^{-1}) as compared with those reported by Cheng et al. [31] and the previous data of the single crystals [134–136], while the frequency of the asymmetric line-shape associated with A_1(LO) mode is significantly softer. Moreover, four additional Raman modes are also observed at ~ 254, 314, 420 and 670 cm^{-1}, all of which are not allowed by the $P6_3mc$ space group in first-order Raman-scattering at the zone center.

It should be noted that the Raman phonon peaks of the GaN nanowires broaden substantially. The line-width increases roughly by a factor of 2 to 5 compared with the single crystals. Possible mechanisms that could bring about these features are discussed here. The low-frequency shifts and broadening are in line with the spatial-correlation or phonon confinement model [134, 135], suggesting that the phonons in nanometric-sized systems can be confined in space by crystallite boundaries or surface disorders, consequently, causing an uncertainty in the wave vector of the phonons. Based on the model proposed by Campbell and Fauchet, the calculated Raman spectra of the E_2(low) and A_1(LO) phonon modes of GaN nanowires with different diameters (L_1) while keeping a fixed length of 1 μm are shown in Fig. 23. Qualitatively, as the size of nanowires decreases, the Raman peak shifts to lower frequency and the line-shape becomes asymmetric with a low-frequency tail. The experimental data for the E_2(low) phonon peak agrees reasonably well with the calculated results with $L_1 = 20$ nm, which is consistent with the size estimated from SEM images. However, this model with diameters ranging from 10 to 50 nm fails to account for the large softening, broadening and unusual asymmetry observed in the A_1(LO) phonon mode. Other factors should be considered, which include a resonantly enhanced behavior through electron-phonon interaction effects, as manifested by the excitation wavelength dependence of the intensity of the Raman peak. The strongly coupled phonons are particularly notable in that they sensitively gauge the electronic transitions in GaN nanowires.

The relaxation of the momentum conserved ($q = 0$) selection rule due to the surface disorders of finite crystallite size not only broadens the Raman-allowed modes, but also causes new modes which correspond to $q \neq 0$ phonons to appear. One extra peak at ~254 cm^{-1} in GaN nanowires could be possibly attributed to the zone-boundary phonons activated by the surface disorders and finite-size effects. The observed peak position is close to the value estimated by lattice dynamical calculations [136, 139, 140] of the phonon-dispersion relations for the wurtzite GaN, in which the E_2 optical branches at the M symmetry point in the Brillouin zone produce peak in the phonon density of states near ~250 cm^{-1}. The other two extra peaks at 314 and 420 cm^{-1} are very close to those of overtone process of acoustic phonons observed in single crystal GaN [136, 140].

5.3.2. Comparison of nanowires and epitaxial films in InN

Fig. 24 shows the room temperature Raman spectra of the Au-catalyzed InN nanowires [28] and the highly oriented MOCVD-grown InN film [141, 142] for comparison. Three prominent Raman peaks at around 445, 489 and 579 cm^{-1} are observed, which can be assigned to the first-order A_1(TO), E_2, and A_1(LO) modes,

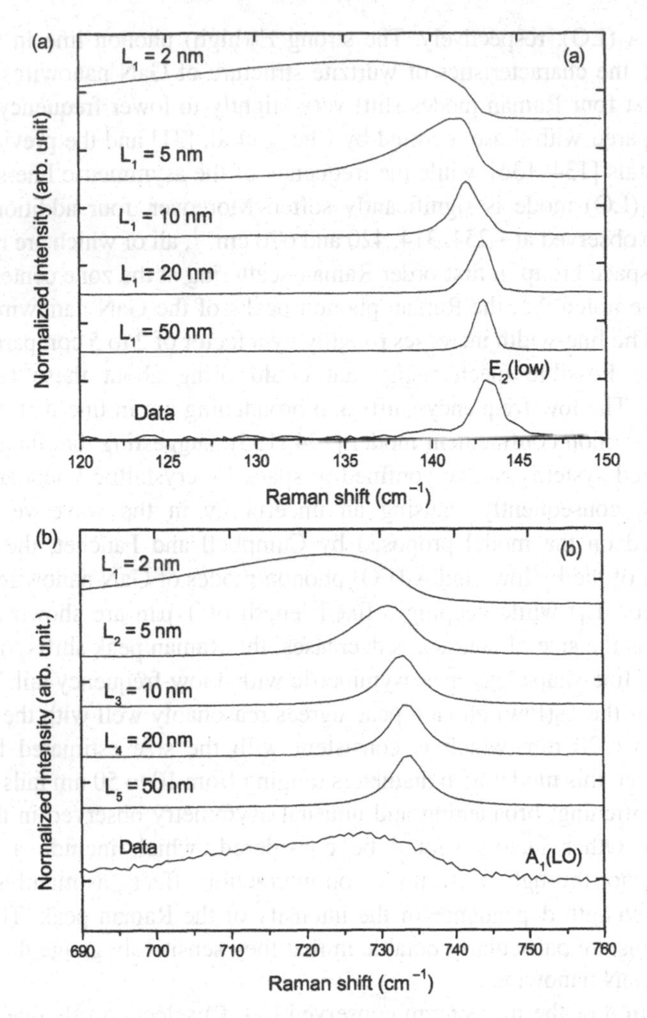

Fig. 23. Calculated Raman spectra of (a) E_2(low) and (b) A_1(LO) phonon modes for the GaN nanowires with different diameters L_1 (solid lines) using the spatial-correlation model described in the text. The filled circles denote the experimental spectra after subtracting the background and other phonon contributions. [With permission from H. L. Liu et al., *Chem. Phys. Lett.* **345** (2001) 245]

respectively. In comparison with the first–order optical phonon peaks of InN film, the corresponding Raman peaks of InN nanowires exhibit broader line width and more asymmetric line shape. It is also intriguing that the A_1(LO) mode exhibits most significant broadening and is the only one that shows observable peak shift by about 14 cm^{-1} toward lower wave numbers. The large softening, broadening and unusual asymmetry observed in the A_1(LO) phonon mode in the InN nanowires is quite similar to that observed in the GaN nanowires described above.

5.3.3. Ternary In-Ga-N nanowires and nanorods

Fig. 25 shows the Raman spectra taken from the GaN, InN, and ternary In-Ga-N nanowires and nanorods, the same four samples as shown in Fig. 21. Consistent with

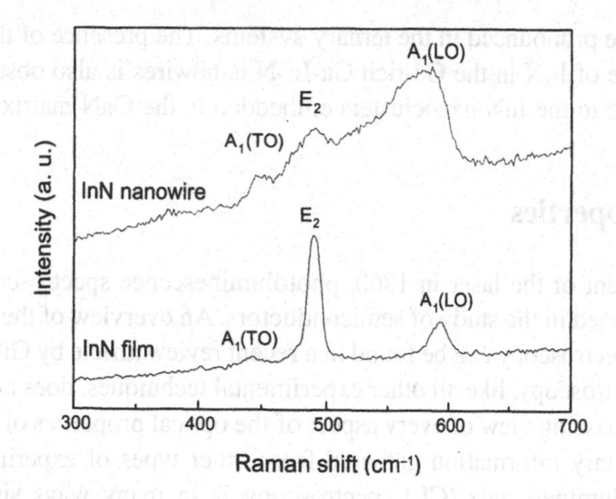

Fig. 24. Room temperature Raman scattering spectra of (a) InN nanowires and (b) InN film, the same set of samples described in Fig. 16. Three prominent modes typical for wurtzite structure are observed. [With permission from C. H. Liang et al., *Appl. Phys. Lett.* **81** (2002) 22]

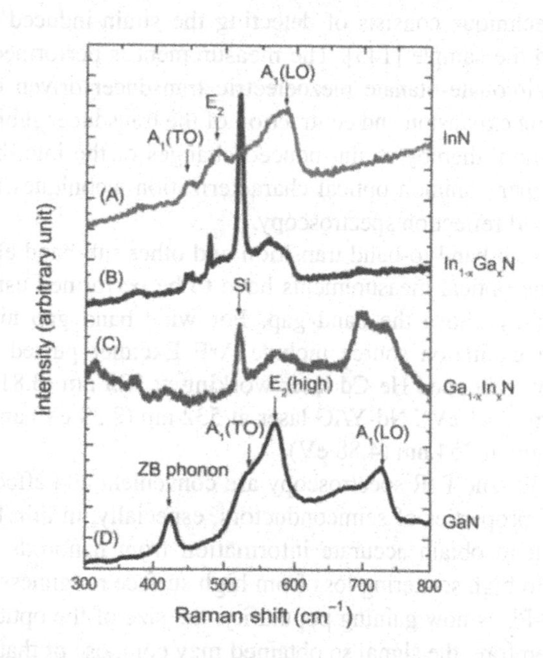

Fig. 25. Room temperature Raman scattering spectra of Ga-rich and In-rich ternary Ga-In-N nanowires, the same set of samples described in Fig. 19. For comparison, data taken from the binary GaN nanowires and InN nanowires are also shown.

the XRD data, the Raman spectra of Ga-rich or In-rich Ga-In-N nanowires and nanorods are quite similar to those of binary GaN or InN counterparts. Although the spectra of ternary systems are predominant by the matrix of nanowires and nanorods, some additional fine features and differences can still be seen, in particular, the A_1(LO) modes. Most notably, the lower wave number side band of A_1(LO) mode appears to be more pronounced in the ternary systems. The presence of the characteristic A_1(LO) mode of InN in the Ga-rich Ga-In-N nanowires is also observable. This could be attributed to the InN nanoclusters embedded in the GaN matrix.

6. Optical Properties

Since the advent of the laser in 1960, photoluminescence spectroscopy (PL) has been extensively used in the study of semiconductors. An overview of the current state of the art of PL spectroscopy can be found in a recent review article by Gilliland [143]. However, PL spectroscopy, like all other experimental techniques, does not give a universal, all-encompassing view of every aspect of the optical properties of the material. Some complimentary information gathered from other types of experiments is also needed. Cathodoluminescence (CL) spectroscopy is in many ways similar to PL, except that the excitation source is electron beam, having excitation energy much higher than the usual optical source [144]. Piezoreflectance (PzR) spectroscopy is another example of the complimentary tool to be described here. The basic mechanism of the PzR technique consists of detecting the strain-induced changes of the reflected light from the sample [145]. The measurement is performed by gluing the sample on a lead-zirconate-titanate piezoelectric transducer driven by a sinusoidal wave. The alternating expansion and contraction of the transducer subjects the sample to an alternating strain, thereby strain-induced changes of the inter-band transitions can be detected. Other common optical characterization techniques include absorption, transmission and reflection spectroscopy.

To obtain complete band-to-band transition and other sub-band electronic transition information, the optical measurements have to be performed using light source with excitation energy above the band gap. For wide band gap nitride materials, popular choices of excitation source include ArF Excimer pulsed laser operating at 193 nm (6.42 eV), chopped He-Cd laser working at 325 nm (3.81 eV), Ar^+ laser operating at 514 nm (2.41 eV), Nd-YAG laser at 532 nm (2.33 eV) and Xe lamp with excitation wavelength at 254 nm (4.88 eV).

Although both PL and PzR spectroscopy are convenient and effective techniques for probing optical properties of semiconductors, especially, in thin film forms, it is often quite difficult to obtain accurate information from nanorods and nanowires. This in part is due to high scattering loss from high surface roughness of the samples. Despite that micro-PL is now gaining popularity, the size of the optical probe is still relatively large; therefore, the signal so obtained may compose of that from a tremendous number of nanowires, with inherently broad size distribution. Single nano-object optical spectroscopy is now under development. In contrast, the CL technique, with an electron beam size comparable to the size of the nanorods and nanowires, can easily offer the opportunity for measuring optical properties of a single nano-object.

6.1. Theory on the size effects of nanorods and nanowires

We shall emphasize that the small size of nanowires and nanorods can generate some unique optical features different from those observed in their bulk counterparts.

One of the key features is the spectral modification in both peak position and intensity due to the quantum confinement effect. Limiting the electron motion to fewer dimensions dramatically modifies the electron energy spectrum, leading to an enhancement of the density of states near the band edge [146]. When the size of the object is comparable to the Bohr radius of an exciton, the exciton transition energies will also be modified. Published values of the excitonic Bohr radius of GaN are in the range between 2.8 and 11 nm [1, 147–149].

In the effective mass approximation, the allowed energy levels of the electrons in a quantum wire can be found by solving the Schrödinger equation, assuming an infinite confining potential. For a relatively long nanowire with a wire diameter of *d*, the energy shift can be expressed as [150]:

$$\Delta E = (\eta^2/2m_{eff})(j_x^2 + j_y^2)\pi^2/d^2 \quad (18)$$

where η is the Planck constant, j_x and j_y are integer numbers and m_{eff} is the effective mass of electron. Published values of the electron effective mass for GaN [1] and InN [151] are $0.22m_o$ and $0.12m_o$ (where m_o is the electron rest mass), respectively. Given the relatively large electron effective mass, a GaN or InN nanowire with a diameter of about 100 nm could only generate a transition energy shift of about 0.5–1 meV. However, when the size of the nanowire is reduced to below 10 nm, significant energy shifts of about several tens to over one hundred meV can be produced.

Besides the dimension and size confinement effect, additional optical features may also be originated from the surface states due to the large surface to volume ratio of the nanowires and nanorods and/or defect states inherently different from those observed in their bulk counterparts. The surface states and/or defect states in the nanowires and nanorods are poorly understood at present time. Nevertheless, it is imperative that what is observed, no matter how incomplete, be discussed for the reader to be abreast of the developments.

6.2. Binary III-nitride nanorods and nanowires

Typical PL spectra from catalytic grown GaN nanowires taken in the temperature range from 300 K to 15 K are shown in Fig. 26. A He-Cd laser with an output power of >10 mW was used. The broad half-width of the PL spectra probably reflects the broad size distribution of the catalytic grown GaN nanowires. Notably, a strong emission at around 3.4 eV, which can be ascribed to the band-edge emission of wurtzite GaN [152], was observed even at room temperature. The band-edge emission intensity at room temperature was about 60% of that obtained at 15 K. In comparison to the data obtained at the same energy range of emission from bulk GaN, the GaN nanowires exhibited the weakest photoluminescence quenching [153–155]. This implies that thermal quenching of PL is much reduced in GaN nanowires. The small shoulder around 3.1 eV is very close to the band-edge emission of cubic GaN [156], consistent with the HRTEM result, which indicated the existence of a small fraction of cubic phase incorporated in the hexagonal matrix, as the former phase was easily present as "stacking faults" in the matrix phase. A weak and broad PL band centered at ~2.3 eV was also observed and can be attributed to the well-known yellow luminescence observed in

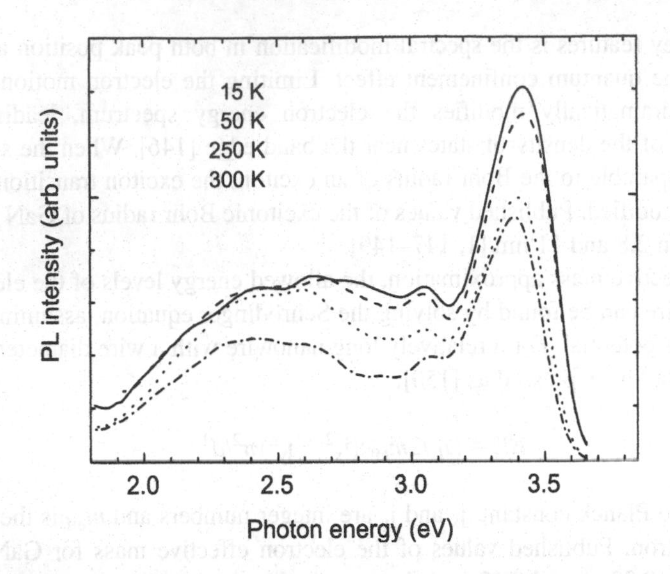

Fig. 26. Temperature dependence of photoluminescence spectra of GaN nanowires produced by VLS growth with iron powder as the catalyst. A He-Cd laser (325 nm) with an output power of >10 mW was used as the excitation source. The band-edge emission intensity at 300 K is ~60% of that at 15 K, suggesting very weak PL quenching. [With permission from C.-C. Chen et al., *J. Am. Chem. Soc.* **123** (2001) 2791]

bulk GaN, which is thought to be originated from the Ga or N vacancies or a related complex [157, 158]. The weak defect-related signal suggests that the catalytically grown GaN nanowires contained few defects and were indeed of high quality. For GaN nanowires obtained by anodic alumina membrane-confined growth, the PL spectra exhibited a much broader feature [53] than those taken from catalytically grown samples. Multi-peak fitting to the broad feature revealed three bands located at 3.4, 2.8 and 2.28 eV, respectively. The first and third peaks are attributed to the band-edge and yellow emission, respectively, whereas the second peak can result from the interaction between the GaN nanowires and the anodic alumina membrane, or the existence of defects or surface states.

PL measurement of InN is rarely reported in the literature. Due to the inherent process difficulties as noted in section 2, the InN sample usually contains extremely high defect density. In the case of InN nanowires, sample with much reduced defect density can be obtained and strong PL signals have been observed. Fig. 27 shows the PL spectra measured (using Ar^+ laser operating at 514 nm as the excitation source) from Au-catalyzed InN nanowires at temperature range of 129–433 K. Very strong PL peaks at room temperature and above were noted, suggesting that thermal quenching of PL in InN nanowires was extremely weak. Interestingly, only one broad peak was observed. The temperature variations of this PL peak can be fitted to a semiempirical relationship described by Varshni [159]:

$$E(T) = E(0) - \alpha T^2/(\beta + T) \tag{19}$$

where E(0) is the direct band gap at absolute zero, and α and β are constants. The constant α is related to the electron-phonon interaction and β is related to the Debye

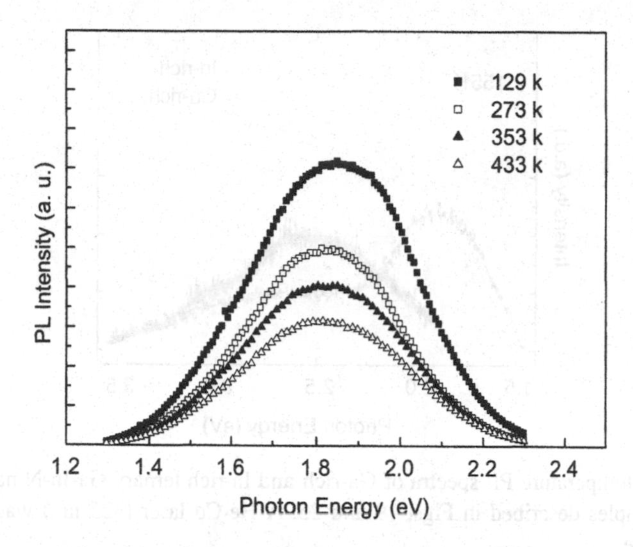

Fig. 27. Temperature dependence of photoluminescence spectra of InN nanowires produced by VLS growth with Au as the catalyst. An Ar^+ ion laser (514 nm) was used as the excitation source. Strong PL intensity was still observed at temperatures much higher than room temperature.

temperature. The estimated values of E(0), α and β for the Au-catalyzed InN nanowires were 1.85 eV, 0.26 meV/K and 597 K, respectively. The corresponding values reported for InN thin films were 1.99 eV, 0.245 meV/K and 624 K, respectively [160]. It should be emphasized that the empirically determined E(0) may not be unambiguously attributed to the band to band transition in InN. Other transitions such as recombination of bound excitons are also possible.

The widely referenced 1.9 eV band gap value (at room temperature) reported by Tansley et al. in 1986 was determined using optical absorption on polycrystalline InN films grown on glass substrate [161]. The film on glass was by no means epitaxial; and, the existence of defects in the films could have affected seriously the correct determination of the band gap value. More recent luminescence and transmission measurements on InN films with higher crystalline quality seemed to agree with a band gap value near 2.1 eV at room temperature [162–164], whereas Davydov et al. suggested a band gap of 0.9 eV [165, 166]. As the value of band gap in InN film or bulk is still controversial, clarification of the origins of the observed PL peak in our InN nanowires as shown in Fig. 28 is beyond the scope of this chapter.

6.2.1. *Band edge emission peak shift in GaN nanowires*

Apart from the apparent broadening of the PL spectra, the band-edge emission peak position observed in GaN nanowires reported in most literatures did not show noticeable shift from that observed in bulk GaN, presumably due to the fact that the majority of the nanowires still have diameters larger than the excitonic Bohr radius [27, 53, 60, 65]. However, Chen et al. has reported a blue shift of about 0.2 eV in the band edge emission peak taken from GaN nanowires, having diameters of 10–40 nm, grown by using NiO nanoparticle catalyst and directed flow of carrier and reaction gas [59]. Kim et al. has also reported a blue shift of 60 meV in the GaN nanorods of

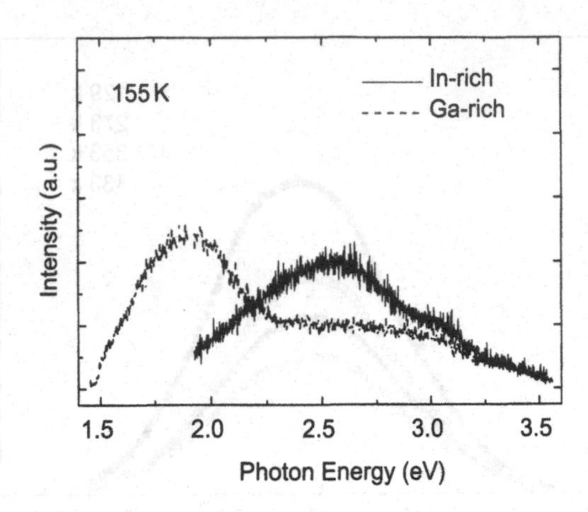

Fig. 28. Room temperature PL spectra of Ga-rich and In-rich ternary Ga-In-N nanowires, the same set of samples described in Figs. 19 and 25. A He-Cd laser (325 nm) was used as the excitation source.

80–120 nm, grown by a hydride vapor phase epitaxy method on sapphire substrate without using any catalyst [167]. Besides the peak shift observed by luminescence spectroscopy methods, transmission spectroscopy study also indicated that the absorption edge of some catalyst-assisted CVD grown GaN nanowires, with diameters ranging from 10 to 50 nm, was blue shifted by 0.2 eV from the bulk GaN [168].

The energy shifts observed by these three groups are apparently larger than the values estimated by effective mass approximation (Eq. 18). Other contribution to the excessive energy shift may include strain experienced in the nanorods. One possible source of strain arises in hetero-epitaxial growth. The GaN nanorods grown by HVPE on sapphire substrate were preferentially oriented along the *c*-axis direction [169]; hence, some strain might likely be present in this system. However, in the experiments performed by Chen et al. [59], the GaN nanorods did not exhibit any orientation alignment, although $LaAlO_3$ substrate (on which epitaxial growth of GaN film has been demonstrated) was used. The GaN nanowires used by Lee et al. [168] were also randomly oriented. The exact origins of the relatively large blue shift in these "free standing" GaN nanorods and nanowires still need to be clarified.

In contrast, Seo et al. reported a red shift of about 0.12 eV in the band edge emission peak taken from the strained GaN nanowires [123], of which the process and structure analyses have been described in section 5.1. As these strained nanowires experience biaxial compressive stresses in the inward direction and uniaxial tensile stresses in the growth direction, the change of lattice constant would lead to band gap shift. Apparently, the decreases of the band gap due to the tensile stresses occurred more significantly than the increases due to the compressive stresses.

6.2.2. *Size-dependent CL: Evidence of quantum confinement*

While the actual values of energy shift may be affected by the strains or defects of other origins, further evidence of quantum confinement in the GaN nanorods is supported by CL measurements performed over individual nanorods with various sizes [90]. Fig. 29 shows a panchromatic CL image and CL spectra of three individual GaN nanorods with a diameter of 120, 100 and 80 nm. Rather broad linewidths were observed even for these individual nanorods, however, distinctive CL peak energies of 3.435, 3.473 and 3.512 eV could still be determined for the GaN nanorods with a diameter of 120, 100 and 80 nm, respectively. It is clearly noted that (i) CL peak energies from all the GaN nanorods were higher than that of bulk GaN, and (ii) the CL peak energy exhibited blue shift with decreasing diameter of the nanorod. The broad line widths may be associated with the carrier diffusion along the rod axes [169–171].

6.2.3. Single GaN nanowire laser

Besides the luminescence property studies, a considerable advance towards the realization of electron-injected nanowire-based UV-blue coherent light sources has been demonstrated recently. Johnson et al. have reported lasing from a single optically pumped GaN nanowire [172]. The GaN nanowires, typically 300 nm in diameter and 40 μm in length, were synthesized using a nickel catalyst on a sapphire substrate under a VLS process. The nanowires were removed from the substrate and dispersed onto a sapphire substrate by drop-casting in a mixture of ethanol for near-field scanning optical microscope (NSOM) study. Laser excitation from the fourth harmonic of an optical parametric amplifier pumped by a regeneratively amplified Ti : sapphire oscillator provides 100–200 fs and 290–400 nm beams of 5–10 mW and 1 kHz repetition rate. Despite catalyst particle on the ends, UV-blue laser action of Fabry-Perot mode with a Q factor above 1000 was observed, which is attributed to an electron-hole plasma mechanism. For a nanowire of 16 μm long and 400 nm diameter, a stimulated emission threshold of 700 nJ/cm^2 was concluded. The threshold value for the nanowires is lower than those reported for other types of GaN microstructures, presumably due to the reduction in mode volume [173]. The emission spectra showed multiple peaks, related to the longitudinal modes of the cavity, and a strong red shift of the center of spectral intensity at higher pumping power. Further optimization and carrier injection pumping to realize an optoelectronic device is suggested.

6.3. Ternary Ga-In-N nanorods and nanowires

Fig. 29 shows the PL spectra measured at 155 K from the ternary In-Ga-N nanowires and nanorods. PL spectra of Ga-rich or In-rich Ga-In-N nanowires and nanorods were very different from those of binary GaN or InN counterparts, which have been shown in Figs. 26 and 27, respectively. The In-rich sample showed a broad band emission with a major peak centered at 2.57 eV and a shoulder at 3.06 eV, whereas the Ga-rich sample exhibited two resolvable bands at 1.91 and 2.63 eV. Note that the main peaks observed in the binary InN (~1.9 eV) or GaN (3.4 eV) nanowires and nanorods were absent in their In-rich or Ga-rich counterparts. It appears that the PL spectra of the ternary systems are likely to be dominated by the minor phases, such as In nanoclusters embedded in the GaN matrix of nanowire or GaN nanoclusters embedded in the InN matrix of nanowire. Further experiments are needed to clarify the exact origins of the emission. Nevertheless, it is worth mentioning here that, for film-based optoelectronic device, carrier captured by InN quantum dots has been attributed to the efficient emission of blue-violet commercial InGaN light emitting diode lasers [174], and the mechanism of the luminescence in InGaN/GaN quantum wells is thought to arise from the radiative recombination within self-organized InN quantum dots [175].

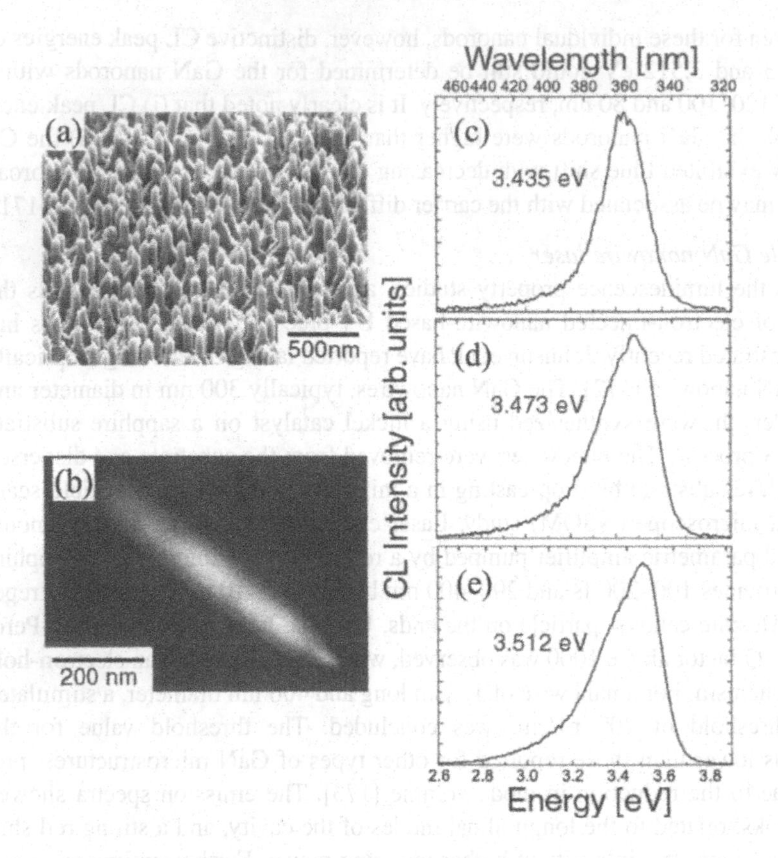

Fig. 29. (a) Tilted side-view SEM image of GaN nanorods grown on sapphire substrate at 478 °C by hydride vapor phase epitaxy. (b) Panchromatic CL image and (c)–(e) CL spectra for three individual GaN nanorods taken at room temperature ($E_b = 10$ keV, $I_b = 220$ pA and CL bandpass = 2.4 nm). The CL peak energies of the individual nanorod with a diameter of (c) 120, (d) 100 and (e) 80 nm are found to be 3.435, 3.473 and 3.512 eV, respectively. The CL image is taken from the GaN nanorod with a diameter of 100 nm. [With permission from H. M. Kim et al., *Appl. Phys. Lett.* **81** (2002) 2193]

6.4. Ternary Si-C-N nanorods

PL spectra of the Si-C-N nanorods were measured using an ArF laser at 193 nm. At room temperature, the Si-C-N nanorods exhibited a broad intense emission centered at 4.2 eV (Fig. 30). From Varshni analysis (equation 18) of the temperature-dependent PL spectra, E(0), α and β values of 4.26 eV, 0.32 meV/K and 313 K, respectively, were deduced. The band gap emission was also confirmed by piezoreflectance spectroscopic technique [176]. As shown in the inset of Fig. 30, a typical PzR spectrum of the Si-C-N nanorods measured at 300 K showed a direct band gap of 4.2 eV, consistent with the PL data. This value is smaller than that of the α-Si_3N_4 (~4.8 eV) [177], but is larger than that for a polycrystalline Si-C-N film (3.81 eV) [21]. The present results indicate a variation of the band gap of Si-C-N phase with its composition and microstructure while their exact relation is yet to be clarified.

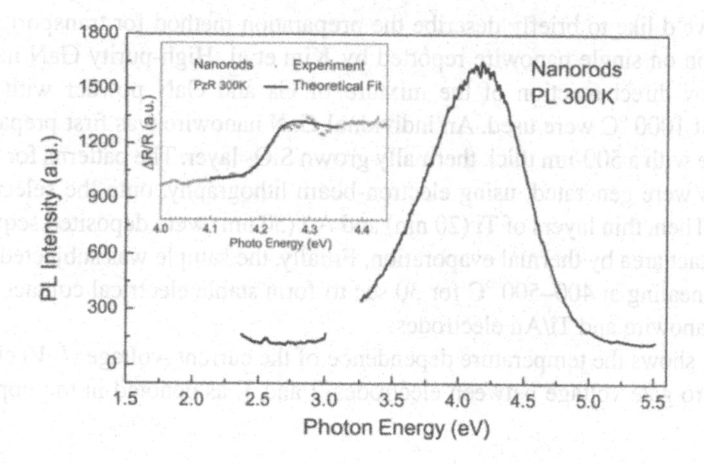

Fig. 30. Experimental PL and PzR (inset) spectra of Si-C-N nanorods at 300 K. The PzR curve has been fitted to the first derivative of a Lorentzian profile, which yields the direct band-to-band transition energy. [With permission from L. C. Chen et al., *J. Phys. Chem. Solids* **62** (2001) 1567]

7. Electrical Properties

Two types of electrical properties (transport and field emission) are discussed here. While various CNT-based molecular-scale devices have been demonstrated recently, there might be some critical limitations difficult to overcome in the CNT-based devices. For instance, precise control of the chiralities, which allows selective growth of semiconducting or metallic CNTs, is still difficult. Controlled and long-term stable doping is yet another challenging topic. In contrast, the semiconducting nanowires, such as Si, GaAs and III-nitrides, appear to be able to overcome the fundamental limitations of the CNTs. These nanowires are inherently semiconducting and relatively easy to dope through conventional semiconductor technology.

Meanwhile, the extremely large field enhancement factor resulting from the sharp curvature of CNTs is one of the key factors for their super electron emission properties. However, poor emission stability of CNTs has somewhat undermined their application in vacuum microelectronics. The sharp decay of the emission current and degradation of the field emission property can be largely accounted for by progressive destruction of CNT during the emission process. In view of this, nanorods and nanowires have the potential as field emitters due to the similarity in their geometric features with those of CNT but with more stable structural properties.

7.1. Electrical transport properties

Recently, Huang et al. reported successful fabrication of logic gates and demonstrated the computation capabilities from assembled *p*-Si and *n*-GaN crossed nanowire junctions [178]. Meanwhile, Kim et al. have performed electrical measurements on individual GaN nanowires with Au/Ti metal electrode [179].

Here, we'd like to briefly describe the preparation method for transport property investigation on single nanowire reported by Kim et al. High-purity GaN nanowires produced by direct reaction of the mixture of Ga and GaN powder with flowing ammonia at 1000 °C were used. An individual GaN nanowire was first prepared on a Si substrate with a 500-nm thick thermally grown SiO_2 layer. The patterns for the electrical leads were generated, using electron-beam lithography, onto the selected GaN nanowire. Then, thin layers of Ti (20 nm) and Au (50 nm) were deposited sequentially on the contact area by thermal evaporation. Finally, the sample was subjected to rapid thermal annealing at 400–500 °C for 30 sec to form stable electrical contact between the GaN nanowire and Ti/Au electrodes.

Fig. 31 shows the temperature dependence of the current–voltage (*I–V*) characteristics at zero gate voltage between electrodes 2 and 3, as denoted in the upper inset,

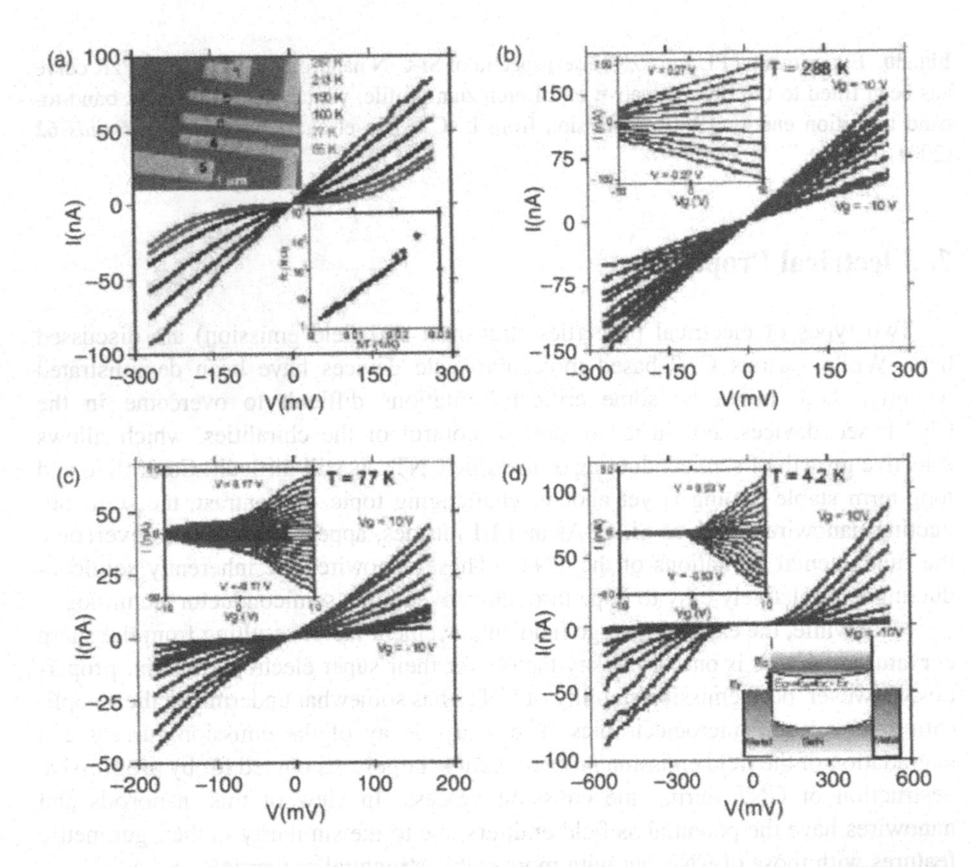

Fig. 31. (a) Temperature-dependent *I–V* curves of a GaN nanowire between electrodes 2 and 3. The measured temperatures were 283, 213, 150, 100, 77 and 55 K. Inset: SEM image of the sample in this study. Numbers in the image represent Ti/Au electrodes for electrical measurement (upper figure). Resistance as a function of inverse temperature (1/*T*) between electrodes 2 and 3 is shown in lower figure. [With permission from J. R. Kim et al., *Appl. Phys. Lett.* **80** (2002) 3548] (b), (c), and (d) show the *I–V* curves as a function of V_g between electrodes 2 and 3 at 288, 77, and 4.2 K, respectively. The gate bias varies from +10 to −10 V. Insets show the corresponding source-drain current changes as a function of V_g. The lower inset in (d) represents the schematic band diagram for the n-type GaN nanowire and metal electrodes.

which depicts the SEM image of an individual GaN nanowire with five Ti/Au electrodes attached to it. The linear resistance increased from 3 to 14 MΩ as the temperature was decreased from 280 to 100 K. A simple thermal activation analysis of this temperature dependence gave activation energy of 20.9 meV at the high temperature region, as shown in the lower inset of Fig. 31. Similar transport measurements were carried out as a function of gate voltage. The gate modulation curves of the other pair of electrodes, such as 3–4, 4–5 and 1–2, also showed similar behaviors as that of 2–3. At ambient temperature, the *I*–*V* curve at room temperature exhibited linear behavior with resistance of 3 MΩ at zero gate voltage. Application of positive gate voltage progressively increased the conductance, which is a signature of an *n*-type field-effect transistor. As the temperature was decreased from ambient to 77 K and further to 4.2 K, the nonlinearity of the resultant *I*–*V* curve and the *n*-type gating effect became more and more pronounced. Such an *n*-type electrical property is known to result from nitrogen vacancies [180]. As illustrated in the schematic band-diagram for an *n*-type GaN nanowire in contact with a metal electrode (the lower inset of Fig. 31d), the energy band of the n-type semiconductor is bent to align its Fermi level to that of the metal [181]. When a positive gate voltage is applied, the conduction band of the *n*-type semiconductor is pulled downward, giving an increasing density of states near the Fermi level, resulting in an enhanced conductance. On the other hand, a negative gate bias voltage raises the bands upward and suppresses the conductance.

The energy difference between conduction band edge of the nanowire and the Fermi level of the metal electrode acts as an energy barrier to overcome for an electron to pass through the nanowire. At high temperature, thermal energy becomes comparable or larger than this energy barrier; therefore, the thermionic emission would be a dominant transport mechanism. At low temperature, on the other hand, the thermionic emission current is suppressed. Other transport mechanism such as quantum tunneling of the electrons through the energy barrier becomes important. The room temperature carrier mobility estimated from the gate modulation characteristics is about 2.15 cm^2/Vs at a bias voltage of 0.27 V. Although this value is comparable to that of B-doped Si nanowire [182], it is considerably smaller than the electron mobility of bulk GaN crystal (380 cm^2/Vs) [181], suggesting a highly diffusive nature of electron transport in the nanowire, presumably due to defects, disorder, interface scattering and possibly enhanced scattering because of a smaller diameter.

7.2. Field emission properties

Three types of samples, GaN nanorods, GaN nanorods encapsulated inside CNTs (GaN@CNT) and Si-C-N nanorods, are presented below. The GaN@CNT exhibiting a core-shell coaxial cable structure is formed by two-step catalytic reactions [183]. First, the GaN nanowires were catalytically grown as described in section 2.2. Then, by changing the reaction gas to methane, CNT were subsequently grown from the residual catalysts at the tips, wrapping around and along the nanowires.

Field emission measurements were performed using parallel-plate configuration with spacing at about 30 μm. An indium-tin-oxide (ITO) glass plate was used as the anode of which the area was restricted to be smaller than that of the sample to avoid the field emission at the sample edges due to their structure enhancement. The field emission *I–V* characteristics were obtained under a base pressure of 1×10^{-7} Torr at room temperature. A Keithely 237 electrometer was employed for sourcing the voltage and measuring the current.

Typical field emission characteristics of the GaN and GaN@CNT along with CNT are shown in Fig. 32. The turn-on field, which is defined as the field required for drawing an emission current density of 0.01 mA/cm^2, of the GaN nanorods, GaN@CNT nanocables and CNT is 12 V/μm, 5 V/μm and 3 V/μm, respectively. A peculiar tenfold increase of the emission current in GaN@CNT was observed at around 4 V/μm, which was reproducible from repetitive runs. Although the mechanism of the jump is still unclear, it might be related to the unique cable-like structure. For an undoped wide band gap material such as GaN, an emission current density of about 100 μA/cm^2 attainable at a field of 15 V/μm is significant. Encapsulating GaN with CNT obviously helps improving the electrical conductivity and thus enhances the overall field emission property.

In Fig. 33, the field emission characteristics of two Si-C-N nanorods samples are depicted. Emission current density at the maximum accessible voltage (i.e., a field of 36.7 V/μm) in excess of 1 mA/cm^2 can be routinely achieved for sample with high rod number density. This material also showed a low turn-on field at 10 V/μm. While achieving highest emission current at lowest applied field is the goal in most field emission study, the long-term stability of the emission is also essential. The temporal stability test for the Si-C-N nanorods was carried out at a constant applied voltage that gave an initial emission current density of about 100 μA/cm^2. As shown in the inset of Fig. 33, there was no significant degradation and fluctuation (±10%) for over 8 h.

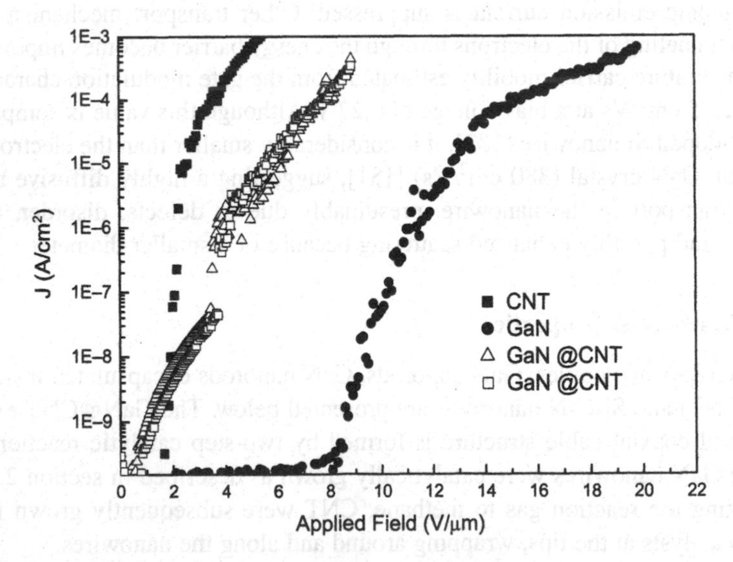

Fig. 32. Comparison of the field emission characteristic of pure GaN nanowires, GaN@CNT nanocables and CNTs is shown. [With permission from C.-C. Chen et al., *J. Phys. Chem. Solids* **62** (2001) 1577]

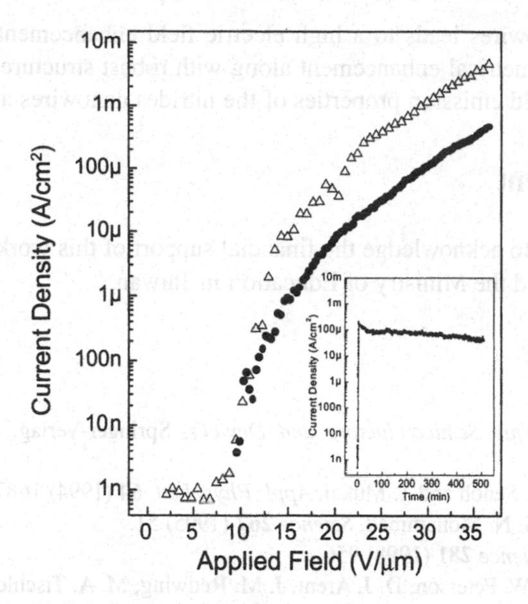

Fig. 33. Field emission J-E curves for two Si-C-N nanorods samples in log-linear scale. Field emissions are characterized using parallel-plate configuration (with spacing at about 30 μm) under a base pressure of 8.0×10^{-8} Torr and over a voltage sweep from 0 to 1100 V. The maximum accessible current density for Si-C-N nanorods varies, presumably, due to the difference in the number density as well as the geometry and orientation of the rods. Typical temporal stability test for Si-C-N nanorods is shown in the inset. [With permission from L. C. Chen et al., *J. Phys. Chem. Solids* **62** (2001) 1567]

8. Summary

We have reviewed the available information about the growth, characterization, as well as the optical and electrical properties for 6 technologically important group III- and IV-Nitrides nanorods and nanowires (AlN, GaN, InN, Si_3N_4, Ga-In-N and Si-C-N alloys). Whereas numerous reports on synthesizing this class of nanomaterials may be found in the literature, no review article has been published previously. As is commonly the case, research articles were either restricted to particular material systems, did not address all parameters of interest, or biased to the results of a specific group of researchers. Our goal has been to provide a comprehensive and even-handed reference source, as free of internal contradiction and significant omissions as is practically possible.

Rapid development in the synthetic techniques allows us to produce nanorods and nanowires with narrow size distribution. Some control over their position and orientation or epitaxial growth has also been demonstrated. Whenever applicable, comparison of the structure and bonding characteristics between the nanowires and films (or bulk) is shown. Optical characterization indicates that quantum confinement effect has occurred in the nanorods and nanowires with a diameter of around 100 nm, which is more pronounced than that predicted by simple theory such as effective mass approximation. Strong surface effects are thus suggested. The high aspect ratio of the nanorods and nanowires leads to a high electric field enhancement in this class of materials. The geometrical enhancement along with robust structure properties gives rise to excellent field emission properties of the nitrides nanowires and nanorods.

Acknowledgment

We would like to acknowledge the financial support of this work by the National Science Council and the Ministry of Education in Taiwan.

References

1. H. Morkoç, *Nitride Semiconductors and Devices*, Springer-Verlag, Berlin Heidelberg (1999).
2. S. Nakamura, M. Senoh and T. Mukai, *Appl. Phys. Lett.* **64** (1994) 1687.
3. H. Morkoç and S. N. Mohammad, *Science* **267** (1995) 51.
4. S. Nakamura, *Science* **281** (1998) 956.
5. S. S. Kocha, M. W. Peterson, D. J. Arent, J. M. Redwing, M. A. Tischler and J. A. Turner, *J. Electrochem. Soc.* **142** (1995) L238.
6. E. N. Matthias and B. M. Allen, *IEEE Trans. Electron Devices* **ED-34** (1987) 257.
7. A. Yamamoto, M. Tsujino, M. Ohkubo and A. Hashimoto, *Sol. Energy Mater. Sol. Cells* **35** (1994) 53.
8. N. Kuramoto, H. Taniguchi and I. Aso, *Am. Ceram. Soc. Bull.* **68** (1989) 883.
9. R. S. Wagner and W. C. Ellis, *Appl. Phys. Lett.* **4** (1964) 89.
10. H. Pierson, *Handbook of Refractory Carbides and Nitrides*, Noyes Publications, Westwood, New Jersey (1996).
11. L. C. Chen, C. Y. Yang, D. M. Bhusari, K. H. Chen, M. C. Lin, J. C. Lin and T. J. Chuang, *Diamond Rel. Mater.* **5** (1996) 514.
12. L. C. Chen, D. M. Bhusari, C. Y. Yang, K. H. Chen, T. J. Chuang, M. C. Lin, C. K. Chen and Y. F. Huang, *Thin Solid Films* **303** (1997) 66.
13. R. Riedel, A. Greiner, G. Miehe, W. Dressler, H. Fuess, J. Bill and F. Aldinger, *Angew. Chem. Int. Ed. Engl.* **36** (1997) 603.
14. A. Badzian and T. Badzian, *Int. J. Refract. Metals Hard Mater.* **15** (1997) 3.
15. A. Badzian, T. Badzian, W. D. Drawl and R. Roy, *Diamond Rel. Mater.* **7** (1998) 1519.
16. G. Lehmann, P. Hess, J.-J. Wu, C. T. Wu, T. S. Wong, K. H. Chen, L. C. Chen, H.-Y. Lee, M. Amkreutz and Th. Frauenheim, *Phys. Rev. B* **64** (2001) 165305.
17. C.-Z. Wang, E.-G. Wang and Q. Dai, *J. Appl. Phys.* **83** (1998) 1975.
18. R. N. Musin, D. G. Musaev and M. C. Lin, *J. Phys. Chem. B* **103** (1999) 797.
19. J. E. Lowther, *Phys. Rev. B* **60** (1999) 11943.
20. L. C. Chen, K. H. Chen, S. L. Wei, J.-J. Wu, T. R. Lu and C. T. Kuo, *Thin Solid Films* **112** (1999) 355–356.
21. L. C. Chen, C. K. Chen, S. L. Wei, D. M. Bhusari, K. H. Chen, Y. F. Chen, Y. C. Jong and Y. S. Huang, *Appl. Phys. Lett.* **72** (1998) 2463.
22. L. C. Chen, K. H. Chen, J. J. Wu, D. M. Bhusari and M. C. Lin, in "*Silicon Based Materials and Devices*", Edited by H. S. Nalwa, Academic Press, San Diego, California (2001) pp. 73–125.
23. S. Chattopadhyay, L. C. Chen, C. T. Wu, K. H. Chen, J. S. Wu, Y. F. Chen, G. Lehmann and P. Hess, *Appl. Phys. Lett.* **79** (2001) 332.

24. R. Riedel, H. Kleebe, H. Schonfelder and F. Aldinger, *Nature* **374** (1995) 526.
25. F. G. Tarntair, C. Y. Wen, L. C. Chen, J.-J. Wu, K. H. Chen, P. F. Kuo, S. W. Chang, Y. F. Chen, W. K. Hong and H. C. Cheng, *Appl. Phys. Lett.* **76** (2000) 2630.
26. L. C. Chen, S. W. Chang, C. S. Chang, C. Y. Wen, J.-J. Wu, Y. F. Chen, Y. S. Huang and K. H. Chen, *J. Phys. Chem. Solids* **62** (2001) 1567.
27. W. Han, S. Fan, Q. Li and Y. Hu, *Science* **277** (1997) 1287.
28. C. H. Liang, L. C. Chen, J. S. Hwang, K. H. Chen, Y. T. Hung and Y. F. Chen, *Appl. Phys. Lett.* **81** (2002) 22.
29. H. Parala, A. Devi, F. Hipler, E. Maile, A. Birkner, H. W. Becker and R. A. Fischer, *J. Cryst. Growth* **231** (2001) 68.
30. A. Hashimoto, T. Motiduki, H. Wada and A. Yamamoto, *Mater. Sci. Forum* **264** (1998) 1129.
31. G. S. Cheng, L. D. Zhang, Y. Zhu, G. T. Fei, L. Li, C. M. Mo and Y. Q. Mao, *Appl. Phys. Lett.* **75** (1999) 2455.
32. H. Dai, E. W. Wong, Y. Z. Lu, S. Fan and C. Lieber, *Nature* **375** (1995) 769.
33. C. J. Frosch and C. D. Thurmond, *J. Phys. Chem.* **62** (1958) 611.
34. C. J. Frosch and C. D. Thurmond, *J. Phys. Chem.* **66** (1962) 877.
35. C. M. Balkas and R. F. Davis, *J. Am. Ceram. Soc.* **79** (1996) 2309.
36. Y. Zhang, J. Liu, R. He, Q. Zhang, X. Zhang and J. Zhu, *Chem. Mater.* **13** (2001) 3899.
37. J. Liu, X. Zhang, Y. Zhang, R. He and J. Zhu, *J. Mater. Res.* **16** (2001) 3133.
38. Y. J. Liang and M. X. Che, *Handbook for Thermodynamic Data of Inorganic Compounds*, Press of Northeast University, Shenyang, China (1993).
39. M. P. Corral, R. Moreno, J. Requena, J. S. Moya and R. Martinez, *J. Eur. Ceram. Soc.* **8** (1991) 229.
40. P. G. Caceres and H. K. Schmid, *J. Am. Ceram. Soc.* **77** (1994) 977.
41. C. R. Martin, *Science* **266** (1994) 1961.
42. M. Nishizawa, V. P. Menon and C. R. Martin, *Science* **268** (1995) 700.
43. K. B. Jirage, J. C. Hulteen and C. R. Martin, *Science* **278** (1997) 655.
44. S. B. Lee, D. T. Mitchell, L. Trofin, T. K. Nevanen, H. Soderlund and C. R. Martin, *Science* **296** (2002) 2198.
45. H. Masuda and K. Fukuda, *Science* **268** (1995) 1466.
46. H. Masuda and F. Hasegawa, *J. Electrochem. Soc. Interface* **144** (1997) L127.
47. M. Saito, M. Kirihara, T. Taniguchi and M. Miyagi, *Appl. Phys. Lett.* **55** (1994) 607.
48. D. Routkevich, A. A. Tager, J. Haruyama, D. Almawlawi, M. Moskovits and J. M. Xu, *IEEE Trans. Electron Devices* **42** (1996) 1646.
49. D. Al-Mawlawi, C. Z. Liu and M. Moskovits, *J. Mater. Res.* **9** (1998) 1014.
50. S. A. Sapp, B. B. Lakshmi and C. R. Martin, *Adv. Mater.* **11** (1999) 402.
51. J. Zhang, X. S. Peng, X. F. Wang, Y. W. Wang and L. D. Zhang, *Chem. Phys. Lett.* **345** (2001) 372.
52. G. S. Cheng, S. H. Chen, X. G. Zhu, Y. Q. Mao and L. D. Zhang, *Mater. Sci. Engineering A* **286** (2000) 165.
53. J. Zhang, L. D. Zhang, X. F. Wang, C. H. Liang, X. S. Peng and Y. W. Wang, *J. Chem. Phys.* **115** (2001) 5714.
54. L. Maya, *Adv. Ceram. Mater.* **1** (1986) 150.
55. I. Kimura, K. Ichiya, M. Ishii, N. Hotta, and T. Kitamura, *J. Mater. Sci. Lett.* **8** (1989) 303.
56. G. Selvaduray and I. Sheet, *Mater. Sci. Technol.* **9** (1993) 463.
57. X. Duan and C. M. Lieber, *J. Am. Chem. Soc.* **122** (2000) 188.
58. C.-C. Chen and C.-C. Yeh, *Adv. Mater.* **12** (2000) 738.
59. X. Chen, J. Li, Y. Cao, Y. Lan, H. Li, M. He, C. Wang, Z. Zhang and Z. Qiao, *Adv. Mater.* **12** (2000) 1432.

60. C.-C. Chen, C.-C. Yeh, C.-H. Chen, M.-Y. Yu, H.-S. Liu, J.-J. Wu, K.H. Chen, L. C. Chen, J. Y. Peng and Y. F. Chen, *J. Am. Chem. Soc.* **123** (2001) 2791.
61. W. Han, P. Redlich, F. Ernst and M. Ruhle, *Appl. Phys. Lett.* **76** (2000) 652.
62. C. C. Tang, S. S. Fan, H. Y. Dang, P. Li and Y. M. Liu, *Appl. Phys. Lett.* **77** (2000) 1961.
63. W. Q. Han and A. Zettl, *Appl. Phys. Lett.* **80** (2002) 303.
64. R. E. Honig and D. A. Kramer, *RCA Rev.* **30** (1969) 285.
65. K. W. Chang and J. J. Wu, *J. Phys. Chem. B* **106** (2002) 7796.
66. T. J. Trentler, K. M. Hickman, S. C. Goel, A. M. Viano, P. C. Gibbons and W. E. Buhro, *Science* **270** (1995) 1791.
67. W. E. Buhro, K. H. Hickman and T. J. Trentler, *Adv. Mater.* **8** (1996) 685.
68. T. J. Trentler, S. C. Goel, K. M. Hickman, A. M. Viano, M. Y. Chiang, A. M. Beatty, P. C. Gibbons and W. E. Buhro, *J. Am. Chem. Soc.* **119** (1997) 2172.
69. S. D. Dingman, N. P. Rath, P. D. Markowitz, P. C. Gibbons and W. E. Buhro, *Angew. Chem. Int. Ed.* **39** (2000) 1470.
70. A. Miehr, O. Ambacher, W. Rieger, T. Metzger, E. Born and R. A. Fischer, *Chem. Vap. Deposition* **2** (1996) 51.
71. A. C. Jones, S. A. Rushworth, D. J. Houlton, J. S. Roberts, V. Roberts, C. R. Whitehouse and G. W. Critchlow, *Chem. Vap. Deposition* **2** (1996) 5.
72. R. A. Fischer, A. Miehr, T. Metzger, E. Born, O. Ambacher, H. Angerer and R. Dimitrov, *Chem. Mater.* **8** (1996) 1356.
73. C. Kawai and A. Yamakawa, *Ceram. Int.* **24** (1998) 135.
74. M. He, I. Minus, P. Zhou, S. N. Mohammad, J. B. Halpern, R. Jacobs, W. L. Sarney, L. Salamanca-Riba and R. D. Vispute, *Appl. Phys. Lett.* **77** (2000) 3731.
75. M. He, P. Zhou, S. N. Mohammad, G. L. Harris, J. B. Halpern, R. Jacobs, W. L. Sarney and L. Salamanca-Riba, *J. Cryst. Growth* **231** (2001) 357.
76. S. T. Lee, Y. F. Zhang, N. Wang, Y. H. Tang, I. Bello, C. S. Lee and Y. W. Chung, *J. Mater. Res.* **14** (1999) 4503.
77. Y. F. Zhang, Y. H. Tang, N. Wang, C. S. Lee, I. Bello and S. T. Lee, *Phys. Rev. B* **61** (2000) 4518.
78. W. Shi, Y. Zheng, N. Wang, C. S. Lee and S. T. Lee, *Adv. Mater.* **13** (2001) 591.
79. H. Y. Peng, X. T. Zhou, N. Wang, Y. F. Zheng, L. S. Liao, W. S. Shi, C. S. Lee and S. T. Lee, *Chem. Phys. Lett.* **327** (2000) 263.
80. J. Zhang, L. Zhang, X. Peng and X. Wang, *J. Mater. Chem.* **12** (2002) 802.
81. J. A. Haber, P. C. Gibbons and W. E. Buhro, *J. Am. Chem. Soc.* **119** (1997) 5455.
82. J. A. Haber, P. C. Gibbons and W. E. Buhro, *Chem. Mater.* **10** (1998) 4062.
83. D. B. Roa and V. V. Dadape, *J. Phys. Chem.* **70** (1966) 1349.
84. K. G. Nickel, R. Riedel and G. Petzow, *J. Am. Ceram. Soc.* **72** (1989) 1804.
85. K. J. Lee, D. H. Ahn and Y. S. Kim, *J. Am. Ceram. Soc.* **83** (2000) 1117.
86. H. Chen, Y. Cao and X. Xiang, *J. Cryst. Growth* **224** (2001) 187.
87. M. Schwartz, *Brazing: For the Engineering Technologist*, Chapman and Hall, London (1995) p. 236.
88. J. Y. Li, X. L. Chen, Z. Y. Qiao, Y. G. Cao and Y. C. Lan, *J. Cryst. Growth* **212** (2000) 408.
89. T. Ogino and M. Aoki, *Jpn. J. Appl. Phys.* **19** (1980) 2395.
90. H. M. Kim, D. S. Kim, D. Y. Kim, T. W. Kang, Y. H. Cho and K. S. Chung, *Appl. Phys. Lett.* **81** (2002) 2193.
91. H. P. Maruska and J. J. Tietjen, *Appl. Phys. Lett.* **15** (1969) 327.
92. M. Yoshizawa, A. Kikuchi, M. Mori, N. Fujita and K. Kishino, *Jpn. J. Appl. Phys.* **36** (1997) L459.
93. C. C. Yu, C. F. Chu, J. Y. Tsai, H. W. Huang, T. H. Hsueh, C. F. Lin and S. C. Wang, *Jpn. J. Appl. Phys.* **41** (2002) L910.

94. R. J. Shul, C. G. Willison, M. M. Bridges, J. Han, J. W. Lee, S. J. Pearton, C. R. Abernathy, J. D. Mackenzie and S. M. Donovan, *Solid State Electron.* **42** (1998) 2269.
95. S. J. Pearton, J. C. Zolper, R. J. Shul and F. Ren, *J. Appl. Phys.* **86** (1999) 1.
96. L. C. Chen, C. H. Liang, Y. T. Hung, Y. F. Chen, C. T. Wu, J. S. Hwang, K. H. Chen, C. W. Hsu and R. H. Lan, IUMRS-ICEM2002, Abstract, p.18.
97. I. Ho and G. B. Stringfellow, *Appl. Phys. Lett.* **69** (1996) 2701.
98. N. Yoshimoto, T. Matsuoka, T. Sasaki and A. Katsui, *Appl. Phys. Lett.* **59** (1991) 2251.
99. S. Nakamura and T. Mukai, *Jpn. J. Appl. Phys.* **31** (1992) L1457.
100. K. Kijima, N. Setaka and H. Tanaka, *J. Cryst. Growth* **24/25** (1974) 183.
101. S. Motojima, T. Yamana, T. Araki and H. Iwanaga, *J. Electrochem. Soc.* **142** (1995) 3141.
102. Y. Inomata and T. Yamana, *J. Cryst. Growth* **21** (1974) 317.
103. P. S. Gopalakrishnan and P. S. Lakshminarsimham, *J. Mater. Sci. Lett.* **12** (1993) 1422.
104. M. J. Wang and H. Wada, *J. Mater. Sci.* **25** (1990) 1690.
105. P. D. Ramesh and K. J. Rao, *J. Mater. Res.* **9** (1994) 2330.
106. M. A. Rodriguez, N. S. Makhonin, J. A. Escrina et al., *Adv. Mater.* **7** (1995) 745.
107. Y. G. Cao, C. C. Ge, Z. J. Zhou and J. T. Li, *J. Mater. Res.* **14** (1999) 876.
108. T. Hashishin, Y. Kaneko, H. Iwanaga and Y. Yamamoto, *J. Mater. Sci.* **34** (1999) 2193.
109. S. C. Zhang and W. R. Cannon, *J. Am. Ceram. Soc.* **67** (1984) 691.
110. S. A. Siddiqi and A. Hendry, *J. Mater. Sci.* **20** (1985) 3230.
111. S. Shimada and T. Kataoka, *J. Am. Ceram. Soc.* **84** (2001) 2442.
112. P. C. Silva and J. L. Figueiredo, *Mater. Chem. and Phys.* **72** (2001) 326.
113. Y. Zhang, N. Wang, R. He, J. Liu, X. Zhang and J. Zhu, *J. Cryst. Growth* **233** (2001) 803.
114. M. J. Wang and H. Wada, *J. Mater. Sci.* **25** (1990) 1690.
115. W. Seo and K. Koumoto, *J. Am. Ceram. Soc.* **79** (1996) 1777.
116. G. J. Jiang, H. R. Zhuang, J. Zhang, M. L. Ruan, W. L. Li, F. Y. Wu and B. L. Zhang, *J. Mater. Sci.* **35** (2000) 63.
117. C. C. Tang, X. X. Ding, X. T. Huang, Z. W. Gan, W. Liu, S. R. Qi, Y. X. Li, J. P. Qu and L. Hu, *Jpn. J. Appl. Phys.* Part 2, **41**(5B) (2002) L589.
118. K. H. Chen, J.-J. Wu, C. Y. Wen, L. C. Chen, C. W. Fan, P. F. Kuo, Y. F. Chen and Y. S. Huang, *Thin Solid Films* **355–356** (1999) 205.
119. O. A. Popov, *Plasma sources for thin film deposition and etching*, Academic Press, San Diego (1994).
120. D. C. Nesting, J. Kouvetakis and D. J. Smith, *Appl. Phys. Lett.* **74** (1999) 958.
121. O. Lagerstedt and B. Monemar, *Phys. Rev. B* **19** (1979) 3064.
122. P. Perlin, C. J. Carillon, J. P. Itie and A. S. Miguel, *Phys. Rev. B* **45** (1992) 83.
123. H. W. Seo, S. Y. Bae, J. Park, H. Yang, K. S. Park and S. Kim, *J. Chem. Phys.* **116** (2002) 9492.
124. S. O. Kucheyev, J. S. Williams and S. J. Pearton, *Mater. Sci. Eng.* **33** (2001) 51.
125. S. O. Kucheyev, J. S. Williams, C. Jagadish, J. Zou and G. Li, *J. Appl. Phys.* **88** (2000) 5493.
126. S. O. Kucheyev, M. Toth, M. R. Phillips, J. S. Williams, C. Jagadish and G. Li, *J. Appl. Phys.* **91** (2002) 3940.
127. S. O. Kucheyev, H. Boudinov, J. S. Williams, C. Jagadish and G. Li, *J. Appl. Phys.* **91** (2002) 4117.
128. T. Nagatomo, T. Kuboyama, H. Minamino and O. Omoto, *Jpn. J. Appl. Phys.* **28** (1989) L1334.
129. A. Mlayah, A. M. Brugman, R. Carles, J. B. Renucci, M. Ya. Valakh and A. V. Pogorelov, *Solid State Commun.* **90** (1994) 567.
130. K. K. Nanda, S. N. Sahu, R. K. Soni and S. Tripathy, *Phys. Rev. B* **58** (1998) 15405.
131. G. H. Li, K. Ding, Y. Chen, H. X. Han and Z. P. Wang, *J. Appl. Phys.* **88** (2000) 1439.
132. K. L. Teo, S. H. Kwok, P. Y. Yu and S. Guha, *Phys. Rev. B* **62** (2000) 1584.

133. H. L. Liu, C. C. Chen, C. T. Chia, C. C. Yeh, C. H. Chen, M. Y. Yu, S. Keller and S. P. DenBaars, *Chem. Phys. Lett.* **345** (2001) 245.
134. T. Azuhata, T. Sota, K. Suzuki and S. Nakamura, *J. Phys.: Condens. Matter* **7** (1995) L129.
135. L. Filippidis, H. Siegle, A. Hoffmann, C. Thomsen, K. Karch and F. Bechstedt, *Phys. Status Solidi B* **198** (1996) 621.
136. V. Yu, Davydov, Yu. E. Kitaev, I. N. Goncharuk, A. N. Smirnov, A. P. Mirgorodsky and R. A. Evarestov, *Phys. Rev. B* **58** (1998) 12899.
137. H. Richter, Z. P. Wang and L. Ley, *Solid State Commun.* **39** (1981) 625.
138. I. H. Campbell and P. M. Fauchet, *Solid State Commun.* **58** (1986) 739.
139. H. Siegle, G. Kaczmarczyk, L. Filippidis, A. P. Litvinchuk, A. Hoffmann and C. Thomsen, *Phys. Rev. B* **55** (1997) 7000.
140. C. Bungaro, K. Rapcewicz and J. Bernholc, *Phys. Rev. B* **61** (2000) 6720.
141. J. S. Hwang, C. H. Lee, F. H. Yang, K. H. Chen, L. G. Hwa, Y. J. Yang and L. C. Chen, *Mater. Chem. Phys.* **72** (2001) 290.
142. F. H. Yang, J. S. Hwang, K. H. Chen, Y. J. Yang, T. H. Lee, L. G. Hwa and L. C. Chen, *Thin Solid Films* **405** (2002) 194.
143. G. D. Gilliland, *Mater. Sci. Eng. R* **18** (1997) 99.
144. B. G. Yacobi and D. B. Holt, *Cathodoluminescence Microscopy of Inorganic Solids*, Plenum Press, New York and London (1990).
145. P. Y. Yu and M. Cardona, *Fundamentals of Semiconductors*, Springer, Berlin (1996) p. 307.
146. A. R. Adams and E. P. O'Reilly, in *Materials for Optoelectronics*, Edited by Maurice Quillec, Kluwer Academic Publishers, Boston/Dordrecht/London (1996) pp. 61–99.
147. B. K. Ridley, *Quantum Processes in Semiconductors*, Clarendon, Oxford (1982) pp. 62–66.
148. X. Shan, X. C. Xie, J. J. Song and B. Goldenberg, *Appl. Phys. Lett.* **67** (1995) 2512.
149. Y. Xie, Y. Quian, W. Wang, S. Zhang and Y. Heng, *Science* **272** (1996) 1926.
150. C. Y. Yek, S. B. Zhang and A. Zunger, *Phys. Rev. B* **50** (1994) 14405.
151. I. Vurgaftman, J. R. Meyer and L. R. Ram-Mohan, *J. Appl. Phys.* **89** (2001) 5815.
152. B. Monemar, *Phys. Rev. B* **10** (1974) 676.
153. O. Lagerstedt and B. Monemar, *J. Appl. Phys.* **45** (1974) 2266.
154. K. Hiruma, M. Yazawa, T. Katsuyama, K. Ogawa, K. Haraguchi, M. Koguchi and H. Kakibayashi, *J. Appl. Phys.* **77** (1995) 447.
155. G. D. Chen, M. Smith, J. Y. Lin, H. X. Jiang, A. Salvador, B. N. Sverdlov, A. Botchkarv and H. Morkoc, *J. Appl. Phys.* **79** (1996) 2675.
156. H. Okumura, H. Hamaguchi, G. Feuillet, Y. Ishida and S. Yoshida, *Appl. Phys. Lett.* **72** (1998) 3056.
157. S. Kim, I. P. Herman, J. A. Tuchman, K. Doverspike, L. B. Rowland and D. K. Gaskill, *Appl. Phys. Lett.* **67** (1995) 380.
158. H. M. Chen, Y. F. Chen, M. C. Lee and M. S. Feng, *Phys. Rev. B* **56** (1997) 6942.
159. Y. P. Varshni, *Physica* (Amsterdam) **34** (1967) 149.
160. I. Vurgaftman, J. R. Meyer and L. R. Ram-Mohan, *J. Appl. Phys.* **89** (2001) 5815.
161. T. L. Tansley and C. P. Foley, *J. Appl. Phys.* **59** (1986) 3241.
162. H. D. Cho, S. H. Park, N. H. Ko, H. Y. Lee, T. W. Kang, S. H. Won, K. S. Chung, G. S. Eom, G. S. Yoon and C. O. Kim, *J. Korean Phys. Soc.* **30** (1997) S58.
163. G. Pozina, J. P. Bergman, B. Monemar, V. V. Mamutin, T. V. Shubina, V. A. Vekshin, A. A. Toropov, S. V. Ivanov, M. Karlsteen and M. Willander, *Phys. Stat. Sol. B* **216** (1999) 445.
164. V. V. Mamutin, T. V. Shubina, V. A. Vekshin, V. V. Ratnikov, A. A. Toropov, S. V. Ivanov, M. Karlsteen, U. Sodervall and M. Willander, *Appl. Surf. Sci.* **166** (2000) 87.
165. V. Y. Davydov, A. A. Klochikhin, R. P. Seisyan et al., *Phys. Stat. Sol. B* **229** (2002) R1.
166. V. Y. Davydov, A. A. Klochikhin, V. V. Emtsev et al., *Phys. Stat. Sol. B* **230** (2002) R4.

167. H. M. Kim, D. S. Kim, Y. S. Park, D. Y. Kim, T. W. Kang and K. S. Chung, *Adv. Mater.* **14** (2002) 991.
168. M. W. Lee, H. Z. Twu, C.-C. Chen and C.-H. Chen, *Appl. Phys. Lett.* **79** (2001) 3693.
169. Y. Nagamune, H. Watabe, F. Sogawa and Y. Arakawa, *Appl. Phys. Lett.* **67** (1995) 1535.
170. K. Hiruma, M. Yazawa, T. Katsuyama, K. Ogawa, K. Haraguchi, M. Koguchi and H. Kakibayashi, *J. Appl. Phys.* **77** (1995) 447.
171. D. Snoke, *Science* **273** (1996) 1351.
172. Justin C. Johnson, Heon-Jin Choi, Kelly P. Knutsen, Richard D. Schaller, Peidong Yang and Richard J. Saykally, *Nature Materials* **1** (2002) 1.
173. R. E. Slusher and U. Mohideen, in *Optical Processes in Microcavities*, Edited by A. J. Campillo and R. Chang, World Scientific, Singapore (1996) pp. 315–338.
174. K. P. O'Donnell, R. W. Martin and P. G. Middleton, *Phys. Rev. Lett.* **82** (1999) 237.
175. H. C. Yang, P. F. Kuo, T. Y. Lin, Y. F. Chen, K. H. Chen, L. C. Chen and J. I. Chyi, *Appl. Phys. Lett.* **76** (2000) 3712.
176. C. H. Hsieh, Y. S. Huang, P. F. Kuo, Y. F. Chen, L. C. Chen, J.-J. Wu, K. H. Chen and K. K. Tiong, *Appl. Phys. Lett.* **76** (2000) 2044.
177. R. D. Carson and S. E. Schnatterly, *Phys. Rev. B* **33** (1986) 2432.
178. Y. Huang, X. Duan, Y. Cui, L. J. Lauhon, K. H. Kim and C. M. Lieber, *Science* **294** (2001) 1313.
179. J. R. Kim, H. M. So, J. W. Park, J. J. Kim, J. Kim, C. J. Lee and S. C. Lyu, *Appl. Phys. Lett.* **80** (2002) 3548.
180. Z. Fan, S. N. Mohammad, W. Kim, O. Aktas, A. E. Botchkarev and H. Morkoc, *Appl. Phys. Lett.* **68** (1996) 1672.
181. S. M. Sze, *Physics of Semiconductor Devices*, Wiley, New York (1981).
182. Y. Cui, X. Duan, J. Hu and C. M. Lieber, *J. Phys. Chem. B* **104** (2000) 5213.
183. C. C. Chen, C. C. Yeh, C. H. Liang, C. C. Lee, C. H. Chen, M. Y. Yu, H. L. Liu, L. C. Chen, Y. S. Lin, K. J. Ma and K. H. Chen, *J. Phys. Chem. Solids* **62** (2001) 1577.

Chapter 10

Template Assisted Synthesis of Semiconductor Nanowires

Dongsheng Xu and Guolin Guo

State Key Laboratory for Structural Chemistry of Unstable and Stable Species, College of Chemistry and Molecular Engineering, Peking University, Beijing 100871, P.R. China

1. Introduction

The synthesis of semiconductor nanowires (SNWs) is critical to work directed towards understanding their fundamental properties and developing nanotechnologies [1–16]. Over the past several years considerable effort has been placed on the bulk synthesis of nanowires, and there are mainly three experimental approaches to fabricate SNWs. One method is laser ablation or thermal evaporation of metal containing semiconductor targets through a vapor-liquid-solid (VLS) growth mechanism [17–24]. Nanowires of element semiconductors (Si, Ge, Se) [17, 18, 22], binary and ternary III-V, II-VI and I-IV group materials [18, 19, 23], and metal oxide semiconductors [9, 21, 24] have been prepared in bulk quantities as high purity single crystals. A critical feature of this method is that the catalyst used to define one-dimensional growth and thus the metal always observed at the nanowire end. The second method is oxide-assisted semiconductor nanowire growth by laser ablation, thermal evaporation, or chemical vapor deposition [25–28]. By means of this approach, each nanowire consists an amorphous oxide shell and a high density of defects has been observed in the crystalline semiconductor core [26]. The third is a template synthesis method, pioneered by Martin [29–32], Moskovits [33], and Searson [34]. It has been demonstrated that template synthesis is a general and versatile method for preparing nanostructural materials including both nanotubles and nanofibrils composed of conductive polymers, metals, carbon, semiconductors and other materials [32], which entails synthesizing the desired material within the pores of a mesoporous template membrane [29–34]. Due to cylindrical pore geometry and monodisperse diameters, corresponding cylindrical and oriented nanostructural materials with a narrow diameter distribution are obtained. More recently, template synthesis of SNWs has been well developed and a broad range of semiconductor materials have been fabricated [31, 35–90].

In this chapter, we summarize the works on template synthesis of SNWs by now. In Section 2, we will describe the types of templates used for synthesis of SNWs. Next, a general outline of six representative chemical strategies that have been used in template synthesis of SNWs will present in Section 3. Finally, we will discuss structural characterization and fundamental properties of these SNWs in the Sections 4 and 5.

2. Templates Used

There are mainly two kinds of templates, i.e. "hard" template and "soft" template. The hard templates include inorganic mesoporous materials such as anodic aluminum oxides (AAO) and zeolites, mesoporous polymer membranes, carbon nanotubes, etc. The soft templates generally refer to surfactant assemblies such as mono-layers, liquid crystals, vesicles, micelles, etc. Most of the works in template synthesis of SNWs, to date, have entailed the use of the hard templates.

2.1. Porous AAO membranes

Porous AAO membrane is formed via the anodization of aluminum metal in acidic solutions, which has been studied in detail over the last five decades [91]. Using two-step or self-organization anodization, porous AAO membranes have a densely regular hexagonal pore structures [92–96], as shown in Fig. 1a–c. The pore densities as high as 10^{11} pores cm^{-2} can be achieved and the pores in these membranes have little or no tilt with respect to the surface normal resulting in an isolating, non-connecting pore structure (Fig. 1d). The pore diameters can be varied from 5 nm to 250 nm in

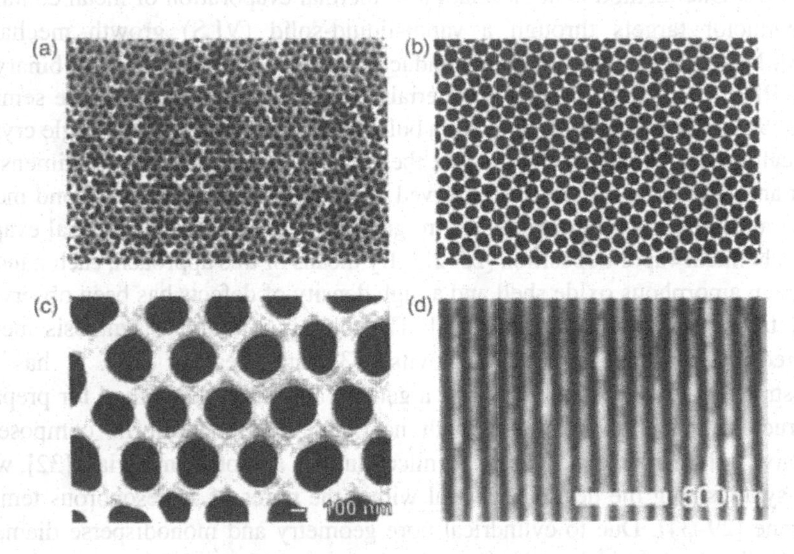

Fig. 1. SEM micrographs of the AAO membranes with various pore sizes: (a) ~30 nm, (b) ~ 90 nm, and (c) ~250 nm [95]. The cross-sectional view in Fig. 1d shows nonintercrossing, parallel cylindrical pores [96].

a narrow distribution, depending on the anodizing condition including the electrolyte compose, voltage, and temperature.

A detailed report on the preparation of through hole AAO membranes can be found in Refs. [39–41, 92–96]. First, high purity aluminum foils were degreased in acetone, cleaned in mixed solution of $HF/HNO_3/HCl/H_2O$, and annealed under nitrogen ambient at 400 °C, and then electropolished in a mixture of $HClO_4$ and C_2H_5OH. Second, the aluminum foils were anodized in aqueous acidic solutions such as sulfuric, oxalic, and phosphoric acids. Third, after the anodization, the remained aluminum was removed by a 20% $HCl-0.1\,mol \cdot l^{-1}$ $CuCl_2$ mixed solution or a saturated $HgCl_2$ solution. Finally, the pore bottoms were opened by chemical etching in 5 wt% aqueous phosphoric or 20 wt% H_2SO_4 acids.

Owing to the regular hexagonal pore structures, higher pore density, highly thermally and chemically stability (i.e., neutral aqueous and organic baths), and easily removing by strong acids or bases, the AAO membranes have been used as an important template material.

2.2. Track-etched polymer membranes

Microporous and nanoporous polymer membranes are commercially available filters, which are prepared by the track-etch method. This method entails bombarding a non-porous sheet of the desired material with nuclear fission fragments to create damage tracks in this material, and then chemically etching these tracks into pores [97]. A broad range of pore diameters (down to 10 nm) is available, and pore densities approaching 10^9 pores cm^{-2} can be obtained. The pores in these membranes are randomly distributed across the membrane surfaces with uniform diameters, as shown in Fig. 2. However, due to the random nature of the pore-production process, the pores have tilt with respect to the surface normal, and a number of pores may actually intersect within the membrane [29].

2.3. Other nanoporous materials

Mesoporous molecular sieves, such as hexagonally ordered MCM-41 and mesoporous aluminosilicates (MAS-5) [98–100], which have been synthesized from assembly of aluminosilicate precursors with surfactants, are taken as potential

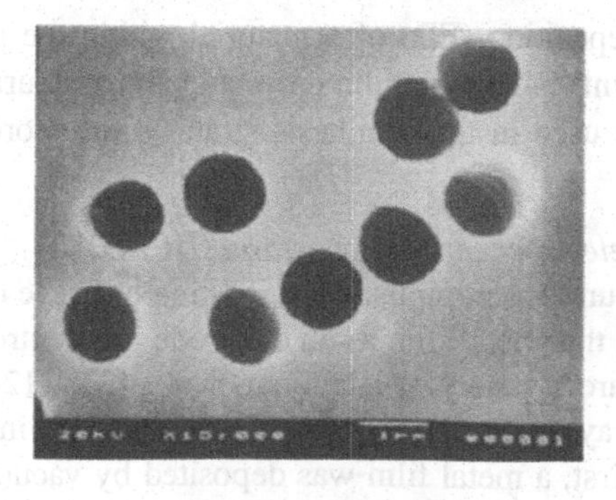

Fig. 2. SEM micrograph of polycarbonate template membrane with 1 μm diameter pores [31].

templates for synthesis of nanowire materials. These new zeolites are highly ordered hexagonal porous structure with uniform pore sizes. Yang et al. have template synthesized densely packed metal nanostructures in functionalized (MCM)-41 and MCM-48 [101]. Beck et al. have template synthesized Poly-aniline and graphitic nanowires within the pores of these new zeolites [102]. However, as compared with AAO membrane, the pores in these materials generally parallel with the surface, which severely hinder their practical applications in template synthesis. Tonucci et al. have described a nanochannel array glass membrane containing pores with diameters as small as 33 nm and densities as high as 3×10^{10} pores/cm^2 [103]. Sun et al. reported that nanopores with lateral dimensions as small as 33 nm have been fabricated by nuclear track etching in single crystal mica wafers [104, 105]. The nanopores have a diamond shape with their axes aligned with the crystal axes of mica as a result of anisotropic etching. Ozin discusses a wide variety of other nanoporous solids that could, in principle, be used as template materials [106]. Carbon nanotubes have been used to synthesis of nanowires through a carbon nanotubes-confined reaction [107–110]. More recently, Thurn-Albrecht et al. described a route to the fabrication of ultrahigh-density arrays of nanopores with high aspect ratios using the equilibrium self-assembled morphology of asymmetric diblock copolymers [111]. In contrast to anodized aluminum, there is no insulating barrier layer at the bottom of the nanopores, such that direct dc electrical contact to the substrate is possible.

3. Template Synthesis Strategies of Semiconductor Nanowires

In general, typical concerns that need to be addressed when developing new template synthesis methods include the following [31]: (1) will the precursor solutions used to prepare the material wet the pore (i.e., hydrophobic/hydrophilic consideration); (2) will the deposition reaction proceed too fast resulting in pore blockage at the membrane surface before tubule/fibril growth can occur within the pores; (3) will the host membrane be stable (i.e., thermally and chemically) with respect to the reaction condition? The following is a general outline of six representative chemical strategies that have been used in template synthesis of SNWs by now.

3.1. Electrochemical deposition

Electrochemical deposition (ED) of a material within the pores is a simple and versatile method for synthesizing one-dimensional nanostructural material. Two main types of templates are used in this synthesis strategy, i.e., porous AAO and porous polymer membranes.

3.1.1. Direct-current electrochemical deposition (DCED)

The DCED procedure is accomplished by coating one face of the membrane with a metal film and using this metal film as an electrode for electroplating. This method has been used to prepare a variety of metal nanowires [34, 112–117]. In general, the synthesis of SNW arrays within the pores of the template involves three steps as shown in Scheme 1: first, a metal film was deposited by vacuum evaporation or ion sputtering onto the back of the template membrane; second, semiconductor materials

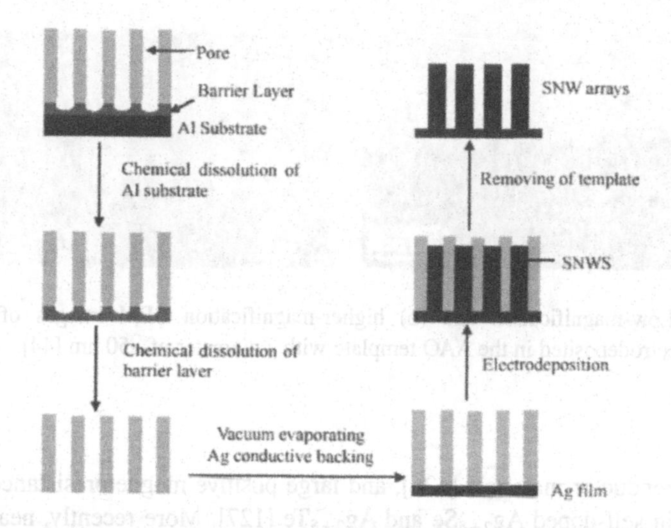

Scheme 1. DC electrochemical synthesis scheme for SNWs [41].

were cathodically deposited on the metal surface at the pore bottom from a solution containing the metal ion (e.g., Cd^{2+}) and another anion (e.g. sulfur and selenium); and finally, the AAO template was dissolved in strong acid or base (i.e., H_3PO_4 and NaOH). The deposition was commonly performed potentiostatically or galvanostatically in a three-electrode configuration. By this strategy, a variety of highly ordered and crystallized metal chalcogenide SWN arrays have been prepared [36–49].

There are some important considerations, however, in the choice of electrochemical reaction routes and the operational parameters of ED synthesis. ED offers a simple and viable alternative to the cost-intensive methods such as laser ablation or thermal evaporation synthesis. In particular, it occurs closer to equilibrium than those high-temperature vacuum deposition methods, and affords precise process control due to its electrical nature. For example, the lengths or aspect ratios of the nanowires can be controlled by the amount of material deposited. However, contamination of the target material with impurity phases is often a problem with ED as is the morphological quality of the product. In particular, both the greater number of active species in the solution and the by-reaction render the composition modulation a little trickier. On the other hand, the important point to emphasize is that the electrochemical reaction routes have a key role in the growth of nanowires.

Significant improvements have also been claimed in the morphological quality of the resultant nanowires using non-aqueous media [37–45]. For example, highly aligned and crystallized CdS and CdSe nanowires have been prepared by DCED in porous AAO template from dimethylsulfoxide (DMSO) solution containing $CdCl_2$ and S (Se) [37–41]. These nanowires have uniform diameters and lengths are up to tens of micrometers. Fig. 3 presents the SEM images of highly ordered single crystal CdTe nanowire arrays prepared by ED from an ethylene glycol bath containing $CdCl_2$, $TeCl_4$ and KI [44]. There has been report for preferred [001] orientation in these CdTe nanowires [44]. Silver chalcogenides (Ag_2Se and Ag_2Te) belong to $A_2^IB^{VI}$ group of compound semiconductors with narrow band-gaps. Their high temperature phases are

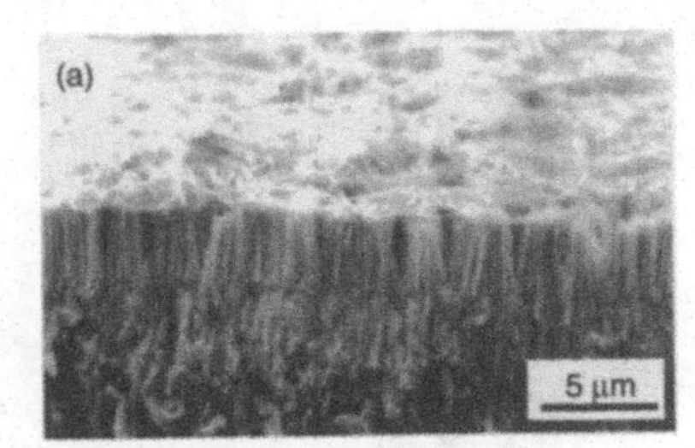

Fig. 3. (a) Low-magnification and (b) higher-magnification SEM images of the CdTe nanowires electrodeposited in the AAO template with a diameter of ~50 nm [44].

superionic conductor materials [126], and large positive magnetoresistance has been discovered in self-doped $Ag_{2+\delta}Se$ and $Ag_{2+\delta}Te$ [127]. More recently, near stoichiometric Ag_2Se [49] and Ag_2Te [45] nanowires have been synthesized by cathodic electrolysis from DMSO solutions containing $AgNO_3$ and $TeCl_4$ (or $SeCl_4$). In these studies, the composition of the nanowires can be controlled continuously from Ag-rich to Te-rich mainly by changing the concentration of $TeCl_4$ in the solutions [45].

There have been proponents of the use of non-aqueous solvents. The proclivity of the chalcogen to exist predominantly in low oxidation states in these baths, and the attendant lack of complications in the ED chemistry have been the principal motivating factors in evaluation of non-aqueous media. For example, CdSe nanowires by ED in porous AAO template from DMSO solution containing $CdCl_2$ and Se are highly (002) oriented, and the atomic composition of Cd and Se is very close to a 1 : 1 stoichiometry [39]. However, in $CdCl_2$ and SeO_2 ammonia alkaline solutions, the deposited CdSe nanowires are randomly oriented and the ratio of Se to Cd depends on the pH of the deposition bath [47]. Another important factor is that interference from solvent electrochemistry is circumvented in aprotic media; this proves to be crucial in the ED synthesis of relatively difficult systems such as ZnX (S, Se, Te).

An attractive feature of the electrochemical synthesis approach is the ease with which alloy nanowires may be generated. Routkevitch et al. have fabricated ternary compound CdS_xSe_{1-x} nanowires in a DMSO solution containing $CdCl_2$, S and Se [33]. We have synthesized ternary compound $Ag_{2+\delta}Se_{1-x}Te_x$ nanowires at the co-deposition potentials of Ag_2Se and Ag_2Te using DMSO as a solvent and $CdCl_2$, $SeCl_4$ and $TeCl_4$ as Cd, Se and Te sources respectively (Fig. 4) [50]. By varying the concentrations of $SeCl_4$ and $TeCl_4$ in DMSO solutions, ternary $Ag_{2+\delta}Se_{1-x}Te_x$ compounds with tunable compositions were prepared.

3.1.2. Alternating-current electrochemical deposition (ACED)

The compact barrier layer between the porous AAO and Al substrate is insulating, which obstructs the passage of direct current. Thus, it is difficult that materials are directly deposited on the AAO/Al substrate by DCED. Routkevitch et al. demonstrated that this problem was resolved by using AC electrolysis [33]. They have fabricated CdS nanowire arrays by a single-step ACED in an electrolyte containing Cd^{2+} and S in DMSO. On average the thicker wires (with diameters larger than 12 nm) consist of a

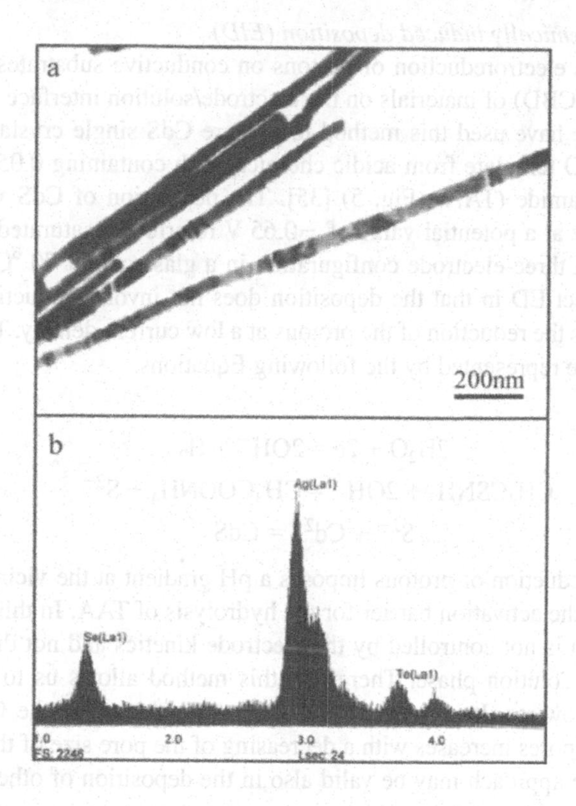

Fig. 4. (a) TEM image and (b) EDS spectrum of the $Ag_{2+\delta}$ Se_{1-x} Te_x nanowires by electrodeposition in DMSO solution.

large number of crystallites in the axial direction and rather few in the radial direction [33, 51–53].

ACED is a simple method to fabricate aligned nanowires in AAO template with retaining the compact barrier layer [33]. A major disadvantage of this synthesis route is that there are large numbers of stacking faults and twinned segments in the deposited nanowire and the structure of the nanowire appears to be mainly the hexagonal form, interleaved with domains of cubic structure [52].

In general, for the case of electrodepositing in the pore of the AAO templates, the direction of the diffusing is limited on one dimension. The diffusing rate may be much slower than the surface electrochemical reaction and the diffusing process would limit the nanowire growth. For AC process, the rate and the direction of the diffusing would vary with the alternation of the electric field, and thus make a high density of defects and small polycrystalline CdS structure in the nanowires [33, 52]. On the other hand, it is ease to reach a steady diffusing process under the DCED condition. Thus, the defects in the nanowires by DCED are largely decreased and produce perfect and highly crystalline nanowires.

3.1.3. Electrochemically induced deposition (EID)

EID involves electroreduction of protons on conductive substrates and chemical bath deposition (CBD) of materials on the electrode/solution interface in the pores of the template. We have used this method to prepare CdS single crystal nanowires in the pores of AAO template from acidic chemical bath containing 0.05 M $CdCl_2$ and 0.10 M thioacetamide (TAA) (Fig. 5) [35]. The deposition of CdS was performed potentiostatically at a potential value of −0.65 V referred to saturated calomel electrode (SCE) in a three-electrode configuration in a glass cell at 70 °C. This method differs from direct ED in that the deposition does not involve reduction of Cd^{2+} or TAA but requires the reduction of the protons at a low current density. The component processes may be represented by the following Equations:

Cathodic

$$2H_2O + 2e = 2OH^- + H_2 \tag{1}$$

$$CH_3CSNH_2 + 2OH^- = CH_3COONH_4 + S^{2-} \tag{2}$$

$$S^{2-} + Cd^{2+} = CdS \tag{3}$$

The electroreduction of protons imposes a pH gradient at the vicinity of the substrate to reduce the activation barrier for the hydrolysis of TAA. In this case, the CdS nanowire growth is not controlled by the electrode kinetics and not disturbed by the reactions in the solution phase. Therefore, this method allows us to deposit single crystal CdS nanowires. It is interesting that the growth rate of the CdS nanowires deposited in the pores increases with a decreasing of the pore size of the template. Of course, the same approach may be valid also in the deposition of other metal sulfide single crystal nanowires.

3.2. Hybrid electrochemical/chemical synthesis (HEC)

Although direct ED has successfully applied in preparation of metal chalcogen nanowires, controlling of the composition is still a problem in these nanowires. In addition, it is difficult to fabricate metal oxide nanowires by this approach. The hybrid electrochemical/chemical (HEC) synthesis of SWNs involves two steps, as shown in Scheme 2: first, metal nanowires are prepared by DCED in the pores of the template; second, the metal nanowires are reacted with other reactants to yield metal compound nanowires. By now, there are three reaction routes for HEC synthesis:

3.2.1. Metal oxide nanowires by thermal oxidation of metal nanowires

This approach involves thermal oxidation of metal nanowires embedded in the pores of the AAO template, resulting in the formation of metal oxide nanowires. A various of metal oxide SNWs, including ZnO, SnO_2, In_2O_3, TiO_2, MgO and NiO, have been fabricated by this electrochemical deposition and thermally oxidizing methods [54–58]. The key point is that the composition of the metal oxide nanowires is decided by the annealing temperature and the oxidation time. Zheng et al. [54] reported that there are three phases (Sn, SnO, and SnO_2) coexisting in the thermally oxidized nanowires at 823 K, however, only the SnO_2 cassiterite phase is detected under annealing at 923 K. Interestingly, we found that the Ti nanowires can be transferred to a TiO_2/Ti coaxial structure by thermally oxidizing in air at room temperature, while TiO_2 nanowires are formed by thermally

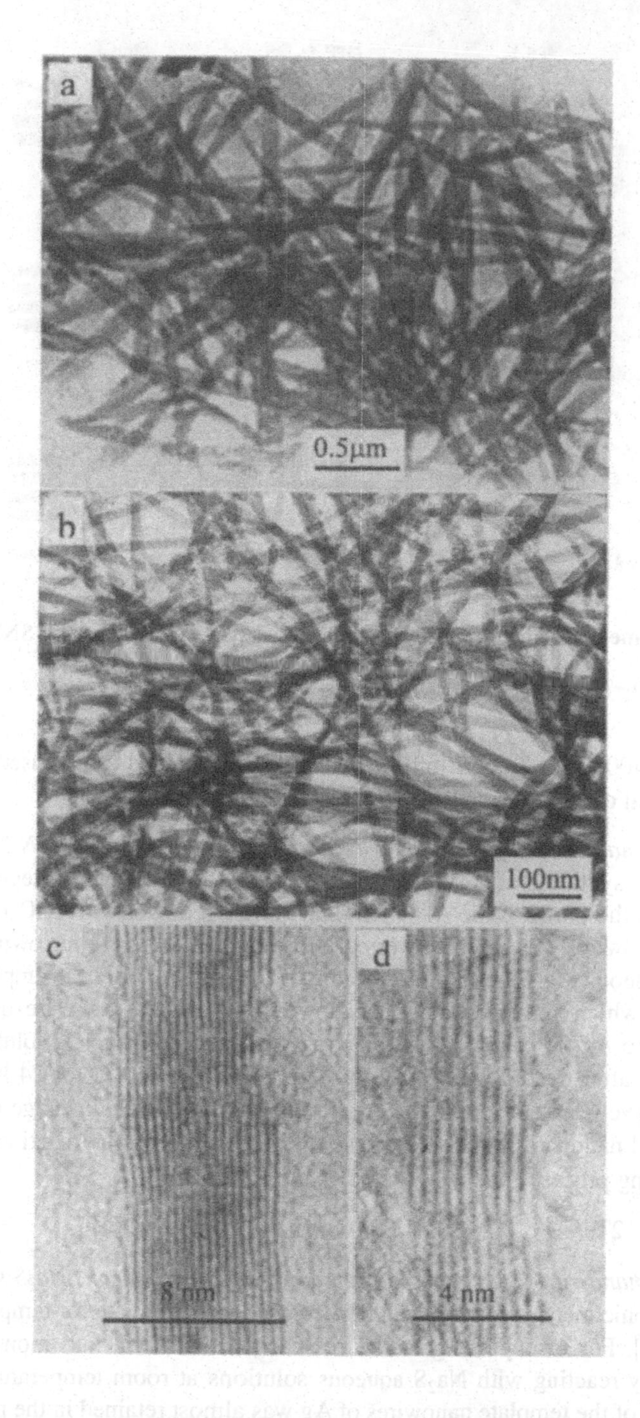

Fig. 5. TEM images of CdS nanowires prepared by electrochemically induced deposition in the AAO templates with diameters of about 90 nm (a) and 20 nm (b), and the HREM images of individual nanowires with different diameters of (c) 8 nm and (d) 4 nm.

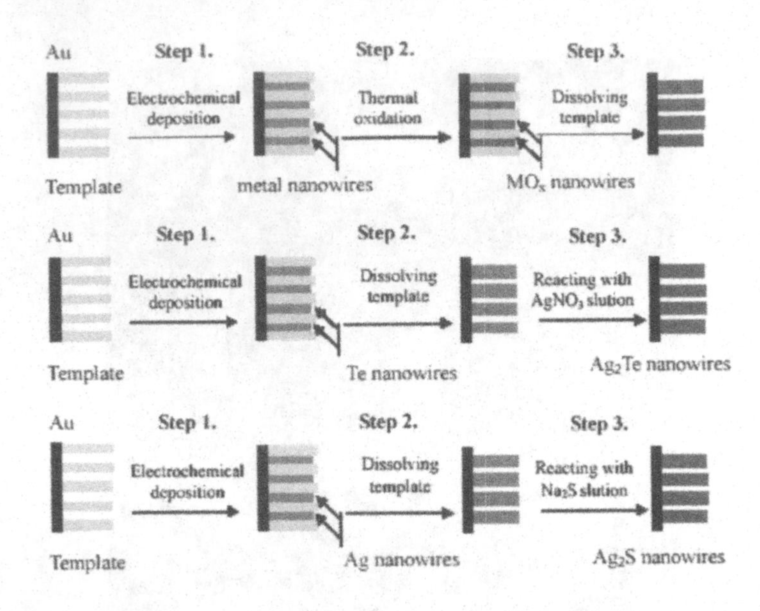

Scheme 2. Hybrid electrochemical/chemical synthesis schemes for SNWs.

oxidizing at 400 °C for 8 hours [58]. In principle, this method can be used to fabricate complex metal oxide nanowires from multi-metal nanowires.

3.2.2. Ag_2Te nanowires by reacting of Te nanowires with aqueous $AgNO_3$ solutions

The HEC synthesis of Ag_2Te nanowires begins with the electrodeposition of Te nanowires in the pores of the AAO template. After removed the AAO template, the metal Te nanowires were then converted into monoclinic Ag_2Te nanowires by reacting with aqueous $AgNO_3$ solutions. Gates et al. [59] reported a template-directed reaction, in which single crystalline nanowires of trigonal Se were quantitatively converted into Ag_2Se nanowires by reacting with aqueous $AgNO_3$ solutions. In our experiments, after treated with 5 mM $AgNO_3$ solution at 60 °C for 24 hours, the Te nanowires have been converted into Ag_2Te nanowires, and the average diameters of the converted nanowires have an increase of 20~30% [60]. The reaction involved in this converting process can be presented as the following:

$$2\mathrm{Te(s)} + 2\mathrm{Ag}^+ + 3\mathrm{H_2O} \rightarrow \mathrm{Ag_2Te} + \mathrm{TeO_3^{2-}} + 6\mathrm{H}^+ \tag{4}$$

3.2.3. Ag_2S nanowires by reacting of Ag nanowires with aqueous Na_2S solutions

The cationic metal nanowires can be also used as a template for template-directed reaction [61]. For example, Ag nanowires can be converted into monoclinic Ag_2S nanowires by reacting with Na_2S aqueous solutions at room temperature [61]. The morphology of the template nanowires of Ag was almost retained in the product Ag_2S nanowires, and the average diameters of the nanowires have an increase of 70~80%. In addition, the fact that excess Ag in the electrodeposited Ag_2Se nanowires were reduced and even disappeared by annealing in Se vapor [48] implies that annealing in Se vapor may be used for the transformation of Ag_2Se nanowires from Ag nanowires.

3.3. Sol-gel deposition

Sol-gel chemistry has evolved into a general and powerful approach for preparing inorganic materials [118, 119]. This method typically entails hydrolysis of a solution of a precursor molecule to obtain first a suspension of colloidal particles (the sol) and then a gel composed of aggregated sol particles. The gel is then thermally treated to yield the desired material. This approach for the synthesis of inorganic materials has advantages in the preparation of both high-purity materials at a lower temperature and homogenous multi- component systems by mixing precursor solutions.

3.3.1. Direct sol filling

Martin's group [62, 63] has first conducted sol-gel synthesis within the pores of the AAO template to create both fibrils and tubules of the desired material, using a simple immersion method. First, the AAO membrane is immersed into a sol for a given period of time, and the sol deposits on the pore walls. After thermal treatment, either a nanotubule or nanofibril is formed within the pores. By this method, nanostructures (nanowires and nanotubules) of TiO_2 [62–67], SiO_2 [63, 68–70], In_2O_3 [71], Ga_2O_3 [71], V_2O_5 [62, 63], MnO_2 [63], WO_3 [63], Fe_2O_3 [72], ZnO [73], SnO_2 [74], CdS [75] and other complex oxide materials [76–80] have been synthesized. More recently, C_{70} nanowires have been fabricated by this method [81].

Whether tubules or fibrils are obtained is determined by the immersion time, the temperature of the sol and the electric properties of the pore walls, Fig. 6. Longer

Fig. 6. SEM images of TiO_2 nanostructures obtained by immersing the template membrane in the sol for (a) 5, (b) 25, and (c) 60s [63].

immersion times yield solid fibrils while brief immersion times produced tubules. However, when the templates with smaller pore sizes were dipped into the sol, solid fibrils are always obtained in the pores [63].

There are some potential limitations to this technique. For example, since the only driving force of this technique is capillary action, for the sol with higher concentration, the filling of the pores would be difficult (at the same time destabilization of the sol remain a big problem), but low concentration leads to nanomaterials with serious shrinkage and cracking.

3.3.2. Sol-gel electrophoretic deposition

Sol-gel electrophoretic growth of nanowires throws some light on the overcoming of the limitations of direct sol filling method [82, 83]. While using this sol-gel electrophoresis in the template-assisted growth of nanorods, an electric field was applied to draw thc charged sol nanoclusters into the template pores (Fig. 7). Both single metal oxides (TiO_2, SiO_2) and complex oxides ($BaTiO_3$, $Sr_2Nb_2O_7$, and $Pb(Zr_{0.52}Ti_{0.48})O_3$) have been grown by this method [83]. Desired stoichiometric chemical composition and crystal structure of the oxide nanorods was readily achieved by an appropriate procedure of sol preparation, with a heat treatment (700 °C for 15 min) for crystallization and densification. But they failed to synthesize nanorodes of <50 nm in diameter. This can be explained by the difficulty of diffusion of clusters in the nano-sized pores.

3.3.3. Electrochemically induced sol-gel synthesis

We have reported an electrochemically induced sol-gel method to prepare TiO_2 single-crystalline nanowire arrays (Fig. 8) [84]. At first, the hydroxyl ion is generated due to the cathodic reduction, and then the generation of OH^- ions increases the local pH at the electrode surface resulting in the titanium oxyhydroxide gel formation in the

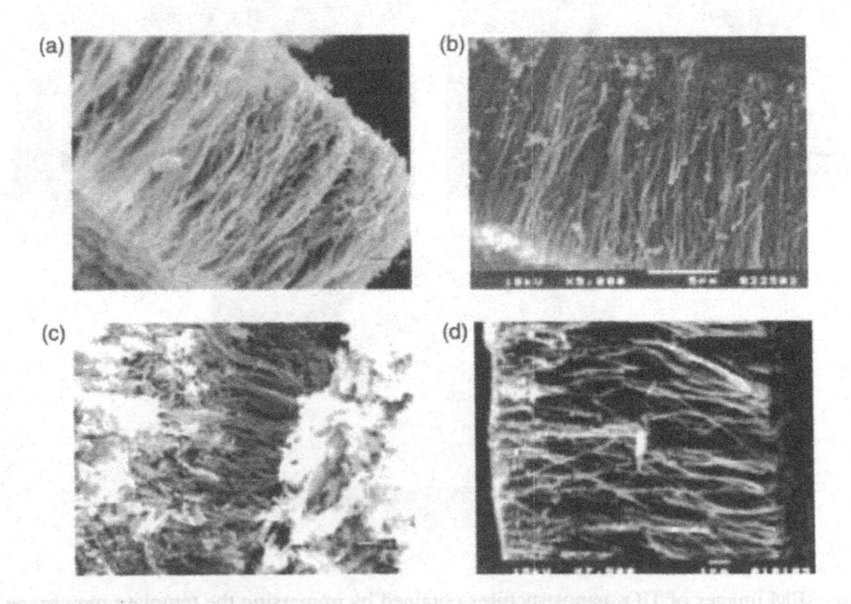

Fig. 7. SEM images of (a) $BaTiO_3$, (b) SiO_2, (c) Sr_2-Nb_2O_7, and PZT nanorods grown in a PC membrane with 200 nm diameter pores by sol-gel electrophoresis [83].

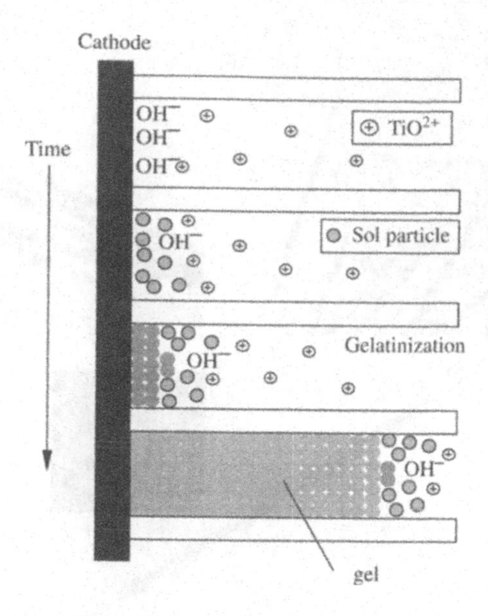

Fig. 8. This schematic demonstrates the progress of the electrochemically induced sol-gel process. It can be seen that both the formation of sol particles and the gelatin process take place in the AAO pores; lastly, the pores are filled with homogeneous titanium oxyhydroxide gel [84].

pores of the template. Finally, subsequent heat-treatment and the removal of the AAO template results in the formation of TiO_2 single-crystalline nanowire arrays. During this electrochemically induced sol-gel process, both the formation of sol particles and the gelation process take place in the AAO pores.

This method offers many advantages for the formation of nanowires and nanotubes. Firstly, the sol-gel preparation of nanowires within the templates that have very small pores (less than 20 nm or even small) would be readily achievable by using this technique (Fig. 9). Secondly, the length of the nanowires can be well controlled by varying the deposition time and potential of the working electrode. Thirdly, less shrinkage and cracking will happen during the heat treatment of nanowires. The transport of ions through the voids of packed sol particles and the tiny pores in the titanium oxyhydroxide gel network courses the expansion of gel in the AAO pores and aging of the gel that has already formed. The AAO template with pores of nanoscaled size can restrict the diffusion of OH^- in solution, so it is easy to construct higher local pH and longer diffusion depth in the AAO pores than on the thin film, resulting a higher packing density gel.

Further, one-dimensional silica nanostructures, such as nanotube, "bamboo-like" structure and nanowire, have also been synthesized by such an electrochemically induced sol-gel deposition (Fig. 10) [85]. It was demonstrated that the growth of these nanostructures is strongly dominated by the electrochemical process. The lower cathode voltage is favor to produce nanotubes while the nanowires were formed at a higher cathode voltage. We believe that the electrochemically induced sol-gel deposition is a general method for the growth of 1D nanostructures of a variety of inorganic oxides (single or complex) materials in small-pore templates.

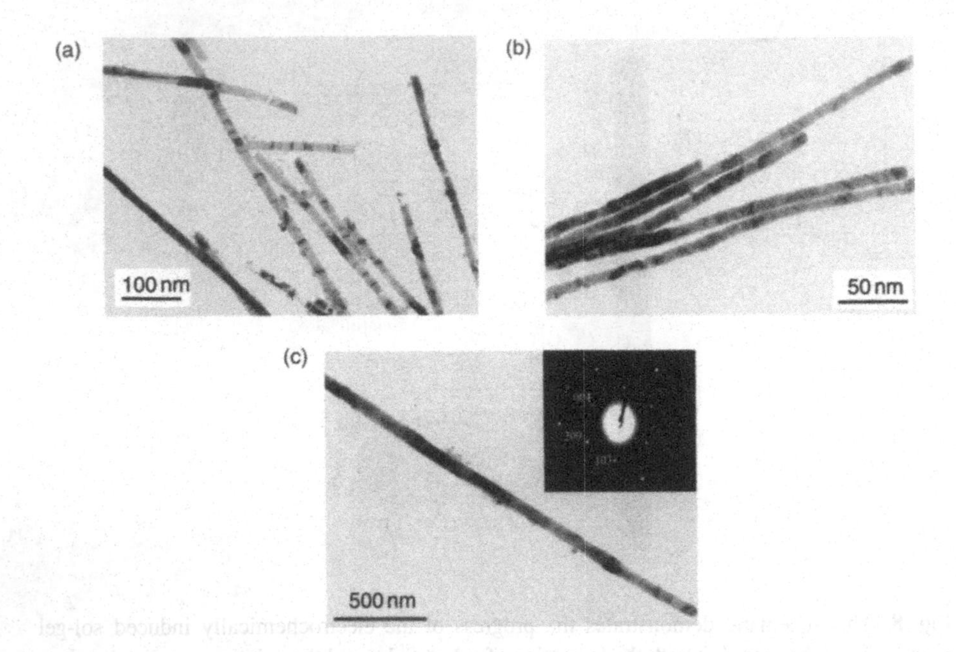

Fig. 9. TEM images of TiO_2 nanowires grown in AAO membranes with pore diameters of (a) 22, (b) 12, and (c) 50 nm by electrochemically sol-gel deposition. The inset plot in Fig. 9(c) gives the corresponding selected area diffraction pattern of the single nanowire in Fig. 9(c) [84].

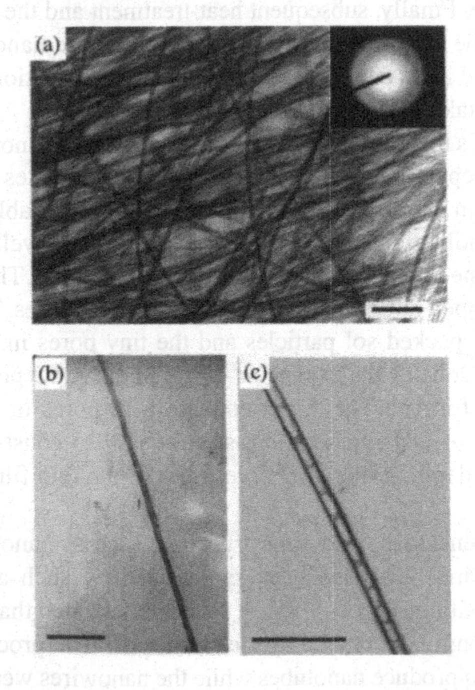

Fig. 10. TEM images of silica nanostructures deposited in 20 nm AAO membranes at different cathode voltages: (a) −1.25, (b) −1.10, and (c) −0.95 V. The scale bars are 100 nm [85].

3.4. Chemical bath deposition (CBD)

CBD involves dispersion of the reactants into the pores of the template membrane and in situ chemical reaction deposition of the desired semiconductors in the inner pores. There are few other reported about the synthesis of one-deminsional nanostrc-tures in AAO membranes in aqueous solutions with methods other than electrochem-ical synthesis. The reason usually is that the in situ formed nanoparticles hinder the dispersion of reactants into the inner part of the pores. For example, using aqueous solutions of Cd^{2+} and H_2S or Na_2S as the reactants, only low aspect ratio nanoparticle were obtained. Li et al. have developed methods by which CdS and other metal sul-fide semiconductor nanowires can be deposited from aqueous solutions into the pores of the AAO template [86]. This method involves applying TAA as the precursor of H_2S, which can gradually liberate H_2S in aqueous solution, and react with Cd^{2+} slowly. First, the membranes were immersed into a cadmium acetate aqueous solution under reduced pressure in a flask. Then, the flask was exposed to the atmosphere to

(a)

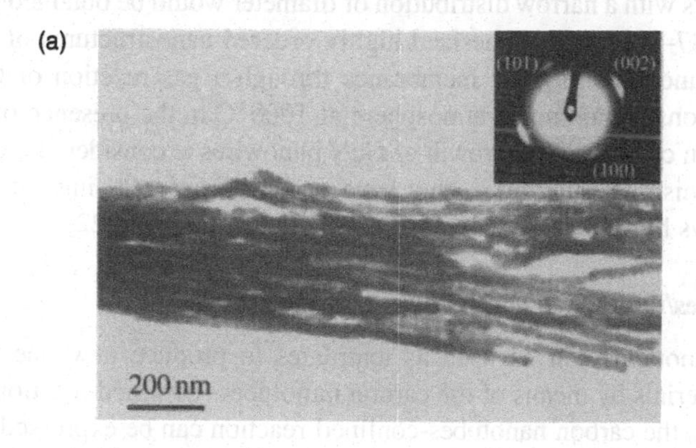

(b)

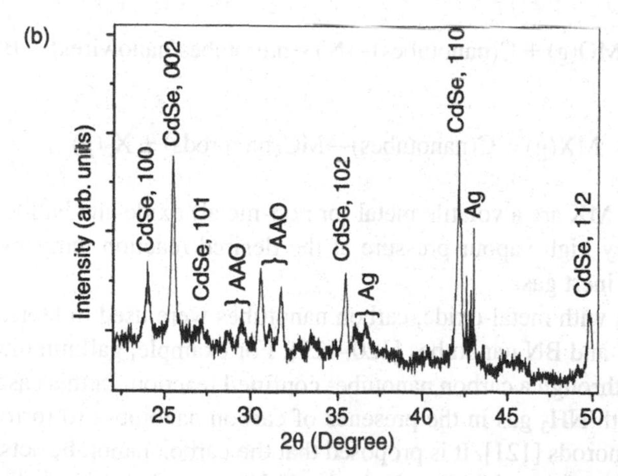

Fig. 11. (a) TEM image and (b) X-ray diffractogram of the CdSe nanowires grown in an AAO membrane with 20 nm pore diameters by dc electrodeposition. The inset plot in Fig. 11(a) is the corresponding selected area diffraction pattern of the nanowires, which can be indexed as the hexagonal CdSe crystal structure [39].

pour the solution into the pores due to the pressure differentiation. After the membranes were dried and washed, the TAA aqueous solution was poured into the membrane similarly. Finally, the membrane was dried very slowly to ensure the CdS nanocrystals could grow well.

3.5. Chemical vapor deposition (CVD)

CVD synthesis typically involves laser ablation or thermal evaporation of the desired metal containing semiconductor targets through a vapor-liquid-solid (VSL) growth mechanism or gas reactions catalyzed on the metal surface. This approach usually results high-purity single crystal semiconductor nanowires [17–24]. Commonly, the template assisted CVD synthesis of semiconductor nanowires begins with the formation of metal nanoparticles within the pores of the template membranes. Then, the membranes containing metal nanoparticles are placed in a high-temperature furnace and a vapor or reactant gases are passed into the pores of the membranes to deposit the desired nanowires.

Applying CVD techniques to template synthesis, highly ordered single crystal nanowire arrays with a narrow distribution of diameter would be obtained [22, 87–90]. Zhang et al. [87–90] have synthesized highly ordered nanostructures of single crystalline GaN nanowires in AAO membranes through a gas reaction of Ga_2O or Ga vapor with a constant ammonia atmosphere at 1000 °C in the presence of nano-sized metallic indium catalysis. The growth of GaN nanowires is considered through a VLS growth mechanism. Furthermore, they have produced ordered single crystal silicon nanowire arrays by catalytic decomposition of silane molecule [22].

3.6. Nanotubes/nanowires confined reaction synthesis

Carbon nanotubes can be used as templates to produce new one dimensional nanoscale materials by means of the carbon nanotubes- confined reaction [120–124]. In this method, the carbon nanotubes-confined reaction can be expressed as [121]:

$$\mathrm{A(g) + MO(g) + C(nanotubes) \rightarrow NTs(nanotubes/nanowires) + B(g)} \tag{5}$$

Or [120]

$$\mathrm{MX(g) + C(nanotubes) \rightarrow MC(nanorods) + X_2(g)} \tag{6}$$

where MO and MX are a volatile metal (or non-metal) oxide and halide, respectively, with a relatively high vapour pressure at the desired reaction temperature; A is the reaction gas or inert gas.

By reaction with metal-oxide, carbon nanotubes were used to fabricate SiC, WC, GaN, TiC, AlN and BN nanotubes [120–125]. For example, gallium nitride nanorods were prepared through a carbon nanotube- confined reaction. In this case, Ga_2O vapor was reacted with NH_3 gas in the presence of carbon nanotubes to form wurtzite gallium nitride nanorods [121]. It is proposed that the carbon nanotube acts as a template to confine the reaction, which results in the GaN nanorods having a diameter similar to that of the original nanotubes (equation 7).

$$\mathrm{MO(g) + C(nanotubes) + NH_3 \rightarrow MN(nanorods) + H_2O + CO + H_2} \tag{7}$$

4. Structural Characterization of the Template-Synthesized SNWs

Using the template synthesis strategies, a variety of SNW materials have been fabricated. These nanowires are size monodisperse and have a highly ordered arrangement in morphology, characterized by SEM and AFM [22, 39–48, 63]. However, crystal structure of the semiconductor nanowires, e.g., single crystal, polycrystalline or amorphous, plays a key role in nanodevices and other applications. This section will discuss the structural characterization of SNWs synthesized by template assisted growth strategies. Electron diffraction, X-ray diffraction and HREM have been employed to determine the crystal structure of the template-synthesized SNWs.

One important feather is that the SNWs prepared by DCED from a non-aqueous bath mostly crystallized as a uniform orientation, i.e. [001], as well as highly ordered arrangement in morphology. Fig. 11(a) shows a TEM image of a bundle of 20 nm diameter CdSe nanowires prepared by dc electrodeposition from a non-aqueous electrolyte bath [39]. These CdSe nanowires have a uniform width and are nearly paralleled to each other. The electron diffraction pattern taken from these nanowires is shown in the inset on the upper right of the micrograph, which are respectively corresponded to (002), (101) and (103) diffraction plane of the hexagonal CdSe single crystal. The diffraction pattern with somewhat dispersed and elongated spots implies that the nanowires grow with dominant direction and uniform crystal structure.

The X-ray diffractogram of the CdSe nanowire arrays is shown in Fig. 11(b), where the diffraction peaks could be assigned to CdSe, Ag and AAO without elemental Se and Cd. These XRD data indicate that the nanowires have diffraction patterns corresponding to the hexagonal phase of CdSe (ASTM standard 8-459). The

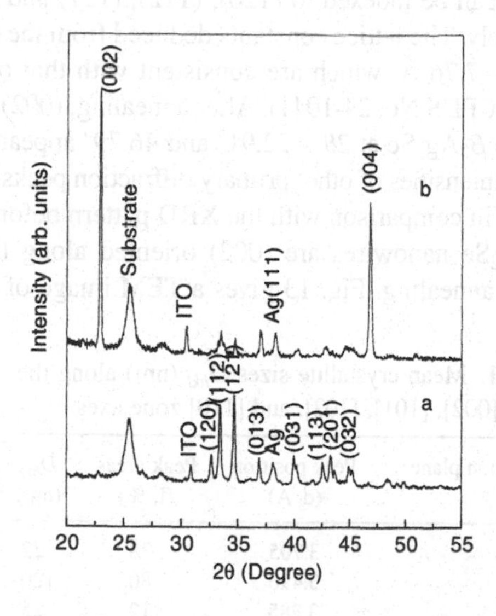

Fig. 12. X-ray diffractogram of the Ag_2Se nanowires embedded in the AAO template with pore diameters of about 50 nm prepared by dc electrodeposition: (a) as-deposited, (b) annealed in Ar atmosphere [48].

interplanar diffraction spacing (d_{hkl}) of CdSe nanowires with diameters of 20 nm differ slightly from those reported for polycrystalline CdSe. The relative intensity of the 002 diffraction peak, which corresponds to interplane distances d = 3.49−3.51(Å), is greater than that of the polycrystalline CdSe powder. This fact indicates that the c-axis of hexagonal crystals is preferentially aligned along the direction normal to the substrate rather than oriented randomly. Furthermore, the dimensions of the crystallites of the CdSe nanowires were estimated from the widths of the major diffraction peaks observed in Fig. 11(b) through the Scherrer formula

$$D_{hkl} = k\lambda/(\Delta_{hkl} \cos\theta) \quad (8)$$

where D_{hkl} is the linear dimension of the coherent diffracting domain along a direction normal to the diffraction plane (*hkl*), λ is X-ray wavelength in angstroms (1.5405 Å in this case), k is a crystal constant (0.9), θ is the angle of reflection of the peak, and is the corrected full width at half-maximum (fwhm) of the peak in radians, which was calculated as the square root of the difference between the squares of the sample's and the reference's fwhm. The dimensions of the CdSe crystallites calculated from the widths of the XRD reflections in Fig. 11(b) are listed in Table 1. These results show that the crystallite domains of the CdSe nanowires are c-axis oriented and the crystallite dimension in the radial direction is very close to the pore sizes of the template used. The aspect ratio of 5 : 1 calculated from the crystallite dimensions along [002] and [100] axes are obtained. The similar crystalline orientations have also observed on other semiconductor nanowires prepared by dc electrodeposition from non-aqueous baths [44, 45].

However, in aqueous baths, the deposited nanowires are randomly oriented [47, 48]. Fig. 12 presents the XRD patterns of the nanowires embedded in the AAO template. For the as-deposited nanowires, the diffraction peaks at $2\theta = 32.66$, 33.40, 34.68 and 36.88° etc. can be indexed to (120), (112), (121) and (013) of orthorhombic β-Ag_2Se respectively. The lattice constants deduced from the diffraction data were $a = 4.46$, $b = 6.96$, $c = 7.76$ Å, which are consistent with that of the polycrystalline bulk Ag_2Se standard (JCPDS No. 24-1041). After annealing, (002) and (004) diffraction peaks of orthorhombic β-Ag_2Se at $2\theta = 22.91°$ and 46.79° appeared with strong intensities. Meanwhile, the intensities of other primary diffraction peaks, e.g., (112) and (121) are almost suppressed in comparison with the XRD pattern before annealed. This fact indicates that the Ag_2Se nanowires are (002) oriented along the direction vertical to the wire axes after annealing. Fig. 13 gives a TEM image of an individual 50 nm

Table 1. Mean crystallite sizes D_{hkl} (nm) along the [100], [002], [101], [102], and [110] zone axes

Diffraction plane (*hkl*)	Peak position (d, A)	Peak areas (I, %)	D_{hkl} (nm)
(100)	3.705	38	22
(002)	3.496	80	120
(101)	3.285	12	15
(102)	2.532	35	58
(110)	2.142	100	30

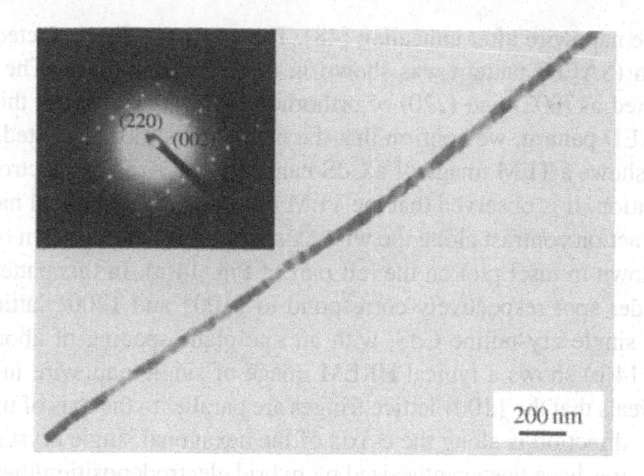

Fig. 13. Typical TEM image of a single nanowire after annealing, (inset) the SAED pattern recorded from this nanowire [48].

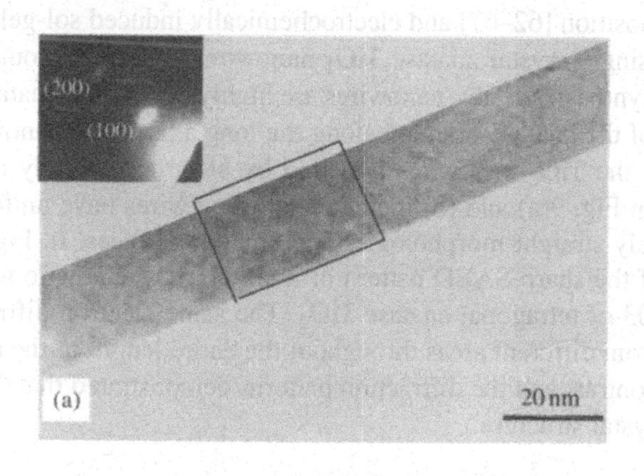

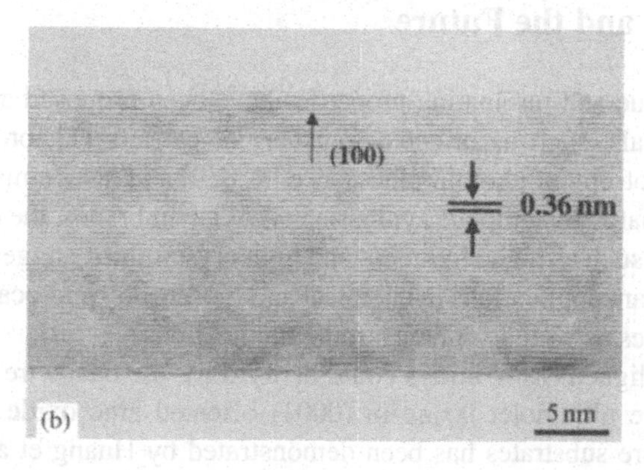

Fig. 14. (a) TEM image of single CdS nanowire with a diameter of 20 nm. (b) HREM image of single nanowire. The inset plot in Fig. 14(a) is the electron diffraction pattern taken from the single nanowire [35].

diameter Ag_2Se nanowire after annealing [48]. The corresponding selected area electron diffraction (SAED) pattern was shown in the inset of Fig. 13. The diffraction spots are indexed as (002) and (220) of orthorhombic β-Ag_2Se. From this morphology and its SAED pattern, we confirm that the nanowire is [002] oriented.

Fig. 14(a) shows a TEM image of a CdS nanowire prepared by electrochemically induced deposition. It is observed that the TEM image of the individual nanowire has a uniform diffraction contrast along the wire axis. The diffraction pattern of the single nanowire is shown in inset plot on the left part of Fig. 14(a). In this pattern, the first and second order spot respectively correspond to {100} and {200} lattice plane of the hexagonal single crystalline CdS, with an interplane spacing of about 0.36 and 0.18 nm. Fig. 14(b) shows a typical HREM image of single nanowire in Fig. 14(a). This image reveals that the {100} lattice fringes are parallel to the axis of the nanowire and the growth direction is along the c-axis of thc hexagonal single crystalline CdS.

TiO_2 nanowires have been synthesized by hybrid electrodeposition/thermal oxidation [58], sol-gel deposition [62–67] or electrochemically induced sol-gel method [84]. The Ti nanowires can be transferred to a TiO_2/Ti coaxial structure by thermally oxidizing in air at room temperature [58]. However, highly crystalline tetragonal anatase TiO_2 nanowires are formed by thermally oxidizing at 400 °C for 8 hours. Both sol-gel deposition [62–67] and electrochemically induced sol-gel strategies [84] have produced single crystal anatase TiO_2 nanowires. Martin's group demonstrated that the sol-gel synthesized TiO_2 nanowires are highly crystalline anatase-phase TiO_2, with the c axis of the anatase oriented along the long axis of the nanowires [62]. The TEM images of the TiO_2 nanowires prepared by electrochemically induced sol-gel process shown in Fig. 9(a) and (b) indicate that these wires have uniform diffraction contrast, relatively straight morphologies and smooth surfaces. In Fig. 9(c), the diffraction spots of the sharp SAED pattern of a 40nm TiO_2 nanowire were indexed as 004, 200 and 103 of tetragonal anatase TiO_2. The same electron diffraction patterns were obtained from different areas throughout the entire length of the nanowire. Both the diffraction contrast and the diffraction patterns demonstrated that these nanowires have a single crystal structure.

5. Properties and the Future

A lot of unique and fascinating properties have been proposed and demonstrated for SNW materials, such as oriented transport of carriers [1], order-of-magnitude polarization anisotropy of photoluminescence [4, 8, 9] and laser emission [9, 10]. For SNW arrays prepared by template synthesis in AAO membranes, the characteristics of these nanowires, such as high alignment and high crystallinity, suggest that they could have very important applications in chemical and biosensors, photocatalysis, and solar conversion devices as well as optoelectronic applications.

First, these aligned SNW arrays could be used for the nanowire lasers. Recently, room-temperature ultraviolet lasing in (0001) oriented zinc oxide nanowire arrays grown on sapphire substrates has been demonstrated by Huang et al. [9], which are expected to serve as miniaturized light sources for microanalysis, information storage, and optical computing. In this case, these single- crystalline, well-faceted nanowires were considered as natural resonance cavities, in which one end is the epitaxial interface between the sapphire and ZnO, whereas the other end is the sharp (0001) plane of the ZnO nanocrystals. The giant oscillator strength effect, which can occur in high-quality nanowire crystals with dimensions larger than the exciton Bohr radius but smaller than the optical wavelength, enables the excitonic stimulated emission in these nanowire arrays. Also, due to their highly crystalline and well-oriented properties, SNW arrays by template assisted growth, such as II-VI compound semiconductors (ZnO, CdS, CdSe, CdTe, etc.) and III-V semiconductors (GaN and InP), are good candidates for excitonic laser action.

Second, these nanowire arrays could also be used as luminescent emitters for the planar displayer. Stronger luminescent emission from a variety of template assisted semiconductor nanowires have also been demonstrated [56, 65, 69, 87–90]. Due to the high density and highly regular arrangement of the nanowire arrays, the size of the emitter would be greatly decreased and the amount of the emitters in the unit would be largely increased, which is expected to increase the resolution and improve the stability of the displayer.

Third, these nanowire arrays could also be used as a new class of semiconductor electrodes for photovoltaics [12]. One of the important applications of these nanowire electrodes could be the "Grätzel solar cell". Conventional cell is based on nanoporous and nanocrystalline TiO_2 on which a monolayer of a photosensitizer is absorbed. Replacing nanoparticles with nanowires could remove one of the electron transport bottlenecks occurring in the conventional Grätzel cell, i.e., the tunneling process between neighboring oxide nanocrystals. With nanowire electrodes, electron can be transport directly to the conducting substrate without going through the multiple steps of the tunneling process, which is expected to be able to greatly decrease the probability of recombination across large heterogeneous interface and improve the overall efficiency of the solar cell. In addition, it has been demonstrated that semiconductor CdSe nanorods can be used to fabricate readily processed and efficient hybrid solar cells together with polymers [16]. As the aspect ratio of the nanorods increases from 1 to 10, the charge transport must improve substantially to yield an external quantum efficiency enhancement by a factor of approximately 3.

Fourth, these nanowire arrays could be used as a photocatalyst for the light-induced redox process. The potocatalytic reaction involves absorption of a UV photon by semiconductor (i.e., TiO_2 and SiO_2) to produce an electron-hole pair and then trapping of the electron or hole by scavengers or surface defects. The valence band holes are powerful oxidants while the conduction band electrons are good reductants. For example, the generated holes have been used to oxidize organic molecules for environmental remediation applications. Also, owing to the increased surface area of the SWN arrays, the decomposition rate of the organic molecules correspondingly increases. Lakshmi et al. [62] have investigated the decomposition of salicylic acid over time on an array of immobilized TiO_2 fibres with exposure to sunlight. They observed a marked increase in decomposition rate of salicylic acid for the template-synthesized TiO_2 fibres.

6. Conclusion

Template assisted growth of SNWs has been well developed and a broad range of semiconductor materials have been fabricated in the recent years. These template-synthesized SWNs are size monodisperse and have a highly ordered arrangement in morphology. The characteristics of these nanowires, such as high alignment and high crystallinity, suggest that they could have very important applications in nanosensor, nanowire laser, photocatalysis, solar conversion devices, and planar displayers.

Acknowledgement

This work was supported by the Major State Basic Research Development Program (Grant No. 2000077503) and the National Science Foundation of China.

References

1. J. T. Hu, T. W. Odom and C. M. Lieber, *Accounts Chem. Res.* **32** (1999) 435.
2. X. F. Duan, Y. Huang, Y. Cui, J. F. Wang and C. M. Lieber, *Nature* **409** (2001) 66.
3. Y. Huang, X. F. Duan, Y. Cui, L. J. Lauhon, K. H. Kim and C. M. Lieber, *Science* **294** (2001) 1313.
4. M. S. Gudiksen, L. J. Lauhon, J. Wang, D. C. Smith and C. M. Lieber, *Nature* **415** (2002) 617.
5. J. F. Wang, M. S. Gudiksen, X. F. Duan, Y. Cui and C. M. Lieber, *Science* **293** (2001) 1455.
6. Y. Cui, Q. Q. Wei, H. K. Park and C. M. Lieber, *Science* **293** (2001) 1289.
7. Y. Cui and C. M. Lieber, *Science* **291** (2001) 851.
8. X. F. Duan, Y. Huang, Y. Cui, J. F. Wang and C. M. Lieber, *Nature* **409** (2001) 66.
9. M. H. Huang, S. Mao, H. Feick, H. Q. Yan, Y. Y. Wu, H. Kind, E. Weber, R. Russo and P. D. Yang, *Science* **292** (2001) 1897.
10. J. C. Johnson, H. Q. Yan, R. D. Schaller, L. H. Haber, R. J. Saykally and P. D. Yang, *J. Phys. Chem. B.* **105** (2001) 11387.
11. H. Kind, H. Q. Yan, B. Messer, M. Law and P. D. Yang, *Adv. Mater.* **14** (2002) 158.
12. Y. Y. Wu, H: Q. Yan and P. D. Yang, *Top. Catal.* **19** (2002) 197.
13. Z. W. Pan, Z. R. Dai and Z. L. Wang, *Science* **291** (2001) 1947.
14. Z. L. Wang, R. P. Gao, P. Poncharal, W. A. de Heer, Z. R. Dai and Z. W. Pan, *Mat. Sci. Eng. C* **16** (2001) 3.
15. Z. L. Wang, R. P. Gao, Z. W. Pan and Z. R. Dai, *Adv. Eng. Mater.* **3** (2001) 657.
16. W. U. Huynh, J. J. Dittmer and A. P. Alivisatos, *Science* **295** (2002) 2425.
17. A. M. Morales and C. M. Lieber, *Science* **279** (1998) 208.
18. X. F. Duan and C. M. Lieber, *Adv. Mater.* **12** (2000) 298.
19. X. F. Duan and C. M. Lieber, *J. Am. Chem. Soc.* **122** (2000) 188.
20. Y. Y. Wu and P. D. Yang, *J. Am. Chem. Soc.* **123** (2001) 3165.
21. Z. W. Pan, Z. R. Dai, C. Ma and Z. L. Wang, *J. Am. Chem. Soc.* **124** (2002) 1817.
22. X. Y. Zhang, L. D. Zhang, G. W. Meng, G. H. Li, N. Y. Jin-Phillipp and F. Phillipp, *Adv. Mater.* **13** (2001) 1238.
23. Y. W. Wang, G. W. Meng, L. D. Zhang, C. H. Liang and J. Zhang, *Chem. Mater.* **14** (2002) 1773.
24. C. H. Liang, G. W. Meng, Y. Lei, F. Phillipp and L. D. Zhang, *Adv. Mater.* **13** (2001) 1330.
25. D. P. Yu, C. S. Lee, I. Bello, X. S. Sun, Y. H. Tang, G. W. Zhou, Z. G. Bai, Z. Zhang and S. Q. Feng, *Solid State Commun.* **105** (1998) 403.
26. S. T. Lee, N. Wang, Y. F. Zhang and Y. H Tang, *MRS Bull.* **24** (1999) 36.
27. Z. G. Bai, D. P. Yu, H. Z. Zhang, Y. Ding, Y. P. Wang, X. Z. Gai, Q. L. Hang, G. C. Xiong and S. Q. Feng, *Chem. Phys. Lett.* **303** (1999) 311.

28. D. P. Yu, Q. L. Hang, Y. Ding, H. Z. Zhang, Z. G. Bai, J. J. Wang, Y. H. Zou, W. Qian, G. C. Xiong and S. Q. Feng, *Appl. Phys. Lett.* **73** (1998) 3076.
29. C. R. Martin, *Science* **266** (1994) 1961.
30. C. R. Martin, *Acc. Chem. Res.* **28** (1995) 61.
31. J. C. Hulteen and C. R. Martin, *J. Chem. Mater.* **7** (1997) 1075.
32. C. R. Martin, *Chem. Mater.* **8** (1996) 1739.
33. D. Routkevitch, T. Bigioni, M. Moskovits and J. M. Xu, *J. Phys. Chem.* **100** (1996) 14037.
34. T. M. Whitney, J. S. Jiang, P. C. Searson and C. L. Chien, *Science* **261** (1993) 1316.
35. Dongsheng Xu, Yajie Xu, Dapeng Chen, Guolin Guo, Linlin Gui and Youqi Tang, *Adv. Mater.* **12** (2000) 520.
36. J. D. Klein et al., *Chem. Mater.* **5** (1993) 902.
37. S. A. Sapp, B. B. Lakshmi and C. R. Martin, *Adv. Mater.* **11** (1999) 402.
38. Y. J. Xu, D. S. Xu, D. P. Chen, G. L. Guo and C. J. Li, *Acta Physico-Chemica Sinica* **15** (1999) 577.
39. D. S. Xu, X. S. Si, G. L. Guo, L. L. Gui and Y. Q. Tang, *J. Phys. Chem. B* **104** (2000) 5061.
40. D. S. Xu, Y. J. Xu, D. P. Chen, G. L. Guo, L. L. Gui and Y. Q. Tang, *Chem. Phys. Lett.* **325** (2000) 340.
41. D. S. Xu, D. P. Chen, Y. J. Xu, X. S. Si, G. L. Guo, L. L. Gui and Y. Q. Tang, *Pure and Appl. Chem.* **72** (2000) 127.
42. A. L. Prieto, M. S. Sander, M. S. Martin-Gonzalez, R. Gronsky, T. Sands and A. M. Stacy, *J. Am. Chem. Soc.* **123** (2001) 7160.
43. M. S. Sander, A. L. Prieto, R. Gronsky, T. Sands and A. M. Stacy, *Adv. Mater.* **14** (2002) 665.
44. D. S. Xu, Y. G. Guo, D. P. Yu, G. L. Guo and Y. Q. Tang, *J. Mater. Res.* **17** (2002) 1711.
45. R. Z. Chen, D. S. Xu, G. L. Guo and Y. Q. Tang, *J. Mater. Chem.* **12** (2002) 2435–2438.
46. X. S. Peng, G. W. Meng, J. Zhang, L. X. Zhao, X. F. Wang, Y. W. Wang and L. D. Zhang, *J. Phys. D-Appl. Phys.* **34** (2001) 3224.
47. X. S. Peng, J. Zhang, X. F. Wang, Y. W. Wang, L. X. Zhao, G. W. Meng and L. D. Zhang, *Chem. Phys. Lett.* **343** (2001) 470.
48. R. Z. Chen, D. S. Xu, G. L. Guo and L. L. Gui, *J. Electrochem. Soc.* **150** (2003) G183.
49. R. Z. Chen, D. S. Xu, G. L. Guo and L. L. Gui (submitted).
50. R. Z. Chen, D. S. Xu, G. L. Guo and Y. Q. Tang (in the preparation).
51. D. Routkevitch, T. L. Haslett, L. Ryan, T. Bigioni, C. Douketis and M. Moskovits, *Chem. Phys.* **210** (1996) 343.
52. J. L. Hutchison, D. Routkevitch, M. Moskovits and R. R. Nayak, *Inst. Phys. Conf. Ser* **157** (1997) 389.
53. J. S. Suh and J. S. Lee, *Chem. Phys. Lett.* **281** (1997) 384.
54. M. J. Zheng, G. H. Li, X. Y. Zhang, S. Y. Huang, Y. Lei and L. D. Zhang, *Chem. Mater.* **13** (2001) 3859.
55. M. J. Zheng, L. D. Zhang, G. H. Li, X. Y. Zhang and X. F. Wang, *Appl. Phys. Lett.* **79** (2001) 839.
56. Y. Li, G. W. Meng, L. D. Zhang and F. Phillipp, *Appl. Phys. Lett.* **76** (2000) 2011.
57. D. S. Xu, Y. X. Yu, G. L. Guo and Y. Q. Tang (in the preparation).
58. M. Zhen, D. S. Xu, Y. X. Yu, G. L. Guo and Y. Q. Tang (unpublished).
59. B. Gates, Y. Y. Wu, Y. D. Yin, P. D. Yang and Y. N. Xia, *J. Am. Chem. Soc.* **123** (2001) 1500.
60. R. Z. Chen, D. S. Xu, G. L. Guo and L. L. Gui (in the preparation).
61. R. Z. Chen, D. S. Xu, G. L. Guo and L. L. Gui (in the preparation).
62. B. B. Lakshmi, P. K. Dorhout and C. R. Martin, *Chem. Mater.* **9** (1997) 857.
63. B. B. Lakshmi, C. J. Patrissi and C. R. Martin, *Chem. Mater.* **9** (1997) 2544.
64. P. Hoyer, *Adv. Mater.* **8** (1994) 857.

65. Y. Lei, L. D. Zhang, G. W. Meng, G. H. Li, X. Y. Zhang, C. H. Liang and S. X. Wang, *Appl. Phys. Lett.* **78** (2001) 1125.
66. M. Zhang, Y. Bando and K. Wada, *J. Mater. Sci.* **20** (2001) 167.
67. S. Z. Chu, K. Wada, S. Inoue and S. Todoroki, *Chem. Mater.* **14** (2002) 266.
68. H. Nakamura and Y. Matsui, *J. Am. Chem. Soc.* **117** (1995) 2651.
69. M. Zhang, E. Ciocan, Y. Bando, K. Wada, L. L. Cheng and P. Pirouz, *Appl. Phys. Lett.* **80** (2002) 491.
70. M. Zhang, Y. Bando and K. Wada, *J. Mater. Res.* **15** (2000) 387.
71. B. Cheng and E. T. Samulski, *J. Mater. Chem.* **11** (2001) 2901.
72. C. R. Gong, D. R. Chen, X. L. Jiao and Q. L. Wang, *J. Mater. Res.* **12** (2002) 1844.
73. Z. Wang and H. L. Li, *Appl. Phys. A-Mater. Sci. Proc.* **74** (2002) 201.
74. N. C. Li and C. R. Martin, *J. Electrochem. Soc.* **148** (2001) A164.
75. H. Q. Cao, Y. Xu, J. M. Hong, H. B. Liu, G. Yin, B. L. Li, C. Y. Tie and Z. Xu, *Adv. Mater.* **13** (2001) 1393.
76. Y. K. Zhou and H. L. Li, *J. Mater. Chem.* **12** (2002) 681.
77. B. A. Hernandez, K. S. Chang, E. R. Fisher and P. K. Dorhout, *Chem. Mater.* **14** (2002) 480.
78. Y. K. Zhou, C. M. Shen and L. H. Li, *Solid State Ion.* **146** (2002) 81.
79. M. Zhang, Y. Bando and K. Wada, *J. Mater. Res.* **16** (2001) 1408.
80. C. Hippe, M. Wark, E. Lork and G. Schulz-Ekloff, *Microporous Mesoporous Mat.* **31** (1999) 235.
81. H. Q. Cao, Z. Xu, X. W. Wei, X. Ma and Z. L. Xue, *Chem. Commun.* (2001) 541.
82. S. J. Limmer, S. Seraji, M. J. Forbess, Y. Wu, T. P. Chou, C. Nguyen and G. Z. Cao, *Adv. Mater.* **13** (2001) 1269.
83. S. J. Limmer, S. Seraji, Y. Wu, M. J. Forbess, T. P. Chou, C. Nguyen and G. Z. Cao, *Adv. Funct. Mater.* **12** (2002) 59.
84. Z. Miao, D. S. Xu, J. H. Ouyang, G. L. Guo, X. S. Zhao and Y. Q. Tang, *Nano Lett.* **2** (2002) 717.
85. Z. Miao, D. S. Xu, G. L. Guo and Y. Q. Tang (in the preparation).
86. Y. Li, D. S. Xu, Q. M. Zhang, F. Z. Huang, D. P. Chen, Y. J. Xu, G. L. Guo and Z. N. Gu, *Chem. Mater.* **11** (1999) 3433.
87. J. Zhang, L. D. Zhang, X. F. Wang, C. H. Liang, X. S. Peng and Y. W. Wang, *J. Chem. Phys.* **115** (2001) 5714.
88. G. S. Cheng, S. H. Chen, X. G. Zhu, Y. Q. Mao and L. D. Zhang, *Mater. Sci. Eng. A-Struct. Mater. Prop. Microstruct. Process.* **286** (2000) 165.
89. G. S. Cheng, L. D. Zhang, S. H. Chen, Y. Li, L. Li, X. G. Zhu, Y. Zhu, G. T. Fei and Y. Q. Mao, *J. Mater. Res.* **15** (2000) 347.
90. G. S. Cheng, L. D. Zhang, Y. Zhu, G. T. Fei, L. Li, C. M. Mo and Y. Q. Mao, *Appl. Phys. Lett.* **75** (1999) 2455.
91. F. Keller, M. S. Hunter and D. L. Robinson, *J. Electrochem. Soc.* **100** (1953) 411.
92. H. Masuda and K. Fukuda, *Science* **268** (1995) 1466.
93. H. Masuda, F. Hasegawa and S. Ono, *J. Electrochem. Soc.* **144** (1997) L127.
94. O. Jessensky, F. Muller and U. Gosele, *Appl. Phys. Lett.* **72** (1998) 1173.
95. A. P. Li, F. Muller, A. Birner, K. Nielsch and U. Gosele, *J. Appl. Phys.* **84** (1998) 6023.
96. J. S. Suh and J. S. Lee, *Appl. Phys. Lett.* **75** (1999) 2047.
97. R. L. Fleisher, P. B. Price and R. M. Walker, *Neclear Tracks in Solids*, University of California Press, Berkeley, CA (1975).
98. A. Corma, *Chem. Rev.* **97** (1997) 2373.
99. P. D. Yang, D. Y. Zhao, D. I. Margolese, B. F. Chmelka and G. D. Stucky, *Nature* **396** (1998) 152.

100. Z. T. Zhang, Y. Han, F. S. Xiao, S. L. Qiu, L. Zhu, R. W. Wang, Y. Yu, Z. Zhang, B. S. Zou, Y. Q. Wang, H. P. Sun, D. Y. Zhao and Y. Wei, *J. Am. Chem. Soc.* **123** (2001) 5014.
101. C. M. Yang, H. S. Sheu and K. J. Chao, *Adv. Funct. Mater.* **12** (2002) 143.
102. J. E. Beck, J. C. Vartuli, W. J. Roth, M. E. Leonowicz, C. T. Kresge, K. D. Schmitt, C. T.-W. Chu, D. H. Olson, E. W. Sheppard, S. B. McCullen, J. B. Higgins and J. L. Schlenker, *J. Am. Chem. Soc.* **114** (1992) 10834.
103. R. J. Tonucci, B. L. Justus, A. J. Campillo and C. E. Ford, *Science* **258** (1992) 783.
104. L. Sun, C. L. Chien and P. C. Searson, *J. Mater. Soc.* **35** (2000) 1097.
105. L. Sun, C. L. Chien and P. C. Searson, *Appl. Phys. Lett.* **74** (1999) 2803.
106. R. Schollhorn, *Chem. Mater.* **8** (1996) 1747.
107. T. Thurn-Albrecht, J. Schotter, C. A. Kastle, N. Emley, T. Shibauchi, L. Krusin-Elbaum, K. Guarini, C. T. Black, M. T. Tuominen and T. P. Russell, *Science* **290** (2000) 2126.
108. M. J. Tierney and C. R. Martin, *J. Phys. Chem.* **93** (1989) 2878.
109. C. J. Brumlik and C. R. Martin, *J. Am. Chem. Soc.* **113** (1991) 3174.
110. C. J. Brumlik, C. R. Martin and K. Tokuda, *Anal. Chem.* **64** (1992) 1201.
111. C. A. Foss Jr., G. L. Hornyak, J. A. Stockert and C. R. Martin, *J. Phys. Chem.* **96** (1992) 7497.
112. C. A. Foss Jr., G. L. Hornyak, J. A. Stockert and C. R. Martin, *Adv. Mater.* **5** (1993) 135.
113. C. A. Foss Jr., G. L. Hornyak, J. A. Stockert and C. R. Martin, *J. Phys. Chem.* **98** (1994) 2963.
114. A. J. Yin, J. Li, W. Jian, A. J. Bennett and J. M. Xu, *Appl. Phys. Lett.* **79** (2001) 1039.
115. P. C. Searson, R. C. Cammarata and C. L. Chien, *J. Electronic Mater.* **24** (1995) 955.
116. G. Sauer, G. Brehm, S. Schneider, K. Nielsch, R. B. Wehrspohn, J. Choi, H. Hofmeister and U. Gosele, *J. Appl. Phys.* **91** (2002) 3243.
117. Y. W. Wang, L. D. Zhang, G. W. Meng, X. S. Peng, Y. X. Jin and J. Zhang, *J. Phys. Chem.* **106** (2002) 2502.
118. C. J. Brinker and G. W. Scherer, *Sol-gel Science*, Academic Press Inc., New York (1990).
119. L. L. Hench and J. K. West, *Chem. Rev.* **90** (1990) 33
120. H. Dai, E. W. Wang, Y. Z. Lu, S. S. Fan and C. M. Lieber, *Nature* **375** (1995) 769.
121. W. Q. Han, S. S. Fan, Q. Q. Li and Y. D. Hu, *Science* **277** (1997) 1287.
122. A. Fukunaga, S. Y. Chu and M. E. McHenry, *J. Mater. Res.* **13** (1998) 2465.
123. W. Q. Han, Y. Bando, K. Kurashima and T. Sato, *Appl. Phys. Lett.* **73** (1998) 3085.
124. Y. J. Zhang, J. Zhu, Q. Zhang, Y. J. Yan, N. L. Wang and X. Z. Zhang, *Chem. Phys. Lett.* **317** (2000) 504.
125. Y. J. Zhang, J. Liu, R. R. He, Q. Zhang, X. Z. Zhang and J. Zhu, *Chem. Mater.* **13** (2001) 3899.
126. V. Damodara Das and D. Karunakaran, *J. Appl. Phys.* **68** (1990) 2105.
127. R. Xu, A. Husmann, T. F. Rosenbaum and M. L. Saboungi, *Nature* **390** (1997) 57.

Chapter 11

Wide Band-Gap Semiconductor Nanowires Synthesized by Vapor Phase Growth

D. P. Yu

School of Physics, Electron Microscopy Laboratory, and State Key Laboratory for Mesoscopic Physics, Peking University, Beijing 100871, P. R. China

1. Introduction

With the rapid development of the contemporary sciences and technologies into nanoscale regime, one needs to synthesize nanostructured materials, to addresses their peculiar physical and chemical properties related to the lower dimensionality, and what is more important is to explore their possible applications. Among those nanostructures, functional nanowire materials have stimulated intensive research interests from fundamental research to application community. A variety of one-dimensional nanostructured materials, such as silicon, germanium [1–6], GaAs, InAs [7, 8], gallium nitride [9], have been prepared. Fig. 1 shows an example that very pure silicon nanowires can be synthesized via simple physical evaporation [2].

Those 1-D structures often show novel physical properties that are different from their bulk [11–15]. Based on the particular structure characteristics and size effects, much progress has been made in nanodevice applications using semiconductor nanowires as the building blocks. For example, GaAs, InAs nanowires have found application in developing high speed field effect transistor (FET), or laser working at low-threshold current density and high gain [7, 8]. Recently, Si and other semiconductor nanowires (Si, GaN, InP, ……) have been used as the building blocks to construct random access memory, nano-sized LEDs, logical and computational circuits [16–22]. Wide band-gap semiconductor nanowires are of particular interests for their wide band-gap which allows the possibility to tune the optoelectronic devices working from infrared to ultraviolet range [23]. Preparation of wide band-gap quantum nanowires in a controllable way thus becomes a challenging task for material scientists, because those nanowires usually possess intriguing magnetic, electronic and nonlinear optical properties, which can find potential applications in optoelectronics, and nanoelectronics. For example, ZnO nanowires were also used to make nano-sized UV lasers [24]. A diverse of other wide-band gap semiconductor nanowires [25–34]

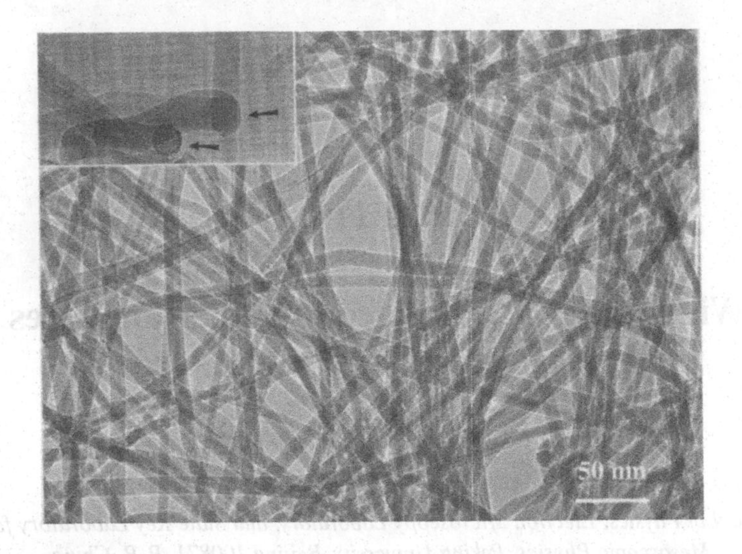

Fig. 1. TEM image showing the homogeneity and smoothness of the silicon nanowires prepared via vapor phase grown [1].

have been synthesized using different methods in recent years, such as physical vapor evaporation (PVD), chemical vapor deposition (CVD), laser ablation and other methods. It is the purpose of this chapter to review some of the recent publications on synthesis and characterization of the wide band-gap nanowires, such as Ga_2O_3, GaN, ZnO and ZnSe. As an addition, highly oriented silicon oxide nanowires and amorphous silicon nanowire films were also summarized. Because most of the above-mentioned nanowires were fabricated using the vapor phase growth, we will first start to give a brief introduction to this growth method.

2. Vapor Phase Evaporation

The vapor phase evaporation approach was used in the early year of 1960's for preparation of micrometer-sized whiskers, which were destined to study the theoretical strength of a single whisker [36]. An abundance of metallic (Ag, Al, Au, Cu, Fe, Ni, Ti, Zn, for example) and oxide whiskers were synthesized [37–39]. These whiskers were prepared either through vapor phase by simple physical sublimation of a source material, or through reduction of a volatile metal halide. In recent years, this method found wide application to prepare a variety of nanometer-sized wires (Si, Ge, ZnO, GeO_2, Ga_2O_3, GaSe) based on vapor phase evaporation [1, 4, 5, 11, 25, 27]. Such a growth process was usually conducted in a programmable tube furnace. The growth chamber was designed such a way that a proper temperature gradient can be obtained which is in favor of the growth of the nanowires, as is depicted schematically in Fig. 2.

The evaporated or reduced atomic species from the target materials transport through vapor phase towards the growth site of lower concentration along the temperature gradient, and subsequent nucleation and growth can occur automatically that is caused by random thermal/concentration fluctuations. The growth of nanowires

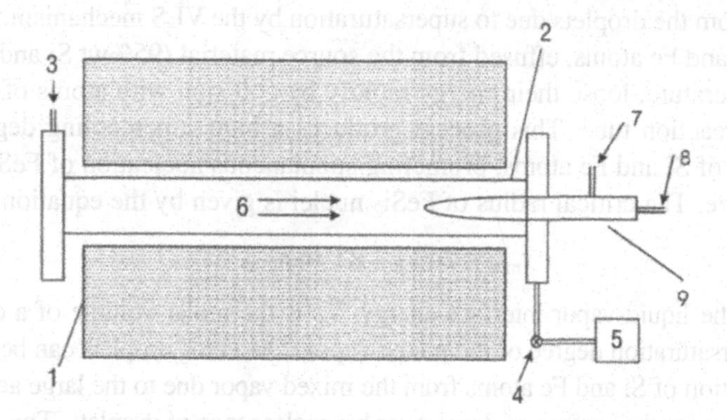

Fig. 2. Schematic depiction of the vapor phase growth system: 1-Tube furnace; 2-Quartz tube; 3-Gas inlet; 4-Needle valve; 5-To mechanical pump; 6-Central zone where source materials are placed; 7-Cooling water in; 8-Cooling water out; 9-Copper collector.

via physical sublimation are usually found involved with impurity elements (the catalysts). In the case of silicon nanowire growth, no nanowire growth can occur if catalysts like Fe, Ni, or Au elements were not added in the source materials, which is indicative of the participation of liquid phase in the whole growth process. The role of the added impurity elements is usually to form with the source material a low temperature eutectic liquid alloy phase, which will reduce considerably the energy barrier of crystalline precipitation via supersaturation.

The central questions in nano-sized wire growth are why matter prefers to grow unidirectionally, and how atoms and other atomic building blocks are stacked into wires of one-dimensionality. Conventional crystal growth was usually explained by the dislocation theory proposed by Frank [37], in which a screw dislocation terminate at the growth surface of the crystal, and the spiral steps around the dislocation core provide energetically favored sites where atoms can stack continuously at a low energy cost. This theory was extended by Sears [38, 39] and Price et al. to explain the preferred growth of whiskers. In their speculations, screw dislocations should emerge at the tips of whiskers, and the axial dislocations provide preferred growth sites, which can account for the unidirectional growth of whiskers. In practice, however, there were few reports on the observations of axial dislocations in whiskers, and in most cases it was revealed that the whiskers were dislocation-free.

Ellis and Wagner [40] proposed a novel model to account for the preferential unidirectional growth process of crystal whiskers (wires). In this model, both the vapor, liquid, and solid phases (VLS) were found involved in the whisker growth, and in particular, the importance of the involved impurity elements was emphasized in the whisker formation. It seems that the VLS model well explained many of the wire growth, especially in case of the existence of catalysts in the wire growth.

According to the VLS growth mechanism, the central idea is the participation of the liquid-forming agents. Another important key is the formation of nanosized catalyst droplets during the VLS growth at proper temperature, which can be chosen rationally from a phase diagram [2, 4] under consideration of the melting effect at nanosize. For the synthesis of Si nanowires, the growth process can be divided into two stages: the nucleation and the growth of the liquid droplets, and the growth of the nanowires from the droplets due to supersaturation by the VLS mechanism. In the first stage, the Si and Fe atoms, effused from the source material (95%wt Si and 5%wt Fe) at high temperature, loose their energy rapidly by collision with atoms of buffer gas (Ar) in the reaction tube. This process produces a high supercooling degree of the mixed vapor of Si and Fe atoms, promoting spontaneous nucleation of $FeSi_2$ droplets with nano-size. The critical radius of $FeSi_2$ nuclei is given by the equation:

$$r_{min} = 2\alpha V_L / RT \ln\sigma \qquad (1)$$

Where α is the liquid-vapor interface energy, V_L is the molar volume of a droplet and σ is the supersaturation degree of the mixed vapor. The $FeSi_2$ droplets can be formed by either absorption of Si and Fe atoms from the mixed vapor due to the large accommodation coefficient for the surface of droplets or by coalescence of droplets. The growth rate of droplets by coalescence process is much higher than that by absorption process. The second growth stage occurred when the liquid droplets become supersaturated by absorbing much more Si atoms from the vapor phase. This led to the growth of Si nanowires by VLS mechanism. Such process would continue until the droplets were carried out of the high temperature region by the gas flow and solidified. The nano particles are observed at the tips of the nanowires, which is nearly a signature of the VLS growth.

Givargizov [41] developed the VLS model and discussed it within the framework of kinetics. Based on the dependence of the growth rate on the whisker diameter, it was found that VLS growth would terminate at some critical diameter $d_c \leqslant 0.1\ \mu m$ according to the Gibbs-Thomson effect. However, Lieber and Yu found that SiNWs with a diameter of less than 20 nm can be synthesized by laser ablation of silicon/metal target and the growth process was ascribed to the VLS mechanism [2–4] too. Experimental evidences for a VLS controlled growth are the existence of solidified particles at the ends of nanowires.

In the synthesis of silicon nanowires, however, Lee et al. found that the adding of silicon oxide into the target materials can also favorize the growth of the nanowires [42]. In fact such a case is an analogy to the VLS growth [40], because the metastable oxide phase can play the role as the catalysts.

Table 1 summarizes the nanowires fabricated via the vapor phase growth. The advantages of the vapor phase growth is manifested by their simplicity, the high quality of the crystals, and the easiness to scale up for commercial production.

Table 1 Nanowires Prepared by Vapor Phase Evaporation or Reduction

Types of Nanowires	Growth Conditions: Source Materials (at%)	Temperature (°C)	Atmosphere	Pressure (Torr)	Average Diameter
Si	Si (95) + Fe (5)	1200	Ar	100	~12 nm
ZnSe	Zn (50) + Se (50)	520	Ar + H_2	200	60 nm
GaSe	Ga (10) + Ga_2O_3(40) + Se(50)	1000	Ar + H_2	100	20~90 nm
SiO_2	Si(10) + SiO_2(85) + Fe	1250	Ar	150	~15 nm
GeO_2	Ge + Fe	900	Ar	150	15~80 nm
Ga_2O_3	Ga + SiO_2		Ar + H_2	100	60 nm
GaN	Ga + Ni	900	NH_3 + H_2	< 1 Torr	~8 nm

However, this method usually needs to proceed at elevated temperatures compared to other methods either from melt (more than 1000 °C), from CVD method (around 500 °C), or from the so-called solution–liquid–solid (SLS) growth which can be grown at a low temperature (≤200 °C) [43].

3. Gallium Oxide Nanowires and Nanoribbons

Monoclinic gallium oxide (β-Ga_2O_3) is a wide band gap semiconductor compound ($Eg = 4.9$ eV) which has long been known to exhibit both interesting conduction and luminescence properties. As a transparent conducting oxide, β-Ga_2O_3 has potential applications in optoelectronic devices [44–46]. For example, Ga_2O_3 thin films are used as one of the most promising materials for stable high-temperature gas sensors.

Gallium oxide nanowires (GaONWs) were first synthesized by vapor phase growth using gallium powder as a source material [25], and later a variety of methods were used to fabricate it [47–53]. The evaporation system via vapor phase growth was the same as previously used [1, 5]. The gallium powder in a quartz boat was placed in the front of the quartz tube. During the experiment, a constant flow of mixture gas (90% Argon and 10% Hydrogen) was maintained at flow rate of 30 sccm. The ambient pressure inside the quartz tube was kept at 100 Torr, and the temperature near the Ga-containing boat is about 300 °C. A white web-like product was collected from the inner wall of the quartz boat. XRD a monoclinic crystalline gallium oxide phase Ga_2O_3. The lattice parameters of the crystalline phase are: $\boldsymbol{a} = 1.223$ nm, $\boldsymbol{b} = 0.304$ nm, $\boldsymbol{c} = 0.580$ nm, and $\boldsymbol{\beta} = 103.7°$, respectively, and its space group is identified as ***C2/m***.

The general morphology of the product is shown in Fig. 3. It is visible from the TEM images that the product consists of pure GaONWs. The length of the wires can be found as long as up to hundreds of microns. SAED analysis and the continuous bend contours visible in each single wire reveal that the nanowires are of single crystal.

In above-mentioned growth conditions, the formation of the GaONWs and nanoribbons was explained as follows. The Ga is a metal element, easier to get oxygen from the silicon oxide on the surface of the quartz boat to form Ga_2O_3 nanowires. The growth of the nanowires were realized through the following reactions:

$$SiO_2(s) + H_2(g) \rightarrow H_2O(g) + SiO(g) \tag{2}$$

$$4Ga(l) + 3SiO(g) + 3H_2O(g) \rightarrow 2Ga_2O_3(s) + H_2(g) + 3Si(g) \tag{3}$$

which is similar to the growth of alumina whisker.

Though the structure defects are crucial in the formation of nanomaterials, their roles were seemingly ignored in now-existing models to explain the preferable growth in nanostructured materials. The HREM images [25, 47] shown in Fig. 4 reveal the existence of coherent symmetric twin in the GaONWs, which is a representative characteristic in microstructure of many GaONWs. The twin boundary (marked with arrow) is parallel to the nanowire axis and highlights the importance of twinning in the growth of the GaONWs, because in this peculiar configuration of twinning, the growing frontier perpendicular to the wire axis keeps zigzag, which is energetically favorable, and serves as the stable sites for the rapid stacking of atoms, resulting in a fast axial growth rate over the radius direction. In many other cases, it was demonstrated

Fig. 3. Low magnification (a) and magnified (b) TEM images showing the morphology of the GaONWs [25].

that structure defects also play an important role in the nucleation, in particular, the triple, quadruple, and quintuplet junction points of the micro twin variants can serve as the center of the nucleation of the silicon nano wires. In fact, in sub-micro Si particles with diameter around 200 nm synthesized by the arc-discharge gas evaporation method reported by Iijima [54], multiple-twins and stacking faults frequently occurred in the center of the spherical Si particles. In the growth of silicon nanowires, the importance of structure defects has been utilized to describe the growth of the nanowires [55]. It is therefore proposed that the GaONWs follow a growth mechanism similar to vapor-solid (VS) model. In this sense, vapor-solid phase (VS) was involved in the growth of GaONWs in a way similar to that of vapor-phase epitaxy.

By modifying the growth conditions used in reference [25], micron-sized dendrite-like Ga_2O_3 complex structures with very peculiar morphology were fabricated on large-area substrate [56]. The morphology of the gallium oxide products, however, were found much dependent on the distance from the source materials. Fig. 5 (a) shows the XRD spectrum of the gallium oxide products, revealing a β-Ga_2O_3 phase. In Fig. 5 (b), the Ga_2O_3 whiskers was collected far away from the gallium source, and they have a very smooth and straight morphology, and average diameter around 500 nm. With the further approaching to the source materials, the collected whiskers become larger, and nano-sized branches were grown out of the whisker stems, as is shown in the SEM image in Fig. 5 (c). In place close to the source materials, the whiskers become much larger, and

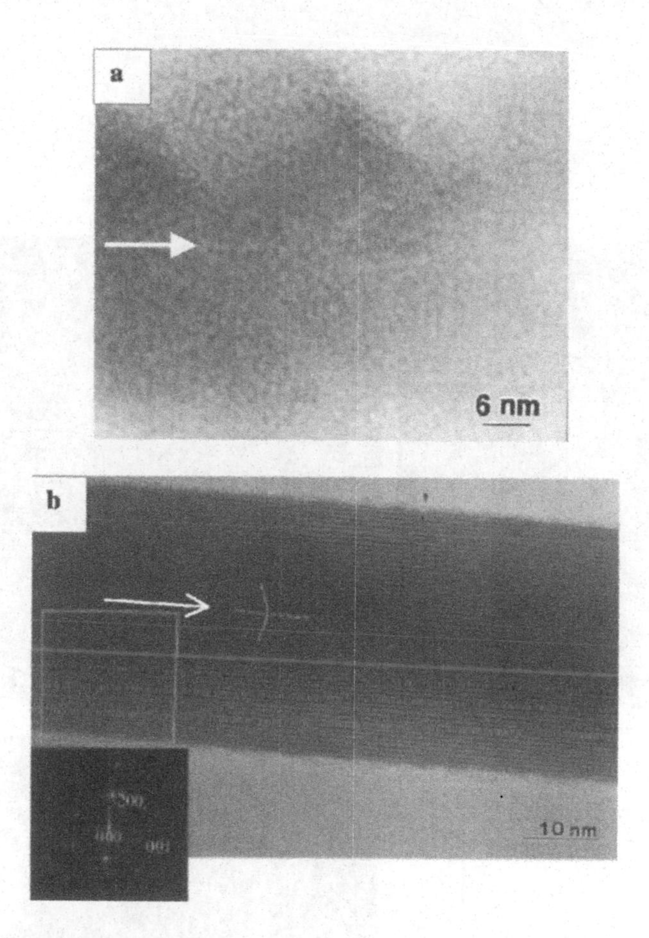

Fig. 4. (a) HREM of a single GaONW. A symmetrical twin boundary which is parallel to the axis of the wire [25]; (b) HREM images showing the symmetrical twin boundary in a single GaONW [47]. Those images provide strong evidence for the important role that defects may play in the growth of nanowires.

very complex dendrite structures (pine tree-like) are formed, which is shown in the SEM image in Fig. 5 (d). The evolution of the whisker morphologies as a function of the distance from the source materials results in the concentration distribution of gallium and oxygen in the vapor phase. The above-mentioned gallium oxide whiskers with particular morphology are advantageous in fabrication of composite materials, because the branches of dendrite whiskers can greatly increase the adhesion force with the matrix material, thus enhance the tensile strength of the composite materials.

By changing the growth parameters such as the temperature and ambient pressure, nanosized Ga_2O_3 ribbons can be grown in large quantity [57]. Fig. 6 (a, b) shows the SEM images revealing the representative morphology of the Ga_2O_3 product. It is visible that the product consists of nanoribbons and very fine nanowires. The TEM images in Fig. 6 (c, d) show the microstructure details of the nanoribbons. Some of the nanoribbons are so thin that they appear transparent to other underlying nanoribbons

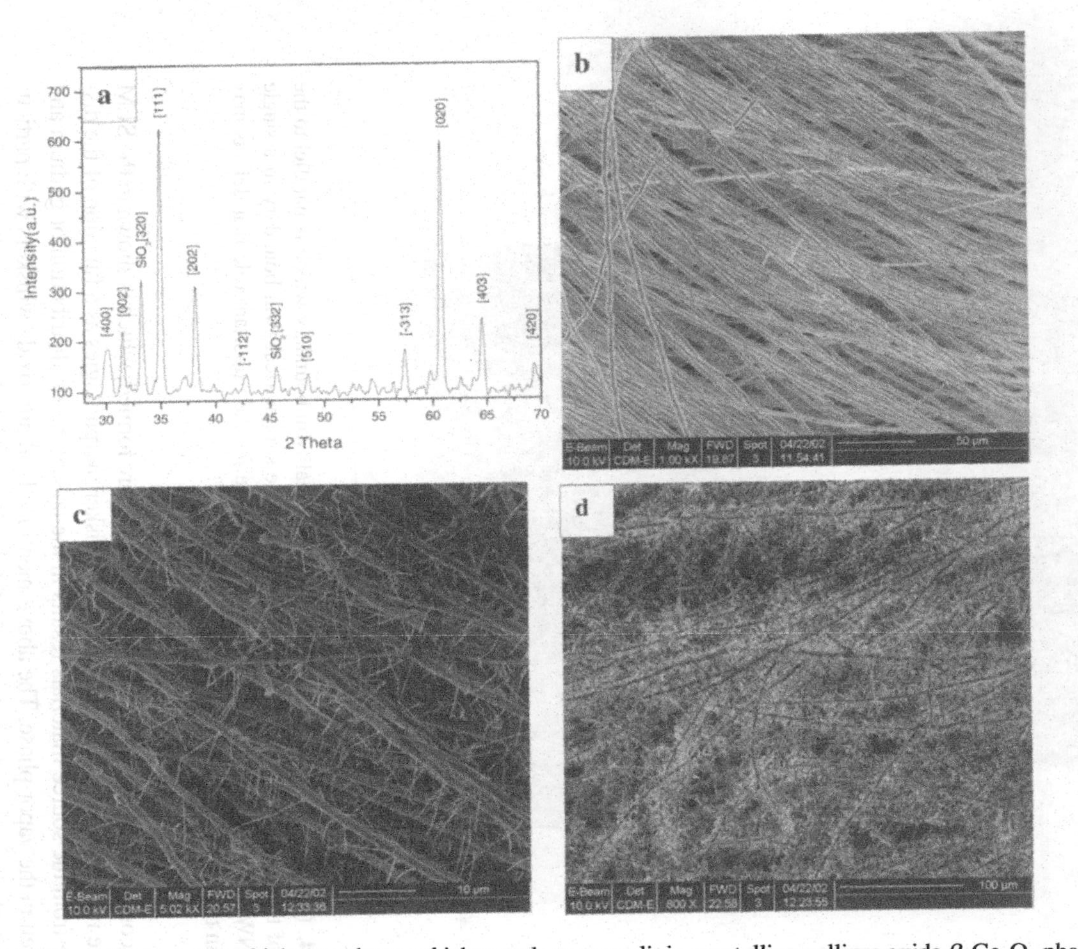

Fig. 5. (a): XRD spectrum of the gallium oxide whisker products, which reveals a monoclinic crystalline gallium oxide β-Ga_2O_3 phase; (b): SEM image showing the smooth morphology of the Ga_2O_3 whiskers; (c,d): Ga_2O_3 whiskers with complex dendrite morphology [56].

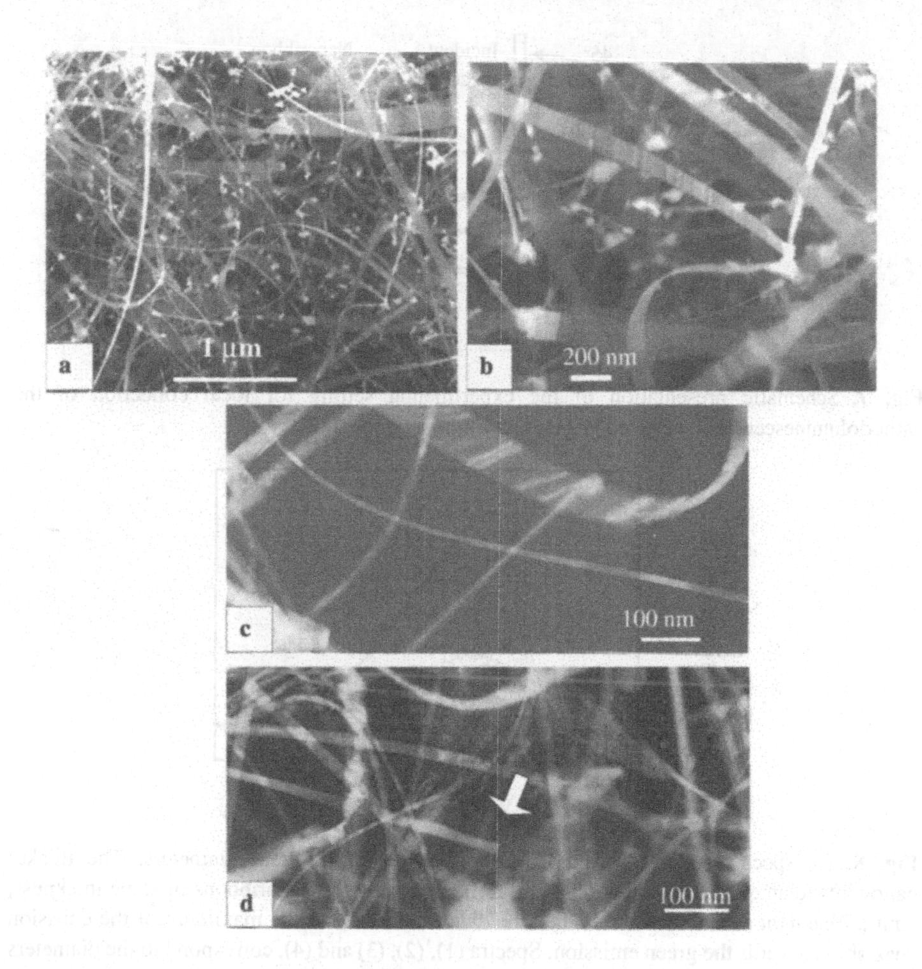

Fig. 6. SEM images of the Ga_2O_3 nanoribbons fabricated via vapor phase growth [57].

under electron beam (marked with arrow head in Fig. 6 (d)). The thinnest nanoribbons were found to have a thickness smaller than 5 nm, whereas their length can vary from 2 to 10 μm. Such thin ribbons are ideal building blocks in optoelectronic nanodevice applications. From the SEM tilting experiments, it was able to measure the thickness of every nanoribbon of interest, and found that the thickness of the nanoribbons can change between 5~500 nm. The diverse distribution of the nanoribbon thickness makes it possible to characterize the CL features in function of their thickness.

Using a new setup combining SEM and near field optical microscope system (SNOM), the localized cathodoluminescence (CL) from one single nanowire/nanoribbons were characterized. The CL was measured with a combined SPM/FE-SEM instrument that has been developed to collect the photons in near-field regime so as to improve the resolution [58] that is generally obtained with conventional CL systems. In the present experiment, the photons excited by the electron from the nanostructure were collected in the far-field regime in order to improve the collection efficiency on consideration of the morphology of our samples (nanoribbons/wires dispersed on the surface).

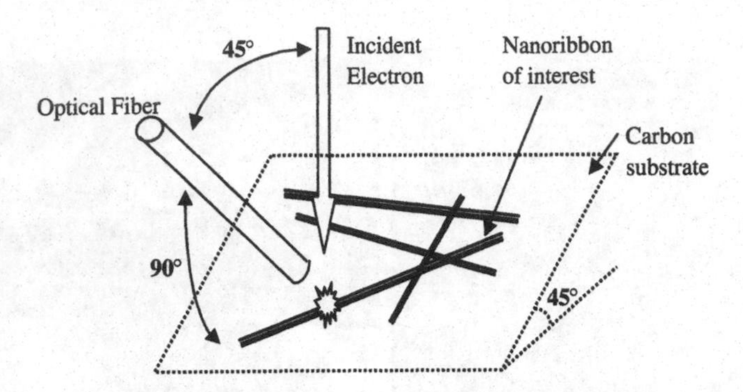

Fig. 7. Schematic presentation of the experimental setting for local collection of the cathodoluminescence on selected Ga_2O_3 nanoribbons/wires [60].

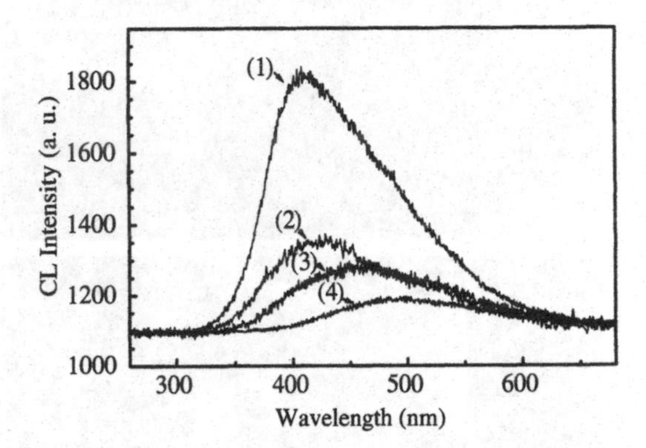

Fig. 8. CL spectra collected from Ga_2O_3 nanowires of different diameters. The thicker nanowires (diameter larger than about 170 nm), as exactly the nanoribbons of same thickness, emit a blue light around 410 nm. When the diameter decreases the maximum of the emission peak shifts towards the green emission. Spectra (1), (2), (3) and (4), correspond to the diameters 180, 120, 70 and 30 nm, respectively [60].

Indeed, since the nanoribbons/wires are sufficiently separated from one another, the CL information is very localized because the electron beam is highly focused into a small probe on the nanostructure of interest. The working principle of the SPM/FE-SEM was described in details elsewhere [59]. The experimental setting is depicted in Fig. 7.

The sample stage is set on the SPM scanning piezotube, which is 45° tilted with respect to the electron beam. The piezotube allows X and Y displacements of 30×30 μm, and the sample is performed with two stepper piezomotors. A multimode optical fiber (~ 50 μm) with a numerical aperture of 0.48 is controlled to approach closely to the single nanowire of interest with a third stepper piezomotor. The high brightness of the field emission gun (FEG) enables a high electron probe current in a very small probe size, which favors the photon yield.

More than 50 nanoribbons/nanowires were selected and their CL spectra were collected from each individual ribbon/wire. Fig. 8 shows the evolution of the CL spectra versus the diameter of the nanowires. In the spectral region between 320 nm and 800 nm of our detection system, only one broad peak was observed for all nanoribbons/wires. As the energy band gap of gallium oxide (around 4.8 eV) is too large for

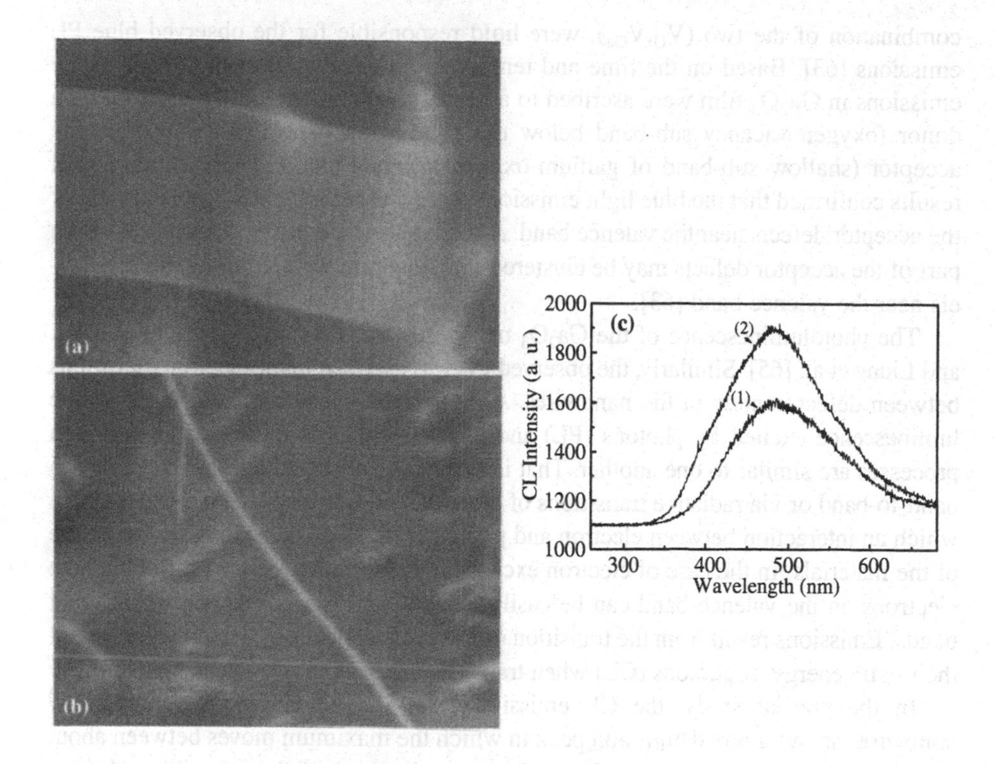

Fig. 9. SEM images of a thin nanoribbons of about 30 nm thickness (a), and a nanowire of 35 nm in diameter (b). The corresponding CL spectra are shown on (c). The spectrum (1) corresponds to the nanowire and the spectrum (2) to the nanoribbon. The peak maximum is around 480 nm [60].

our system, we can only characterize the defect levels, such as oxygen or gallium vacancies. Interestingly, the position of the peak maximum depends on the thickness/diameter of the nanoribbons/nanowires. The thicker nanoribbons with a thickness > 170 nm present a blue emission peaked around 410 nm (~3.00 eV). A shift towards the green emission occurs for thinner nanoribbons/wires to progressively attain an emission maximum around 490 nm (~ 2.53 eV) for the thinnest ones. The nanoribbons behave the similar way.

Figs 9 (a) and (b) show the SEM images of a thin nanoribbon (about 30 nm in thickness) and a nanowire (35 nm in diameter). The corresponding CL spectra are shown in Figure 9 (c). It is well revealed that those thin Ga_2O_3 nanoribbons/wires have a CL emission peak around 480 nm corresponding to a blue-green dominant light [60].

There are little reports on the optical properties of the Ga_2O_3. Harwig et al. [61] studied the luminescence of the Ga_2O_3 crystals and powders, and ultra violet emission was observed in all samples. Besides the UV emission, blue and green emissions were also observed from Ga_2O_3 crystals. Binet et al. [62] investigated in detail the photoluminescence (PL) of the Ga_2O_3 film, and a blue PL emission peaked at 2.95 eV (~420 nm) was observed upon excitation of 4.67 eV (~266 nm). It was further revealed that either the oxygen vacancy (V_O) or gallium vacancy (V_{Ga}), or a combination of the two (V_O,V_{Ga}), were hold responsible for the observed blue PL emissions [63]. Based on the time and temperature dependence of the PL, the blue emissions in Ga_2O_3 film were ascribed to a tunnel recombination of an electron on a donor (oxygen vacancy sub-band below the conduction band) with a hole on the acceptor (shallow sub-band of gallium-oxygen vacancy pairs). Those experimental results confirmed that the blue light emissions were closely related to the excitation of the acceptor defects near the valence band. A model was recently put forward in which part of the acceptor defects may be clustered into quantum wells forming discrete levels near the valence band [63].

The photoluminescence of the Ga_2O_3 nanowires was analyzed by Wu et al. [64], and Liang et al. [65]. Similarly, the observed PL was ascribed to the electron transitions between defect centers in the nanowires. Although the excitation mechanism of the luminescence excited by photons (PL) and electrons (CL) is different, the emission processes are similar to one another. That is to say, the observed emissions are due to band-to-band or via radiative transitions of electron and hole between defect bands, in which an interaction between electron and photon is involved depending on the nature of the materials. In the case of electron excitation under high accelerating voltage, the electrons on the valence band can be easily excited, and can be trapped by different bands. Emissions result from the transition of the excited electrons, which will give out their extra energy as photons (CL) when transitioning back to the valence band.

In the present study, the CL emission from individual Ga_2O_3 nanoribbons/ nanowires shows a broad emission peak in which the maximum moves between about 410 and 490 nm, depending on the thickness or diameter of the nanoribbons/wires, respectively. The detailed origin of the different emission peaks has yet to be understood. It is well know that oxygen vacancy is formed in Ga_2O_3 under a reduction growth conditions, forming an n-type semiconductor. Harwig et al. [61] showed that samples annealed in reduction atmosphere is in favor of the formation of oxygen vacancies and blue emission was enhanced, while sample heated in O_2 atmosphere is in favor of the formation of Ga vacancies, which shows a dominant green emission. Villora et al. [66] revealed recently very interesting results via CL investigation of the Ga_2O_3 single crystals grown under different oxygen partial pressure, which confirms the results of Harwig et al. It was found that with the decrease of the oxygen partial pressure, the CL emission decreased in intensity, and shifted from green to blue, and finally to UV when grown in N_2 ambient. Based on those experimental evidences, we conclude that the observed CL difference in thin and thick nanoribbons/wires result from the difference of defect structures. There is certainly a different amount of oxygen vacancies, and the thick nanoribbons are probably rich in oxygen vacancies while there are less oxygen vacancies in the thin nanoribbons/wires. This is reasonable because the thin and thick nanoribbons/wires can have a different growth rate, which can result in a preferential distribution of defects and thus in different emission properties.

The β-Ga_2O_3 compound have very interesting conductivity behavior from room temperature to high temperature, which makes it advantageous in nanodevice applications. It appears insulating at room temperature. At high temperature [67, 68], however, it becomes *n-type* semiconductor, and its conductivity is highly environment dependent. Between 600~700 °C, surface absorbed molecules react directly with the oxygen in the lattice, forming oxygen vacancies at the surface, causing a dramatic change in the conductivity, which is the main working mechanism of the gas sensor. At high temperature above 700 °C, the oxygen atoms comes out of the lattice, forming Schottckey defects, and the major carriers are Ga^{2+}. Now β-Ga_2O_3 compound are excellent candidates for oxygen gas sensors working at high temperature [69–72]. Compared to other gas sensors, the Ga_2O_3 based sensors have the following advantages: (1) high stability; (2) very small differences between products; (3) low sensitivity to moisture; (4) quick response time; (5) self-cleaning function; (6) no pre-aging process is needed.

3. ZnO Nanowires and Nanotubes

ZnO is a wide band-gap semiconductor with a band gap energy of 3.37 eV [73]. ZnO films [74] and disordered nanoparticles [75], and nanowires [27] exhibited interesting UV photoluminescence emission, and very intense stimulated ultraviolet lasing action was observed at room temperature. In addition, the low growth temperature, high chemical stability, and low threshold intensity make the ZnO an excellent candidate for commercial UV laser [19]. Nanowire nano lasers based on array of ZnO nanowires were fabricated by P. D. Yang et al. [76], and ZnO nanobelts were synthesized by Z. W. Pan, Z. L. Wang et al [29]. There are already a lot of progress in the synthesis and characterization of the ZnO nanowire using various methods [77–117]. In this section, the synthesis and characterization of UV emitting ZnO nanowires and nanotubes fabricated using PVD and other methods were summarized.

ZnO nanowires were first reported by Kong et al. using vapor phase deposition [27] Powder of 70 wt% zinc and 30 wt% selenium were well mixed in a quartz boat, and placed in a quartz tube. The quartz tube was heated in a tube furnace to 1100 °C for 10 hours at a wet oxidation atmosphere under 100 Torr.

Fig. 10 (a) shows a transmission electron microscope (TEM) image at low magnification, revealing the representative morphology of the ZnO nanowires. The diameter of the ZnO nanowires varies from 30 nm to 100 nm, averaged around 60 nm. Selected-area transmission electron diffraction (SAED) taken from one single ZnO nanowire shows a complete diffraction pattern (inset), indicating that the ZnO nanowires are single crystalline. The particles (as indicated by arrows) were revealed to be ZnSe particles. In the growth of ZnO nanowires, it was observed that ZnO nanowires are distributed radically about large ZnSe droplets, implying that the growth of the ZnO nanowires were controlled by the VLS mechanism [40] with the Se serving as a liquid-forming agent in the growth of the nanowires.

The room temperature photoluminescence spectrum of the ZnO nanowires is shown in Fig. 11. A broad intense emission peaked at 3.2 eV was observed at low excitation intensity. Two factors are responsible for the enhanced emission intensity of the room temperature PL and the increased gain of lasing from lower dimensional ZnO structures. Bagnall et al. [73] demonstrated that improvement of crystal quality can cause a high NBE emission to deep-level emission ratio, resulting in detectable UV emissions at room temperature. In addition, the quantum confinement effect related to the nanostructures also contributes to the increase of the emission intensity of the UV emission. Considering that the diameter of the ZnO nanowires are comparable to that of the ZnO

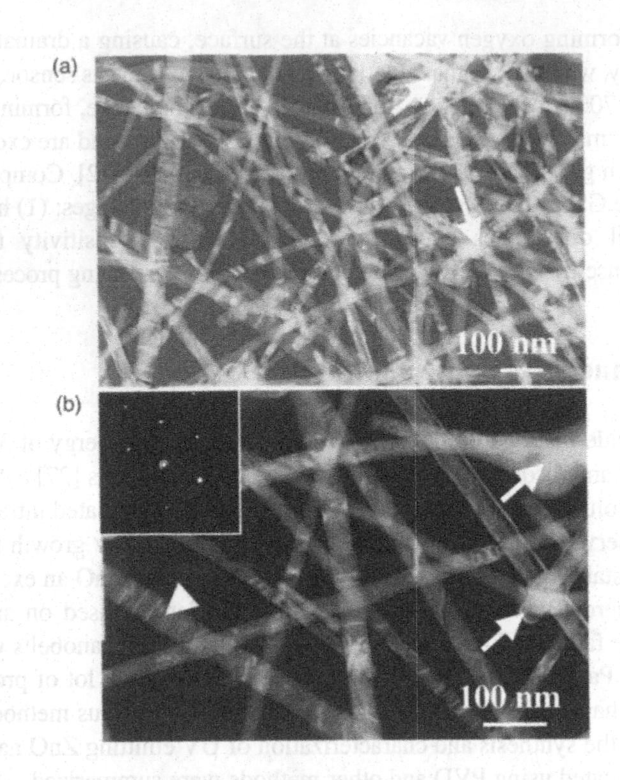

Fig. 10. TEM image at low (a) and high (b) magnification showing the representative morphology of the ZnO nanowires, with an average diameter around 60 nm. Inset shows the SAED patterns taken on different single ZnO nanowire, revealing the monocrystalline nature of the nanowires [27].

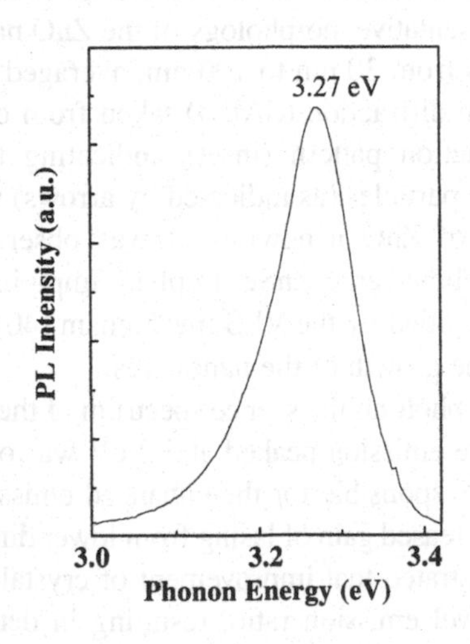

Fig. 11. Room temperature photoluminescence of the ZnO nanowires. A broad (FWHM = 142 meV) emission peak at 3.2 eV was observed at low excitation intensity, which was assigned to emission from free excitons [27].

film reported, we ascribe the PL band at 3.2 eV observed from the ZnO nanowires to the emission from free excitons. Those results confirm that fabrication of nano-sized materials is an important way to tailor the physical properties of semiconductor materials.

Recently, Y. J. Xing [79] et al. succeeded in large-scale synthesis of ZnO nanowires using simple physical evaporation of the zinc oxide powder. The experimental details used in this work was described as follows. Powders of zinc oxide was put into the central zone of alumina tube in a tube furnace. The whole system was evacuated to about 300 Torr. The furnace was heated to about 1300 °C under flowing Ar atmosphere (40 sccm) for 1 hour. Si substrates (5×20 mm) were placed in sequence to collect the products. From the low magnified SEM image as shown in Fig. 12 (a), the nanowires self-organize into very beautiful regular arrays which is a mimic the snow flames. Those flames grow radically from one common central nucleus, as are revealed from the magnified SEM images in Fig. 12 (b). There exists well defined crystallographic relationships between the branches of the nanowires. For example, the orientation of the three branches of the dendrite in Fig. 12 (c, d) have crystallographic direction of <101>, <011>, and <1 1–20>, respectively, which are identical to each other, and the angle between them are 80.75°. Such a kind of self-assembly originates indeed from the intrinsic microstructure of the hexagonal ZnO crystal. Detailed TEM analysis (Fig. 12e) revealed that the dendrite ZnO nanorods are nucleated from a 4-fold twin center, and coalescence can occur between two nanorods (Fig. 12f).

By optimal control of the growth parameters, ZnO nanorods with homogeneity in diameter along their length can be achieved, as is shown in the SEM image in Fig. 13. It is visible that the ZnO nanorods have a perfect orientation perpendicular to the Si substrate.

ZnO nanorod arrays with amazing peculiar morphologies were synthesized recently by H. Z. Zhang, X. C. Sun et al. on (111)-oriented Si substrate and glass via a vapor phase growth [80]. The morphology of the individual nanorod is variable from flat-headed, bottle-like, to needle-like, which depends on the deposition positions relative to the source materials in the presence of a controlling element Se. Fig. 14 shows typical SEM images revealing three different morphologies of the ZnO nanorod arrays, which were deposited on substrates with different positions relative to the Zn source. First of all, all the nanorods exhibit perfect orientation perpendicular to the substrates and the coverage ratios of the three situations on the substrates are estimated to be over 70%. Fig. 14A reveals a peculiar morphology which represents nano-sized cavities, because it is visible that each nanorod consists a gradual pillar and a faceted head. Such morphology was observed on substrates that were placed 2 cm away from the Zn source. Top view SEM image shown in the inset further reveals that all the nanocavities have very flat hexagonal faceted heads. The average diameter of the nanocavities is about 240 nm, while the lengths of the nanorods are uniform and over 0.5~1.5 μm. X-ray diffraction reveals a very intense (002) peak. The intensity of the (002) peak surpasses others, which demonstrates the c-oriented nature of the as grown arrays. The degree of the orientation is illustrated by the relative texture coefficient which is given by equation (4)

$$TC_{002} = \frac{I_{002} \geqslant I^0_{002}}{I_{002} \geqslant I^0_{002} + I_{101} \geqslant I^0_{101}} \tag{4}$$

Fig. 12. SEM and TEM images of the ZnO nanowires that self-assembly to form arrays. (a) Low magnified and (b) magnified SEM images. The selected magnified SEM images in (c,d) reveal that the ZnO nanowires grow from one common central nucleus, and the dendrites have exact crystallographic orientation relationship. TEM images reveal the X-shaped dendrite (e) and erminal-coalescence of two nanowires (f). TEM images of the dendrite ZnO nanorods provide evidence that the dendrite structure is nucleated from a 4-fold twin center [79].

Fig. 13. SEM image showing the perfect orientation and homogeneity in diameter of the ZnO nanorods grown on Si substrate [79].

where TC_{002} is the relative texture coefficient of diffraction peaks (002) over (101), I_{002} and I_{101} are the measured diffraction intensities due to (002) and (101) planes, respectively, I^0_{002} and I^0_{101} are the corresponding values of PDF. For a random oriented material, the texture coefficient is 0.5, while the value of our sample is 0.968, which indicates an extremely high c-orientation. Fig. 14B shows an SEM image which corresponds to products deposited on the substrates placed 3 cm away from the Zn source. It reveals that the ZnO nanorods have a bottle-shaped morphology. Each individual nanorod consists of a well-faceted base stem and a small faceted head. The top-view SEM image in inset reveals a 6-fold symmetry for both the stems and the heads, indicating that they grow along the same direction, which is parallel to the *c*-axis. The average diameters of the base stems and the heads are around 350 nm and 60 nm, respectively. The average length of the stems of ZnO nanorods is around 3 μm. It is also noted that most of the ZnO nanorods not only have the same ***c***-axis growth direction, but also exhibit almost the same in-plane orientations. Fig. 14C shows ZnO nanorods with needle-like morphology deposited on substrate placed 1 cm away from the Zn source. The nanoneedles are well aligned too. The diameters of the nano-needle stems are below 100 nm, and each one has a very sharp tip. Some tips are magnified in the inset of Fig. 14C to highlight the sharpness of the tips. The lengths of the nanoneedles usually exceed 3μm. The microstructures of the ZnO nanorods with peculiar morphologies were further analyzed using TEM, as shown in Fig. 15. Fig. 15A is a magnified TEM image of the nanocavities, from which the hexagonal heads of the nanocavities are clearly visible. The selected-area electron diffraction (SAED) pattern corresponding to the one of the nanocavity is shown in inset. The diffraction spot (001) in the pattern confirmed that the growth of the observed nanocavities is along their ***c***-axis. The single-crystal nature of the stripe was verified by tilting SAED experiments over each nanocavity of the whole stripe. It was found that all the nanocavities on the stripe have exactly the same crystallographic orientation both in plane and along the ***c***-direction, which is an important factor for optical device applications. The crystal habit planes of the ZnO nanocavities are depicted schematically

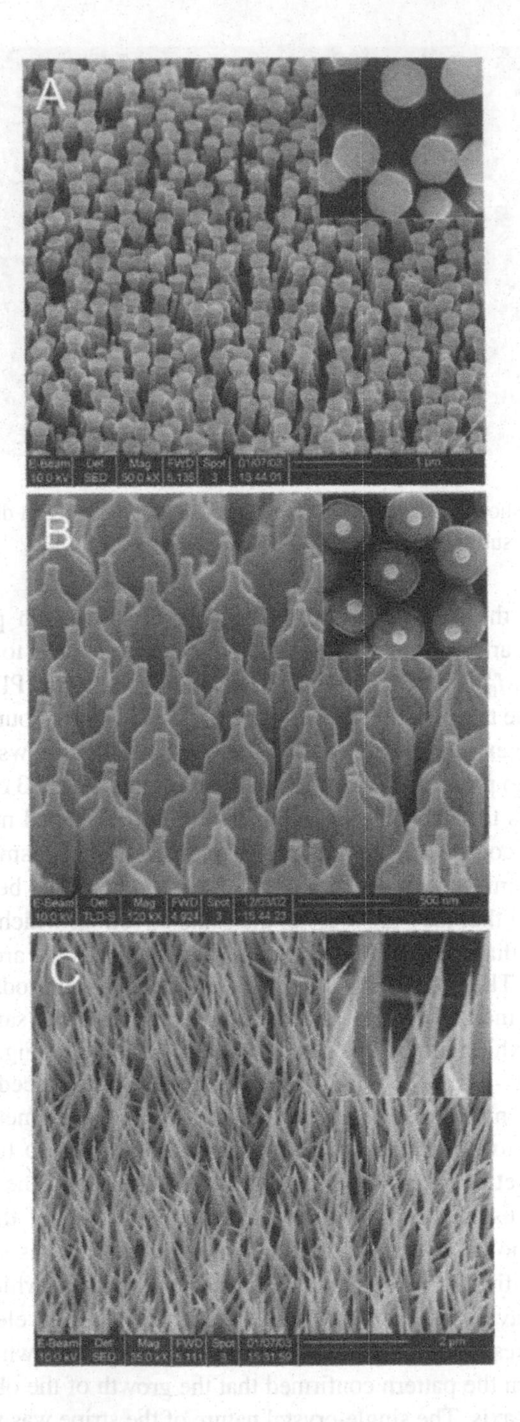

Fig. 14. SEM images of the ZnO nanorods with peculiar morphologies. (A) Nanocavities. Inset: top view of the hexagonal facet of the head (B) Bottle-shaped nanorods. Inset: the top view of the hexagonal facets of both the stem and the tips. (C) Nanoneedles. Inset: magnified view revealing the sharp tips [80].

in 15B. EDS analysis was conducted on single nanocavity which confirmed that the nanorods are composed of Zn and O elements, as shown in Fig. 15C. The TEM image in Fig. 15D shows the typical morphology of two bottle-like ZnO nanorods with gradual diameters along the nanorod axis. The nanorods thicken gradually from the root to the shoulder. The HRTEM image in Fig. 15E represents the atomic details of the head of a single ZnO nanorod. The lattice constant measured from the lattice fringes showing that the nanorod grows along the [001] *c*-axis direction. The continuous lattice fringes demonstrate the head of the nanorod is defect free. It is noted that the sizes of the roots for different ZnO nanorods vary between 100 ~ 400 nm, while the tip has an average diameter around 60 nm. Fig. 15F shows the TEM image revealing the general morphology of the needlelike ZnO nanostructures, which indicates gradual decrease of the diameters of the nanoneedles from the roots to form very sharp tips. The magnified HRTEM image in Fig. 15G reveals that the diameters of the tips of the nanoneedles are only several nanometers, and its growth direction is parallel to the ***c***-axis. Due the very sharp tips, the ZnO nanoneedles show very high field emission current density comparable to that of the carbon nanotubes, which will be reported elsewhere.

Detailed cross-sectional SEM analysis of the ZnO nanorods grown on different substrate provides further insight into the nucleation and growth mechanism of the nanorods. Fig.16 shows cross-sectional SEM view of the ZnO nanorods grown on the Si substrate. This image sheds light on how these nanorods nucleated and grew from the substrate. There is a near continuous layer composed of many smaller grains just above the substrate, as indicated by the white small rectangle in Fig. 16A, revealing that the density of ZnO nuclei is very high at the nucleation stage. Since the growth rate for (001) lattice plane of ZnO is much higher than other lattice planes, for example, it is about twice as fast as that for (100). Among grains in the film, there must be quite large number of ZnO grains with (001) ***c***-axis perpendicular to the substrate, in the following process such grains can grow epitaxially much faster than any others, leading to the formation of c-axis well-oriented ZnO nanorod arrays. The area in the dash-lined rectangle is magnified to show the detail between the substrate and the nanorods. It is clearly seen from Fig. 16B that the nanorods are grown directly from the substrate. It is revealed that the ZnO nanorods were nucleated directly on the substrate.

The well-aligned ZnO nanoneedles with tips as small as several nanometers make them excellent candidate for effective field emission [81]. TEM analysis provides insight into the microstructure details of the ZnO nanoneedles. Fig. 17a indicates that the diameters of the nanorods are gradually decreased from the root to form a very sharp tip, and the diameters of the tips are several nanometers. The sharpness of the tip is manifested in the inset, which is the boundary contour of the nanoneedle. The dimension of its apex is about 7 nm, and the initial half-angle θ (as defined in the figure) is as small as 4°, showing excellent configuration of the as-grown nanoneedles for field-emission applications. As is shown in Fig. 17b, the HRTEM image discloses the detailed lattice structure of the nanoneedles. The continuous lattice fringes indicate the single crystal nature of the nanoneedles. The lattice distance measured is 0.52 nm, which is consistent with the distance of (001) lattice planes of ZnO. The growth direction of the nanoneedle is along (001), as is indicated by the white arrow. The chemical composition and crystal structures of the sample are clarified by EDS analysis (the inset of Fig. 17b), and it turns out that the samples are well- crystallized ZnO nanoneedle arrays.

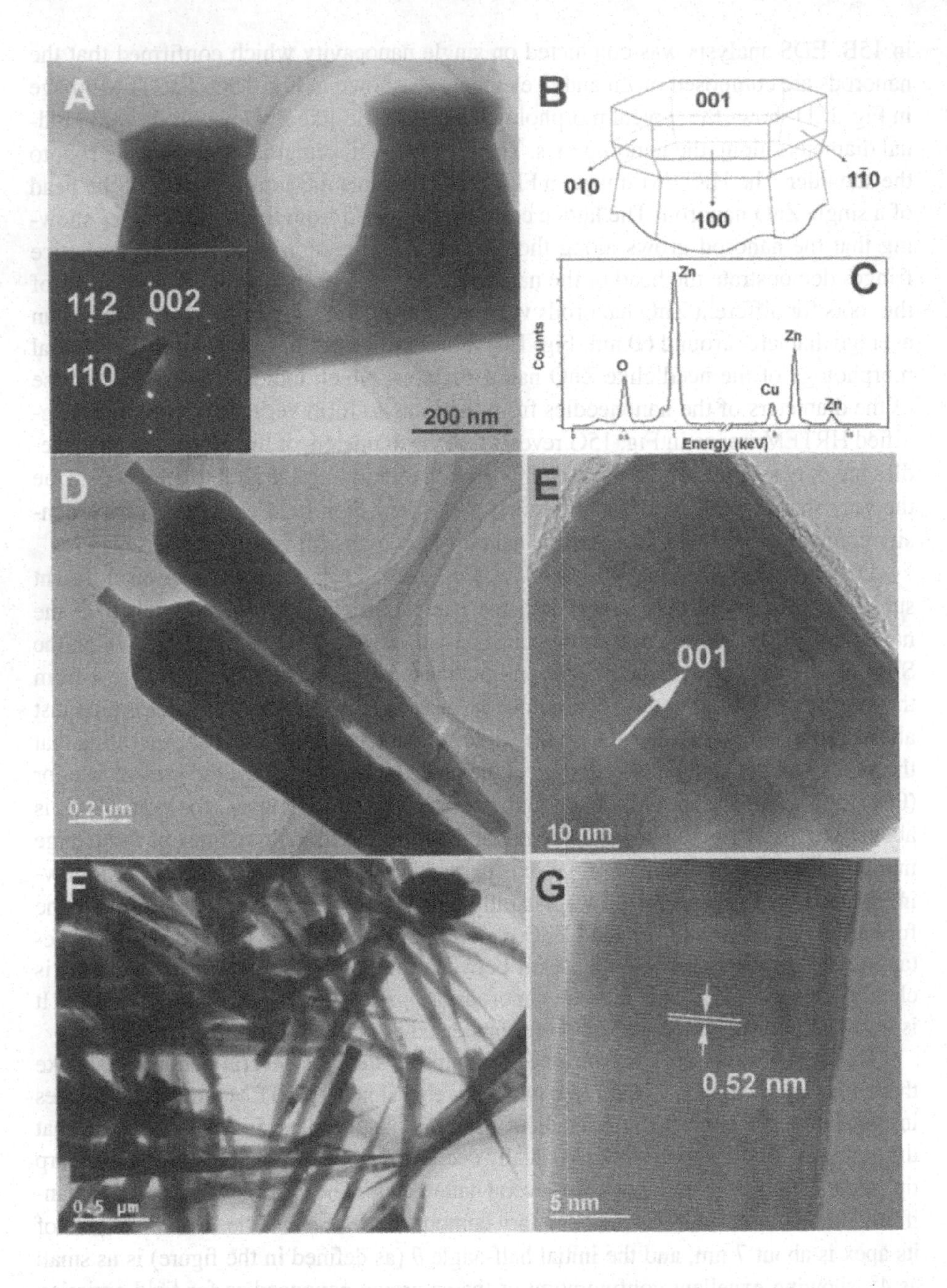

Fig. 15. (A) TEM image of the nanocavities. Inset: SAED pattern of a single nanocavity. The growth direction is parallel to the c-axis. (B) Schematic depiction showing the crystal habit planes of the ZnO nanocavities. (C) EDS analysis showing the nanorods are composed of Zn and O elements. (D) Bottle-shaped nanorods with nonuniform diameters. (E) HRTEM image shows the lattice fringes of a head of bottle-like nanorod. (F) TEM image showing the morphology of the nanoneedles. (G) HRTEM image showing the atomic structure of a single nanoneedles [80].

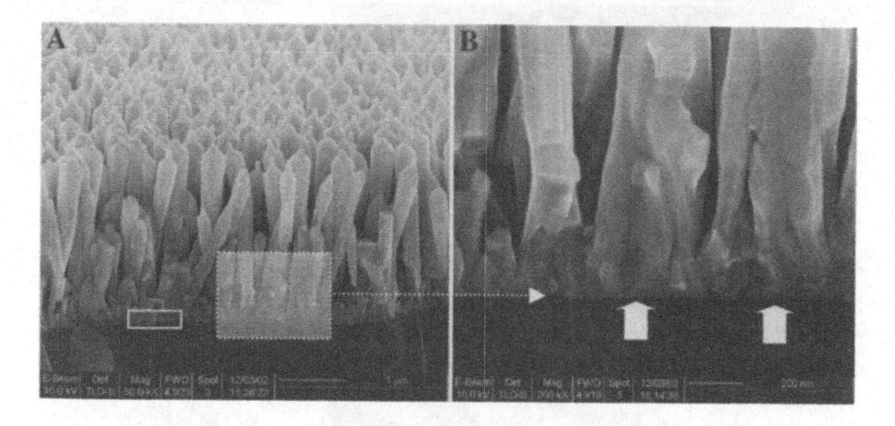

Fig. 16. (A) Cross-sectional SEM view of the substrate-nanorod interface, which provide insight into the growth mechanism of the nanorods. The area inside the rectangle is magnified in (B) The white arrow indicates that the nanorods grow directly from the substrate [80].

Field emission measurement was carried out by a two-parallel-plate configuration in a vacuum chamber with pressure $<3\times10^{-7}$ Pa. The sample was attached to a stainless-steel plate using conducting glue as cathode with the other stainless-steel plate as anode. The distance between the electrodes was 510 μm. The sample was first heated for 15 min to degas to further improve the vacuum. The emission current was measured by applying a voltage increasing from 1 kV to 4 kV with a sweep step of 50 V. The results are shown in Fig.18. The F-N plot of the data is also shown in the inset. It can be seen that the curve is consistent with the F-N mechanism by exhibiting linear dependence. The results show that the field emission of the ZnO nanoneedles can be substantially enhanced by their small geometry. In fact, the emission current density of our sample can be as high as 2.4 mA/cm^2 at the field of 7 V/μm [81], and the turn-on field is about 2.4 V/μm, that is defined as the field where the emission current density can be districted from the background noise. Both values are comparable to those of the CNTs, which makes the ZnO nanoneedle arrays a prospective candidate for the field-emission devices.

Lao, Wen and Ren fabricated a novel kind of hierarchical ZnO nanostructures by a vapor phase transport and condensation technique with the assistance of In_2O_3 [82]. The synthesis process was described as follows: ZnO, In_2O_3, and graphite powders were well mixed and placed at a quartz tube in a tube furnace, which was evacuated around 0.5 to 2.5 Torr. The mixed powders were heated to 950 ~ 1000 °C and for 30 min. The temperature where the nanostructures were collected on a graphite foil was about 820–870 °C.

Abundant dendrite structures were produced via such a growth route. An extremely interesting thing is that the nanosized dendrites appear a ZnO-In_2O_3 complex. It was revealed in combination of SEM and TEM analysis that the major cores of the dendrites (the stem) are single-crystal In_2O_3 with 6-fold, 4-fold, and 2-fold facets. The branches of the dendrites are single-crystal hexagonal ZnO and grow either perpendicular on or slanted to all the facets of those core In_2O_3 nanowires.

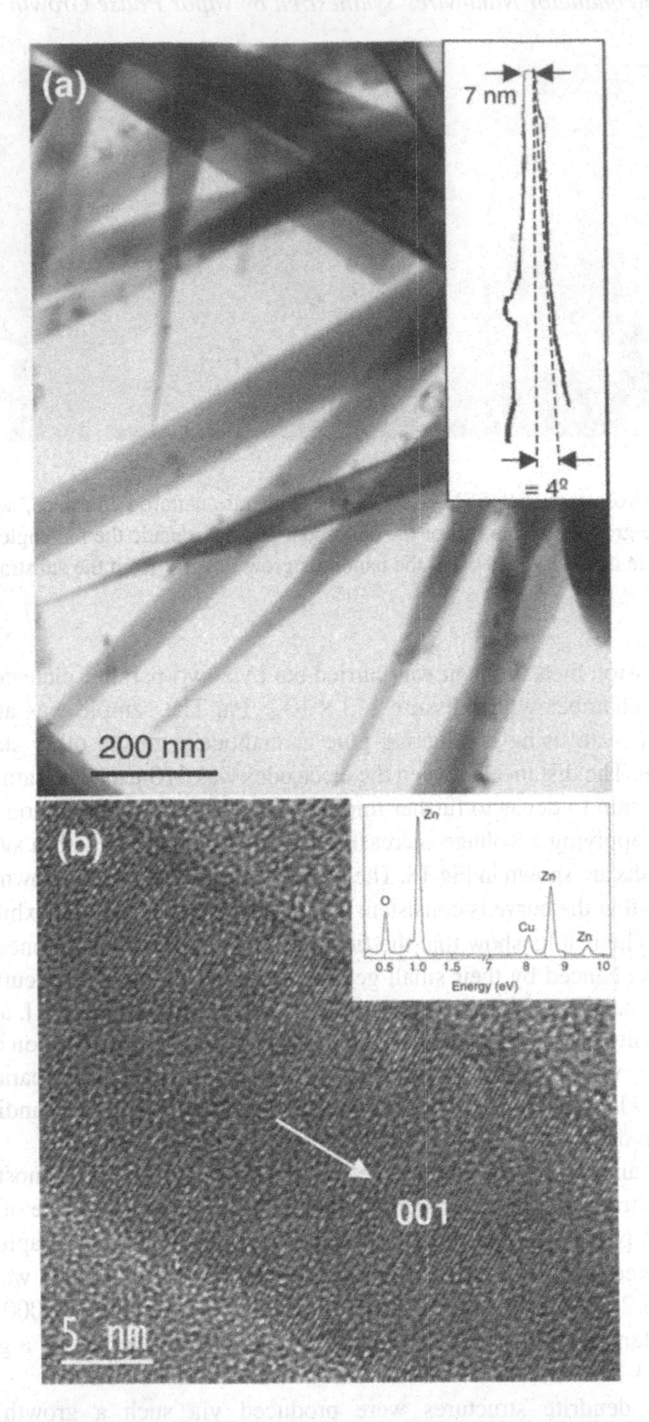

Fig. 17. (a) TEM image of the ZnO nanoneedles. It is visible that the needles have a tiny tip of several nanometers. The inset is the boundary contour of a nanoneedle, with a tip size as small as 7 nm, revealing that the nanoneedles have excellent configuration for field -emission applications. (b) HRTEM image of a tip of a nanoneedle. The growth direction is along (001), and the distance of lattice plane is 0.52 nm. The inset shows the EDS results, indicating the nanoneedles consist of Zn and O [81].

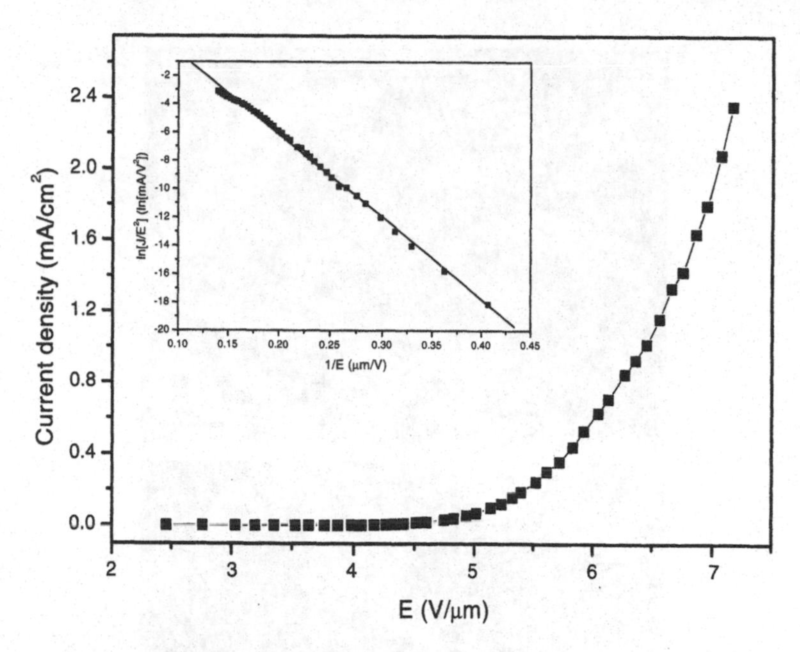

Fig. 18. (a) Field emission I–V curves from the ZnO nanoneedle arrays at working distance of 510 μm. The emission current density can be as high as 2.4 mA/ cm^2 at the field of 7 V/μm, and the turn-on field is about 2.4 V/μm. The inset depicts the F-N plot, which indicates that the emission satisfies the F-N mechanism by showing linear dependence [81].

The SEM image shown in Fig. 19 (a) represents a typical morphology of the dendrite-shaped products. In the SEM image of medium magnification in Fig. 19 (b), it is further visible that the dendrites have perfect 2-fold, 4-fold and 6 fold symmetry, as are marked in the image. Fig. 20 (a–f) provides detailed insight into the microstructures of the nanostructures with 4-fold symmetry. Cross-sectional TEM image in Fig. 20 (h) combining diffraction and compositional analysis revealed that the core is In_2O_3 and the four branches are ZnO nanowires, which was further confirmed by the selected-area electron diffraction patterns taken on the cross-sectional sample in Fig. 20 (i). The core In_2O_3 nanowires have diameters of about 50–500 nm, whereas the branch ZnO nanowires have diameters of about 20–200 nm. Interestingly, the branch ZnO nanowires grow either as a single row or multiple rows, depending on the diameter of the core In_2O_3 nanowires, and they exists a strict epitaxy relationship between the In_2O_3 core nanowires and the ZnO branches. Fig. 21 shows the structure details of the 6-fold dendrites, which are similar to the 4-fold ones. The authors are very optimistic about their ZnO/In_2O_3 hierarchical heteronanostructures, and believe that those complex nanostructures may find applications in a variety of fields such as field-emission, photovoltaics, supercapacitors, fuel cells, high strength and multifunctional nanocomposites, etc. that require not only high surface area but also structural integrity.

S. C. Lyu, Y. Zhang [83] et al. succeeded recently in large-scale synthesis of well-aligned single-crystalline ZnO nanowires with high density on nickel monoxide (NiO) catalyzed alumina substrate by using a simple metal vapor deposition method at a very low temperature around 450 °C. The ZnO nanowire arrays were fabricated by the

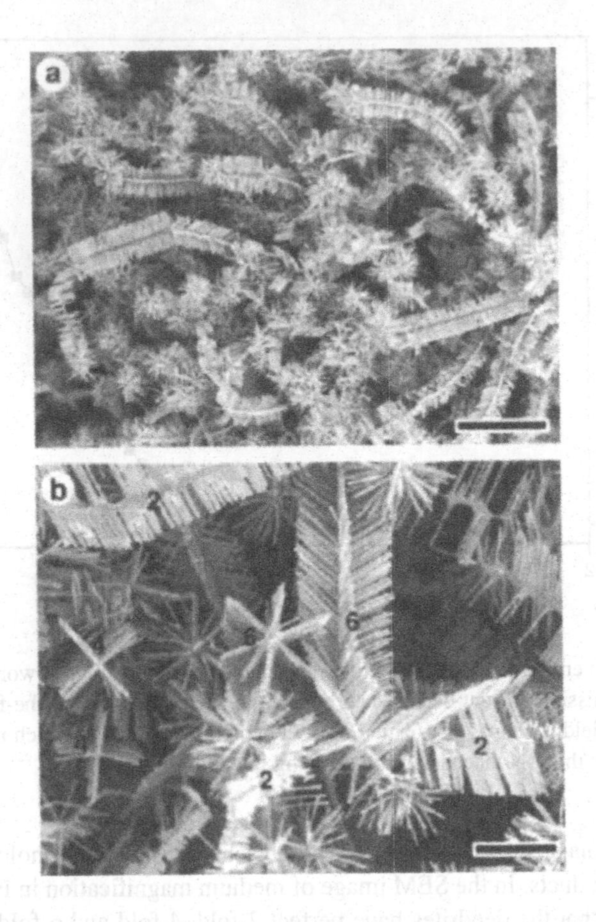

Fig. 19. SEM images of the ZnO nanostructures synthesized by the vapor transport and condensation technique. (a) Low magnification SEM image of the ZnO nanostructures to show the abundance. Scale bar, 10 μm. (b) Medium magnification SEM image of ZnO nanostructures to show the various structural symmetries. Three major basic symmetries of 6-, 4-, 2-fold were clearly seen. Scale bar, 3 μm [82].

following procedure. A nickel nitrate/ethanol solution was spread onto an alumina substrate (10 mm 5 mm in size), and was placed on a quartz boat filled with metal zinc powder. The quartz boat was then heated to 450 °C for 60 min under a constant flow of argon (flow rate: 500 sccm). XRD analysis revealed that the products are of crystalline ZnO phase.

Fig. 22 shows the SEM images of the low temperature produced ZnO nanowires. In the low magnified SEM images in (a, b), a high density of well oriented ZnO nanowires were grown perpendicular to the surface of the substrate. It is demonstrated the diameter of the ZnO nanowires increased with increasing the growth temperature. Those ZnO nanowires were grown at 450 °C, and they appear free-standing with an average diameter around 55 nm. When the growth temperature increased to 500 °C, the average diameter increased to 190 nm, as are shown in Fig. 22 (c, d). The quality of the ZnO nanowires thus prepared seems very high, which were indicated in the

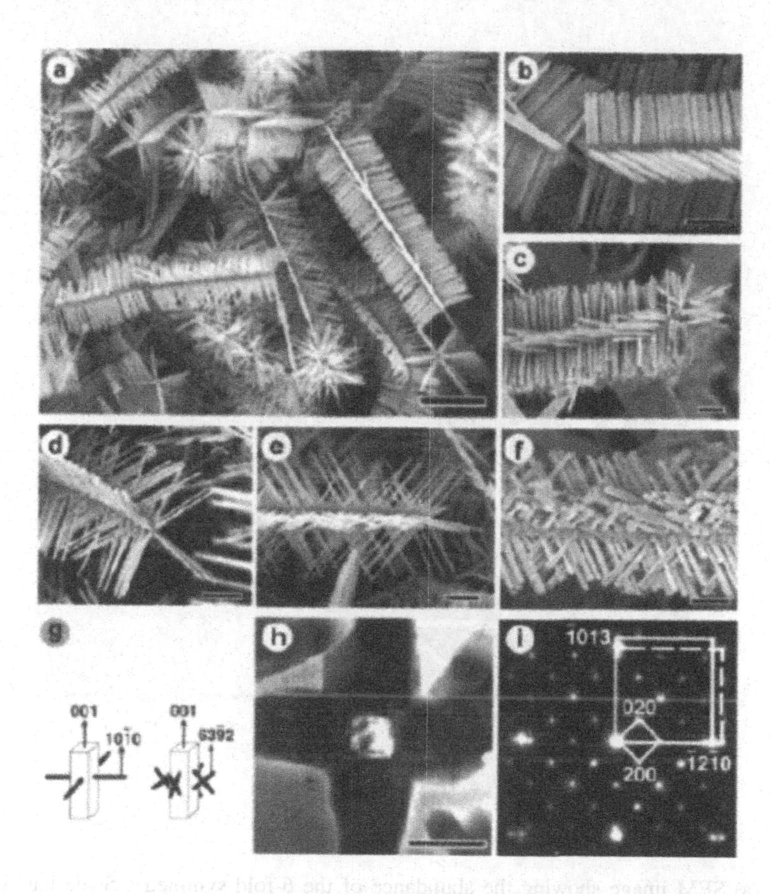

Fig. 20. (a) SEM image showing the abundance of the 4-fold nanostructures. Scale bar, 5 μm. (b~e) High magnification SEM image of the 4-fold symmetry. Scale bars are 1 μm. (f) High magnification SEM image of the 4-fold symmetry. Scale bar, 500 nm (g) Schematic model of the 4-fold symmetry. (h) Cross-sectional bright-field TEM image of 4-fold symmetry to show the four facets of the central core. Scale bar, 200 nm. (i) Selected-area electron diffraction pattern taken from (h), corresponding to the In_2O_3 core (the [001] zone axis of In_2O_3) and the branch ZnO nanorods (the dashed rectangle) [82].

detailed TEM analysis in Fig. 23. The low magnified TEM image in Fig. 23 (a) shows the well-aligned ZnO nanowires appear single crystalline, and the HREM image in Fig. 23 (b) further revealed that the single crystalline ZnO nanowires have their wire axis parallel to the longitudinal *c*-axis. This work offers the audience a quick and easy way to prepare ZnO nanowires at a much lower temperature, which is obviously advantageous in future nanoelectronic applications.

ZnO-Pb nanocable structures were recently prepared by Kong et al. [84] in a similar way to that reported in reference [25]. Fig. 24 (a) shows a TEM image revealing the typical morphology of the ZnO-Pb nanocables, which appear very straight and have an average diameter around 80 nm. Energy dispersive spectroscopy (EDS) analysis shows that the ZnO nancables consist a Pb-ZnO core around 10 nm, and a ZnO sheathing layer. The Pb-containing core appears much darker in the TEM image in

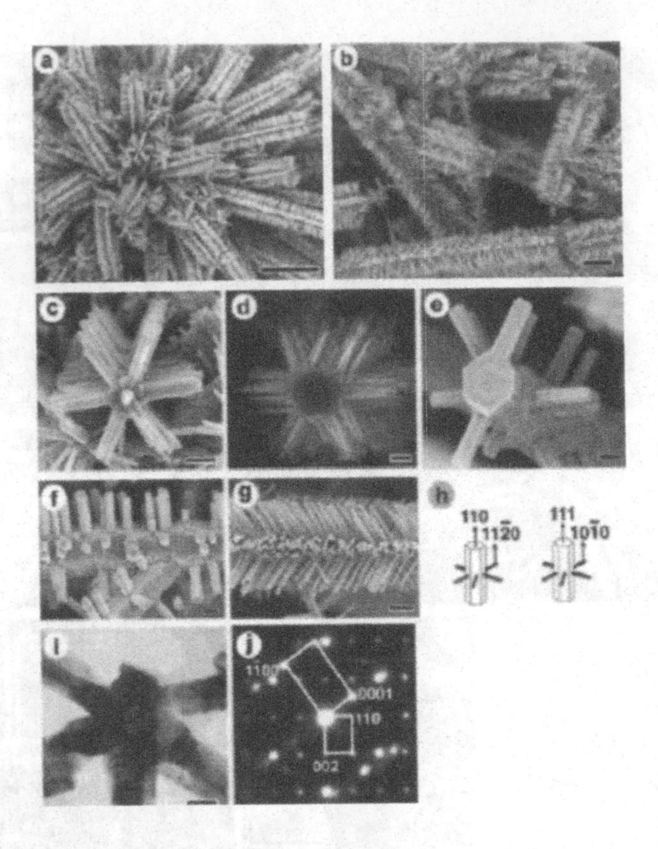

Fig. 21. (a) SEM image showing the abundance of the 6-fold symmetry. Scale bar, 10 μm. (b~d) SEM images of the 6-fold symmetry of the ZnO dendrites. Head-on look (e) and (f) Side view showing the hexagonal nature of the In_2O_3 core nanowires. The branch ZnO nanorods have hexagonal symmetry, and the same growth orientations with the In_2O_3 core nanowire. (g) The branch nanorods are not perpendicular to the In_2O_3 core. (h) Schematic orientation relationships between the In_2O_3 core nanowire and the branch ZnO nanorods. (i) Cross-sectional TEM image of 6-fold symmetry to show the six facets of the central core. (j) Selected-area electron diffraction pattern of (i) corresponding to the In_2O_3 core and the branch ZnO nanorods. Scale bar, (b), (c), (g), 1 μm; (d–f), (i), 200 nm [82].

Fig. 24 (b) because of the higher atomic scattering factor arising from the heavy Pb atoms in the core. The peculiar ZnO-Pb nanocable structures may be used as the building blocks for future nanotechnology.

Semiconductor ZnO nanotubes with a regular polyhedral shape, hollow core, and wall thickness as small as 4 nm, have been prepared in large-area substrate by vapor phase growth by Y. J. Xing, D. P. Yu, et al. [85]. X-ray diffraction analysis of the nanotubes revealed a hexagonal wurtzite ZnO phase with lattice parameters $\boldsymbol{a} = 0.325$ nm, and $\boldsymbol{c} = 0.521$ nm. SEM analysis revealed that the whole Si substrate (5×20 mm) was covered with nanoscale hollow tubes. The amount of the tubular structures is estimated more than 90 wt%. The formed ZnO nanotubes, randomly distributed on the substrate, have a diameter distribution between 30~100 nm, with a mean value ~ 60 nm.

Fig. 22. SEM micrographs of ZnO nanowires vertically aligned on alumina substrates. (a, b) ZnO nanowires grown at 450 °C have a diameter of 55 nm. (c, d) ZnO nanowires grown at 500 °C have a diameter of 190 nm [83].

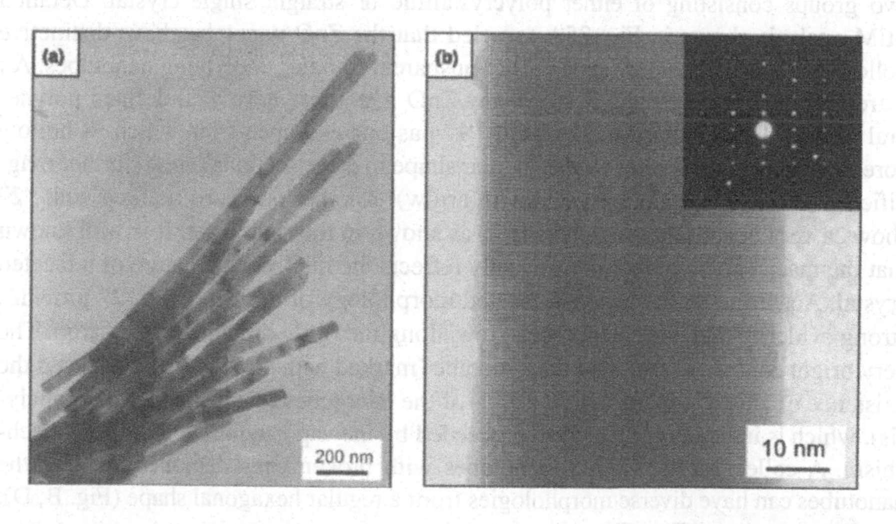

Fig. 23. TEM micrograph of ZnO nanowires: (a) Low magnified TEM image and (b) HRTEM image of a single ZnO nanowire typical with its ***c*** axial growth direction. The SAED shown in inset reveals the single-crystalline structure of the ZnO nanowire [83].

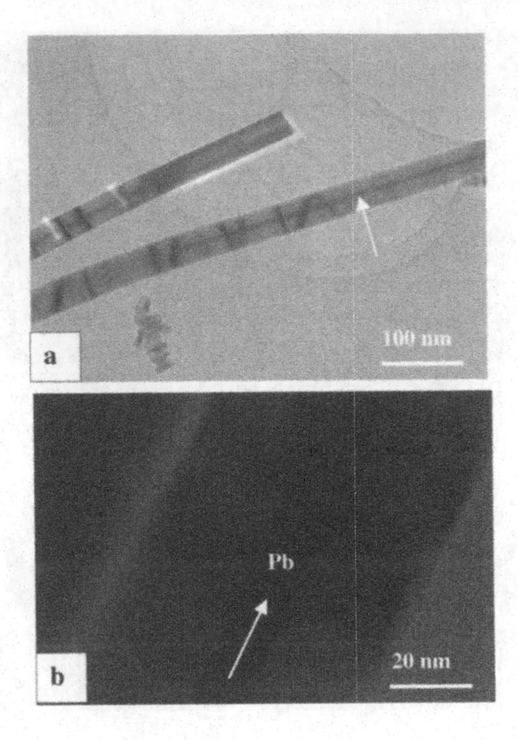

Fig. 24. Low (a) and magnified (b) TEM images of the ZnO-Pb nanocables. The core consists Pb and ZnO, and was sheathed with ZnO layer. (c) SEM image revealing abundance of hollow ZnO nanotubes on Si substrate [84].

Their length can be over a few tens micrometers. The nanotubes can be classified into two groups consisting of either polycrystalline or straight single crystal. Detailed SEM analysis shown in Fig. 25A revealed that the ZnO nanotubes have distinctive hollow cores, which appear apparently transparent to other underlying nanotubes. An extremely surprising feature is that many ZnO nanotubes have well defined polyhedral shapes. The nanotube marked with "**1**" has one end open from which its hollow core is visible, and possesses a rectangular shape in cross sectional view (further magnified in the left inset, and marked with arrow). Another nanotube marked with "**2**" shows a very beautiful hexagonal shape, as shown in the right inset. It is well known that the macroscopic morphology usually reflects the microscopic nature of a faceted crystal. Accordingly, the hexagon faceted morphology of the nanotube "**2**" provides strong evidence that those nanotubes grow along the (001) direction (the ***c***-axis). The very bright contrast at the tip of the nanotube (marked with arrow head) evidenced the existence of zinc nanoparticle at the ends of the nanotubes (confirmed by EDS analysis), which is indicative of a growth proceeded by the vapor-liquid-solid (VLS) mechanism. A collection of different nanotubes with broken ends demonstrates that the nanotubes can have diverse morphologies from a regular hexagonal shape (Fig. B, D), a to a square shape (Fig. C).

TEM provides further insight into the microstructure of ZnO nanotubes. As revealed in the TEM investigation in Fig. 26, all ZnO nanotubes manifest their hollow

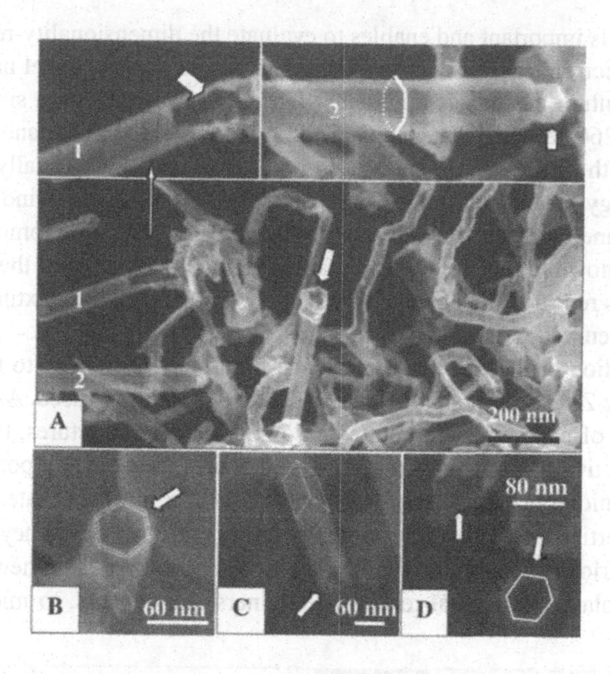

Fig. 25. (A) Magnified SEM image of the ZnO nanotubes. The broken end of the tube "1" reveals the hollow core with a rectangular shape (left inset). The tube "2" has a well-defined hexagonal shape (right inset), and appears transparent to other underlying nanotubes. The very bright-contrast at the tip of the nanotube (marked with arrow head) shows the existence of zinc-rich nanoparticle at the ends of the nanotubes. (B–D) A collection of the SEM images revealing the open tips of the ZnO nanotubes, showing that many of the ZnO nanotubes have a regular polyhedral shape [85].

core in contrast. The side walls (parallel to the incident electron beam) of the ZnO nanotubes appear much dark in the TEM contrast due to the relative larger thickness than the other part (perpendicular to the electron beam) of the nanotubes, which is independent of the focus degree. Based on the difference in morphology and microstructure, the ZnO nanotubes can be classified into two groups. As revealed from the TEM image in Fig. 26A, the first group appears "regular," and usually has a straight morphology, smooth surface, and a homogenous wall contrast. The inset corresponds to fast Fourier transform of the high resolution TEM along the [011] zone axis. Detailed TEM analysis also revealed that some ZnO nanotubes have longitudinal axis parallel to the [001] zone axis. By setting the focus point near to the Gaussian value, both the top and bottom walls are visible at the broken end of a single nanotube (indicated by two white arrows in the HREM image in Fig. 26B. The lattice fringes at the side wall (dark contrasted) corresponding to the (100) plane are well resolved (magnified and shown in inset), and reveals that the walls of the ZnO nanotubes are as solid as the bulk, that is distinct to the multi-walled graphite-like nanotubes (C, BN, BCN, and WS_2), in which there exist spacings between wall sheets with interlayer van der Waals interaction. It is worth noting that the ZnO nanotubes have a very fine wall thickness as small as 4 nm. Such a wall thickness is unique for the ZnO nanotubes compared to ZnO nanowires, and makes them a quantum-confined structure. Such a unique feature is important and enables to evaluate the dimensionality-related properties such as electron transport, optical behavior, and to explore novel nanodevices in which the quantum mechanics dominates. The EDS spectrum on a single nanotube shown in Fig. 26C revealed that chemical compositions of the ZnO nanotubes are zinc and oxygen with Zn/O ~ 1. The second group of ZnO nanotubes usually has an irregular morphology. Compared to the "regular" ZnO nanotubes, this kind of nanotubes is polycrystalline, and has a thicker wall (>20 nm), as well as inhomogeneous wall thickness, as shown in Fig. 26D. The variable bright/dark contrast of the grains on the nanotube walls revealed that the nanotubes is polycrystalline. The textured nanotubes can have very curved morphology (Fig. 26E).

The formation of the ZnO nanotubes was found closely related to the hexagonal structure of the ZnO crystal and the peculiar growth conditions used. As an extension to the concept of nanotubular structures based on lamellar structures, the ZnO nanotubes is distinctive from the nanotubes and nanowires previously reported. The novelty and the unique feature of the ZnO nanotubes offer a model material to address physical properties related to dimensionality confinement effects. They can be easily doped with various elements, and increase the flexibility to design new nanodevices from UV nanolasers, chemical reaction containers, gas sensors, to microscope tips.

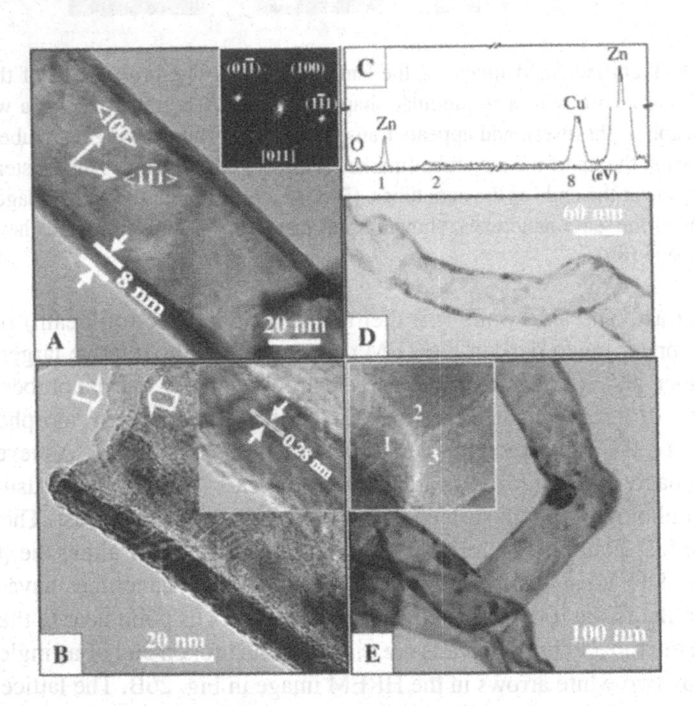

Fig. 26. (A) HREM image of a single ZnO nanotube with wall thickness around 8 nm. The electron diffraction in inset reveals a single crystalline nature of the regular ZnO nanotubes. (B) Magnified HREM image showing the details of the open end of a single ZnO nanotube. The top and bottom walls of the nanotube are well resolved and marked with two arrows. The (100) lattice plane image of the wall is shown in inset. (C) EDS analysis showing that the nanotubes consist of zinc and oxygen. (D) Low magnified TEM image of the irregular nanotube morphology. (E) TEM image of highly curved ZnO nanotubes [85].

Zinc compounds are also regarded as the spark of human life, so the ZnO nanotubes an be also used as safe nanobiological carriers for drug delivery and other biological applications due to their characteristic hollow structure. This can be achieved by first opening the tube's ends by simple mechanical grinding, and followed by filling their hollow cores with targeted substances via capillarity suction. Besides the ZnO nanotubes reported here, we also prepared bismuth nanotubes and nanotubular structures using C60 as the building blocks. The successful preparation of these tubular nanostructures from 3-D materials strongly suggests the generality to roll a range of non-lamellar structures into the distinctive nanotubes. Therefore, the synthesis of ZnO nanotubes via simple method offers a new and quick way to produce solid-state nanotubes of non-lamellar semiconductor compounds.

ZnO–Zn coaxial cable and ZnO nanotube heterojunctions were prepared by Wu et al. by pyrolysis of zinc acetylacetonate hydrate [$Zn(C_5H_7O_2)_2.xH_2O$] at around 500°C [86] in a two temperature zone furnace. Zinc acetylacetonate hydrate was first vaporized in the low temperature zone (130 ~ 140°C). The evaporated species were blown down stream to the higher temperature zone (500°C) of the furnace. The growth zone temperature is around 230°C from where the products were collected. The SEM image shown in Fig. 27 (a) and the low-magnification TEM image in (b) indicate the general morphology of the products. Detailed TEM (Fig. 27 (c)) including high resolution TEM analysis revealed that the heterostructures consist of coaxial nanocables and the ZnO nanotubes. The nanocables have a zinc core and a ZnO sheath shell. There exists an epitaxial relationship between the Zn core and the ZnO sheathing layer

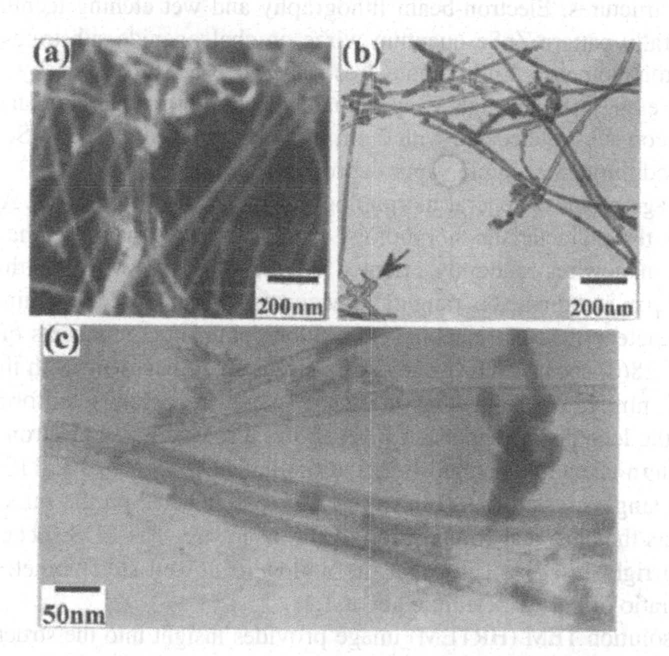

Fig. 27. SEM (a) and low magnification TEM (b) images of the heterostructures of the coaxial ZnO nanocables and the nanotubes by pyrolysis of zinc acetylacetonate. The magnified TEM image in (c) clearly shows the details of the ZnO nanotubes and heterojunctions, which are marked with arrowheads [86].

along their longitudinal *c*-axis. The ZnO nanotubes have a smallest wall thickness around 4 nm, which is comparable to the Bohr radius of the bulk ZnO, and makes the materials a quantum confined structure. Such a unique structure feature compare to ZnO nanowires is important in device applications at nanometric regime in which quantum mechanics dominates. Since ZnO compounds are regarded as the sparkle of human life, the hollow ZnO nanotubes can also be used as a safe carrier for biological applications.

4. ZnSe Nanowires

With the development of contemporary semiconductor technology, desire for nano-electronic device applications and nanometric manipulation technology is considerably increasing. Preparation of quantum nanowires becomes a challenge for scientists of a wide community. ZnSe (and ZnS) are wide-band gap semiconductor compounds, which have stimulated intensive research interests for possible applications in green-blue light emitting opto-electronic devices. ZnSe based photonic devices [87–90] have been realized recently. The miniaturization of the optical systems will eventually encounter now-existing planar chips-based fabrication limitation, and the quantum effects are of importance to understand fundamental physical problems and quantum effect based new devices as well. Though the ZnSe quantum dots have been hitherto well investigated [91–95], there are little results about one-dimensional (1D) ZnSe structures. Electron-beam lithography and wet etching techniques were used to laterally pattern ZnSe quantum wires on chalcogenide substrates [96–101]. Recently, stoichiometric ZnSe nanowires have been synthesized by B. Xiang, H. Z. Zhang et al. [102] through a vapor phase reaction of powder zinc and selenium on (100) silicon substrate coated with 2 nm thickness gold film. The ZnSe nanowires was fabricated through a simple vapor deposition method.

SEM image provides general morphology of the as-grown products. As is shown in Fig. 28(a), the characteristic morphology of the nanowires manifests the abundance of ultra-thin nanowires randomly oriented on the substrate, with lengths normally exceeding 2 μm and diameters ranging between 5 ~ 50 nm. The ultra-thin nanowires with the diameters of several nanometers are the dominant components of the whole sample. Fig. 28(b) shows a TEM image of a single ZnSe nanowire with the diameter of about 50 nm. It is visible that the ZnSe nanowire exhibits uniform diameter throughout the length. The upper-left inset shows a selected-area electron diffraction (SAED) pattern taken on the nanowire, and the lattice spots reveal a [110] zone axis of the ZnSe single crystal. EDS analysis was also conducted on the same nanowire, which depicts the chemical composition of the nanowire. The EDS spectrum shown in the lower-right inset indicates that the nanowire is well stoichiometric with the component ratio of zinc to selenium about 1:1.

High resolution TEM (HRTEM) image provides insight into the structure details of the ZnSe nanowires. As shown in Fig. 29(a), the HRTEM image represents the atomic details of a single ZnSe nanowire with diameter as thin as 5 nm. It is visible from the image that the growth direction of the nanowire is along the <111> direction. Fig. 29(b) is an HRTEM image of another single ZnSe nanowire with a

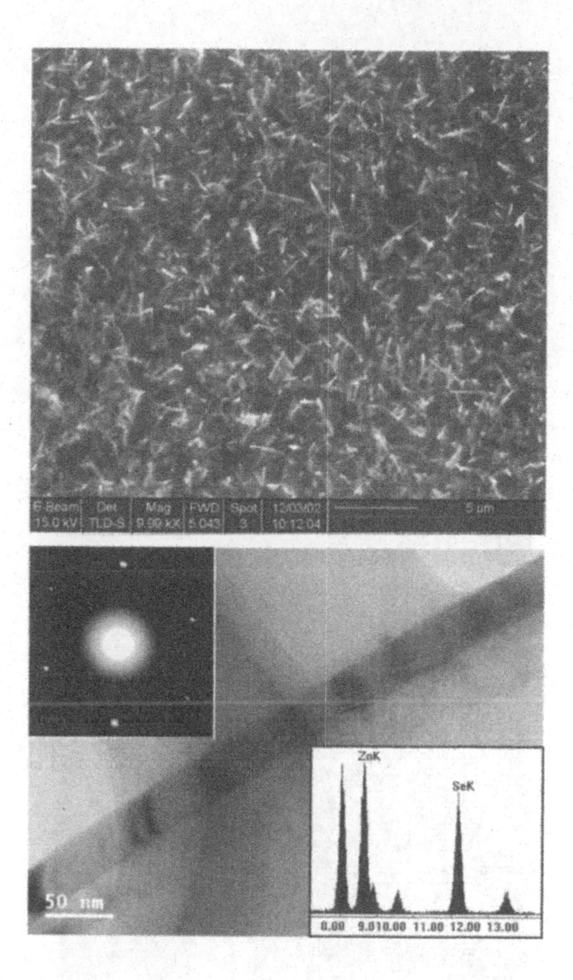

Fig. 28. (a) SEM image showing the representative morphology of the as grown ZnSe nanowires on a (100) silicon substrate. (b) TEM image of a single ZnSe nanowire. The upper-left inset shows that the nanowire is single crystal, and the lower-right inset reveals the chemical composition of the nanowire is composed of ZnSe [102].

diameter around 30 nm, and the two sets of the {220} lattice fringes are visible, revealing that the nanowire is of perfect single crystal. The corresponding fast Fourier transform (FFT) shown in the inset reveals that the incident electron beam is parallel to the <111> zone axis, with the wire axis along the [1$\bar{1}$2] direction.

Raman scattering spectroscopy was employed to confirm the structures of the ZnSe nanowires. As shown in Fig. 30(a), two Raman peaks were observed which are positioned at 204.7 and 247.2 cm^{-1}. Those are well assigned respectively to the longitudinal optical phonon mode (LO) and the transverse optical phonon mode (TO) of the ZnSe crystal, confirming that the as-grown nanowires are ZnSe crystal. The temperature dependent photoluminescence (PL) spectra were analyzed under 325 nm excitation from room temperature down to 10 K. As is shown in Fig. 30(b), a very broad PL peak (from 450 to 570 nm) is dominant centering at ~505 nm (2.46 eV),

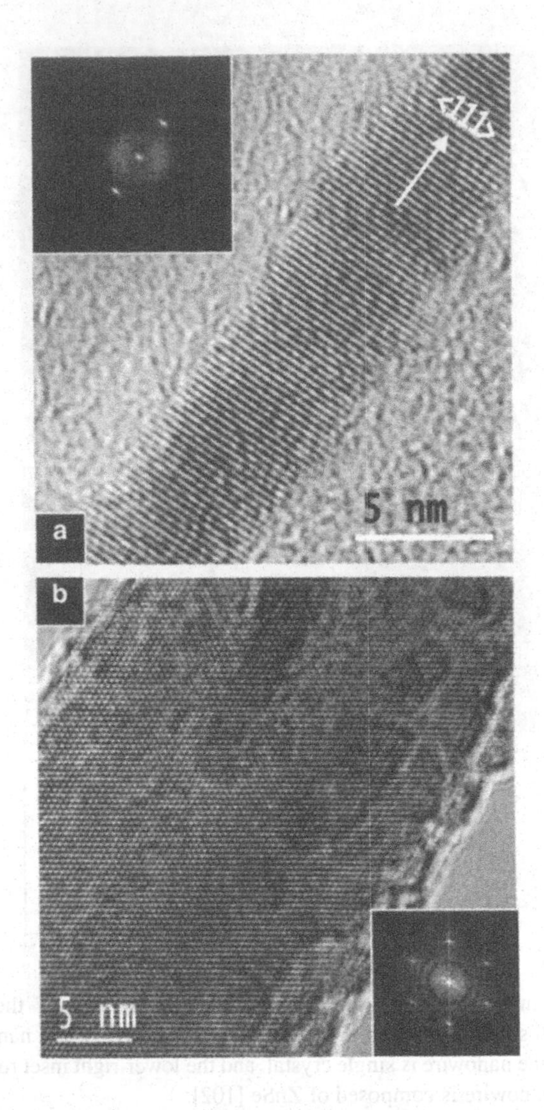

Fig. 29. (c) HRTEM images of the ultrafine ZnSe nanowires. (a) A single ZnSe with the diameter around 5 nm, and the growth direction is along the < 111 > zone axis. (b) Atomic structural details of a single ZnSe nanowire with the incident electron beam parallel to the [111] zone axis. Both insets show the FFT patterns of the corresponding HRTEM images [102].

and an increase of the emission intensity was observed as the temperature decreased from 300 K to 10 K. It is well known that the band gap energy of the bulk ZnSe crystal E_g is 2.70 eV (459 nm) and the blueshift is usually expected either due to the quantum confinement as the size decreases or because of the thermal expansion effects and the broadening of the band gaps as the temperature decreases. For example, when the diameters of the ZnSe nanoparticles decrease below 10 nm, the E_g can be as large as 3.27 eV (379 nm) [95]. On the other hand, the E_g can be increased to 2.8 eV (443 nm) as the temperature is lowered to ~50 K solely [103]. In the present case, however, the

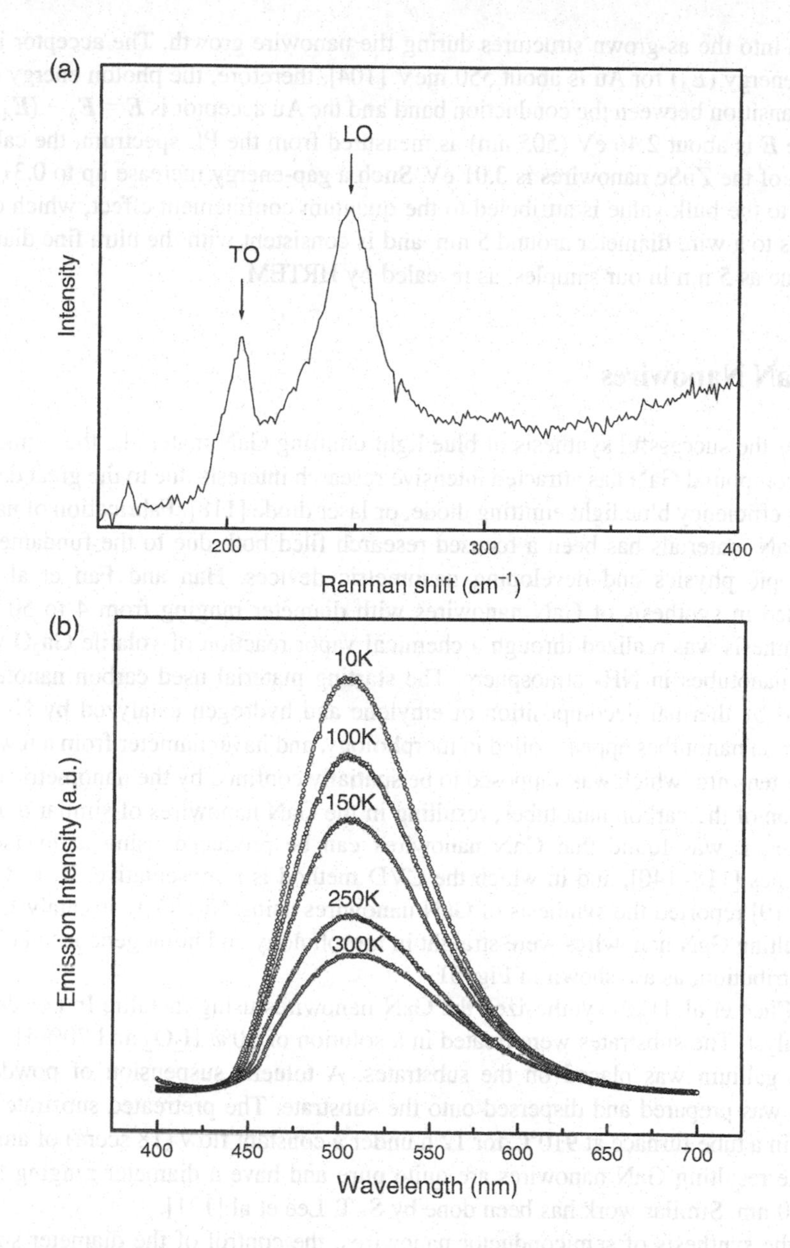

Fig. 30. (a) Raman spectrum of the as-grown ZnSe nanowires. The LO and TO Raman modes are assigned to the ZnSe crystal. (b) PL spectrum of the ZnSe nanowires. The green emission peaks around 505 nm are attributed to the Au acceptors [102].

dominant green emission relating to the recombination from defect centers (as discussed below) is so intensive that the emission corresponding to the band-edge emission was much suppressed and did not show up in the PL spectra. The most possible origin of the observed PL emission is ascribed to the medium deep acceptor of Au as the defect levels [104]. Because Au was used as the catalyst, the Au atoms can easily diffused into the as-grown structures during the nanowire growth. The acceptor ionization energy (E_A) for Au is about 550 meV [104], therefore, the photon energy due to the transition between the conduction band and the Au acceptor is $\boldsymbol{E} = \boldsymbol{E}_g - (\boldsymbol{E}_A)_{Au}$. Because $\boldsymbol{E}$ is about 2.46 eV (505 nm) as measured from the PL spectrum, the calculated $\boldsymbol{E}_g$ of the ZnSe nanowires is 3.01 eV. Such a gap-energy increase up to 0.31 eV relative to the bulk value is attributed to the quantum confinement effect, which corresponds to a wire diameter around 5 nm, and is consistent with the ultra fine diameter as fine as 5 nm in our samples, as revealed by HRTEM.

5. GaN Nanowires

After the successful synthesis of blue light emitting GaN materials, the semiconductor compound GaN has attracted intensive research interests due to the great desire for high efficiency blue light emitting diode, or laser diode [118]. Fabrication of nanosized GaN materials has been a focused research filed both due to the fundamental mesoscopic physics and developing nanometric devices. Han and Fan et al. [9] succeeded in synthesis of GaN nanowires with diameter ranging from 4 to 50 nm. The synthesis was realized through a chemical vapor reaction of volatile Ga_2O with carbon nanotubes in NH_3 atmosphere. The starting material used carbon nanotubes prepared by thermal decomposition of ethylene and hydrogen catalyzed by Ni, Fe. The carbon nanotubes appear coiled in morphology, and have diameter from a few nm to a few tens nm. which was supposed to be spatially confined by the nanometric configuration of the carbon nanotubes, resulting in the GaN nanowires of similar diameter. Later, it was found that GaN nanowires can be produced using a diverse of approaches [118–140], and in which the CVD method is representative. X. L. Chen et al. [119] reported the synthesis of GaN nanowires using $Ni(NO_3)_2$ as catalyst, and the resulting GaN nanowires were straight in morphology and homogeneous in diameter distribution, as are shown in Fig. 31.

C. Chen et al. [120] synthesized the GaN nanowires using metallic In powder as the catalyst. The substrates were treated in a solution of 30% H_2O_2 and 70% H_2SO_4. Molten gallium was placed on the substrates. A toluene suspension of powdered indium was prepared and dispersed onto the substrate. The pretreated substrate was heated in a tube furnace at 910°C for 12 h under a constant flow (18 sccm) of ammonia. The resulting GaN nanowires are quite pure and have a diameter ranging from 20 to 50 nm. Similar work has been done by S. T. Lee et al [121].

In the synthesis of semiconductor nanowires, the control of the diameter size is very important, otherwise, if the scale of the nanomaterials is much larger than the Bohr radius, the materials will behavior similarly as the bulk material, and no quantum confinement effects can be expected. A typical example of a CVD growth of GaN nanowires was introduced in details below, by which very fine GaN nanowires can be produced [124]. A conventional hot-filament CVD system was modified to grow the GaN nanowires. A p-type Si (111) wafer of 5 mm × 5 mm deposited with 10 nm Ni film were ultrasonically cleaned in acetone and ethanol for 3–5 min each to collect the final products. A drop of liquid Ga was placed on a Si substrate as Ga source, which were then loaded into the CVD chamber. The CVD chamber was first filled with H_2 gas. The temperature of the substrate was raised to about 700°C. High-purity

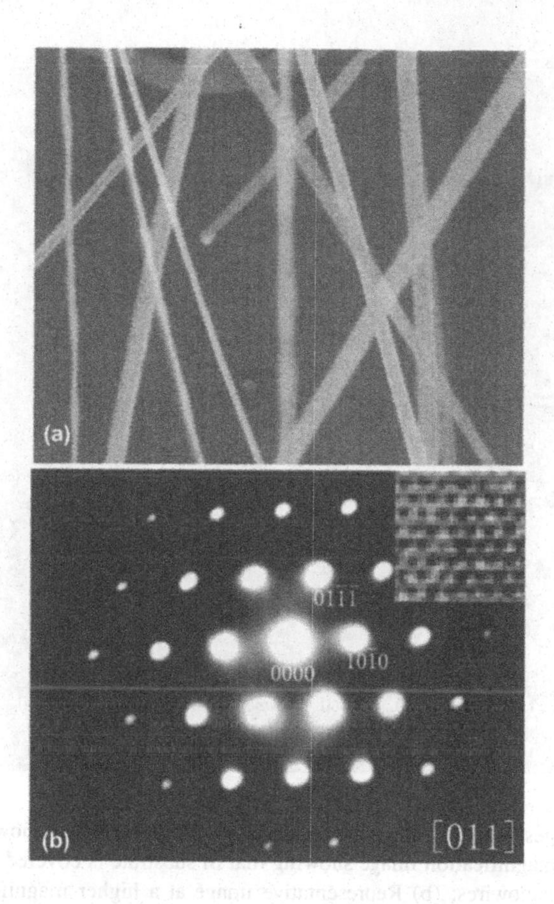

Fig. 31. (a) SEM image of GaN nanowires. (b) The selected area diffraction dots of GaN nanowires with its lattice image (inset) [119].

NH_3 with was introduced into the reaction chamber at a flow rate of 100 sccm. The reaction time was kept for 5 min under a pressure of about 100 mTorr.

Bulk quantity of the product was used to characterize structure using the X-ray diffractometer, and the diffraction spectrum. All the peaks can be indexed to the hexagonal wurtzite GaN phase.

The general morphology of the products was first analyzed using SEM. Surprisingly, it was found that the entire substrate (~1 cm^2) was covered with a thick layer of wool-like nanowire film, as shown in the SEM image at low magnification in Fig. 32(a). This image was taken near the edge of the Si substrate, and it reveals that the GaNWs can grow on area where there is thin nickel layer of about 5 nm. This implies the importance of the catalyst in the growth of the GaN nanowires. The magnified SEM image in Fig. 32 (b) shows the representative morphology, and it reveals that the film consists of pure and high density of GaN nanowires. In some area, as shown in Fig. 32(c), the nanowires tend to grown oriented. It is not yet well understood how this orientation is realized and what are the conditions to determine the orientation, and we think that the topological morphology of the substrate play a critical role. Such an orientation is interesting and important to control the growth of nanowires.

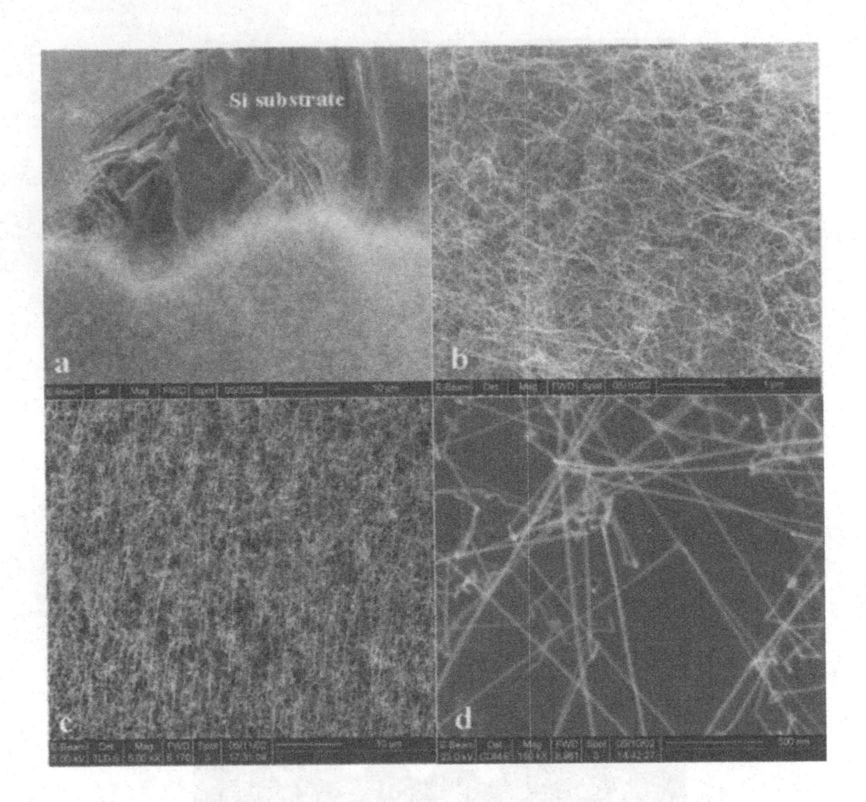

Fig. 32. SEM images revealing the general morphology of the GaN nanowires grown on Si substrate. (a) Low magnification image showing that Si substrate is covered with a thick layer of wool-like GaN nanowires; (b) Representative image at a higher magnification shows the high density of the GaN nanowires; (c) In some area the nanowires tend to grown oriented; (d) Magnified SEM image showing the uniformity and narrow distribution in diameter of the nanowires ranging from 4 to 12 nm [124].

Detailed SEM analysis shows that the Gnaws have very even distribution of diameters. The magnified SEM image shown in Fig. 32(d) reveals the smoothness and uniformity of the nanowires with diameter around 10 nm. This image was taken near the edge of the substrate, and the observed straight morphology of the nanowires is great in contrast to curved nanowires observed in the center region where the nanowires are in a high density. This indicates that the spatial confinement has significant influence on morphology of the nanowires, i.e., the nanowires grown near the edge have enough space to grow straight, while the nanowires grown at the high density regions interact with each other and eventually form a curved morphology.

Most nanowires have a very fine diameter, and based on statistic analysis of many TEM images, the diameters range is from 4 to 12 nm, with a mean diameter of about 8 nm. Such an average diameter is in general smaller than the Bohr radius of the GaN (~11 nm), and thus make the material a quantum confined structure, which is essential to further investigations of the quantum confinement effects related to materials of lower dimensionality.

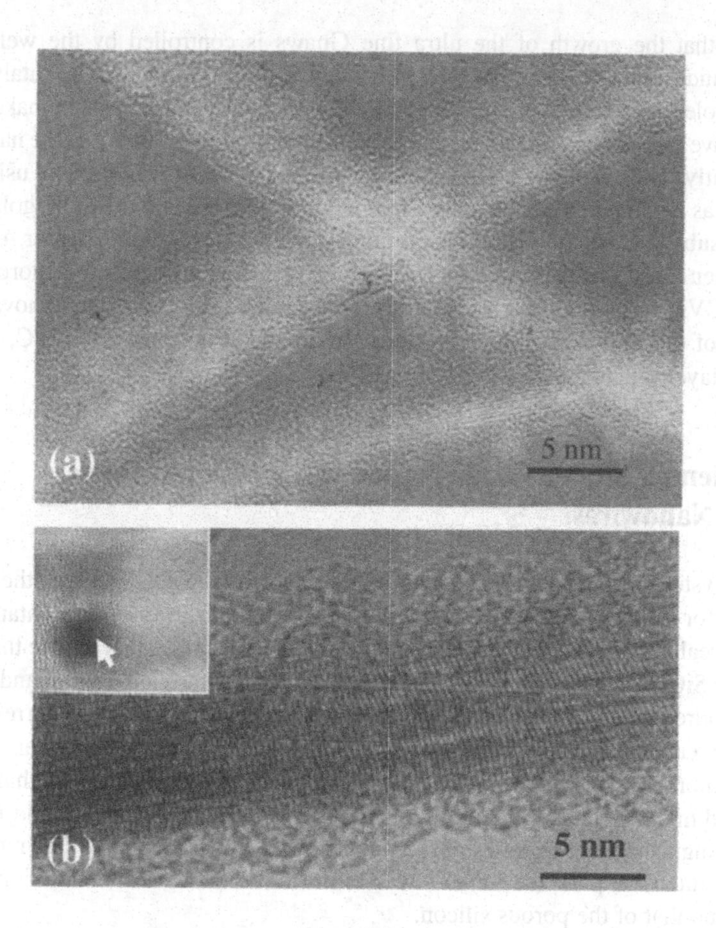

Fig. 33. HREM image of the GaN nanowires. (a) HREM image of three GaN nanowires with diameter around 5, 6, and 8 nm, respectively. (b) HREM image revealing the structure details of a single GaN nanowire. The core diameter of the nanowire is about 7 nm, and stacking faults and microtwins are visible with their fault plane parallel to the axis of the nanowire. A sheathing layer of 2 nm was observed outside the nanowire [124].

The high resolution electron microscopy (HREM) images in Fig. 33 further reveal the structure details of the GaN nanowires. The HREM image in (a) shows three single GaN nanowires with diameter around 4 ~ 8 nm. The lattice fringes are clearly visible. The GaN nanowire in the lower part of the image has the finest diameter as small as 4 nm, and one set of the lattice plane is found parallel to the axis of the wire. The HREM image in (b) reveals the existence of structure defects in a single GaN nanowire. Its core diameter is about 7 nm, and stacking faults and microtwins are visible with their fault plane (the (101) plane) parallel to the axis of the nanowire. According to our former work on Ga_2O_3 nanowires, such a configuration of structure defects play an important role in the unidirectional growth of the nanowires. A sheathing layer of about 2 nm is observed outside the core of the nanowires. One nanowire is capped with a catalyst nanoparticle (indicated with an arrow), and electron diffraction spectroscopy (EDS) analysis reveals that the nanoparticle is composted of Ga–N–Ni. This is an evidence that the growth of the ultra fine Gnaws is controlled by the well-known vapor–liquid–solid (VLS) mechanism, in which the liquid phase forming catalysts play a crucial role to the growth of the nanowires. The VLS growth mechanism makes it possible to have a controlled growth (orientation, size, morphology, et al.) of the nanowires.

Recently, GaN nantubes were synthesized via a MO-CVD growth using ZnO nanorods as a template [141]. Firstly ZnO nanorod arrays were grown on gold-coated sapphire substrate via thermal evaporation of ZnO powder at 900°C under Ar atmosphere. Then GaN was deposited on the surface of the as-grown ZnO nanorod arrays via MO-CVD technique at a temperature around 600 ~ 700°C. After removal of the template of the ZnO nanorod arrays under hydrogen atmosphere at 600°C, the GaN overcoat layer remains as GaN nanotubes.

6. Oriented Silicon Oxide Nanowires and Amorphous Silicon Nanowires

The system for synthesis of silicon oxide nanowires (SiONWs) was the same as that used for synthesis of silicon nanowires [1]. Fig. 34(a) shows a representative TEM image revealing the general morphology of the SiONWs [142]. It is visible that the as-produced SiONWs have uniform distribution in diameter around 15 nm, and a length up to hundreds micrometers. The highly diffusive ring pattern in the corresponding selected-area electron diffraction shown in inset reveals that SiONWs are of completely amorphous state. The most striking property of the SiONWs is that it emits stable and high brightness blue light. As is shown in Fig. 34(b), two broad PL peaks were distinguishable at energies of 2.65 eV and 3.0 eV, respectively. The intensity of the more intensive peak (at 3.0 eV) was found more than two orders of magnitude higher than that of the porous silicon.

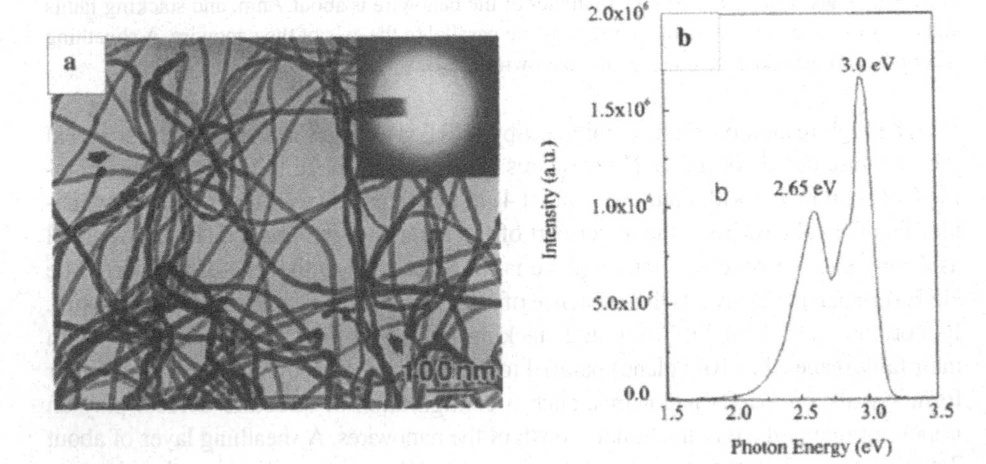

Fig. 34. (a) TEM image revealing the general morphology of the SiONWs. (b) PL of the SiONWs. Extremely intensive blue light emission was revealed peaking at 2.65 eV, and 3.0 eV [142].

Recently, super long, ultra fine silicon oxide nanowires were synthesized as a by-product by accident in the synthesis of GaN nanowires via CVD method by Yang [143], Wang [144] and Yu et al. [145]. In this method, cetain amount of liquid gallium was placed on a silicon wafer, which was in turn put into a tube furnace. When the temperature incresed to about 950 °C, amounia was introduced into the system to react with Ga. Surprisingly, millimeter-sized spheres made of silicon oxide nanowires were formed on the surface of the substrate. Those spheres appear opal or nearly transparent to naked eyes. Detailed SEM and TEM revealed that those spheres are made of self-assembly of silicon oxide nanowires. Fig. 35(a) shows a SEM image of the sphere.

From Fig. 35(a) it is visible that the sphere is multi-shelled, and the inset reveal the diameter of the sphere is around 5 millimeters. The length of the silicon oxide nanowires is >0.1 mm. From the TEM image of the SiONWs in Fig. 35(b), it is found that the nanowires are parallel to each other, and have a super fine diameter <10 nm. SAED revealed that the SiONWs are completely amorphous. The formation of the self-organized SiONWs in the presence of gallium as catalyst was explained as follows: the metallic Ga dissolved gradually Si wafer to form Si-Ga liquid phase, and the dissolved silicon was finally oxidized by the water vapor through the following reaction:

$$Si + xH_2O \rightarrow SiOx + x\ H_2$$

Those SiONWs show extremely intensive blue light emission too. The blue-light emission materials have long been of great interest for full-color display. In the field of

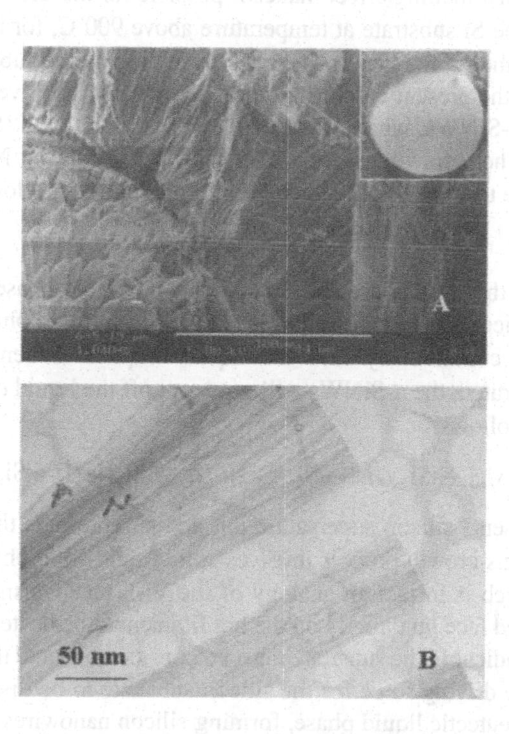

Fig. 35. (A) SEM image of the sphere-shaped self-assembly of SiONWs. Inset shows a millimeter-sized sphere which is multi-shelled, and made of pure SiONWs. (B) TEM image of the super long SiONWs [145].

recently developed SNOM, the spatial resolution depends both on the diameter of the tip and the tip-sample separation. The use of the SiONWs as the optical head in the new generation SNOM may give an opportunity to improving the resolution. It was discovered recently that unusually high transmittivity of visible light was obtained through regular arrays of holes [27] (about 150 nm in diameter, which are much smaller. The SiONWs is helpful for studying the optical phenomenon beyond the diffraction limit, e.g., the behavior when visible light propagates through nano optical wires whose diameter is far smaller than the wavelength of the propagating light.

Besides the amorphous SiONWs as discussed above, amorphous silicon nanowires were also produced by simple evaporation method [146–148]. A thin layer of 40 nm nickel was thermal deposited on heavy-doped ($1.5 \times 10^{-2}\Omega$/cm) *n*-type Si (111) wafer. The substrate was placed in a quartz tube which was heated in a tube furnace at 950°C under flowing Ar.

Fig. 36(a) shows a SEM image morphology of the amorphous SiNWs. The nanowires are highly oriented perpendicular to the substrate. As is shown in the SEM images in Fig. 36(b, c), the amorphous SiNWs are oriented perpendicular to the substrate, forming a dense films, with a length >30 μm. TEM analysis shows that the diameter of the SiNWs is between 20 ~ 80 nm, as is shown in the TEM image in Fig. 36(d).

It was found that the growth of the SiNWs here is different from the VLS mechanism for conventional whiskers [146–148]. Though the eutectic point of Si_2Ni is 993 °C, the small-size-melting effect makes it possible for the deposited Ni nanoparticles to react with the Si substrate at temperature above 900°C, forming Si_2Ni eutectic liquid droplets. So the source materials comes from directly the substrate instead from the vapor phase in the present case. Therefore, we proposed a novel model to explain the growth of the a-SiNWs, which involves the solid-liquid-solid (SLS) phases in the growth process of the nanowires. In the SLS growth, the catalyst Ni dissolve directly the silicon substrate to form Si_2Ni eutectic liquid phase in the following reaction:

$$2\text{Si}\ (s) + \text{Ni}\ (s) \rightarrow \text{NiSi}_2\ (l) \qquad (5)$$

where ***s*** represents the solid phase and ***l*** represents the liquid phase.

Because the silicon has a large solubility in the Si_2Ni liquid phase, more Si atoms will be dissolved continuously into the liquid droplets. When the liquid phase becomes supersaturated, the a-SiNWs will grow out off the liquid droplets, which can be represented as follows:

$$\text{Si}\ (s) + \text{NiSi}_2\ (l) \rightarrow \text{Si}_{super}\ \text{Ni}\ (l) \rightarrow \text{Si}_2\text{Ni}\ (l) + \text{Si}_w\ (s) \qquad (6)$$

where ***super*** represents silicon supersaturation and ***w*** represents the finally solidified SiNWs. Because this growth process involves solid-liquid-solid phases, it is named as a SLS growth, which is in fact an analogy of the VLS mechanism. Since the silicon substrate was placed face up directly on the hot filament to be heated from the back, so the temperature gradient in the substrate must be considerable, and the temperature gradient should be the driving force for the silicon substrate to be dissolved to form low temperature Ni-Si eutectic liquid phase, forming silicon nanowires at the cooler side.

Two more questions shall be addressed here. Firstly, it is not fully well understood why the final nanowires are in amorphous state instead of a crystalline phase in the

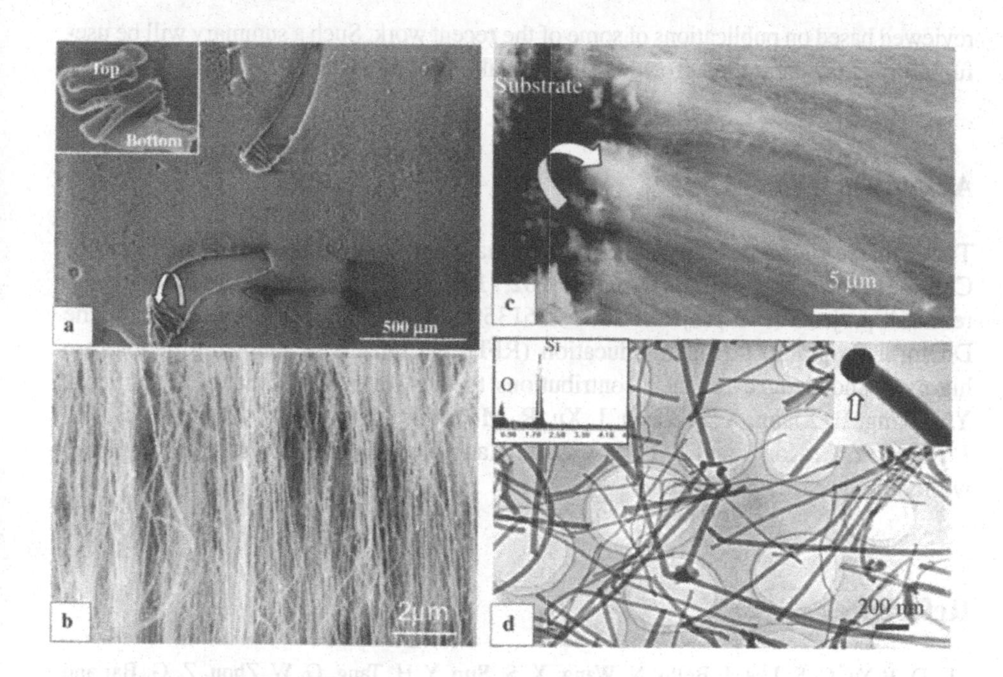

Fig. 36. (a) SEM image revealing a wool-like film on large area. Inset shows one sheet of the film scratched off the substrate. (b) SEM images showing the edge of the nanowire film showing a wide-spread orientation of the nanowires. (c) SEM image showing the details of the highly oriented nanowire film. (d) TEM image showing the morphology of the silicon nanowires. The inset in the left shows the EDS spectrum and Si and O are visible, where oxygen comes from the surface oxidation of the nanowires. The inset in the right shows a Si-Ni nanoparticle capped at the end of the nanowire [146].

present case. The most possible explanation is due to the unusual high growth rate. The estimated growth rate is about 100 nm/sec. Such a high growth rate may explain why the resulting nanowires are amorphous instead of crystalline, because the growth is so rapid that the atoms have no time to stack themselves into crystalline order. Secondly, we think the driving force for the formation of the oriented nanowire film is similar to that of the orientation of carbon nanotubes, and the crowding effect between the very dense nanowires plays an important role to keep the nanowires staying oriented. From the SEM images in the previous sections, it is evident that the density of the a-SiNWs are very high, as a result the van der Waals force between nanowires should be large. This interaction force is another important factor to keep the nanowire grow forth ward to be oriented from each other.

7. Summary

The synthesis and structure characterization of wide band-gap semiconductor nanowires (Ga_2O_3, ZnO, ZnSe and GaN) and silicon oxide, and amorphous silicon nanowires were reviewed based on publications of some of the recent work. Such a summary will be useful to direct the controlled growth of functional nanowire materials.

Acknowledgement

This project was financially supported by national Natural Science Foundation of China (NSFC, No. **50025206**, 50228202, **19834080**), national key fundamental research projects (973 projects: 2002CB613505), and by the Research Fund for the Doctoral Program of Higher Education (RFDP), China. The following people are highly acknowledged for their contributions to the work: H. Z. Zhang, S. Q. Feng, Y. J. Xing, B. Xiang, X. H. Chen, J. Xu, R. M. Wang, X. C. Sun, Y. Zhang, H. B. Jia, J. Xiang, Y. C. Kong, and Z. G. Bai, Q. L. Hang. This chapter can not be completed without their participation.

References

1. D. P. Yu, C. S. Lee, I. Bello, N. Wang, X. S. Sun, Y. H. Tang, G. W. Zhou, Z. G. Bai and S. T. Lee, *Solid State Communications* **105** (1998) 403.
2. M. Morales and C. M. Lieber, *Science* **279** (1998) 208.
3. Westwater, J., Gosain, D. P., Tomiya, S., Usui, S. and Ruda, H., *J. Vac. Sci. Technol. B* **15** (1997) 554.
4. D. P. Yu, Z. G. Bai, Y. Ding, Q. L. Hang, H. Z. Zhang and S. Q. Feng, "Silicon Nano-wires synthesized using simple physical evaporation", *Appl. Phys. Letters* **72** (1998) 3458.
5. N. Wang, D. P. Yu, et al., "Transmission electron microscopy evidence of the defect structure in Si nanowires synthesised by laser ablation", *Chem. Phys. Letters* **283** (1998) 368.
6. Dai, H., Wong, E. W., Lu, Y. Z., Fan, S. and Lieber, C. M., *Nature* **375** (1995) 769.
7. Hiruma, K., Katsuyama, T., Ogawa, K., Morgan, G. P., Koguchi, M. and Kakibayashi, H., *Appl. Phys. Letters* **59** (1991) 431.
8. K. Hiruma, M. Yazawa, T. Katsuyama, K. Ogawa, K. Haraguchi, M.Koguchi and H. Kakibayashi,. *J. Appl. Phys.* **77** (1995) 447.
9. W. Q. Han, S. S. Fan, Q. Q. Li and Y. D. Hu, *Science* **277** (1997) 1287.
10. Zhang, L. D. Zhang, X. F. Wang, C. H. Liang, X. S. Peng and Y. W. Wang, "Fabrication and photoluminescence of ordered GaN nanowire arrays", *J. Chem. Phys.* **115** (2001) 5714.
11. L. Pavesi, L. Dal Negro, C. Mazzoleni, G. Franzo and F. Priolo, "Optical gain in silicon nanocrystals", *Nature* **440** (2000) 408.
12. M. P. Zach, K. H. Ng and R. M. Penner, *Science* **290** (2000) 2120.
13. D. K. Ferry and S. M. Goodnick, Transport in Nanodtructures, Cambridge University Press, 1997.
14. Y. Kanemitsu, *Phys. Rev. B* **48** (1993) 4883.
15. Y. Makhlin, G. Schon and A. Shnirman, *Reviews of Modern Physics* **73** (2001) 357.
16. A. G. Cullis and L. T. Canham, *Nature* **353** (1991) 335.
17. J. T. Hu, M. Ouyang, P. Yang and C. M. Lieber, *Nature* **399** (1999) 48.
18. X. F. Duan, Y. Huang, Y. Cui, J. F. Wang and C. M. Lieber, *Nature* **409** (2001) 66.
19. C. M. Lieber, *Solid State Communications* **107** (1998) 607.
20. T. Rueckes, K. Kim, E. Joselevich, G. Y. Tseng, C. L. Cheung, C. M. Lieber, Y. Cui and C. M. Lieber, *Science* **289** (2001) 94.

21. M. Kociak, K. Suenaga, K. Hirahara, Y. Saito, T. Nakahira and S. Iijima, "Linking Chiral Indices and Transport Properties of Double-Walled Carbon Nanotubes", *Phys. Rev. Letters* **89** (2002) 155501.
22. A. Tilke, R. H. Blick, H. Lorenz and J. P. Kotthaus, "Coulomb blockade in quasimetallic silicon-on-insulator nanowires", *Appl. Phys. Lett.* **75** (1999) 3704.
23. J. L. Costa-Kramer, N. Garcýa, P. G. Mochales, P. A. Serena, M. I. Marques and A. Correia, "Conductance quantization in nanowires formed between micro and macroscopic metallic electrode", *Phys. Rev. B* **55** (1997) 5416.
24. M. H. Huang, S. Mao, H. Feick, H. Q. Ya, Y. Y. Wu, H. Kind, E. Weber, R. Russo and P. D. Yang, *Science* **292** (2001) 1897.
25. H. Z. Zhang, Y. C. Kong, Y. Z. Wang, X. Du, Z. G. Bai, J. J. Wang, D. P. Yu, Y. Ding, Q. L. Hang and S. Q. Feng, "Nano-scale Ga_2O_3 wires synthesized using physical evaporation", *Solid State Commun.* **109** (1999) 677.
26. C. C. Chen, C. C. Yeh, C. H. Chen, M. Y. Yu, H. L. Liu, J. J. Wu, K. H. Chen, L. C. Chen, J. Y. Peng and Y. F. Chen, "Catalytic growth and characterization of gallium nitride nanowires", *J. Am. Chem. Soc.* **123** (2001) 2791.
27. Y. C. Kong, D. P. Yu, B. Zhang, W. Fang and S. Q. Feng, "UV-emitting ZnO nanowires synthesized by a PVD approach", *Appl. Phys Lett.* **78** (2001) 407.
28. D. H. Cobden, "Nanowires begin to shine", *Nature* **492** (2001) 32.
29. Z. W. Pan, Z. R. Dai and Z. L. Wang, *Science* **291** (2001) 1947.
30. Y. Li, G. S. Cheng and L. D. Zhang, "Fabrication of highly ordered ZnO nanowire arrays in anodic alumina membranes", *J. Mater. Res.* **15** (2000) 2305.
31. Y. Li, G. W. Meng, L. D. Zhang and F. Phillipp, "Ordered semiconductor ZnO nanowire arrays and their photoluminescence properties", *Appl. Phys. Lett.* **76** (2000) 2011.
32. X. F. Duan and C.M. Lieber, *J. Am. Chem. Soc.* **122** (2000) 188.
33. T. W. Ebbesen, H. J. Lezec, H. F. Ghaemi, T. Thio and P. A. Wolff, *Nature* **391** (1998) 667.
34. B. Zheng, Y. Wu, P. Yang and J. Liu, *Adv. Mater.* **14** (2002) 122.
35. E. Betzig and K. Trautman, *Science* **257** (1992) 189.
36. A. P. Levitt, Whisker Technology, *John Wiley & Sons, Inc.* (1970) 25.
37. F. C. Frank, *Discussions Faraday Soc.* **5** (1949) 48.
38. G. W. Sears, *Acta Met.* **1** (1953) 367; G. W. Sears, *Acta Met.* **3** (1955) 457.
39. Melmed, A. J. and Gomer, R., *J. Phys. Chem.* **34** (1961) 1802.
40. R. S. Wagner and W. C. Ellis, *Appl. Phys. Letters* **4** (1964) 8.
41. E. I. Givargizov, *J. of Crystal Growth* **20** (1973) 217.
42. S. T. Lee, N. Wang, Y. F. Zhang and Y. H. Tang, *Mater. Res. Soc. Bull.* **24** (1999) 36.
43. T. J. Trentler, K. M. Hickman, S. C. Goel, A. M. Viano, P. C. Gibbons and W. E. Buhro, *Science* **270** (1995) 1791.
44. S. Geller, *J. Chem. Phys.* **33** (1960) 676.
45. T. Sasaki and K. Hijikata, *Proc. Inst. Nat. Sci. Nibon Univ.* **9** (1974) 29.
46. T. Harwig and J. Schoonman, *J. Solid. State Chem.* **23** (1978) 205.
47. G. S. Park, W. B. Choi, J. M. Kim, et al., "Structural investigation of gallium oxide (beta-Ga_2O_3) nanowires grown by arc-discharge", *J. Cryst. Growth* **220** (2000) 494.
48. Y. C. Choi, W. S. Kim, Y. S. Park, S. M. Lee, D. J. Bae, Y. H. Lee, G. S. Park, W. B. Choi, N. S. Lee and J. M. Kim, "Catalytic growth of beta-Ga_2O_3 nanowires by arc discharge", *Adv. Mater.* **12** (2000) 746.
49. J. Y. Li, X. L. Chen, G. Zhang and J. Lee, "Synthesis and structure of Ga_2O_3 nanosheets", *Mod. Phys. Lett. B* **16** (2002) 409.
50. G. R. Patzke, F. Krumeich and R. Nesper, "Oxidic nanotubes and nanorods—Anisotropic modules for a future nanotechnology", *Angew Chem Int Edit* **41** (2002) 2446.
51. G. Gundiah, A. Govindaraj and C. N. R. Rao, "Nanowires, nanobelts and related nanostructures of Ga_2O_3", *Chem. Phys. Lett.* **351** (2002) 189.

52. C. H. Liang, G. W. Meng, G. Z. Wang, Y. W. Wang, L. D. Zhang and S. Y. Zhang, "Catalytic synthesis and photoluminescence of beta-Ga_2O_3 nanowires", *Appl. Phys. Lett.* **78** (2001) 3202.
53. Z. R. Dai, Z. W. Pan and Z. L. Wang, "Gallium Oxide Nanoribbons and Nanosheets", *J. Phys. Chem. B.* **106** (2002) 902.
54. S. Iijima, *Jpn. J. Appl. Phys.* **26** (1987) 357.
55. G. W. Zhou, H. Li, H. P. Sun, D. P. Yu, Y. Q. Wang, L. Q. Chen and Ze Zhang, "Controlled Li-doping of Si nanowires by electrochemical insertion method", *Appl. Phys. Lett.* **73** (1998) 677.
56. Y. Wang, D. P. Yu et al., *to be published*, 2002.
57. J. Xiang, D. P. Yu et al., *unpublished results*, 2002.
58. D. Pastré, M. Troyon, T. Duvaut and J. L. Beaudoin, *Surf. Interface Anal.* **27** (1999) 495.
59. M. Troyon, D. Pastré, J. P. Jouart and J. L. Beaudoin, *Ultramicroscopy* **75** (1998) 15.
60. D. P. Yu, J-L. Bubendorff, J. F. Zhou, Y. Leprince-Wang and M. Troyon, *Solid State Communications* **124** (2002) 417.
61. T. Harwig, F. Kellendonk and S. Slappendel, *J. Phys. Chem. Solids* **39** (1978) 675.
62. L. Binet and D. Gourier, *J. Phys. Chem Solids* **59** (1998) 1241.
63. L. Binet and D. Gourier, *Appl. Phys. Letters* **77** (2000) 1138.
64. X. C. Wu, W. H. Song, W. D. Huang, M. H. Pu, B. Zhao, Y. P. Sun and J. J. Du, *Chem. Phys. Letters* **328** (2000) 468.
65. C. H. Liang, G. W. Meng, G. Z. Wang, Y. W. Wang and L. D. Zhang, *Appl. Phys. Lett* **78** (2001) 3202.
66. E. G. Villora, T. Atou, T. Sekiguchi, T. Sugawara, M. Kikuchi and T. Fukuda, *Solid State Communications* **120** (2001) 455.
67. M. Le Blanc and H. Sachse, *Physik. Z.* **32** (1931) 88.
68. M. R. Lorentz, J. A. Woods and R. J. Gambino; *J. Physic. Chem. Solids* **28** (1967) 403.
69. M. Fleischer and H. Meixner, *Sensors and Actuators B* **4** (1991) 437.
70. M. Fleischer and H. Meixner, *Sensors and Actuators B* **6** (1992)257.
71. M. Fleischer and H. Meixner, *J. Vac. Sci. Technol. A* **17** (1999) 1866.
72. M. Fleischer and H. Meixner, *J. Appl. Phys.* **74** (1993) 300.
73. D. M. Bagnall et al, *Appl. Phys. Lett.* **70** (1997) 2230.
74. R. F. Service, *Science* **276** (1997)895.
75. H. Cao, J. Y. Xu, D. Z. Zhang and C. Q. Cao, "Spatial confinement of laser light in active random media", *Phys. Rev. Lett.* **84** (2000) 5584.
76. M. H. Huang, Y. Y. Wu, H. Feick, N. Tran, E. Weber and P. D. Yang, *Adv. Mater.* **13** (2001) 113.
77. C. C. Chen and C. C. Yeh, *Adv. Mater.* **12** (2000) 738.
78. X. L. Chen, J. Y. Li, Y. G. Cao, Y. C. Lan, H. Li, M. He, C. Y. Wang, Z. Zhang and Z. Y. Qiao, *Adv. Mater.* **19** (2000) 1432.
79. Y. J. Xing, Z. H. Xi, D. P. Yu, Z. Q. Xue et al., *to be published*, 2003; Y. Zhang, H. B. Jia et al., to be published, 2003.
80. X. C. Sun, H. Z. Zhang, J. Xu, Q. Z., B. Xiang, S. Q. Feng, R. M. Wang and D. P. Yu, to be published, 2003.
81. Y. W. Zhu, H. Z. Zhang, X. C. Sun, S. Q. Feng, J. Xu, Q. Zhao, B. Xiang, R. M. Wang and D. P. Yu, Appl. Phys. Letters, accepted, 2003.
82. J. Y. Lao, J. G. Wen and Z. F. Ren, *Nano Letters* **2** (2002)1287.
83. S. C. Lyu, Y. Zhang, H. R. Lee, H. W. Shim, E. K. Suh and C. J. Lee, *Chem. Phys. Lett.* **363** (2002) 134.
84. Y. C. Kong, H. Z. Zhang and D. P. Yu, *unpublished* 1999.
85. Y. J. Xing et al., Appl. Phys. Letters, to be published, 2003; Y. J. Xing, D. P. Yu, *to be published*, 2003.
86. J. J. Wu, S. C. Liu, C. T. Wu and K. H. Chen, L. C. Chen, "Heterostructures of ZnO–Zn coaxial nanocables and ZnO nanotubes", *Appl. Phys. Letters* **81** (2002) 1312.

87. T. Y. Tsai and M. Birnbaum, *J. Appl. Phys.* **87** (2000) 25.
88. H. Ishikura, T. Abe, N. Fukuda, H. Kasada and K. Ando, *Appl. Phys. Lett.* **76** (2000) 1069.
89. K. Katayama, H. Yao, F. Nakanishi, H. Doi, A. Saegusa, N. Okuda and T. Yamada, *Appl. Phys. Lett.* **73** (1998) 102.
90. M. Kuhnelt, T. Leichtner, S. Kaiser, B. Hahn, H. P. Wagner, D. Eisert, G. Bacher and A. Forchel, *Appl. Phys. Lett.* **73** (1998) 584.
91. Z. H. Ma, W. D. Sun and G. K. L. Wong, *Appl. Phys. Lett.* **73** (1998) 1340.
92. D. Sarigiannis, J. D. Peck, G. Kioseoglou, A. Petrou and T. J. Mountziaris, *Appl. Phys. Lett.* **80** (2002) 4024.
93. T. Tawara, S. Tanaka, H. Kumano and I. Suemune, *Appl. Phys. Lett.* **75** (1999) 235.
94. H. Tho, H. E. Jackson, S. Lee, M. dobrowolska and J. K. Furdyna, *Phys. Rev.* B **61** (2000) 15641.
95. C. A. Smith, H. W. H. Lee, V. J. Leppert and S. H. Risbud, *Appl. Phys. Lett.* **75** (1999) 1688.
96. O. Ray, A. A. Sirenko, J. J. Berry, N. Samarth, J. A. Gupta, I. Malajovich and D. D. Awschalom, *Appl. Phys. Lett.* **76** (2000) 1167.
97. H. P. Wagner, H. P. Tranitz, R. Schuster, G. Bacher and A. Forchel, *Phys. Rev.* B **63** (2001) 155311.
98. X. F. Duan, Y. Huang and C. M. Lieber, *Nano Lett.* **2** (2002) 487.
99. W. Z. Wang, Y. Geng, P. Yan, F. Y. Liu, Y. Xie and Y. T. Qian, *Inorg. Chem. Commun.* **2** (1999) 83.
100. X. F. Duan and C. M. Lieber, *Adv. Mater.* **12** (2000) 298.
101. R. Solanki, J. Huo, J. L. Freeouf and B. Miner, *Appl. Phys. Lett.* **81** (2002) 3864.
102. B. Xiang, H. Z. Zhang, G. H. Li, R. M. Wang, J. Xu, G. W. Lu, X. C. Sun, Q. Zhao and D. P. Yu, *Appl. Phys. Letters* **82** (2003) 3330.
103. L. Malikova, W. Krystek, F. H. Pollak, N. Dai, A. Cavus and M. c. Tamargo, *Phys. Rev.* B **54** (1996) 1819.
104. P. J. Dean, B. J. Fitzpatrick and R. N. Bhargava, *Phys. Rev.* B **26** (1982) 2016.
105. M. Godlewski, E. M.Goldys, M. R. Phillips, R. Langer and A. Barski, *Appl. Phys. Lett.* **73** (1998) 3686.
106. R. Konenkamp, K. Boedecker, M. C. Lux-Steiner, M. Poschenrieder, F. Zenia, C. L. Clement and S. Wagner, *Appl. Phys. Lett.* **77** (2000) 2575.
107. Y. Li, G. W. Meng, L. D. Zhang and F. Phillipp, *Appl. Phys. Lett.* **76** (2000) 2011.
108. Y. W. Wang, L. D. Zhang, G. Z. Wang, ct al., "Catalytic growth of semiconducting zinc oxide nanowires and their photoluminescence properties", *J. Cryst. Growth* **234** (2002) 171.
109. Y. C. Wang, I. C. Leu and M. H. Hon, "Effect of colloid characteristics on the fabrication of ZnO nanowire arrays by electrophoretic deposition", *J.Mater. Chem.* **12** (2002) 2439.
110. B. D. Yao, Y. F. Chan and N. Wang, "Formation of ZnO nanostructures by a simple way of thermal evaporation", *Appl. Phys. Lett.* **81** (2002) 757.
111. Y. Y. Wu, H. Q. Yan and P. D. Yang, "Semiconductor nanowire array: potential substrates for photocatalysis and photovoltaics", *Top Catal.* **19** (2002) 197.
112. P. D. Yang, H. Q. Yan, S. Mao, R. Russo, J. Johnson, R. Saykally, N. Morris, J. Pham, R. R. He and H. J. Choi, "Controlled growth of ZnO nanowires and their optical properties", *Adv. Funct. Mater.* **12** (2002) 323.
113. J. C. Johnson, H. Q. Yan and R. D. Schaller, "Near-field imaging of nonlinear optical mixing in single zinc oxide nanowires", *Nano Letters* **2** (2002) 279.
114. Y. Y. Wu, H. Q. Yan, Huang M, B. Messer, J. H. Song and P. D. Yang "Inorganic semiconductor nanowires: Rational growth, assembly and novel properties", *Chem-Eur. J.* **8** (2002) 1261.
115. Z. X. Zheng, Y. Y. Xi, Dong P, H. G. Huang, J. Z. Zhou, L. L. Wu and Z. H. Lin, "The enhanced photoluminescence of zinc oxide and polyaniline coaxial nanowire arrays in anodic oxide aluminium membranes", *Phys. Chem. Comm.* **9** (2002) 63.

116. H. Kind, H. Q. Y an, B. Messer, M. Law and P. D. Yang, "Nanowire ultraviolet photodetectors and optical switches", *Adv. Mater.* **14** (2002) 158.
117. J. Wallace, "Single ZnO nanowires lase", *Laser Focus World* **38** (2002) 15.
118. S. Nakamura, *Science* **281** (1998) 956.
119. X. L. Chen, J. Y. Li, Y. G. Cao, Y. C. Lan, H. Li, M. He, C. Y. Wang, Z. Zhang and Z. Y. Qiao, *Adv. Mater.* **19** (2000) 1432.
120. C. C. Chen and C. C. Yeh, *Adv. Mater.* **12** (2000) 738.
121. H. Y. Peng, X. T. Zhou, N. Wang, Y. F. Zheng, L. S. Liao, W. S. Shi, C. S. Lee, S. T. Lee, *Chem. Phy. Letters* **32** (2000) 263.
122. G. S. Cheng, L. D. Zhang, Y. Zhu, et al., "Large-scale synthesis of single crystalline gallium nitride nanowires", *Appl. Phys. Lett.* **75** (1999) 2455.
123. X. F. Duan and C. M. Lieber, *J. Am. Chem. Soc.* **122** (2000) 188.
124. X. H. Chen, R. M. Wang, J. Xu and D. P. Yu, *Advanced Materials* **15** (2003) 419.
125. F. L. Deepak, A. Govindaraj and C. N. R. Rao, "Single crystal GaN nanowires", *J Nanosci. & Nanotechno.* **1** (2001) 303.
126. J. Y. Li, X. L. Chen, Z. Y. Qiao, Y. G. Cao and H. Li, Synthesis of GaN nanotubes", *J. Mater. Sci. Lett.* **20** (2001) 1987.
127. M. W. Lee, H. Z. Twu, C. C. Chen and C. H. Chen, "Optical characterization of wurtzite gallium nitride nanowires", *Appl. Phys. Lett.* **79** (2001) 3693.
128. Zhang, X. S. Peng, X. F. Wang, Y. W. Wang and L. D. Zhang, "Micro-Raman investigation of GaN nanowires prepared by direct reaction Ga with NH_3", *Chem. Phys. Lett.* **345** (2001) 372.
129. W. S. Shi, Y. F. Zheng, Wang N, C. S. Lee and S. T. Lee, "Microstructures of gallium nitride nanowires synthesized by oxide-assisted method", *Chem. Phys. Lett.* **345** (2001) 377.
130. C. C. Chen, C. C. Yeh, C. H. Liang, C. C. Lee, C. H. Chen, M. Y. Yu, H. L. Liu, L. C. Chen, Y. S. Lin, K. J. Ma and K. H. Chen, "Preparation and characterization of carbon nanotubes encapsulated GaN nanowires", *J. Phys. Chem. Solids* **62** (2001) 1577.
131. M. Q. He, P. Z. Zhou, S. N. Mohammad, G. L. Harris, J. B. Halpern, R. Jacobs, W. L. Sarney and L. Salamanca-Riba, "Growth of GaN nanowires by direct reaction of Ga with NH_3", *J. Cryst. Growth* **231** (2001) 357.
132. Z. J. Li, X. L. Chen, H. J. Li, Q.Y. Tu, Z. Yang, Y. P. Xu and B. Q. Hu, "Synthesis and Raman scattering of GaN nanorings, nanoribbons and nanowires", *Appl. Phys. A-Mater.* **72** (2001) 629.
133. K. W. Chang and J. J. Wu, "Low-temperature catalytic synthesis gallium nitride nanowires", *J. Phys. Chem. B* **106** (2002) 7796.
134. H. M. Kim, D. S. Kim and Y. S. Park, "Growth of GaN nanorods by a hydride vapor phase epitaxy method", *Adv. Mater.* **14** (2002) 991.
135. H. Y. Peng, N. Wang and X. T. Zhou, "Control of growth orientation of GaN nanowires", *Chem. Phys. Lett.* **359** (2002) 241.
136. H. W. Seo, S. Y. Bae and Park J, "Strained gallium nitride nanowires", *J. Chem. Phys.* **11** (2002) 9492.
137. J. R. Kim, H. M. So and J. W. Park, "Electrical transport properties of individual gallium nitride nanowires synthesized by chemical-vapor-deposition", *Appl. Phys. Lett.* **80** (2002) 3548.
138. H. Li, J. Y. Li, M. He et al., "Fabrication of bamboo-shaped GaN nanorods", *Appl. Phys. A-Mater* **74** (2002) 561.
139. L. X. Zhao, G. W. Meng, X. S. Peng, X. Y. Zhang and L. D. Zhang, "Large-scale synthesis of GaN nanorods and their photoluminescence", *Appl. Phys. A-Mater* **74** (2002) 587.
140. L. X. Zhao, G. W. Meng, X. S. Peng, X. Y. Zhang and L. D. Zhang, "Synthesis, Raman scattering and infrared spectra of large-scale GaN nanorods", *J. Cryst. Growth* **235** (2002) 124.
141. J. Goldberger, R. R. He, Y. F. Zhang, S. Lee, H. Q. Yan, Heon-Jin Choi and P. Yang, Single-crystal gallium nitride nanotubes, *Nature* **422** (2003) 599.

142. D. P.Yu, Q. L. Hang, Y. Ding, H. Z. Zhang, Z. G. Bai, J. J. Wang, Y. H. Zou, W. Qian, G. C. Xiong and S. Q. Feng, "Amorphous silica nano wires: Intensive blue light emitters", *Appl. Phys. Letters* **73** (1998) 3076.
143. Z. W. Pan, Z. R. Dai, C. Ma and Z. L. Wang, *J.A.C.S.* **124** (2002) 1817.
144. B. Zheng, Y. Wu, P. Yang and J. Liu, *Adv. Mater.* **14** (2002) 122.
145. D. P. Yu et al., *unpublished results*, 2001.
146. X. H. Chen D. P. Yu et al., *Chemical Physics Letters, accepted*, 2003.
147. Y.J. Xing, D. P. Yu, Q. L. Hang, H. F. Yan, J. Xu, Z. H. Xi and S. Q. Feng, *Materials Research Society Symposium—Proceedings* **581** (1999) 231.
148. D. P. Yu, Y. J. Xing, Q. L. Hang, H. F. Yan, J. Xu, Z. H. Xi and S. Q. Feng, *Physica E* **9** (2001) 305.

Chapter 12

Semiconducting Oxide and Nitride Nanowires

Lide Zhang (responsible for nitride nanowires) and Guowen Meng (responsible for oxide nanowires)

Institute of Solid State Physics, Chinese Academy of Sciences, P.O. Box 1129, Hefei 230031, P.R.China

1. Synthesis of Oxide Nanowires (written by Guowen Meng)

Binary semiconducting oxides, such as Ga_2O_3, In_2O_3, SnO_2, and ZnO have distinctive properties and are now widely used as transparent conducting oxide materials and sensors [1]. For example, β-Ga_2O_3, with direct band gap energy $Eg \approx 4.9$ eV, exhibits conduction and luminescence properties, and thus has potential applications in optoelectronic devices and high-temperature stable gas sensors. In_2O_3 ($Eg = 3.55$–3.75 eV) has been widely used in microelectronic field as window heaters, solar cells, and flat-panel display materials. SnO_2, a very important n-type semiconductor with a wide band gap ($Eg = 3.6$ eV at 300 K), is well known for its potential applications in gas sensors, transparent conducting electrodes and transistors. ZnO ($Eg \approx 3.2$ eV) is ideal for low-voltage and short wavelength (green or green/blue) electro-optical devices such as light emitting diodes and diode lasers [2]. On the other hand, one-dimensional silica nanostructures have attracted considerable attention because of their potential future applications in high-resolution optical heads of scanning near-field optical microscope or nanointerconnections in integrated optical devices [3]. Rationally controlled growth of nanowires consisting of the above-mentioned materials is important for their applications in nanoscale electronics and photonics. In this section we present the rational growth of these nanowires based on the results from our research group by using physical evaporation method. In addition, a novel synthetic method for semiconducting oxide nanowire arrays will be demonstrated by using electrochemical deposition of metal nanowires in the nanochannels of porous anodic aluminum oxide templates and subsequent oxidization in oxygen atmosphere.

1.1. Vapor–liquid–solid growth of oxide nanowires from vapor species generated by physical evaporation

As described in the previous chapter of this book, a well-accepted mechanism of nanowire growth from vapor species is the so-called Vapor–Liquid–Solid (VLS) process, which was proposed by Wagner and Ellis in 1964 for silicon whisker growth [4, 5]. In this mechanism, the liquid surface has a large accommodation coefficient and is therefore a preferred site for the absorption of vapor species of nanowire materials. After the liquid alloy becomes supersaturated with the gas-phase reactants of nanowire materials, the nanowire growth occurs by precipitation at the solid–liquid interface. Conceivably, nanowires of different compositions can be synthesized by choosing suitable catalysts (or solvents) and growth temperatures. A good catalyst should be able to form a liquid alloy with the desired nanowire material; ideally they should be able to form a eutectic alloy. Meantime, the growth temperature should be higher than the eutectic point, but lower than the melting point of the nanowire material. Generally speaking, the diameters of the nanowires generated from VLS process will be determined by the size of the eutectic alloy droplets, therefore desired diameters of nanowires could be obtained by controlling the size of the eutectic alloy droplets. Both physical method (including thermal evaporation, laser ablation and arc discharge) and chemical method can be used to generate the vapor species required during the nanowire growth.

Here, we will demonstrate how to use VLS mechanism to synthesize oxide (β-Ga_2O_3, In_2O_3 and ZnO) nanowires using Au as catalysts during rapid thermal evaporation in dry oxygen atmosphere.

1.1.1. β-Ga_2O_3 nanowires from VLS growth

For the growth of β-Ga_2O_3 nanowires [6], a thin film of gold (c.a.15 nm) was evaporated on gallium arsenide (GaAs) substrate under high vacuum as the catalyst/initiator for nanowire growth. The GaAs substrate with Au thin film was rapidly (about 7 min) heated up to 1240 °C in a horizontal tube furnace and kept at this temperature for 1 hour in a constant flow of mixture gas ($Ar/O_2 = 4:1$, 200 sccm) atmosphere. Upon heat-up, the Au thin film will break up and self-aggregate into large quantities of Au nanoclusters on the surface of GaAs substrate. The Au nanoclusters etch GaAs substrate to form ultrafine Au-Ga binary eutectic liquid droplets at about 550 °C, and thus create a dense vapor of Ga species around the GaAs substrate region, which will act as Ga source for the growth of β-Ga_2O_3 nanowires. Subsequently, the ultrafine Au-Ga droplets absorb both Ga vapor (evaporated from GaAs) and O_2 (from the mixture gas) to form eutectic Au-GaO_x alloy liquid droplets rather than Au-Ga-O, because Au, Ga and O atoms can't form a miscible Au-Ga-O eutectic alloy. When Ga and O atoms in the Au-GaO_x liquid droplets become supersaturated, solid crystal β-Ga_2O_3 will precipitate and grow anisotropically in the form of nanowire with a preferential direction [6].

X-ray diffraction spectrum (Fig. 1) taken from the as-synthesized products demonstrates that the sharp diffraction peaks in the pattern can be indexed to a monoclinic gallium oxide structure, in good agreement with the reported data of bulk β-Ga_2O_3 crystals ($a_0 = 5.80$ Å, $b_0 = 3.04$ Å, $c_0 = 12.23$ Å, $\beta = 103.42°$, JCPDS 11-370). However, in the XRD spectrum of the as-synthesized β-Ga_2O_3 products the strongest peak is only (200) rather than (004), (104) and (200) of the bulk β-Ga_2O_3 powder. Therefore the as-synthesized β-Ga_2O_3 products may grow preferentially along the [100] axis.

Transmission electron microscopy (TEM) observation reveals that the as-synthesized β-Ga_2O_3 products are nanowires with diameters of about 20–50 nm, and

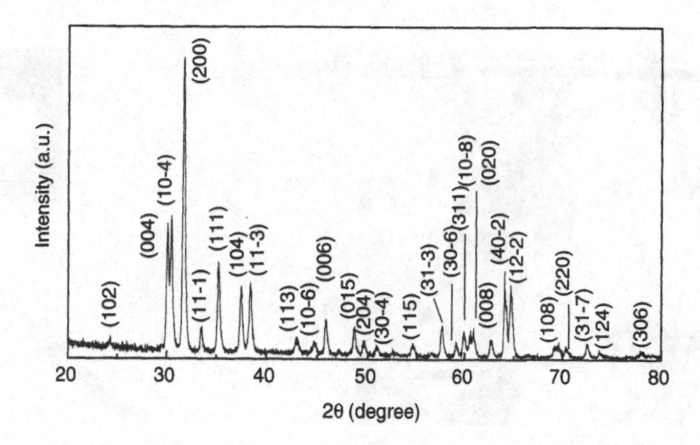

Fig. 1. X-ray diffraction pattern of the as-synthesized β-Ga_2O_3 nanowires. [Reprinted with permission from [6]. Copyright [2001]; American Institute of Physics.]

lengths of several micrometers (Fig. 2a). The detailed structure and composition of the individual β-Ga_2O_3 nanowires have been characterized using TEM, selected area electron diffraction (SAED), and energy dispersive x-ray fluorescence (EDX) analysis. Fig. 2b shows a representative TEM image of two straight nanowires. The SAED pattern (inset) recorded perpendicular to the axis of one nanowire can be indexed for the <010> zone axis of β-Ga_2O_3. Fig. 2c shows a single β-Ga_2O_3 nanowire terminated with a nanoparticle, indicating that the nanowires grew through VLS mechanism. EDX measurements (Fig. 2d) made on the terminal nanoparticle (upper, indicated by A) and the nanowire stem (lower, indicated by B) reveal that the nanoparticle is composed mainly of Au and Ga, with a small amount of O; while the nanowire stem is stoichiometric Ga_2O_3 calculated from the quantitative analysis data within experimental error, in agreement with the morphological and microstructural features of the nanowires generated from VLS process. Fig. 3 is a typical high-resolution TEM (HRTEM) image of a single nanowire. The lattice plane of (100) with interplanar spacing of 5.60 Å is clearly displayed. The angle between the (100) plane and the long axis of the nanowire is about 76°, indicating that the nanowire axis is parallel to the [100] crystalline orientation of β-Ga_2O_3, and further confirming that the nanowire preferably grew along the [100] direction. In addition, there exists a very thin amorphous layer on the surface of the nanowire.

Room temperature photoluminescence (PL) measurements (Fig. 4) demonstrate that two apparent PL peaks at about 330 and 475 nm are related to the β-Ga_2O_3 nanowires, while two peaks at about 380 nm (with weak intensity) and 525 nm are related to the β-Ga_2O_3 powders, respectively. Compared with the PL feature of β-Ga_2O_3 powder, the PL of β-Ga_2O_3 nanowires increases significantly in intensity with a blue-shift about 50 nm. The stable blue emission at 475 nm and an ultraviolet emission at 330 nm may be related to the defects such as the oxygen vacancy and the gallium–oxygen vacancy pair. In our VLS growth of β-Ga_2O_3 nanowires, large quantities of O vacancies and (V_O, V_{Ga}) may be easily produced [6]. Furthermore, just as revealed in Fig. 3, a very thin layer of amorphous gallium oxide has been formed on the nanowire surface. The β-Ga_2O_3 nanowires may have possible applications in optoelectronic nanodevices.

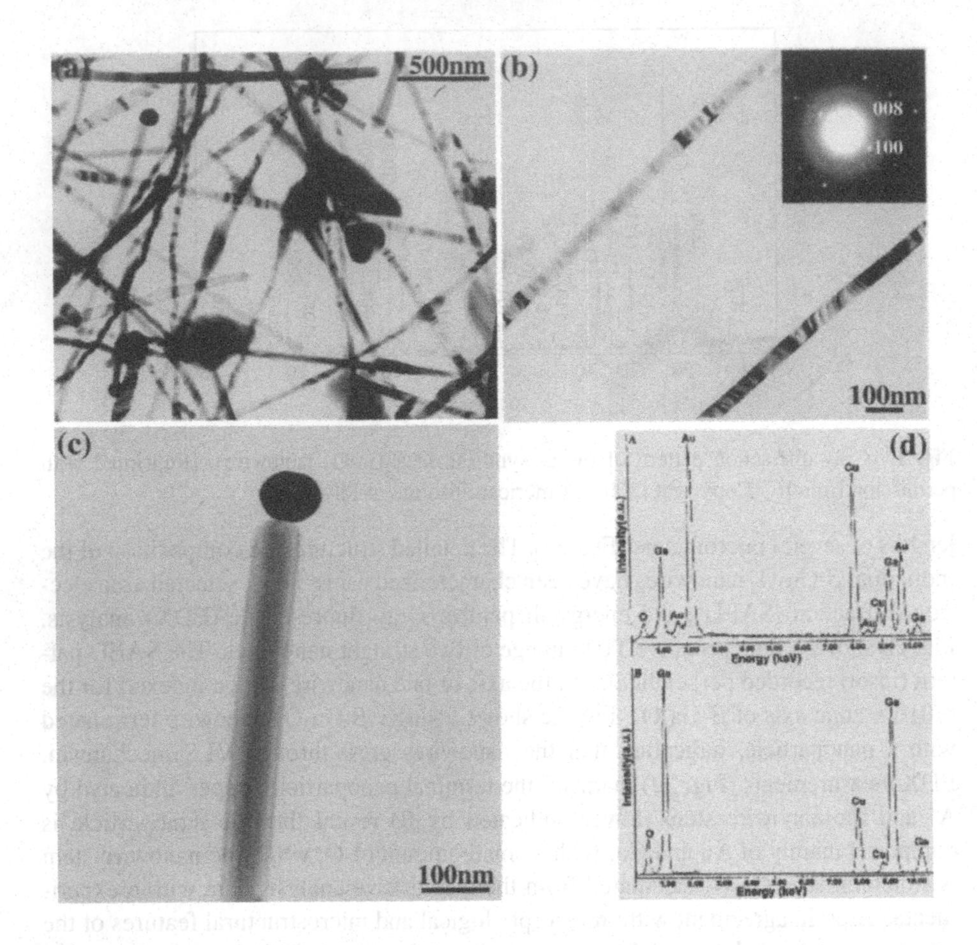

Fig. 2. (a) A typical TEM image shows the general morphology of β-Ga_2O_3 nanowires. (b) A TEM image of two individual β-Ga_2O_3 nanowires, the inset (up right) shows a [010] SAED pattern recorded perpendicular to the long axis of one nanowire. (c) A TEM image of a single nanowire terminated with a nanoparticle. (d) EDX spectra of the nanoparticle (A) and the nanowire stem (B), Cu signals are generated from microgrid mesh that supports the nanowires. [Reprinted with permission from [6]. Copyright [2001]. American Institute of Physics.]

1.1.2. In_2O_3 nanowires from VLS growth

As for the growth of In_2O_3 nanowires, a thin film of Au (c.a. 30 nm) was evaporated on indium phosphoride (InP) substrate. Except rapidly heated to 1080 °C and kept at this temperature for 1.5 h, the other parameters are the same as those used for the growth of β-Ga_2O_3 nanowires. In this case, Au nanoclusters absorb both In vapor (evaporated from InP) and O_2 (from the mixture carrier gas) to form Au-InO_x alloy droplets, which direct the growth of In_2O_3 nanowires through VLS mechanism [7].

It should be pointed out that the diameters and the amount of Au nanoclusters are dependent on the thickness of the Au thin film and the growth temperature. Thus it is possible to control both the diameters and the amount of nanowires by modifying the thickness of the thin Au film.

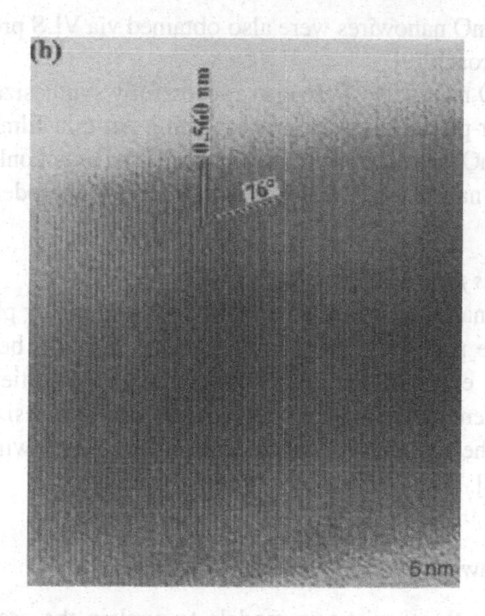

Fig. 3. A typical HRTEM image of the β-Ga_2O_3 nanowire. The (100) lattice planes are clearly resolved. [Reprinted with permission from [6]. Copyright [2001]. American Institute of Physics.]

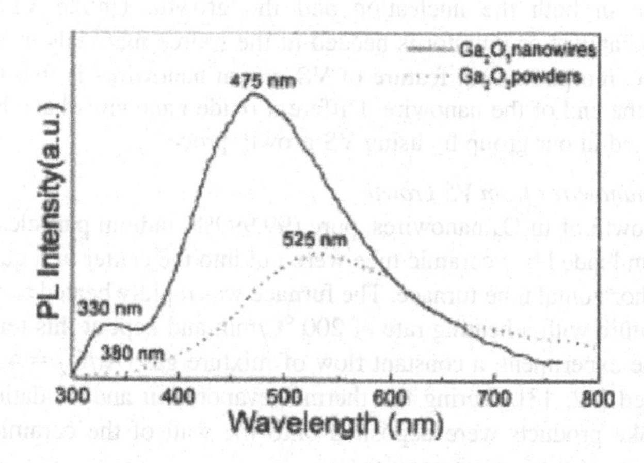

Fig. 4. Room temperature PL spectra of β-Ga_2O_3 nanowires and β-Ga_2O_3 powders with UV fluorescent light excitation of 250 nm and filter wavelength of 310 nm. [Reprinted with permission from [6]. Copyright [2001]. American Institute of Physics.]

1.1.3. ZnO nanowires from VLS growth

For the synthesis of ZnO nanowires, instead of using Au thin film as catalyst we utilized self-made Au nanoparticles as catalyst/initiator to produce ZnO nanowires by rapidly evaporating Zn powders mixed with Au nanoparticles at 900 °C in a constant flow of mixture gas ($Ar/O_2 = 9:1$, 100 sccm) atmosphere [8]. Additionally, ultraviolet-emitting ZnO nanowires were also obtained via VLS process by a physical vapor deposition approach [9].

Furthermore, ZnO nanowires were also successfully synthesized via VLS mechanism through a vapor-phase transport process using Au thin film on Si substrate as catalyst [10]. These ZnO nanowires have already been used as not only room-temperature ultraviolet nanowire nanolasers [2] but also ultraviolet photodetectors and optical switches [11].

1.1.4. SiO_x nanowires from VLS growth

The above oxide nanowires generated from VLS process by physical evaporation are crystalline. Oxide nanowires of amorphous state can also be obtained via VLS process by physical evaporation. For example, large quantities of silicon oxide nanowires with different morphologies were successfully synthesized by simple physical evaporation of the mixture of silicon and silica xerogel with Fe nanoparticles loaded in its pores [3].

1.2. Vapor-solid growth of oxide nanowires

It is accepted that there exist two models to explain the growth mechanism of whiskers and nanowires: VLS growth and vapor-solid (VS) growth. We have already demonstrated VLS growth of nanowires in detail. In this section, we will show VS growth of oxide nanowires. In VS growth of nanowires, structural defects play an important role in both the nucleation and the growth. Unlike VLS growth of nanowires, no catalyst or initiator is needed in the source materials in VS growth of nanowires. The morphological feature of VS-grown nanowires is that there exists a conical tip at the end of the nanowire. Different oxide nanowires have been successfully synthesized in our group by using VS growth process.

1.2.1. In_2O_3 nanowires from VS growth

For VS growth of In_2O_3 nanowires, pure (99.999%) indium particles with diameters of 1–3 mm loaded in a ceramic tube were put into the center of a quartz tube in a conventional horizontal tube furnace. The furnace was rapidly heated to 1030 °C from room temperature with a heating rate of 200 °C/min and kept at this temperature for 2 hours. In the experiment, a constant flow of mixture gas (Ar/O_2 = 4:1, 100 sccm) was maintained [12, 13]. During the thermal evaporation and oxidation, the light-yellow woollike products were deposited onto the wall of the ceramic tube at the downstream side of the indium powder.

Powder X-ray diffraction measurements show that the as-synthesized product is a cubic structure of In_2O_3 with a cell constant of a = 1.012 nm (JCPDS File No. 6-0416). No unreacted elemental indium was detected. Scanning electron microscopy (SEM) observations reveal that the products consist of a large quantity of uniform nanowires with typical lengths in the range of 15–25 micrometers, and diameters from 40 to 100 nanometers (Fig. 5a).

TEM observations reveal that there were no particles at the ends of the nanowires, instead there exists a conical tip at the end of each nanowire (Fig. 5b). A magnified TEM image of a single In_2O_3 nanowire with a cone-tip end and its corresponding selected area electron diffraction (SAED) pattern are shown in Fig. 5c. The SAED

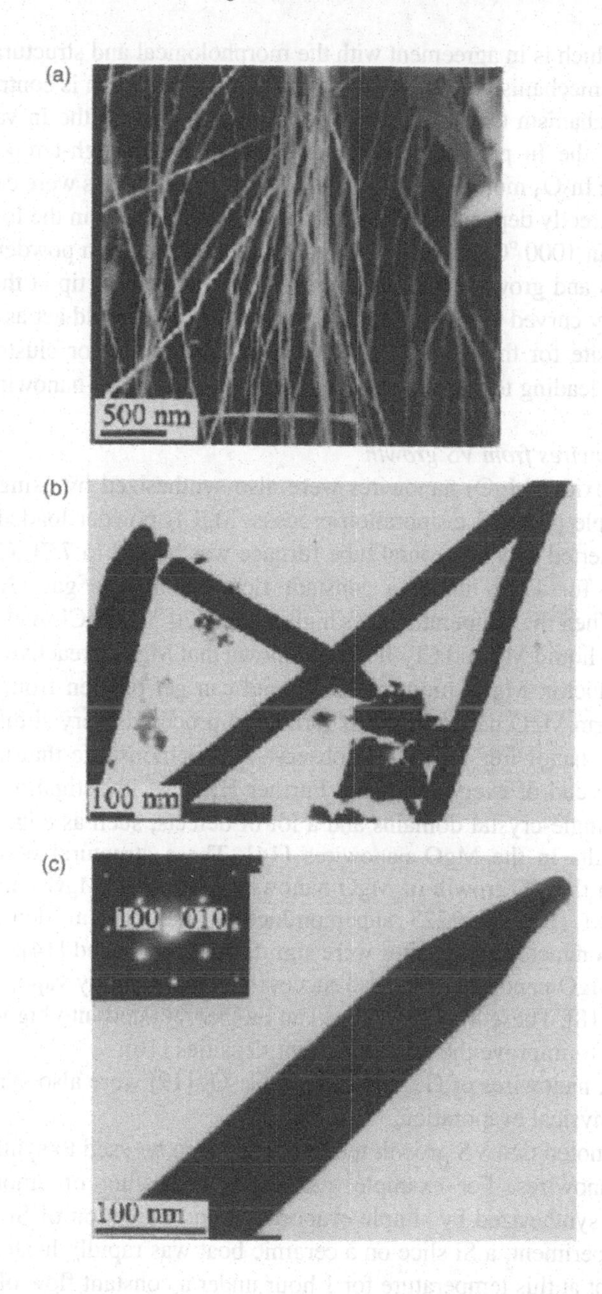

Fig. 5. (a) A typical SEM image of the as-synthesized In_2O_3 nanowires. (b) Typical TEM image of several In_2O_3 nanowires, each nanowire with a conical end. (c) Typical morphology of one In_2O_3 nanowire with a cone-tip end, the *insert* is the SAED pattern taken from this nanowire. [These figures are taken from [12].]

pattern reveals that the nanowires are single-crystalline and grow along <100>, with the surfaces being enclosed by {001} similar to the growth of the In_2O_3 nanobelts [1].

In our VS growth of In_2O_3 nanowires, the only source material used is indium powder without any other intermediates. Virtually, a conical tip (Fig. 5b) is at the end of each nanowire, which is in agreement with the morphological and structural feature of the spiral-growth mechanism. Therefore it is likely that the growth is controlled by the spiral-growth mechanism through a vapor–solid process. Herein the In vapor species (evaporated from the In particles) combines with O_2 at the high-temperature zone (1030 °C) to form In_2O_3 molecules. After that, the In_2O_3 molecules were carried by the flowing Ar and directly deposited on the wall of the ceramic tube in the lower-temperature region (about 1000 °C) at the downstream side of the indium powder, nucleating at the defect sites and growing into nanowires. Since the conical tip at the end of the nanowire is highly curved with a reduced melting point, this would act as the energetically favorable site for the adsorption of the In_2O_3 molecules or clusters from the vapor, eventually leading to the one-dimensional growth of In_2O_3 nanowires [13].

1.2.2. MgO nanowires from VS growth

Magnesium oxide (MgO) nanowires were also synthesized by using VS growth method in a simple physical evaporation process. $MgCl_2$ powder loaded in a quartz boat that was inserted in a horizontal tube furnace was heated to 750 °C and kept at this temperature for 1.5 h under a constant flow of mixture gas ($Ar/H_2 = 9:1$, 30 sccm) [14]. When the temperature was higher than 720 °C, $MgCl_2$ melted, and then decomposed into liquid Mg and Cl_2. It is well known that Mg is a reactive metal, while Si is a semiconductor. Mg is more reductive and can get oxygen from SiO_2 of the quartz boat to form MgO nanowires; this formation process is very similar to the VS growth of Ga_2O_3 nanowires [15]. TEM observations demonstrate that there exists a conical tip at the end of every nanowire. Further HRTEM investigations reveal that there are some single-crystal domains and a lot of defects, such as edge dislocations and stacking faults in the MgO nanowires [14]. These structural defects play an important role in the VS growth of MgO nanowires. After the MgO nanowires were incorporated into (Bi, Pb)-2223 superconductors, the critical densities of the nanowire-superconductor composites were significantly enhanced [14].

Previously, MgO nanorods were also successfully prepared by vapor-solid growth mechanism [16–18]. These MgO nanorods can be incorporated into high-temperature superconductors to improve the critical current densities [16].

Additionally, nanowires of Ga_2O_3 [15] and Ge_2O_3 [19] were also synthesized via VS growth by physical evaporation.

It should be noted that VS growth method could also be used to synthesize amorphous oxide nanowires. For example, nanoropes consisting of amorphous SiO_x nanowires were synthesized by simple evaporation and oxidation of Si [20]. During the synthesis experiment, a Si slice on a ceramic boat was rapidly heated to 1050 °C in 6 min and kept at this temperature for 1 hour under a constant flow of mixture gas ($Ar/O_2 = 9:1$, 200 sccm) atmosphere. As a result, high-yield of amorphous silicon oxide nanoropes was achieved on the Si slice. More recently, aligned SiO_x nanowire arrays standing on Si substrate were also successfully synthesized through a simple method by heating single-crystalline Si slice covered with SiO_2 nanoparticles at 1000 °C in flowing Ar (160 sccm) atmosphere [21]. SEM observations reveal that all the SiO_x nanowires are parallel to each other and perpendicular to the Si substrate, and form self-oriented, regularly aligned arrays (Fig. 6). Furthermore, the diameter of each nanowire becomes smaller and smaller from the bottom to the top of the SiO_x

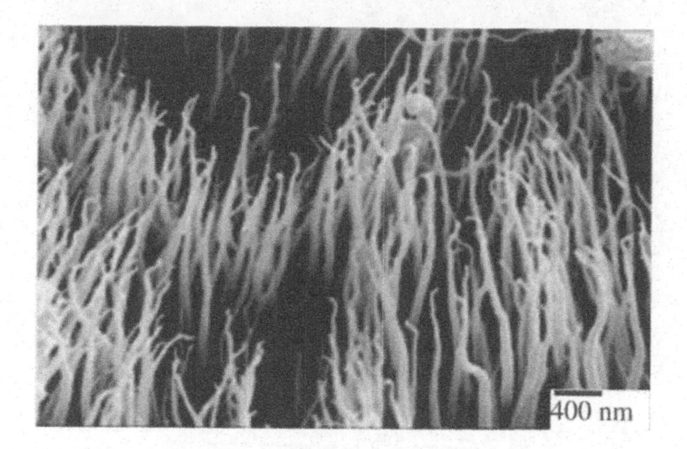

Fig. 6. A typical low-magnification SEM morphology of the aligned SiO_x nanowire arrays. [This figure is taken from [21].]

nanowire, which may be attributed to the decrease of SiO_x vapor in the later period of heating process.

1.3. Oxide nanowires from the oxidization of metal nanowires

It is well known that porous anodic aluminum oxide (AAO) formed in acid electrolytes possesses hexagonally ordered porous structures with channel diameters ranging from below 10 to 200 nm, channel lengths from 1 to over 100 μm, and channel density in the range 10^{10}–10^{12} cm^{-2} [22]. These unique structural features and its thermal and chemical stability make AAO ideal templates for the fabrication of ordered nanowire arrays.

Based on AAO templates, a novel synthetic method for semiconducting oxide nanowire arrays has been developed in our research group. The method involves a three-step process: (1) electrochemical generation of AAO templates with highly ordered hexagonal array of nanochannels; (2) electrochemical deposition of pure metals in the nanochannels of AAO template; (3) oxidation of metal nanowire arrays embedded in AAO template in oxygen or air atmosphere.

Firstly, we tried to produce semiconductor ZnO nanowire arrays using this method. Through-hole AAO templates were fabricated by a two-step anodization process as described previously [23]. A layer of Au was sputtered onto one side of the AAO templates, serving as the working electrode in a standard three-electrode electrochemical cell. The electrolyte contained 80 g/l $ZnSO_4 \cdot 7H_2O$ and 20 g/l H_3BO_3. The electrochemical deposition was performed at 1.25 V relative to the Ag^+/AgCl reference electrode at room temperature, with carbonate serving as the counter electrode. After electrochemical deposition, the Zn nanowire arrays embedded in AAO templates were heated in air at 300 °C for different periods of time (from 0 to 35 h) [24].

X-ray diffraction (XRD) spectra (Fig. 7) were taken from Zn nanowire arrays with different heat treatment conditions. No peak associated with ZnO was found when the spectrum was taken immediately after Zn was electrochemically deposited into the nanochannels of AAO (curve *a*, Fig. 7). Even after being exposed to air at room

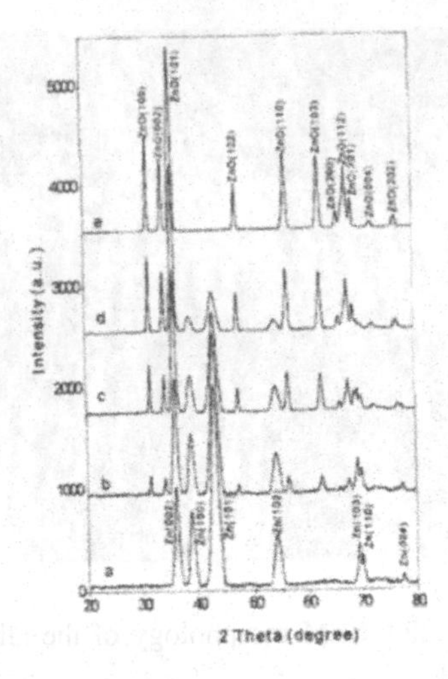

Fig. 7. XRD patterns of the Zn nanowire (40 nm) arrays embedded in AAO templates with different heat treatment conditions. (a) As-electrodeposited, (b) after heated at 300 °C for 8 h, (c) for 15 h, (d) for 25 h, and (e) for 35 h, respectively. [Reprinted with permission from [24]. Copyright [2001]. American Institute of Physics.]

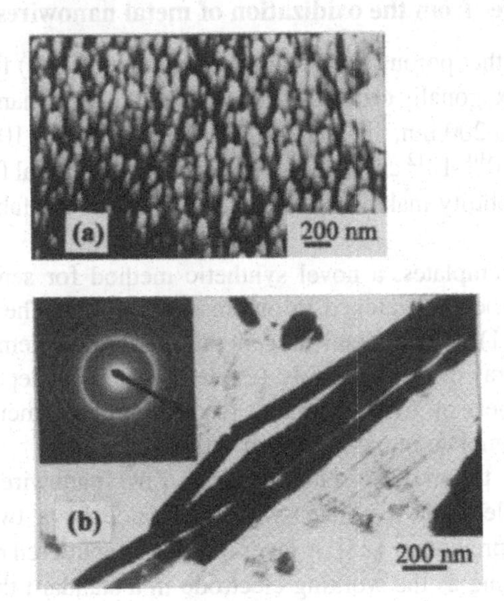

Fig. 8. (a) SEM image of In_2O_3 nanowire arrays after partially removing AAO. (b) TEM image of individual In_2O_3 nanowires and their corresponding SAED pattern, revealing cubic polycrystalline structures of the In_2O_3 nanowires. [Reprinted with permission from [26]. Copyright [2001]. American Institute of Physics.]

temperature for one week, there is still no ZnO peak in XRD spectrum. It can be seen clearly that with the increase of heat treatment time, the intensities of the Zn peaks become weaker and weaker, while those of ZnO peaks become stronger and stronger (from curve *b* to *d*, Fig. 7). After heated at 300 °C for 35 h, all of the Zn peaks disappeared (curve *e*, Fig. 7), indicating that Zn deposited in the nanochannels of AAO had been oxidized completely. In addition, the peak positions and their relative intensities (curve *e*, Fig. 7) are consistent with the standard powder diffraction patterns of ZnO, indicating that there is no preferred orientation and that the ZnO nanowires are polycrystalline. The broadening of ZnO peaks is due to the small particle size. Photoluminescence measurements on the ZnO nanowires show a blue PL band in the wavelength range of 450–650 nm caused by the singly ionized oxygen vacancy in ZnO nanowires [24].

Semiconducting oxide nanowire arrays of In_2O_3 [25, 26] and SnO_2 [27] have also been successfully synthesized by using the electrochemical deposition and subsequent oxidization method. Fig. 8a shows an SEM image of In_2O_3 nanowire arrays after the AAO template was partially removed. Fig. 8b is a TEM image of individual In_2O_3 nanowires and their corresponding SAED pattern, indicating the cubic polycrystalline structures of the In_2O_3 nanowires [26].

2. Synthesis of Nitrides Nanowires (written by Lide Zhang]

According to functional properties, one-dimensional nanoscale nitrides can be mainly divided into two categories. One is the group III-V nitride, such as BN, AlN, GaN and InN etc. They have stimulated great interest due to their importance in basic scientific research and their potential as high-temperature electronic and visible/near-UV optoelectronic devices. The other one is the one-dimensional nanostructured materials with unique mechanical properties (high hardness, high modulus, surprisingly high strength, good flexibility and good resistance to thermal shock etc.). Si_3N_4, CN_x and TiN are representative candidates. These one-dimensional nanostructures can be used as micro- or nano-tools, nano-lever, and nano-axletrees etc in micro- and nanomechanical systems, nano-robots and nanomotors etc. Therefore, studies of this kind of one-dimensional nanostructures have received extensive attention. In the following we will introduce separately the synthetic methods, the growth mechanisms of one-dimensional nanostructures, advantages and disadvantages of various synthetic methods, characterization of morphologies and structures, and properties for InN, Si_3N_4, GaN, and AlN in the above-mentioned two categories of one-dimensional nanostructures.

GaN, a wide direct-bandgap semiconductor, is an ideal material for the fabrication of blue light-emitting diodes and laser diodes [28, 29], which has been the subject of great attention recently. It is well known, not only theoretically but also experimentally, that one-dimensional GaN (nanowires, and nano-ribbons) can improve the performance of blue/green and ultraviolet optoelectronic devices [30–34]. Therefore, the development of new methods for synthesizing one-dimensional nano-GaN has attracted considerable attention. The fact that InN has promising transport and optical properties and its large drift velocity at room temperature could render it better than GaAs and GaN for field effect transistor [35]. Understanding of the optical and structural properties of InN will speed up the development of light-emitting group III nitride devices. Ceramic Si_3N_4 whiskers have a broad range of applications in numerous industries because of its interesting properties, such as high strength, light weight, good resistance to thermal shock and oxidation. Because of quantum confinement effect for nanowires with diameters < 100 nm it can be predicted that Si_3N_4 nanowires may exhibit some new properties. Some results for Si_3N_4 nanowires have proved that Si_3N_4 nanowires have surprisingly high strength with good flexibility [36, 37]. In the following, GaN, InN, and Si_3N_4 nanowires growth will be described in detail, whereas, AlN, BN, and CN_x nanowires growth will be introduced briefly.

2.1. Porous anodic aluminum oxide template method for growth of nitride nanowires

2.1.1. GaN nanowires

In the following, the synthetic method of GaN nanowires using porous anodic aluminum oxide as confinement templates will be introduced. This method was first reported by Cheng et al. [38]. The mixture of Ga and Ga_2O_3 powders with a molar ratio of 4 : 1 was put at the bottom of an alumina crucible, which was located in a quartz tube of a tube furnace. A porous Mo network was placed on the mixture, on which a porous alumina membrane was laid. The quartz tube was evacuated by a mechanic vacuum pump, and then, a N_2 flow passed through the quartz tube. After this degassing and filling NH_3 was repeated for several times, the furnace was heated to 900 °C. The NH_3 flow was held at 300 mL/min. At this time, the following reactions took place in the furnace:

$$Ga_2O_3(S) + 4\ Ga\ (L) \rightarrow 3\ Ga_2O\ (V) \qquad (1)$$

$$Ga_2O\ (V) + 2\ NH_3\ (G) \rightarrow 2\ GaN\ (S) + H_2O\ (G) + 2\ H_2\ (G) \qquad (2)$$

where, S, L, V and G are solid, liquid, vapor, and gas, respectively. When the furnace temperature decreased to room temperature, GaN nanowires were taken off from the surface of porous alumina membrane.

Fig. 9 gives the high resolution TEM images and electron diffraction pattern of a 30 nm diameter GaN nanowire whose growth axis corresponds to the (100) direction of single crystal hexagonal (wurtzite). The lattice constant of the 2-D mesh shown in (c) is 2.76 Å in excellent agreement with the lattice constant in the (100) plane of hexagonal GaN.

The formation of GaN nanowires is through chemical vapor reaction in pores of the porous alumina template. The pores of porous alumina template play a role of confinement growth of GaN nanowires, and the template does not take part in the chemical reaction process. An advantage of this method is that we may prepare arrays of GaN nanowires.

As pointed out by Cheng et al. [39], twisted and curved GaN nanorods or nanowires have been produced by chemical reactions in confined space of templates. This is a common disadvantage of chemical reactions in confined spaces to prepare GaN nanowires. The chemical reactions in confined spaces can be used to synthesize not only GaN nanowires, but also various other one-dimensional materials and even their arrays [40].

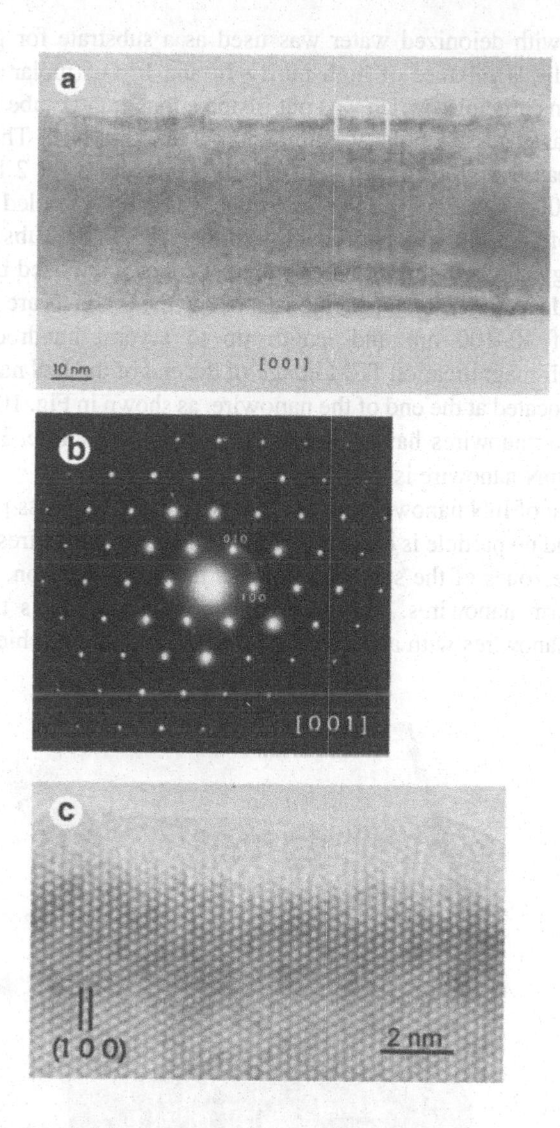

Fig. 9. High-resolution TEM images and electron diffraction pattern of a 30 nm diameter GaN nanowire whose growth axis corresponds to the (100) direction of single crystal hexagonal (wurtzite) GaN. (a) Lower magnification image showing the entire cross-section of the nanowire. (b) Selected-area electron diffraction pattern of hexagonal [001] GaN. (c) The portion of the image outlined in (a) enlarged to a magnification showing atomic resolution. The lattice constant of the 2-D mesh shown in (c) is 2.76 Å in excellent agreement with the lattice constant in the (100) plane of hexagonal GaN.

2.1.2. InN nanowires

Until very recently, only a few reports on InN nanowires have been found [40, 41]. Zhang et al. [40] first achieved large-scale synthesis of hexagonal wurtzite structure single-crystalline InN nanowires with diameters ≤ 100 nm through a gas reaction from mixtures of In metal and In_2O_3 powders with flowing NH_3 at 700 °C via a vapor-solid (VS) process. The main experimental process is as follows. Firstly, an alumina membrane rinsed with deionized water was used as a substrate for growth of InN nanowires. Secondly, a mixture of high purity In and In_2O_3 (molar ratio 4:1) was placed in an alumina crucible, which was put inside a long quartz tube of the furnace. The quartz tube was degassed under vacuum and purged with NH_3. The furnace temperature was increased to 700 °C and kept at this temperature for 2 h and the NH_3 flowing rate was 400 sccm. Finally, the quartz tube was quickly cooled down with the NH_3 flow turned off. A light yellow product was observed on the substrate.

The morphology and structure of the products were characterized by XRD, SEM, TEM, HRTEM and EDX. The results demonstrate that the products are InN nanowires with diameters of 10–100 nm and length up to several hundred micrometers (Fig. 10a). The high-magnification TEM image of the end of the InN nanowires shows that no particle is located at the end of the nanowire, as shown in Fig. 10b. XRD analysis reveals that the nanowires have a hexagonal wurtzite structure. HRTEM image indicates that the InN nanowire is single crystal (see Fig. 10c).

The mechanism of InN nanowire growth is possibly a VS process [42] because no catalyst is used, and no particle is observed at the ends of the nanowires. Possibly, InN nucleated on some zones of the substrate through the vapor reaction, and then these nuclei grew to form nanowires. The advantage of this method is that large-scale synthesis of InN nanowires with a single crystal structure can be achieved.

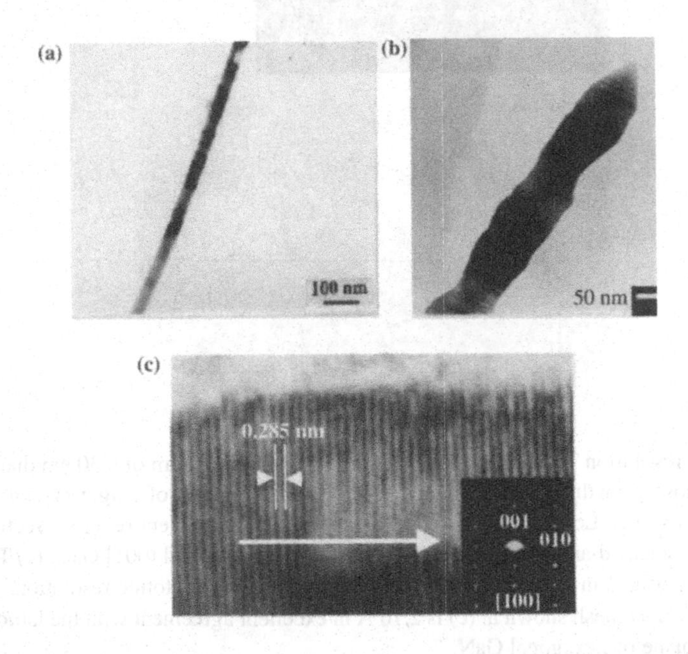

Fig. 10. (a) A TEM image of a single InN nanowire. (b) A typical high-magnification TEM image of the end of the InN nanowire. (c) HRTEM image of an InN nanowire, showing a clean and structurally perfect surface. In this image, an arrow indicates the growth direction [001] of the nanowire. The space of about 0.285 nm between arrowheads corresponds to the distance between (002) planes. The inset at the right-hand bottom of the image is the SAED patterns of the nanowire. The nanowire grows along [001] and it is enclosed by ± (001) and ± (010) crystallographic facets. [These figure are taken from [40].]

H. Parata et al. [41] prepared InN whiskers by chemical vapor deposition (CVD) using [bis (3-dimethylamino) propyl] indium as growth precursor.

Finally, it should be pointed out that the diameters of InN whiskers were distributed in a very wide range (10–200 nm).

2.2. Vapor-solid growth of nitride nanowires

2.2.1. Si_3N_4 nanowires

Si_3N_4 nanowires were prepared by carbothermal nitridization of SiO_2 xerogel with carbon nanoparticles in its pores at 1430 °C for 6 hours in N_2 atmosphere [43].

The products were characterized by TEM and HRTEM. TEM observations show that the products are nanowires with diameters of 20–70 nm and lengths of 20–50 μm (Fig. 11). HRTEM observation shows that all the α-Si_3N_4 nanowires are perfect single crystals (see Fig. 12).

The SEM observation exhibits that the products are fairly pure nanowires with several tens of micrometers in lengths. The crystalline nanowires are characterized as hexagonal α-Si_3N_4 with lattice constants of a = 0.78 nm and c = 0.56 nm. The HRTEM image of a Si_3N_4 nanowire indicates that the Si_3N_4 nanowire is a single crystal. The axis of the nanowire is normal to $[1\,\bar{2}\,\bar{2}]$ lattice plane. It should be noted that the axes of the nanowires might also be normal to {100} or {001} lattice planes.

The little effect of the metal catalyst on the growth of the nanowires and absence of the metal/Si alloy droplets on the tips of the nanowircs, which is proved by TEM observations, suggest that VSL mechanism does not work in the present synthesis. The most possible mechanism is a VS process.

2.2.2. AlN nanowires

AlN, a wide-band-gap semiconductor, is an important substrate and packaging material. It has high thermal conductivity, good electrical resistance, low dielectric loss, and low coefficient of thermal expansion that closely matches that of silicon, good mechanical strength, and excellent chemical stability [44, 45]. Therefore, AlN is

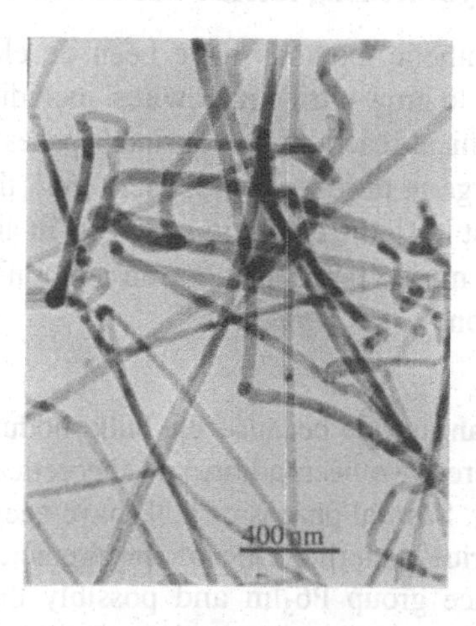

Fig. 11. A TEM image of Si_3N_4 nanowires.

Fig. 12. A HRTEM image of an individual Si_3N_4 nanowire.

the ideal material for high density, high-power and high-speed integrated circuit applications and for applications as electrical packaging material and as components in structural composites.

Most recently, AlN whiskers and nanowires have attracted much attention because they have a perfect or nearly perfect crystal structure and a higher thermal conductivity (320 W $\cdot$ (m $\cdot$ K)$^{-1}$) [46] than ordinary polycrystalline AlN ceramics (30–260 W $\cdot$ (m $\cdot$ K)$^{-1}$) [46, 47], and they could provide better reinforcement in the polymer composites for electronic packaging [48]. Direct reaction of Al with N_2 etc has been employed in the fabrication of AlN nanowires [49].

2.3. Other methods for synthesizing nitride nanowires

A wide variety of synthetic methods have been developed besides the above-mentioned two approaches to grow nitride nanowires, including laser-assisted catalytic growth (LCTG) [50], chemical reactions in confined spaces provided by carbon nanotube [51, 52], arc discharge in nitrogen atmosphere [53], the direct reaction of NH_3 with metal without catalyst [54], the catalytic synthesis in an electric heating furnace [55–56], a sublimation of nitride powder under an ammonia flow [57], and radical-source molecular deposition.

2.3.1. CN_x nanowires

CN_x is also an important nitride because the bulk modulus of CN_x is higher than that of diamond [58]. There have been additional theoretical investigations confirming the prediction [59, 60]. Several phases of CN_x have been considered as possibilities for the superhard nitride material. One of the predicted phases, β-C_3N_4, has a hexagonal form with space group $P6_3lm$ and possibly the highest bulk modulus. Another form that has been considered is cubic C_3N_4 with space group $P\,\bar{4}3m$.

This form has a bulk modulus marginally less than that of the β-phase. These theoretical predictions lead scientists to give intense efforts to synthesize this material and several synthetic methods were developed to prepare CN_x one-dimensional nanostructures. As a result, different synthetic methods gave different phases, such as amorphous or disordered phases and CN_x only contain very low amounts of N, i.e. $x < 1$. In other words, the composition of the CN_x is obviously different from the theoretically predicted value, therefore, it is necessary to develop new synthetic methods to prepare the one-dimensional carbon nitride with composition of the theoretical value.

In summary, we have described briefly the synthesis and growth mechanism of several oxide and nitride nanowires mainly achieved in our group. And a lot of research has been done in the recent years and a wide variety of materials have been grown in one-dimensional form by many synthetic approaches. However, the studies of nitride and oxide one-dimensional nanostructures are sill in the starting stage. The synthetic methods, characterization and property measurements should be continuously developed. The dynamic growth processes for controlled structures should be understood further. The applications of nitride and oxide one-dimensional nanostructures need to be opened.

References

1. Z. W. Pan, Z. R. Dai and Z. L. Wang, *Science* **291** (2001) 1947.
2. M. Huang, S. Mao, H. Ferck, H. Yan, Y. Wu, H. Kind, E. Weber, R. Russo and P. Yang, *Science* **292** (2001) 1897.
3. C. H. Liang, L. D. Zhang, G. W. Meng, Y. W. Wang and Z. Q. Chu, *J. Non-Cryst. Solids* **277** (2000) 63.
4. R. S. Wagner and W. C. Ellis, *Appl. Phys. Lett.* **4** (1964) 89.
5. R. S. Wagner, in *Whisker Technology*, Edited by A. P. Levitt, Wiley, New York (1970) p. 47.
6. C. H. Liang, G. W. Meng, G. Z. Wang, Y. W. Wang, L. D. Zhang and S. Y. Zhang, *Appl. Phys. Lett.* **78** (2001) 3202.
7. C. H. Liang, G. W. Meng, Y. Lei, F. Phillipp and L. D. Zhang, *Adv. Mater.* **13** (2001) 1330.
8. Y. W. Wang, L. D. Zhang, G. Z. Wang, X. S. Peng, Z. Q. Chu and C. H. Liang, *J. Crystal Growth* **234** (2002) 171.
9. Y. C. Kong, D. P. Yu, B. Zhang, W. Fang and S. Q. Feng, *Appl. Phys. Lett.* **78** (2001) 401.
10. M. Huang, Y. Wu, H. Feick, N. Tran, E. Weber and P. Yang, *Adv. Mater.* **13** (2001) 113.
11. H. Kind, H. Yan, B. Messer, M. Law and P. Yang, *Adv. Mater.* **14** (2002) 158.
12. X. S. Peng, Y. W. Wang, J. Zhang, X. F. Wang, L. X. Zhao, G. W. Meng and L. D. Zhang, *Appl. Phys. A Mater. Sci. & Proc.* **74** (2001) 437.
13. X. S. Peng, G. W. Meng, J. Zhang, X. F. Wang, Y. W. Wang, C. Z. Wang and L. D. Zhang, *J. Mater. Chem.* **12** (2002) 1602.
14. Z. Cui, G. W. Meng, W. D. Huang, G. Z. Wang and L. D. Zhang, *Mater. Res. Bull.* **35** (2000) 1653.
15. H. Z. Zhang, Y. C. Kong, Y. Z. Wang, X. Du, Z. G. Bai, J. J. Wang, D. P. Yu, Y. Ding, Q. L. Hang and S. Q. Feng, *Solid State Commun.* **109** (1999) 677.
16. P. D. Yang and C. M. Lieber, *Science* **273** (1996) 1836.
17. P. D. Yang and C. M. Lieber, *J. Mater. Res.* **12** (1997) 2981.
18. P. D. Yang and C. M. Lieber, *Appl. Phys. Lett.* **70** (1997) 3158.

19. Z. G. Bai, D. P. Yu, H. Z. Zhang, Y. Ding, Y. P. Wang, X. Z. Gai, Q. L. Huang, G. C. Xiong and S. Q. Feng, *Chem. Phys. Lett.* **303** (1999) 311.
20. X. S. Peng, X. F. Wang, J. Zhang, Y. W. Wang, S. H. Sun, G. W. Meng and L. D. Zhang, *Appl. Phys. A* **74** (2002) 831.
21. G. W. Meng, X. S. Peng, Y. W. Wang, C. Z. Wang, X. F. Wang and L. D. Zhang, *Appl. Phys. A* **74** (2003) 119.
22. H. Masuda and K. Fukuda, *Science* **268** (1995) 1466.
23. H. Masuda and M. Satoh, *Jpn. J. Appl. Phys. Part I* **35** (1996) 1126.
24. Y. Li, G. W. Meng and L. D. Zhang, *Appl. Phys. Lett.* **76** (2000) 2011.
25. M. J. Zheng, L. D. Zhang, X. Y. Zhang, J. Zhang and G. H. Li, *Chem. Phys. Lett.* **334** (2001) 298.
26. M. J. Zheng, L. D. Zhang, G. H. Li, X. Y. Zhang and X. F. Wang, *Appl. Phys. Lett.* **79** (2001) 839.
27. M. J. Zheng, G. H. Li, X. Y Zhang, S. Y. Huang, Y. Lei and L. D. Zhang, *Chem. Maters.* **13** (2001) 3859.
28. S. Nakamura, *J. Vac. Technol. A* **13** (1995) 307.
29. S. Nakamura, *Science* **281** (1998) 956.
30. F. A. Ponce and D. P. Bour, *Nature* **386** (1997) 351.
31. H. Morkoc and S. N. Mohammad, *Science* **267** (1995) 51.
32. A. Hashimoto, T. Motiduki and A. Yamamoto, *Mater. Sci. Forms* **264–268** (1998) 1129.
33. Y. Arajawa and H. Sakaki, *Appl. Phys. Lett.* **40** (1982) 490.
34. J. Li, Z. Y. Qiao, X. L. Chen, Y. G. Cao and M. He, *J. Phys.: Condens. Mater.* **13** (2001) L285.
35. S. K. O'Leary, E. Foutz, M. S. Shur, U. V. Bhapker and L. F. Eastman, *J. Appl. Phys.* **83** (1998) 826.
36. M. J. Wang and H. Wada, *J. Mater. Sci.* **25** (1990) 1690.
37. P. D. Ramesh and K. J. Rao, *J. Mater. Res.* **9** (1994) 2330.
38. G. S. Cheng, L. D. Zhang, Y. Zhu, G. T. Fei, L. Li, C. M. Mo and Y. Q. Mao, *Appl. Phys. Lett.* **75** (1999) 2455.
39. X. L. Chen, J. Y. Li, Y. G. Cao, Y. C. Lan and H. Li, *Adv. Mater.* **12** (2000) 1432.
40. J. Zhang, L. D. Zhang, X. S. Peng and X. F. Wang, *J. Mater. Chem.* **12** (2002) 802.
41. H. Parata, A. Devi, F. Hipler, E. Maile, A. Birkner, H. W. Becher and R. A. Fisher, *J. Cryst. Growth* **231** (2001)·68.
42. P. Yang and C. M. Lieber, *J. Mater. Res.* **12** (1997) 2981.
43. L. D. Zhang, G. W. Meng and F. Phillip, *Mater. Sci. and Eng. A* **286** (2000) 34.
44. S. M. Bradshaw and J. L. Spicer, *J. Am. Ceram. Soc.* **82** (1999) 2293.
45. Y. J. Zhan, J. Liu, R. R. He, Q. Zhang, X. Z. Zhang and J. Zhu, *Chem. Mater.* **13** (2001) 3899.
46. L. M. Sheppard, *Am. Ceram. Soc. Bull.* **69** (1990) 1801.
47. G. A. Slack and T. F. Mcnelly, *J. Cryst. Growth* **34** (1976) 263.
48. S. M. L. Sastry, R. J. Lederick and T. C. Peng, *J. Met.* **40** (1988) 11.
49. J. A. Haber, P. C. Gibbons and W. E. Buhro, *Chem. Mater.* **10** (1998) 4062.
50. X. F. Duan and C. M. Lieber, *J. Am. Chem. Soc.* **122** (2000) 188.
51. W. Q. Han, S. Fan, Q. Li and Y. Hu, *Science* **277** (1997) 1287.
52. J. Zhu and S. Fan, *J. Mater. Res.* **14** (1999) 1175.
53. W. Q. Han, P. Redlich, F, Ernst and M. Riihle, *Appl. Phys. Lett.* **76**(5) (2000) 652.
54. M. He, I. Minus, P. Zhou, S. N. Mohammed and J. B. Halpern, *Appl. Phys. Lett.* **77** (2000) 3731.
55. X. L. Chen, J. Y. Li, Y. G. Cao, Y. C. Lan and H. Li, *Adv. Mater.* **12** (2000) 1432.
56. C. C. Chen and C. C. Yeh, *Adv. Mater.* **12** (2000) 738.

57. J. Y. Li, X. L. Chen, Z. Y. Qiao, Y. G. Gao and Y. C. Lan, *J. Cryst. Growth* **213** (2000) 408.
58. A. Y. Liu and M. L. Cohen, *Phys. Rev. B* **41** (1990) 10727.
59. A. M. Liu and R. M. Wentzcovitch, *Phys. Rev. B* **50** (1994) 10362.
60. D. M. Teer and K. J. Henley, *Science* **271** (1996) 53.

Chapter 13

Silicon-Based Nanowires

S. T. Lee, R. Q. Zhang and Y. Lifshitz

Center of Super-Diamond and Advanced Films (COSDAF) &
Department of Physics and Materials Science,
City University of Hong Kong, Hong Kong SAR, China

1. Introduction

Since the discovery of carbon nanotubes by Iijima [1] in 1991, one-dimensional (1D) nanomaterials including tubes, wires, cables and ribbons have attracted an explosive interest. A great deal of work has been done on exploring the various fundamental properties and potential applications of carbon nanotubes [2–6]. There are two major reasons that carbon nanotubes have captivated such interest. The first relates to their many potential applications in nanotechnology, such as high-strength materials [7], electronic components [8], sensors [9, 10], field emitters [11, 12] and hydrogen storage materials [13]. The second relates to the nanometer size of carbon nanotubes and the dependence of their electronic properties on diameter and orientation that provide unprecedented and exciting opportunities for the study of size- and dimensionality-dependent chemical and physical phenomena [14–20].

For the same reasons, research on conventional one-dimensional (1D) semiconducting materials has been equally intense. Moreover, there are important limitations of nanotubes. Particularly, both the selective growth of metallic or semiconducting tubes and controlled doping of semiconducting nanotubes are difficult to achieve. Nanowires do not seem to face these problems, which make them much more adaptable for volume fabrication of nanodevices. Among them, silicon nanowires (SiNWs) are of special interest because silicon is the most widely used and studied semiconducting material. In 1998, Morales and Lieber [21] and the CityU team [22] independently reported the bulk-quantity synthesis of SiNWs. At CityU, we proposed an oxide-assisted growth (OAG) model to explain the growth of SiNWs [23–28], while Lieber et al. advocated the laser-assisted metal-catalyst vapor–liquid–solid (VLS) growth [21]. In contrast to the conventional metal-catalyst VLS growth [21], the OAG does not require a catalytic metal nanoparticle tip, thus providing a much "cleaner" method for the 1D material fabrication.

With this OAG approach, highly pure, ultra-long and uniform-sized SiNWs in bulk-quantity could be synthesized by either laser ablation or thermal evaporation of silicon powders mixed with silicon oxide or using silicon monoxide only [23–28]. Section 2 discusses the oxide-assisted growth. Transmission electron microscopic data and theoretical calculations are used to describe the nucleation and the growth of SiNWs via the OAG process.

In the further efforts to achieve control growths, SiNWs of varying diameter, phase purity, morphology, defect density and doping have been obtained. This was achieved by varying the deposition parameters including growth temperature, carrier gas composition, carrier gas flow, and target composition. For example, we have obtained SiNWs of different diameters by varying the carrying gas [29], and we found that the diameters had a wide distribution in each batch of growth. These art contrasted with the work of Cui et al. [30] and Wu and Yang [31], which used metal nanoparticles of uniform size to control the diameter of SiNWs via the laser-assisted catalytic VLS growth. With our OAG method, Si 1D nanostructures of different morphologies [32, 33] including silicon nanoribbon [34] were obtained by adjusting the deposition temperature and pressure, yet the phase purity was not as good. A simple thermal annealing process can reduce stacking faults in the as-grown SiNWs [35], but the crystal perfection needs further improvement. The work on the control of the SiNW structure and size will be reviewed in Section 3.

Future applications of SiNWs require the production of hybrid structures made of SiNWs integrated with nanostructures of other materials. We addressed this issue by first using multi step processes to grow nanocables. Metallization of the SiNW surfaces is another example of hybrid configurations. We addressed it by two approaches: (1) ion implantation of metal ions, (2) Solution methods in which the metallization was done via redox reactions of metal ions on hydrogen-terminated and oxide-removed SiNW surfaces. The later was performed in a liquid solution of the relevant metal. An additional route of addressing the issue of incorporation of SiNWs with other materials/structures is the formation of nanostructures of Si compounds. We demonstrate this approach by our work of SiC nanowires. Section 4 summarizes our works on hybrid structures.

The motivation for studying nanoscience and nanotechnology stems from the exciting properties predicted for nanomaterials due to size effects. The characterization of the SiNWs is the topic of section 5. The optical and electrical properties of the nanowires have been characterized systematically by Raman scattering, photoluminescence and field emission [36, 37]. Understanding the atomic structure and electronic properties of SiNWs including the dopant-induced conductivity is an essential step towards the application of the nanowires. Although the structures and electronic properties of boron-doped silicon wafers have been investigated extensively, the corresponding study for SiNWs is relatively lacking, due to the insulating nature of the oxide sheath on most semiconductor nanowires and the difficulty in dispersing them. We have succeeded in removing the oxide layer of the SiNWs, obtaining atomically resolved STM images of H terminated surfaces of SiNWs with diameters ranging from 1 to 7 nm. This enabled reliable scanning tunneling spectroscopy (STS) measurements of these wires from which the electronic density of states and energy band gaps could be derived. The energy band gaps indeed increase from 1.1 eV for a 7 nm-diameter SiNW to 3.5 eV for a 1.3 nm-diameter SiNW, which are in accord with theoretical prediction demonstrating the quatum size effect in SiNWs. In Section 5 we review our scanning tunneling microscopy (STM)/STS study on boron-doped and undoped SiNWs [38] and on the quantum size effect in SiNWs as well as our characterization work on other SiNWs.

Modeling of SiNW structures, nucleation and growth processes and properties was done parallel to the experimental work. The modeling work is most valuable in providing additional insight into the nature of the OAG and in explaining our experimental results. Our modeling efforts are described in Section 6.

Finally, we show that the OAG is capable of producing different nanostructures based on a variety of semiconducting materials. We have extended the growth approach to synthesize successfully a host of semiconducting materials, including Ge [39], GaN [40, 41], GaAs [42, 43], SiC [44], GaP [45], and ZnO [46]. Section 7 will review these works.

2. Oxide-Assisted Growth of Nanowires

2.1. General

In this section we describe the discovery of the new OAG method, which is distinguishable from the classical metal-catalyst VLS method. We point out how our attempts to join the international effort to produce 1D nanomaterials (nanowires) using laser ablation yielded two types of nanomaterials growth: (1) the conventional metal-catalyst VLS growth, (2) the OAG. We follow by detailing new experimental techniques developed for OAG, simpler and less expensive than laser ablation, and we finish by discussing the nucleation and growth in the OAG process.

2.2. Discovery of oxide-assisted growth [23–25]

Si nanowires were first produced using the classical metal-catalyst VLS approach. [21, 22, 47]. Laser ablation of a metal-containing Si target produces metal/metal silicide nanoparticles that act as the critical catalyst needed for the nucleation of SiNWs. The wires further grow by dissolution of silicon in the metallic nano-cap and concurrent Si segregation from the cap. In a typical experiment, an excimer laser is used to ablate the target placed in an evacuated quartz tube filled with an inert gas, e.g. argon (see Fig. 1) [24].

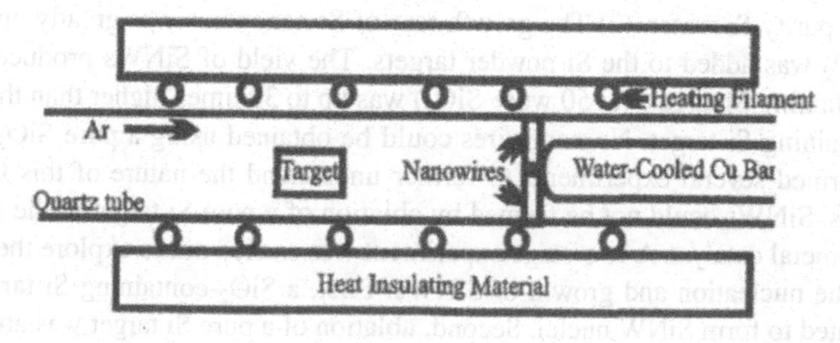

Fig. 1. Experimental set-up for the synthesis of Si nanowires by laser ablation [24].

We followed this idea by ablating a solid composite target of highly pure Si powder mixed with metals (Fe, Ni, or Co). The target temperature was 1100–1400 °C and the nanowire growth temperature was selected as 900–1100 °C. Si wires with a typical diameter of 100 nm were formed (see Fig. 2a) at the high furnace temperature zone (around 1100 °C). The SiNWs were millimeters long and straight with metallic (Fe) spheres at the wire tip, indicating the growth via a metal-catalyst VLS reaction. High-resolution transmission electron microscopy (HRTEM) observations showed that the growth direction of these Si wires was predominantly <111> (Fig. 2b). The formation of such Si wires only at the relatively higher temperature was clearly due to the high melting temperature of Fe-silicides, e.g. $FeSi_2$.

Entirely different SiNWs grew at the lower temperature zone (~900 °C) (Fig. 2c). TEM investigations showed that SiNWs obtained in this region were extremely long and highly curved with a typical smaller diameter of ~20 nm (see Fig. 2c). Each wire consisted of a crystalline Si core in a sheath of Si oxide. The crystalline Si core had a high density of defects, such as stacking faults and micro-twins. HRTEM and electron diffraction showed that the most frequent axis of the SiNWs was along the <112> direction with the {111} surfaces of Si crystalline cores parallel to the nanowire axis [48]. This is in contrast to the <111> growth direction common for the metal-catalyst VLS growth. Most surprisingly, no evidence for metal particles was found either on the SiNWs tips or in the wires themselves regardless of the metal used in the target (Fe, Ni, or Co), in sharp contrast to those SiNWs grown in the high-temperature region. The SiNW tips were generally round and covered by a relatively thick Si oxide layer (2–3 nm) and no other component rather than Si or O was detected by electron energy dispersive spectroscopy (EDS). The Si crystal core near the tip contained a high density of stacking faults and micro-twins [27] generally along the nanowire axis in the <112> direction.

Similar metal-free SiNWs were obtained for other metal catalysts using laser ablation or even by thermal evaporation of either mixed powders of silicon dioxide and silicon or a pure silicon monoxide powder [23, 24, 27]. The morphology and structure of SiNWs obtained from thermal evaporation of SiO were similar to those grown from a (Si + SiO_2) solid source. The yield of SiNWs increased with increasing thermal evaporation temperature (see Fig. 3). Using highly pure SiO powders, we obtained a high yield of SiNWs at temperatures ranging from 1130 to 1400 °C. This provided the direct evidence for the OAG process. These observations led us to propose that the SiNW growth at lower temperatures was induced by the oxide and not by the metal catalyst. This proposition was further substantiated by the observation that: (1) A limited quantity of SiNWs was obtained by laser ablation of pure Si powders (99.995%) or a high-purity Si wafer. (2) The growth rate of Si nanowires was greatly enhanced when SiO_2 was added to the Si powder targets. The yield of SiNWs produced from SiO_2-containing Si targets (at 50 wt % SiO_2) was up to 30 times higher than that from a Fe-containing Si target. No nanowires could be obtained using a pure SiO_2 target. We performed several experiments to further understand the nature of this OAG of nanowires. SiNWs could not be formed by ablation of a pure Si target in the absence of a pure metal catalyst. A two-stage experiment was carried out to explore the role of SiO_2 in the nucleation and growth of SiNWs. First, a SiO_2-containing Si target was laser ablated to form SiNW nuclei. Second, ablation of a pure Si target was attempted

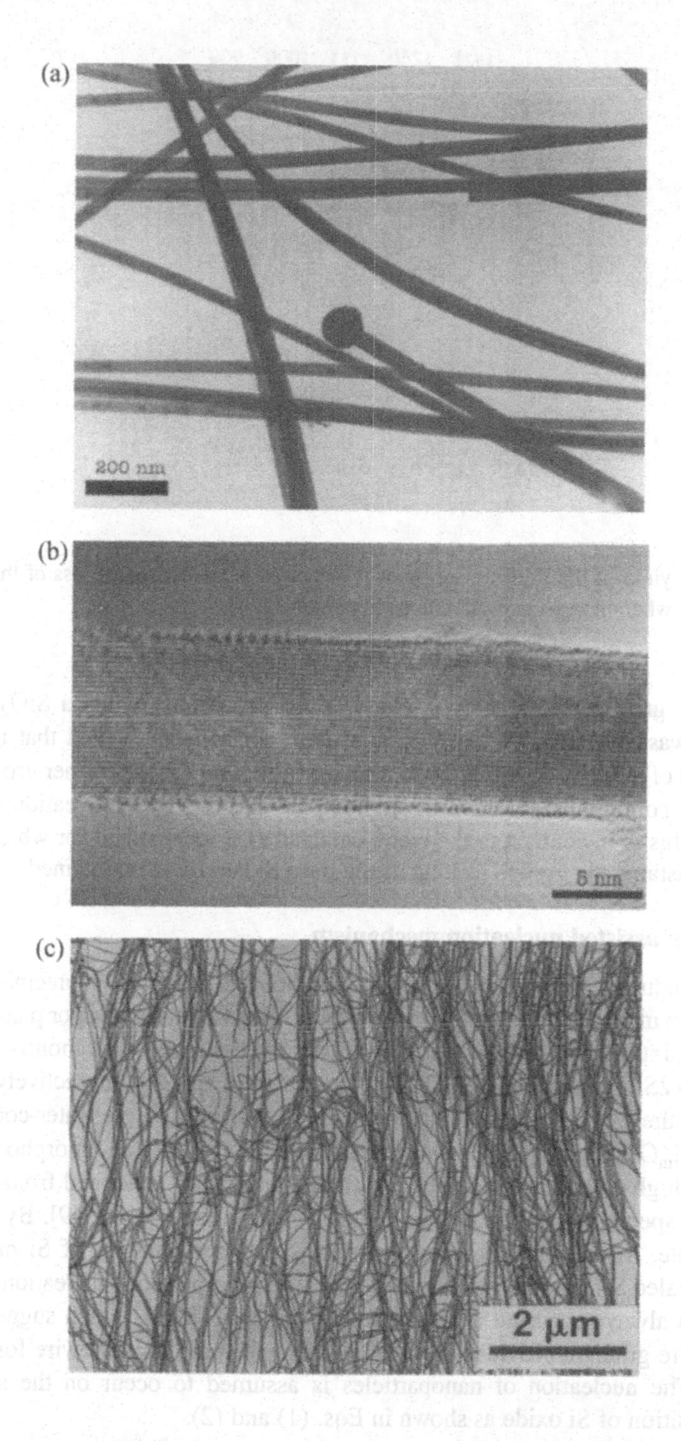

Fig. 2. TEM images of SiNWs from laser ablation [24]: (a) at high temperature (1100 °C); (b) a typical HRTEM image of SiNWs formed at high temperature (1100 °C), note the metallic tip typical for the metal catalyst growth; and (c) TEM micrograph of Si nanowires formed at the low temperature zone (900 °C).

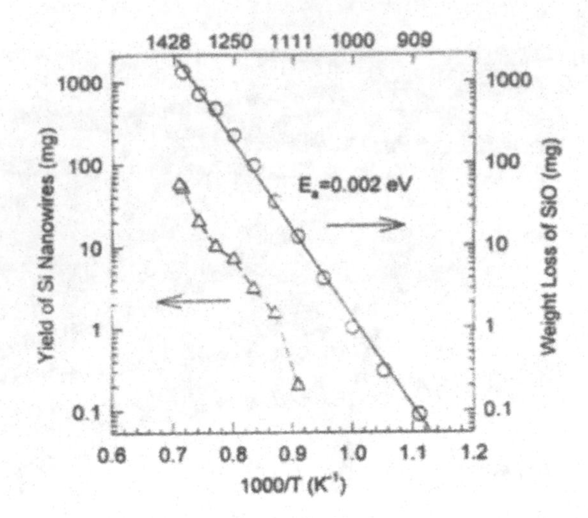

Fig. 3. The yield of the Si nanowire product increases with the weight loss of the SiO target temperature when increasing evaporation temperature [24].

for further growth. SiNW growth could be observed only when a SiO_2-containing Si target was ablated in the second stage. The experiment showed that in the OAG nucleation of a SiNW a pure Si target was not sufficient for the further growth and the oxide was continuously needed throughout the entire SiNW nucleation and growth process. This is in contrast to the metal-catalyst VLS mechanism for which the metal catalyst sustains the growth as long as the pure Si supply is maintained.

2.3. Oxide-assisted nucleation mechanism

We conducted experiments to reveal the nature of the Si core precipitation in the SiO_2 sheath in this new growth process. In these experiments, the vapor phase generated from the mixture of Si and SiO_2 at 1200 °C mainly consisted of Si monoxide [$Si(s) + SiO_2(s) \rightarrow 2SiO(g)$, where (s) and (g) represent solid and gas, respectively]. This was proven by the EDS observation that the material collected on the water-cooled Cu finger was Si_mO_n (m = 0.51, n = 0.49). Si monoxide (SiO) is an amorphous semiconductor of high electrical resistivity, which can be readily generated from the powder mixture (especially in 1 : 1 ratio) of Si and SiO_2 by heating [22, 49]. By heating the SiO sample, Si precipitation was observed. Such precipitation of Si nanoparticles from annealed SiO is quite well known [50]. The precipitation, nucleation and growth of SiNWs always occurred at the area near the cold finger, which suggests that the temperature gradient provided the external driving force for nanowire formation and growth. The nucleation of nanoparticles is assumed to occur on the substrate by decomposition of Si oxide as shown in Eqs. (1) and (2).

$$Si_xO(s) \rightarrow Si_{x-1}(s) + SiO(s) \quad (x > 1) \tag{1}$$

$$2SiO(s) \rightarrow Si(s) + SiO_2(s) \tag{2}$$

Our TEM data suggested that this decomposition results in the precipitation of Si nanoparticles, which are the nuclei of SiNWs, clothed by shells of silicon oxide.

We further initiated theoretical studies to explore the role of the oxide species in the OAG process. The gas-phase composition of silicon oxide clusters evaporated by laser ablation or thermal treatment should be considered to be important in the SiNW synthesis. We first used density functional theory (DFT) calculations to study the nature of the Si_nO_m (n, m = 1–8) clusters formed in the gas phase during OAG [51]. Our calculations show that silicon suboxide clusters are the most probable constituent of the vapor, and they have an unsaturated nature and are highly reactive to bond with other clusters. Moreover, a silicon suboxide cluster prefers to form a Si-Si bond with other silicon oxide clusters as shown in Fig. 4 [52], while an oxygen-rich silicon oxide cluster prefers to form a Si-O bond with other clusters. Based on these calculations we proposed the following SiNW nucleation scheme [53]: First, a silicon suboxide cluster is deposited on the substrate and some of its highly reactive silicon atoms are strongly bonded to the substrate (silicon) atoms limiting the cluster motion on the substrate. Non-bonded reactive silicon atoms in the same cluster are now exposed to the vapor with their available dangling bonds directed outward from the surface. They act as nuclei that absorb additional reactive silicon oxide clusters and facilitate the formation of SiNWs with a certain crystalline orientation. The subsequent growth of the silicon domain after nucleation may be crystallographic dependent. Oxygen atoms in the silicon suboxide clusters might be expelled by the silicon atoms during the growth of SiNWs and diffuse to the edge forming a chemically inert silicon oxide sheath [52]. In a certain orientation, e.g. [112], the diffusion might be lower and the highly reactive silicon oxide phase can still be exposed to the outside and facilitate the continuous growth of the wire in such a direction. The oxygen-rich sheath formed in other directions may however possess lower reactivity and thus does not favor further stacking of silicon oxide clusters from the gas-phase, leading to the growth suppression in such directions.

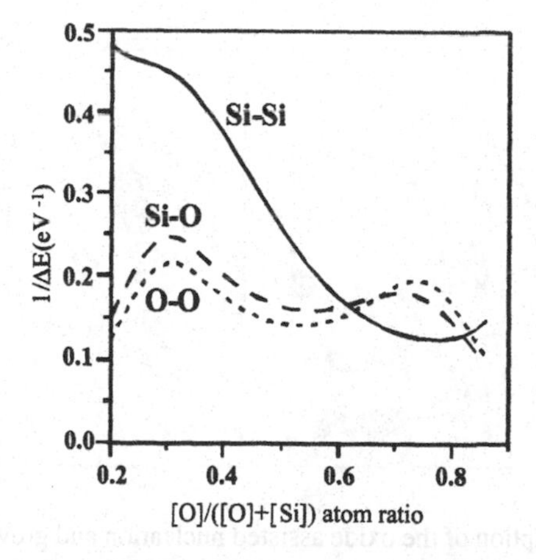

Fig. 4. The inverse of the energy difference ΔE = LUMO (electron acceptor)-HOMO (electron donor) and thus the reactivity (proportional to the inverse of the energy difference) for the formation of a Si-Si bond, a Si-O bond, or an O-O bond between two silicon oxide clusters as a function of Si : O ratio [52].

The reactivity of silicon atoms in oxygen-rich clusters becomes very low at a Si : O ratio of 1 : 2 [52], while the reactivity of oxygen atoms changes to a lesser extent. The overall reactivity for Si : O = 1 : 2 is low. 1D growth in a specific direction is thus facilitated. In summary, the highly reactive SiO_x layer (x > 1) at the tip of nanowires acts as a collector of the vaporized silicon oxide, while the outer SiO_2 layer of the SiNWs stops the nanowires from the diameter growth.

2.4. Oxide-assisted growth mechanism

A schematic description of the OAG for SiNWs is given in Fig. 5, where nanowire nuclei with a preferred growth direction normal to the substrate surface grow fast while those with non-preferred orientations do not grow or form chains or nanoparticles. The Si nanowire growth is determined by four factors: (1) The high reactivity of the Si_xO (x > 1) layer on nanowire tips, (2) the SiO_2 component in the shell, which is formed from the decomposition of SiO and retards the lateral growth of nanowires, (3) defects (e.g. dislocations) in the Si nanowire core, (4) the formation of {111} surfaces, which have the lowest energy among the Si surfaces, parallel to the axis of the growth direction. The first two factors were discussed earlier. As far as the first factor is concerned we would like to add to the previous discussion that the melting temperature of nanoparticles can be much lower than that of their bulk materials. For example, the difference between the melting temperatures of 2 nm-Au nanoparticles and Au bulk material is over 400 °C [47, 54]. The materials in the SiNW tips (similar to the case of nanoparticles) may be in or near their molten states, thus enhancing atomic absorption, diffusion, and deposition.

We suggest that the defects of SiNWs are one of the driving forces for the 1D growth. The main defects in Si nanowires are stacking faults along the nanowire growth direction of <112> (see Fig. 6), which normally contain easy-moving 1 : 6[112] and non-moving 1 : 3[111] partial dislocations, and micro-twins. The presence of these defects at the tip areas should result in the fast growth of Si nanowires

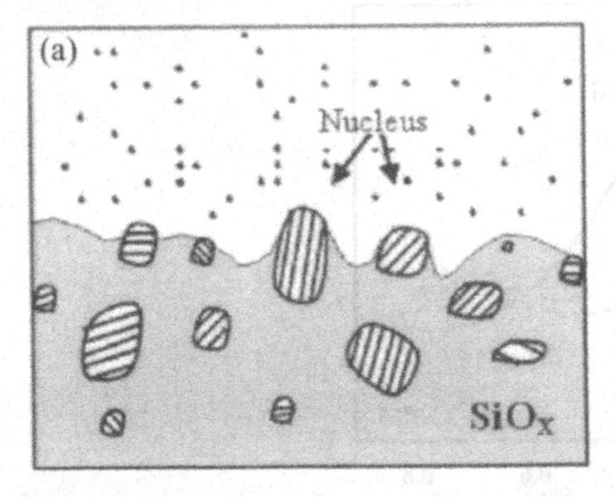

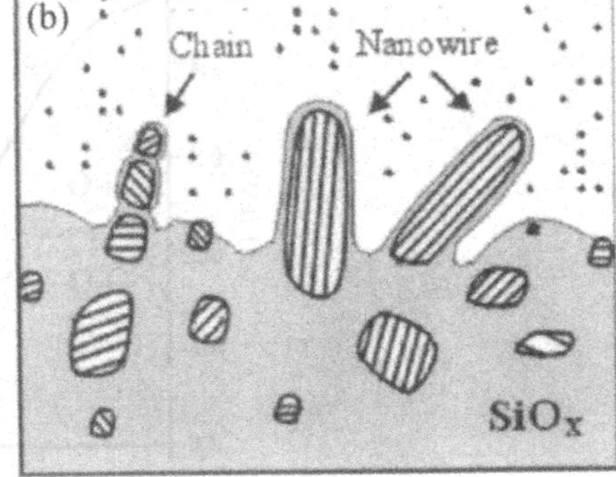

Fig. 5. Schematics description of the oxide assisted nucleation and growth of Si nanowires. The parallel lines indicate the <112> orientation, (a) Si oxide vapor is deposited first and forms the matrix within which the Si nanoparticles are precipitated. (b) Nanoparticles in a preferred orientation grow fast and form nanowires. Nanoparticles with nonpreferred orientations may form chains of nanoparticles.

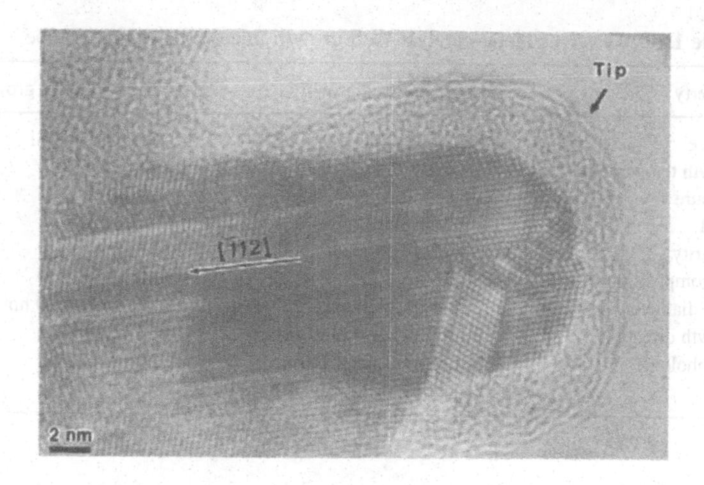

Fig. 6. HRTEM image of a Si nanowire tip. Note the SiO_x tip, the SiO_2 sheath, the stacking faults and the <112> growth direction.

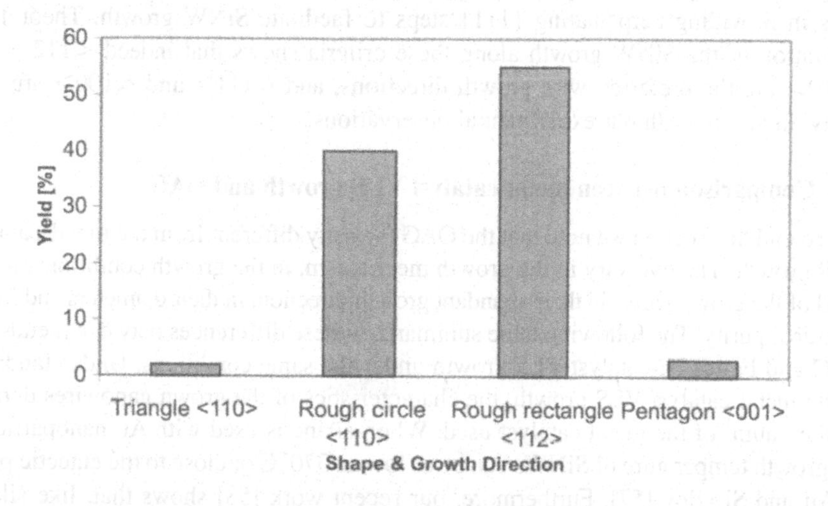

Fig. 7. The distribution of shapes, sizes, and growth directions of SiNWs [55].

since dislocations are known to play an important role in crystal growth. The SiNWs growth rate in certain crystallographic directions is enhanced not only by existing dislocations in the growth direction but also by the formation of facets with a low surface energy (Si {111} facets have the lowest surface energy). Fig. 7 presents the statistical data of SiNW growth directions and shapes derived from the cross-sectional TEM images showing that SiNWs grown by OAG technique are primarily oriented in the <112> and <110> directions, and rarely in the <100> or <111> directions [55]. The cores of SiNWs are bounded by well-defined low-index crystallographic facets with a variety of shapes that can be circular, rectangular and triangular. We found a correlation between the cross sectional shape and the growth direction, and proposed

Table 1. OAG versus metal-catalyst VLS growth in a 3-inch-diameter tube

Property	Oxide-assisted growth	Metal-catalyst growth
Source	SiO; Si + SiO_2	(Fe or Ni) + Si
Growth temperature	850–900 °C	>1100 °C
Pressure	10–800 Torr	10–800 Torr
Yield	3 mg/hr	<0.1 mg/hr
Impurity	None	Metal
Tip composition	SiO_x	Metal
Wire diameter	Typically 2–20 nm	Typically >10 nm
Growth direction	Mostly <112> & <110>	<111>
Morphologies	Nanowires, nanoribbons, nanochains	Nanowires

a model to explain these findings [56]. We suggest that the SiNW growth direction and cross section are determined by four factors: (i) the stability of a Si atom occupying a surface site; (ii) the Si {111} surface stability in the presence of oxygen; (iii) the stepped Si {111} surface layer lateral growth process; and (iv) the effect of dislocations in providing perpetuating {111} steps to facilitate SiNW growth. Theoretical evaluation of the SiNW growth along these criteria shows that indeed <112> and <110> are the preferred wire growth directions, and <111> and <100> are less likely, in accord with our experimental observations.

2.5. Comparison between metal-catalyst VLS growth and OAG

To end this section we note that the OAG is vastly different from the metal-catalyst VLS growth. The two vary in the growth mechanism, in the growth conditions, in the yield of the grown wires, in their abundant growth direction, in their diameters and in the chemical purity. The following table summarizes these differences between metal-free OAG and Fe- or Ni-catalyst VLS growth under the same conditions. Understandably in the metal-catalyst VLS growth, the characteristics of the grown nanowires depend on the nature of the metal catalyst used. When silane is used with Au nanoparticles, the growth temperature of SiNWs can be as low as 370 °C or close to the eutectic point of Au and Si alloy [57]. Furthermore, our recent work [58] shows that, like silane, SiO or other Si-containing vapor in the presence of Au nanoparticles or films could also decompose and lead to the growth of SiNWs, and the growth temperature was as low as 700 °C. We anticipate nanoparticles of other kinds of metals can also induce the deposition of SiNWs from SiO vapor, providing the metal can induce SiO decomposition and form a eutectic alloy with Si.

3. Controlled Synthesis

3.1. General

One of the most important issues of nanomaterials growth is the control of their shape and their size which is the focus of the present section. This can be done by varying different process parameters. We first start by our studies of the effect of the growth temperature on the structure of the Si nanowires. Then we show how we can control the wire diameter by varying the carrier gas. We follow by detailing our study in which we use the flow control to grow long, aligned SiNWs. We then discuss the formation of 2D rather than 1D structures, i.e. Si nano-ribbons. We end this section by discussing the morphologies obtained when B doping of the SiNWs is performed.

3.2. Temperature effects on SiNW shape [32, 33]

The substrate temperature substantially affects the SiNW growth in several ways: (1) determination of the growth process (metal-catalyst VLS growth or OAG), (2) determination of the SiNW shape in the growth process itself (e.g. the SiNW diameter), (3) annealing effects that change the structure and morphology of the SiNW after its formation. Examples of these effects were given in our study of SiNWs deposited by laser ablation of a mixture of Si, SiO_x and metals (the metallic constituent introduced either intentionally or as impurities or contamination of the system). The study was focused on the nature of nanostructures produced on the substrate at temperatures ranging from 850 to 1200 °C.

The Si nanostructures produced in this temperature region can be divided to two groups: (1) Region I (1200–1100 °C) forming nanostructures by the metal-catalyst VLS process (indicative by metallic caps), (2) Region II (1100–850 °C) where the OAG process is dominant (the nanostructures do not have metallic caps). HRTEM analysis of samples prepared at region I (temperatures of 1190, 1160, and 1130 °C denoted I1, I2 and I3) and region II (temperatures 1050, 950, and 900 °C denoted II1, II2 and II3) revealed the structure evolution at different temperatures.

In region I the diameter of SiNWs (all single-crystalline SiNWs) decreases with temperature (200, 80 and 50 nm for 1190, 1160 and 1130 °C respectively) as shown in Fig. 8. While the first two types of nanowires are straight, continuous ones, the third one has a tadpole-like shape and appears to be broken into short Si rods. The head of the tadpoles is crystalline Si, while the tail of the tadpoles is amorphous SiO_x.

In region II the diameter of the Si nanostructures is constant (~20 nm), but their shape changes from tadpole-like through chain-like to wire-like as the temperature decreased from 1050, through 950, to 900 °C respectively. The wires have SiO_x caps rather then metallic caps, and are encapsulated in a SiO_2 sheath, all indicative of an OAG process.

We will now explain these results as a combination of: (1) the dominant growth process (metal-assisted VLS or OAG) and (2) extended annealing of Si nanowires at high temperatures.

Region I is characterized by a metal-catalyst VLS growth as indicated by the metal caps on top of the nanostructures. The diameter of the nanowires in this growth process is determined by the diameter of the liquid alloy droplet at their tips. Metal silicide clusters of different sizes are present in the flowing gas above the substrate leading to condensation of droplets when the substrate temperature is lower than the melting point of the metal silicide clusters. The melting point of nanoclusters decreases with decreasing size in the nanometer region. Large droplets will melt and serve as SiNW nucleation sites at higher temperatures than small droplets, explaining

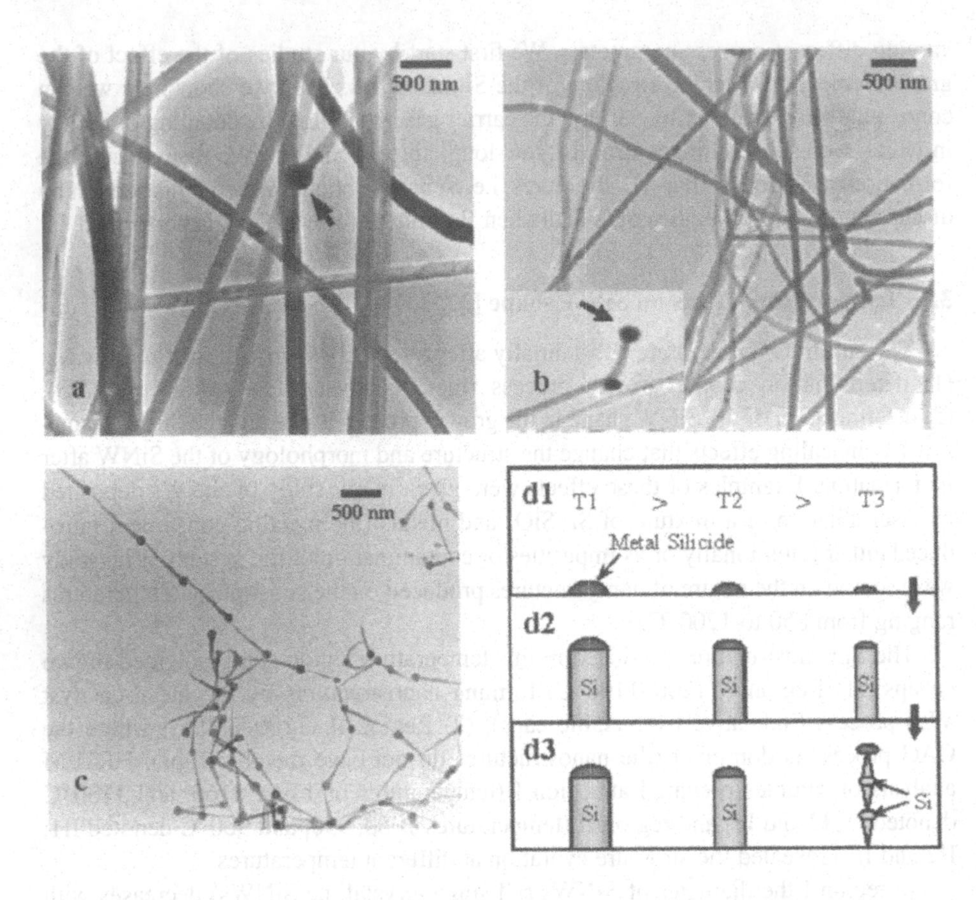

Fig. 8. Bright-field TEM images showing the typical morphology of Si nanowires grown at: (a) 1190, (b) 1160, and (c) 1130 °C. The arrows reveal the metal catalyst present at the tip of nanowires. The diameter of Si nanowires can be seen to decrease with decreasing growth temperature. (d) Diagram showing the morphology evolution of Si nanowires with time in region I: (1) nucleation, (2) growth, and (3) annealing [33].

the decrease of the SiNW diameter with decreasing temperature. Fig. 8d gives a schematic diagram of the SiNW evolution at different temperatures. The size of the molten droplets decreases with temperature (d1). The droplet absorbs Si-containing clusters from the vapor and becomes supersaturated with Si. The excessive Si precipitates out resulting in 1D growth of crystalline SiNWs shown in Fig. 8d2 the diameter of which follows that of the initial droplet. The formation of the SiNWs is restricted to the temperature region for which the temperature is high enough so that the solid particle melts and forms a liquid droplet (lower temperature of region I) on one hand, and the temperature is low enough to melt and condense from the vapor (upper temperature region) on the other hand. Previous work on Si whiskers revealed that there is a critical whisker diameter at which growth stops completely, due to the Gibbs-Thomson effect [59]. This may be the reason why the smallest nanowires obtained in region I have diameters larger than 50 nm (Fig. 8). Fig. 8d3 shows a spherodization effect for the smaller diameter nanowires, which is attributed to annealing of the nanowires as will be described later.

In region II, the OAG region, the diameters of Si nanowires are quite uniform, irrespective of the substrate temperature. This may be explained by a vapor-solid (VS) process governing the initial nucleation of the OAG rather than a VLS process in which the size of the liquid droplet decreases with decreasing energy leading to a respective decrease of the SiNW diameter as discussed above (for the metal-catalyst growth). This would mean that the small nuclei of crystalline Si nanowires were directly solidified from SiO in the vapor phase. This explanation is however not in accord with the formation of a SiO_x cap on the top of the SiNW and the alternative proposition that the oxide-assisted nucleation and growth is occurring due to: (1) the lower melting point of SiO_x compared to that of SiO_2, (2) the high reactivity of the molten SiO_x cap to Si containing clusters in the vapor, and (3) the decomposition of Si sub-oxide to Si and SiO_2. We proposed that the Si_nO_m clusters react with the SiO_x cap, the crystalline Si core precipitates below and the excess oxygen diffuses to the sides forming a SiO_2 amorphous outer layer which solidifies due to its higher melting point and limits the further lateral growth of the nucleus. The lateral growth results from the energetically favorable adsorption of vapor clusters by the highly curved SiO_x molten tip on one hand and the lateral restriction imposed by the solid SiO_2 sheath on the other hand (as shown in Fig. 9d2). The SiNW diameter may be determined not only by the diameter of the initial SiO_x droplet, but also by the equilibrium between the condensation and the disproportionation of SiO to Si and SiO_2 and by diffusion of the excess O to the sides. It is still not completely understood why this equilibrium is not temperature-dependent under our experimental conditions. It could be that the dependence is weak in this limited temperature region and will be revealed if we enlarge this region by different experimental conditions.

The SiO vapor phase is stable at a high temperature, so that the condensation and disproportionation of the SiO vapor into Si + SiO_2 occur only below a certain substrate temperature (the upper limit of region II). On the other hand, below the lower limit in region II the SiO vapor condenses directly to form a SiO solid [60], with no preferential adsorption nor disproportionation so that the 1D growth is suppressed. This explains why the OAG of SiNWs was restricted to the temperature range of 1100–850 °C (region II).

Finally, we discuss the formation of tadpole-like and chain-like Si nanostructures from the metal-catalyst VLS and the OAG (Figs. 8 and 9) processes. Both can be described in terms of a spheroidization mechanism. One-dimensional SiNWs are less stable than the three-dimensional bulk Si, since the wire has a much larger surface area and thus higher surface energy. Annealing of SiNWs for a sufficient time results in spheroidization as schematically shown in Fig. 10. The chemical potential of the SiNWs varies with the local curvature so that small variations in their diameter generate a driving force for diffusive transport between different chemical potentials. The Si nanowire would convert into a nanospherircal chain first (as shown schematically in Fig. 10 and experimentally in Fig. 9b). Later, with further diffusion, the inner crystalline Si core would break up and the spheroidization would become faster due to the larger variations increasing the driving force. The amorphous SiO_x nanorods connecting the Si nanospheres then become thinner, and eventually break up. This is why

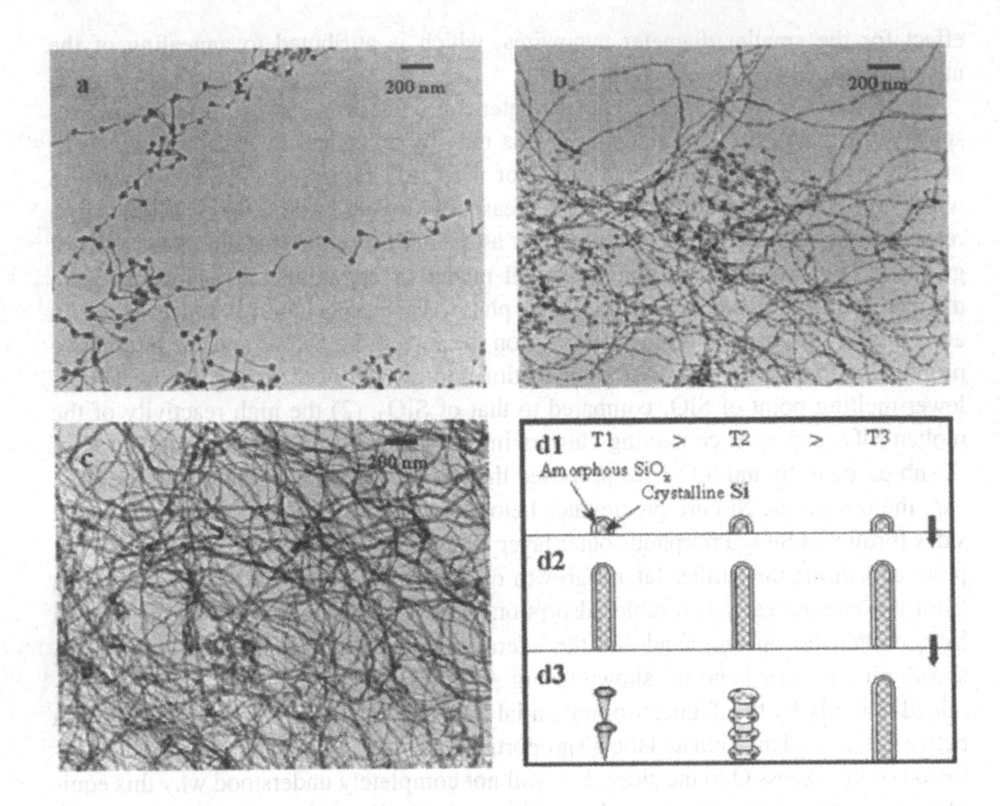

Fig. 9. Bright-field TEM images showing the typical morphology of Si nanowires grown at: (a) 1050, (b) 950, and (c) 900 °C. The SiNW diameter is independent of the growth temperature. (d) Diagram showing the morphology evolution of Si nanowires with time in region II: (1) nucleation, (2) growth, and (3) annealing [33].

the Si nanosphere chains convert into tadpole-like Si nanorods (as shown schematically in Fig. 10 and experimentally in Fig. 9a and Fig. 8c). Eventually, the amorphous SiO_x tails would disappear and perfect Si/Si-oxide spheres might also be formed. Note now the difference between Region I, in which the spherodization occurs only at the lower temperature (Fig. 8), and region II, where spherodization is more significant at higher temperatures (Fig. 9). In region I the first two SiNWs grown at 1190 and 1160 °C are too thick (200 and 80 nm respectively) for spherodization, which is observed only at the lower temperature of 1130 °C (Si nanowire diameter 50 nm), in spite of the higher diffusion rate at higher temperatures. In region II, however, the SiNW diameter is relatively constant with temperature, so the diffusion rate increases with temperature and the spherodization becomes more significant at higher temperatures.

We discussed until now the spherodization process in the context of the formation of different Si nanostructures by varying the substrate temperature at which these structures evolve. One of the structures reported was nanochains of Si nanospheres enclosed in and linked together forming a chain by SiO_2. Similar silicon nanochains (SiNC), this time of boron-doped Si, were grown [61, 62] by laser ablation of a compressed target of a mixed powder of SiO and B_2O_3 at 1200 °C. TEM analysis including imaging (Fig. 11), diffraction and EDS indeed verifies a structure of nanochains with uniform diameters of the Si nanospheres connected by amorphous SiO_2, like a chain of pearls. The SiNC consists of knots and necks with equal distances between them. The average diameters of the knots and necks are 15 and 4 nm, respectively and the thickness of the SiO_2 sheath surrounding the Si spheres is about 2 nm. The product is uniform with no isolated particles and consisting of 95% SiNC and 5% SiNWs.

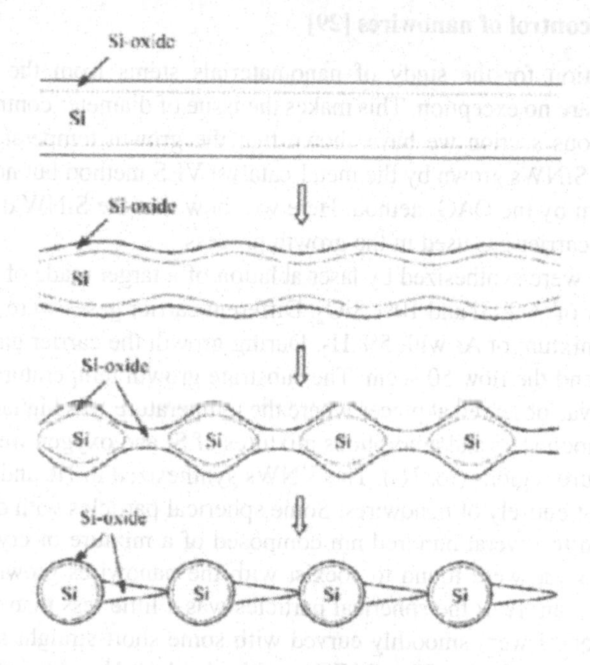

Fig. 10. Diagram showing the spheroidization process of onc-dimensional Si nanowires [33].

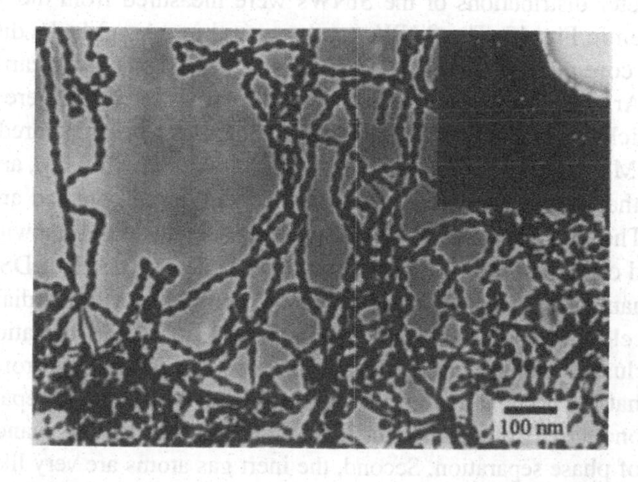

Fig. 11. TEM image showing the morphology of Si nanochains.

To summarize, both the growth process and the morphology of SiNWs can be controlled by the growth temperature. The temperature required for a metal-catalyst growth (using iron as the catalyst, for gold the temperature would be much lower) is higher than that required for the OAG. Post-growth annealing results in spherodization and structural changes which occur faster at higher temperatures. The diameter control by temperature is readily possible for the metal-catalyst VLS method but appears to be difficult for the OAG.

3.3. Diameter control of nanowires [29]

The motivation for the study of nanomaterials stems from the expected size effects; SiNWs are no exception. This makes the issue of diameter control very important. In a previous section we have shown that the growth temperature affects the diameter of the SiNWs grown by the metal-catalyst VLS method but not the diameter of SiNWs grown by the OAG method. Here we show that the SiNW diameter can be affected by the carrier gas used in the growth process.

The SiNWs were synthesized by laser ablation of a target made of compaction of a mixed powder of 90% Si and 10% SiO_2. Different carrier gases were used including He, N_2, and a mixture of Ar with 5% H_2. During growth the carrier gas pressure was about 300 torr and the flow 50 sccm. The substrate growth temperature was ~930 °C. No deposition was observed at places where the temperature was higher than ~950 °C, while some nanoclusters and amorphous mixtures of Si and oxygen were deposited at lower temperature regions (for N_2). The SiNWs synthesized in He and in Ar (5% H_2) consisted almost entirely of nanowires. Some spherical particles with diameters ranging from ~9 nm to several hundred nm composed of a mixture of crystalline Si and amorphous Si oxide were found to coexist with the nanowires grown in N_2 atmosphere, and the quantity of the spherical particles was a little less than that of SiNWs. Most of the SiNWs were smoothly curved with some short straight sections, A few possessed bends and kinks. The SiNWs synthesized in He atmosphere possessed many more bends and curves.

The diameter distributions of the SiNWs were measured from the TEM micrographs as given in Fig. 12. The SiNWs (up to several mm long) had a distribution of diameters (Si core plus SiO_2 sheath) peaked at 13, 9.5 and 6 nm for carrier gas mixtures of He, Ar (5% H_2) and N_2 respectively. The smallest wires were mixed with spherical particles with diameters ranging from ~9 nm to several hundreds nm. High-resolution TEM images of several SiNWs produced in He, Ar (5% H_2), and N_2 atmosphere shows that every nanowire consists of a crystalline Si core and an amorphous SiO_2 sheath. The crystalline Si core has many lattice defects. The nanowires consisted of only Si and oxygen with no ambient gas atoms as determined by EDS and XPS.

The mechanism by which the carrier gas affects the growth and diameter of the SiNWs is not clear. We propose that the ambient atoms affect the formation and transport of nanoclusters and the phase separation process at the growth front of SiNWs. We suggest that the dominant effect of the ambient is on the phase separation. First, the thermal conductivity of the ambient affects the cooling rate of the nanoclusters and thus the rate of phase separation. Second, the inert gas atoms are very likely incorporated in the deposited matrix at the growing tips of SiNWs, and their presence would

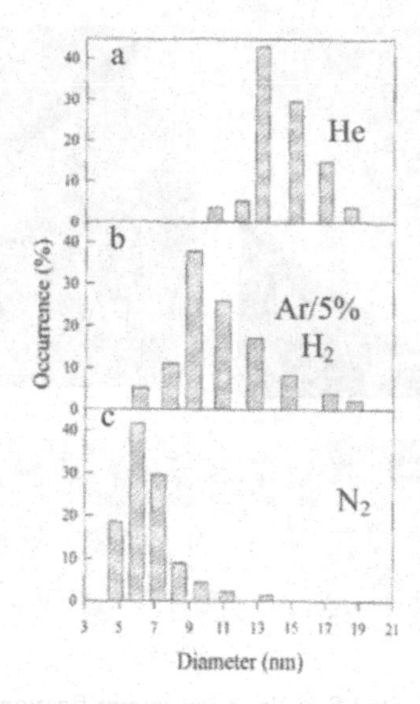

Fig. 12. Size distribution of SiNWs synthesized by laser ablation using different carrying gases [29]: (a) He, (b) Ar(5% H_2), and (c) N_2.

influence the phase separation process despite their eventual outdiffusion. Helium is smaller, faster moving, and more thermally conducting than N_2. The collective effect of these processes leads to a slower supply of heat by N_2 and thus faster cooling of nanoclusters. This in turn results in faster phase separation and the formation of thinner SiNWs. The fast cooling and elemental incorporation in the nanoclusters formed in an N_2 environment probably leads to incomplete phase separation and to the formation of the remnant spherical particles. The Ar (5% H_2) mixture is suspected to exert an intermediate effect between N_2 and He, thus giving rise to nanowires of intermediate diameters.

3.4. Large-area aligned and long SiNWs via flow control [63]

Until now we discussed the effect of two parameters on the SiNW growth: temperature and carrier gas. Now we show that the carrier gas flow can be exploited as well. In this particular example we use it to grow millimeter-area arrays of highly oriented, crystalline silicon nanowires of millimeter length.

The growth in this particular case was performed by thermal evaporation of SiO powder. A carrier gas of argon mixed with 5% H_2 was used with a flow of 50 sccm at 400 Torr. The furnace temperature was 1300 °C, while the growth temperature was about 930 °C.

Large area (about 2 mm × 3 mm) of highly oriented, long (up to 1.5–2 cm) nanowires were grown on the surface of the silicon substrate under these conditions, as detected by SEM (Fig. 13). The thickness of the oriented nanowire product was about 10 μm, as estimated from the cross-sectional image (Fig. 13b) of the sample prepared by focused ion beam cutting. EDX shows that the nanowires are composed

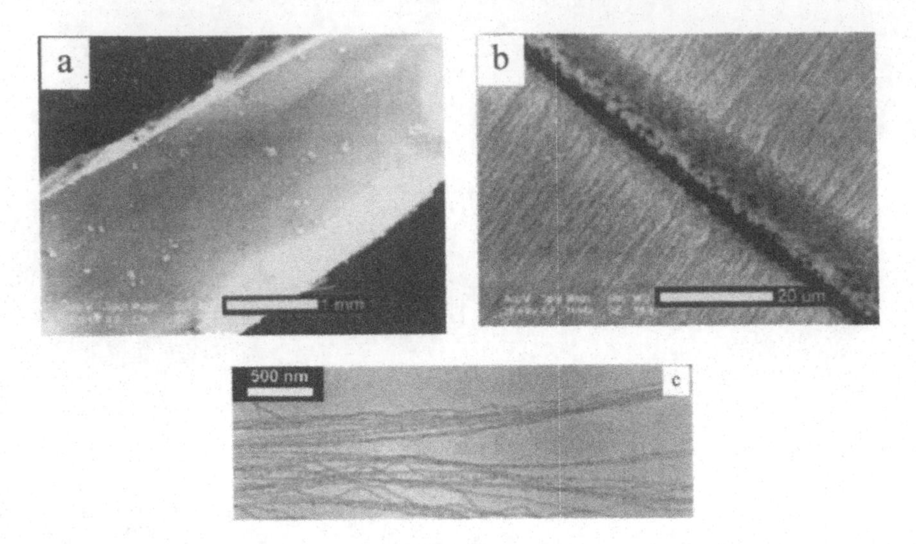

Fig. 13. a) SEM image of oriented SiNWs at low magnification; (b) The cross section of the SiNW in (a) cut by a focused ion beam; (c) TEM image of the oriented silicon nanowires [63].

of silicon and oxygen with no metal impurities, suggesting an oxide assisted growth. Fig. 13c shows the typical morphology of SiNWs. TEM shows that the SiNWs are quite clean, with very few particles attached to their surfaces, and are relatively homogeneous. The SiNW diameters vary from 18 to 46 nm, and the mean value is about 30 nm.

HRTEM shows the typical SiNW Si core encapsulated by a SiO_2 sheath and the {111} planes of crystalline silicon. The diameters of the crystalline silicon core varied from 13 to 30 nm, and the mean value was about 20 nm. The thickness of the amorphous silicon oxide shell varied from 2 to 10 nm, and the mean value was about 5 nm.

The growth of the oriented SiNWs may be related to the flow of the carrier gas, because it was found that the orientation direction of the SiNWs is parallel to the direction of flow of the carrier gas in the alumina tube. A mechanism of the growth of the oriented silicon nanowires is illustrated below. First, the nucleation of silicon nanowires from silicon oxide (SiO_x) vapor started at the proper position of the substrates. Because there was a temperature gradient along the alumina tube, and the planes with the same temperature were perpendicular to the axial direction of the tube, only some particular positions with the appropriate temperature may be suitable for the nucleation of SiNWs [21, 27]. These positions should be located on a line of equal temperature on the substrate and will also be perpendicular to the flow direction of the carrier gas. Once initial nucleation is established, nanowire growth will tend to continue on the substrate. Secondly, the strength of the flow of the carrier gas will force the growing nanowires to grow in the direction of the flow. At the same time, overcrowding of the nanowires will limit the possibility of nanowire propagation in other directions [64]. In addition, it was found that a smooth plane substrate is very helpful for oriented growth.

3.5. Si nanoribbons

SiNWs are 1D nanostructure. A distinct feature of the OAG process revealed in our studies is the variety of different configurations it can form. Some were discussed in the previous sections. Now we present an exciting and unexpected 2D configuration, nanoribbons, discovered in the course of studying OAG. A 2D configuration is not expected to exhibit the magnitude of size effects as a 1D structure, but it may be advantageous in processing and in obtaining signals with more measurable intensities in single object characterization.

The single-crystal silicon nanoribbons were grown by simple thermal evaporation of silicon monoxide (SiO) heated to 1150 °C. No templates or catalysts were used. The nanoribbons have a thickness of only about 10–20 (average 15) nanometers, widths of several hundreds of nanometers (50–450 nm), and lengths of many micrometers [34]. Most of the ribbons have rippling edges (Fig. 14a), and a small portion of the ribbons has smooth edges (Fig. 14b). Due to their small thickness the ribbons seem transparent in TEM imaging (Fig. 14b) using 200 keV electrons. Nanoribbons of different width and morphology have a similar thickness which is constant throughout each individual ribbon. In the typical nanoribbon shown in (Fig. 14b) the thickness is about 14 nm and the width is about 370 nm. The thickness to width ratio of the SiNWs varies from 4 to 22. The rippling and curling features at the edge of most ribbons also confirm that the nanoribbons are quasi-2D structures distinctly different in shape from the 1D SiNWs.

HRTEM imaging of a single nanoribbon revealed that the ribbon has a crystal core nipped by amorphous layers with atomically sharp interfaces. The in-plane layers of the nanoribbon were determined to be silicon (110) facet with a perfect atomic, defect-free, single-crystal structure grown along the <111> direction. This direction is different from the predominant <112> and <110> direction of SiNWs synthesized by the OAG method. It is the same as the most abundant growth direction of Si nanowires synthesized by the metal-catalyzed VLS method. The wide part of the

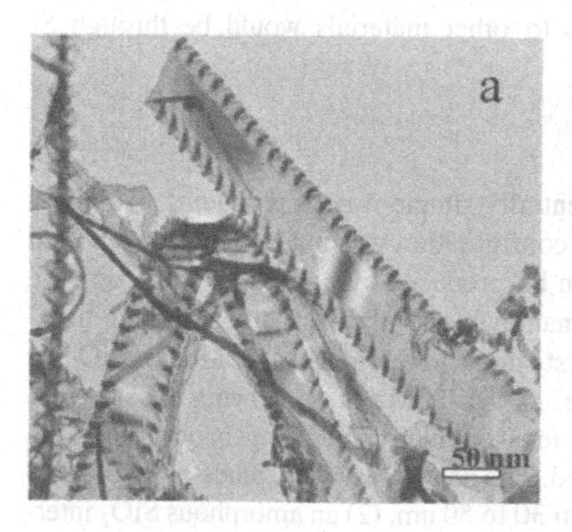

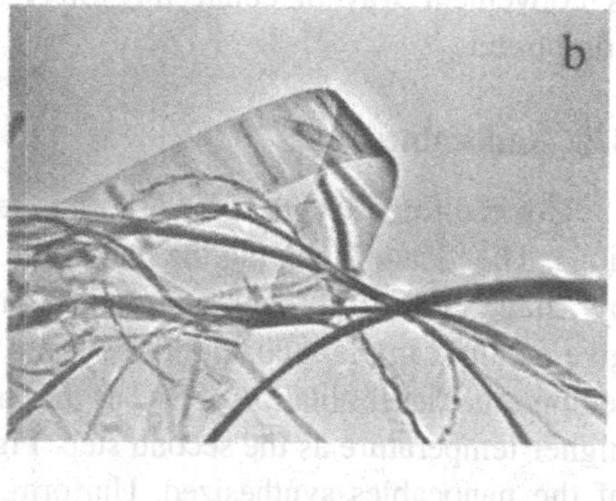

Fig. 14. TEM images (TEM, Philips CM 20 TEM at 200 kV) of a) rippling-edge nanoribbons, and (b) a smooth-edge nanoribbon. The thickness of the ribbons is about 14 nm [34].

ribbon is along the <112> direction. The amorphous edges of the ribbon consist of silicon oxide (SiO_x) as determined by EDS and EELS attached to the TEM. The width of the amorphous edge is about 10 nm. Analysis of a number of nanoribbons with different widths shows that the width of the oxide edges varies from 3 to 25 nm, similar to the thickness of the amorphous silicon oxide shell of the nanowires synthesized by OAG. The thickness of the oxide layer covering the flat surfaces of the ribbons is much less than the width of the oxide edges; that is, the thickness of the oxide layers is anisotropic. This result may be understood in terms of the OAG process, in which the silicon oxide shell was formed by the reaction of $SiO + SiO = Si + SiO_2$, and subsequently separated from the silicon core during growth. According to this reaction, the amount of segregated silicon oxide is proportional to the amount of the silicon at the same place.

The growth of Si nanoribbons cannot be via the twin-plane growth mechanism, (i.e. controlled by a twin plane parallel to the flat surface of the nanoribbons) as suggested for microribbons [65]. A twin-plane mechanism is impossible from crystallographic considerations taking into account the observed structure of the nanoribbons and indeed, no twins were observed. The precise growth mechanism of the nanoribbons is not yet clear, but it is likely that it is governed by anisotropic growth kinetics along different crystallographic directions (i.e. two fast growing directions).

4. Si Hybrids and Compounds

4.1. General

Until now we have discussed the formation of pure silicon nanostructures for which the only (inevitable) additional component was the external SiO_2 sheath. Now we discuss the extension of the work to include hybrids based on SiNWs or alternatively, nanostructures made of Si compounds. Such nanostructures are necessary for the development of SiNWs for applications such as electronic devices (that are usually hybrid devices that include metallic contacts or Si/non-Si material junctions). A convenient way of connecting SiNWs to other materials would be through Si compounds.

4.2. Nanocables [66]

Coaxial three-layer cables offer a potentially simple way of producing nanojunctions. The idea of applying the nanocable configuration was indeed recently reported by Zhang et al. [67]. Such nanocables can be produced by a multi-step process. We give as an example a coaxial three-layer nanocable synthesized by combining high-temperature laser ablation of SiC as the first step, and thermal evaporation of SiO at a higher temperature as the second step. Fig. 15 is a TEM image showing the structure of the nanocables synthesized. Uniform, tens of micrometer long nanocables with diameters smaller than 150 nm were formed. The nanocables were made of: (1) a crystalline Si core with a diameter ranging from 30 to 50 nm, (2) an amorphous SiO_2 interlayer (second layer) 12–23 nm thick, (3) an amorphous carbon sheath (external third layer) 17–31 nm thick. The average dimensions of the nanocable are: core 43 nm in

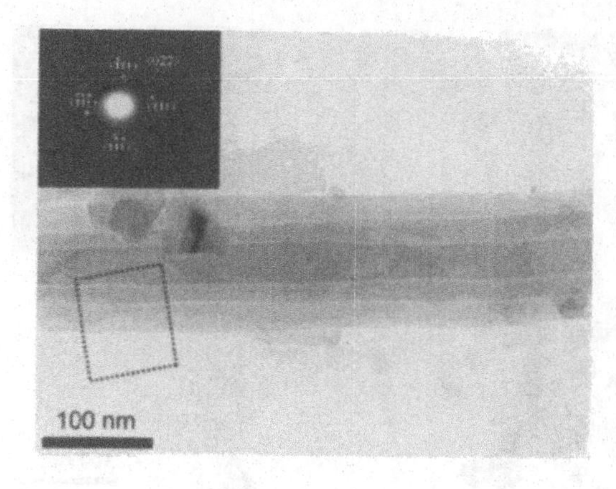

Fig. 15. A magnified image of the coaxial nanostructure, showing a crystalline core and two additional amorphous layers (a-SiO_2 and a-C). The inset shows the selected area diffraction pattern [66].

diameter, second layer 16 nm thick, and third layer 24 nm thick. The interfaces between the layers are sharp. Most of the products were nanocables but small amounts of Si and SiC nanowires were detected as well.

4.3. Metal silicide/SiNWs from metal vapor vacuum arc implantation [68]

A conventional method to produce contacts in the semiconductor industry is ion implantation. The possible advantages of ion implantation for SiNWs are the control in adding precise (and small) amounts of metal atoms to SiNWs, which might be difficult for bulkier techniques. Here we describe ion implantation of Ni and Co into 20 nm-diameter SiNWs produced by thermal decomposition of SiO. The SiNWs were mounted on copper folding grids and directly implanted by Metal Vapor Vacuum Arc (MEVVA) implantation with a 5 keV Ni^+ or Co^+ dose of 1×10^{17} cm^{-2} at room temperature. The implanted samples were annealed afterwards in argon.

Ni implantation results in the formation a Ni silicide layer on the implanted SiNW surface. The layer contains lots of defects. Rapid thermal annealing (RTA) at 500 °C smoothened the surface of the Ni-implanted SiNWs, which was transformed to a continuous outer layer with a typical thickness of about 8 nm, as shown in Fig. 16.

The Co-implanted SiNWs surface is much rougher than that of the Ni-implanted surface with isolated $CoSi_2$ particles 2–40 nm in diameter (Fig. 17).

The generation of $NiSi_2$ and $CoSi_2$ is schematically described in Fig. 18. Room temperature Ni^+ or Co^+ implantation of the as-grown SiNWs (Fig. 18a) results in the formation of a metal/Si mixture (Fig. 18b). The energy of the ion beam should be optimized (5 keV in the present experiment) to avoid excessive damage of the SiNWs. Post-implantation annealing was found efficient in reducing the ion implantation damage. The metal silicides (MS) are expected to give an improved electrical conductivity of the SiNWs and provide electrical contacts to the SiNWs. The structure of the MS/SiNWs layer is sensitive to annealing treatment. Under proper annealing conditions, the MS layer can exhibit a highly oriented relationship to the SiNW core.

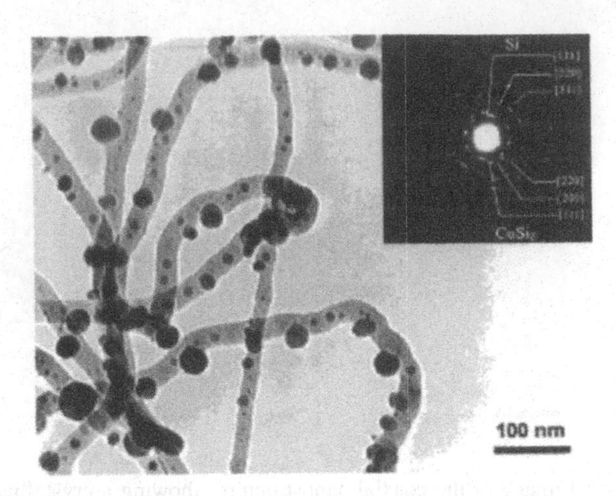

Fig. 16. Ni implanted SiNWs annealed at 500 °C. The inset shows the TED pattern of the Ni layer [68].

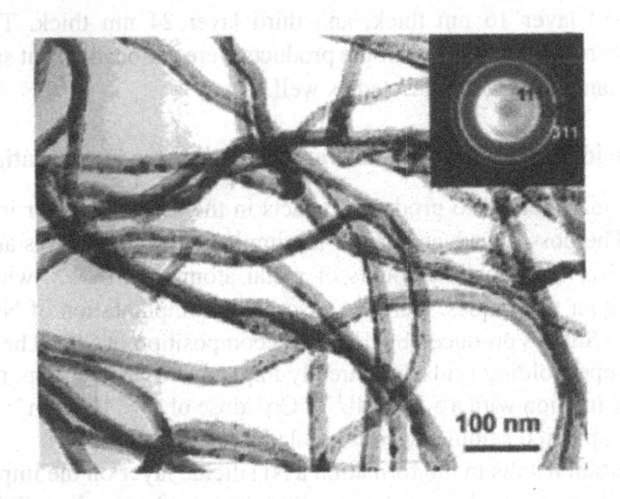

Fig. 17. Co implanted SiNWs annealed at 900 °C. The inset shows the TED of the Co polycrystals [68].

4.4. Metal deposition of SiNWs by solution methods

The reductive deposition of silver and copper ions on (oxygen-removed and hydrogen-terminated) SiNW surfaces in a solution was investigated [69] as an alternative method to ion implantation. The SiNWs surface is indeed capable of reducing silver and copper ions to metal aggregates of various morphologies at room temperature.

Laser ablation was used [22] to produce SiNWs ~20 nm in diameter with a polycrystalline silicon core in a thin silicon oxide sheath with 1/4-1/3 of the nominal diameter and 1/3 of the weight of the SiNW. The oxide layer (which makes the SiNWs surfaces inert) was removed by a 5% HF dip for 5 min resulting in smooth, stable,

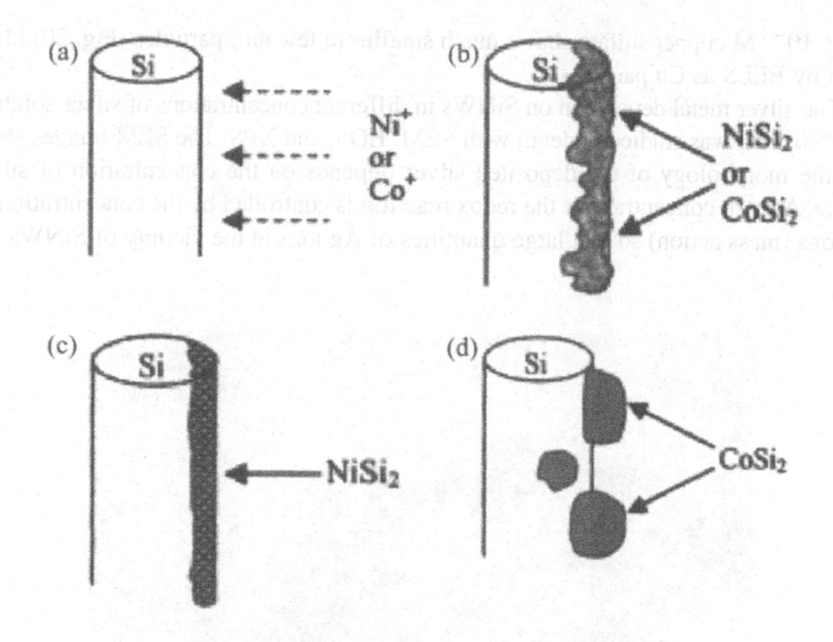

Fig. 18. The formation mechanism of $NiSi_2$/Si and $CoSi_2$/Si on the surface of bare SiNWs [68]. (a) Bare SiNWs implanted with metal ions. (b) Formation of the metal/Si mixture layer on one side of SiNW. (c) The $NiSi_2$/Si nanowire after low temperature annealing. (d) $CoSi_2$ nanoparticles formed by coarsening at high temperature annealing.

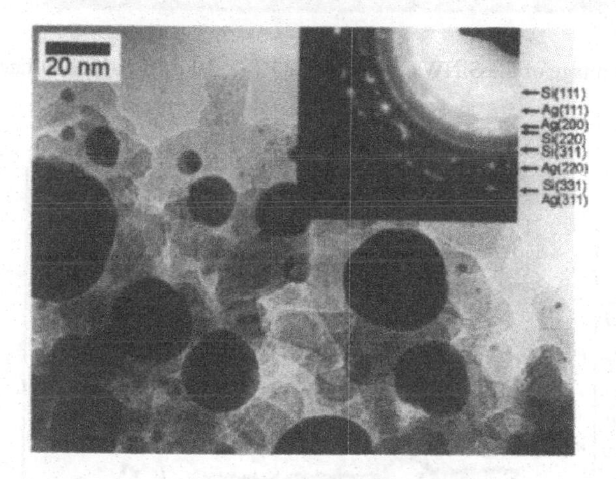

Fig. 19. A TEM image and the corresponding electron diffraction pattern (inset) of the SiNWs treated by a 1.0×10^{-4} M silver nitrate solution [69].

H-terminated SiNW surfaces [70]. The etched SiNWs were immersed into solutions of silver nitrate and copper sulfate of different concentrations. Silver and copper ions were reduced to metallic aggregates deposited onto the surface of SiNWs. The TEM image of the sample treated with a 10^{-4} M silver nitrate solution (Fig. 19) shows dark, round silver particles 5–50 nm in diameter. The HF-etched SiNWs treated with

1.0×10^{-3} M copper sulfate shows much smaller (a few nm) particles (Fig. 20) identified by EELS as Cu particles.

The silver metal deposition on SiNWs in different concentrations of silver solution (10^{-6}–0.1 M) was studied in detail with SEM, EDS, and XPS. The SEM images show that the morphology of the deposited silver depends on the concentration of silver nitrate. At high concentrations the redox reaction is controlled by the concentration of Ag ions (mass action) so that large quantities of Ag ions in the vicinity of SiNWs are

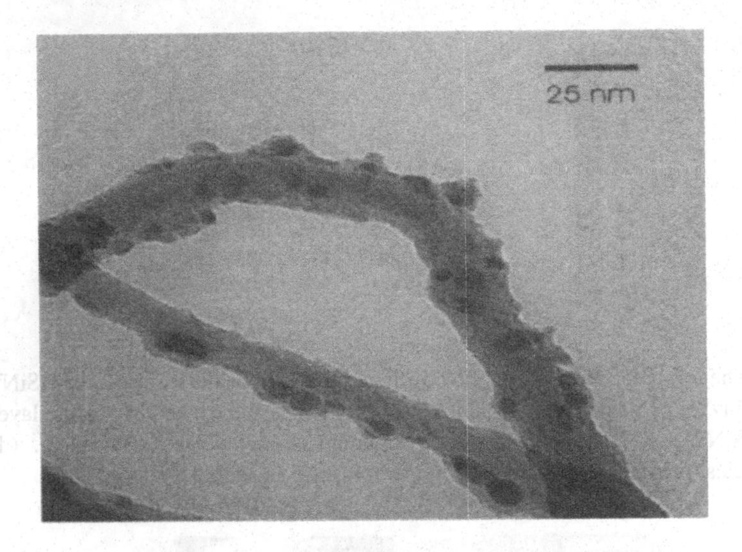

Fig. 20. A TEM image of the SiNWs treated by a 1.0×10^{-3} M copper sulfate solution [69].

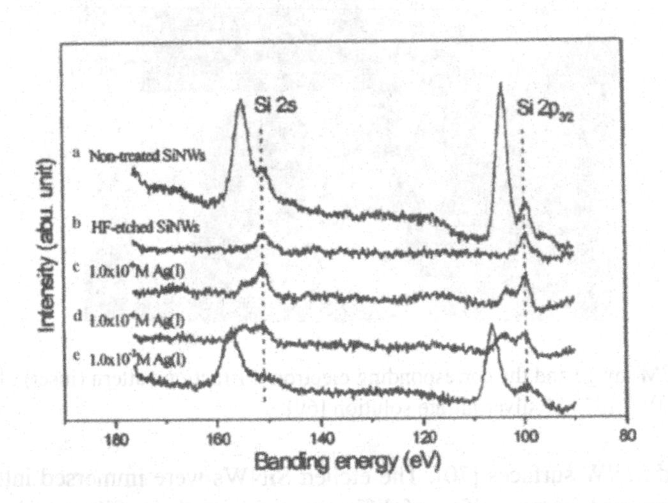

Fig. 21. Si 2p and 2s core level XPS spectra of (a) nontreated SiNWs, (b) HF etched SiNWs, (c) 1.0×10^{-6} M silver nitrate treated SiNWs, (d) 1.0×10^{-4} M silver nitrate treated SiNWs, and (e) 1.0×10^{-2} M silver nitrate treated SiNWs. Note the oxidation of the Si associated with the reduction of the silver ions [69].

reduced and aggregated as dendrites. At low concentrations, the conditions required for dendrite formation are not reached and silver is deposited as clusters or small aggregates on the SiNW surfaces. XPS analysis verifies the redox reaction (Fig. 21). Most of the Si in the initial SiNWs is oxidized (see the Si2s and Si2$p_{3/2}$ lines in Fig. 21) and becomes elemental Si after the HF dip that removed the oxygen. The immersion in the Au solution results in the oxidation of Si, which increases with increasing Au concentration, as expected for a redox reaction of the Au ions. The metal concentration in the solution thus controls the size and the morphology of the metal deposited by the SiNWs from small clusters to dendrites. This can be used for controlled deposition of metal nanoparticles on SiNWs on one hand and larger self-supported metal configurations on the other hand.

4.5. Synthesis of oriented SiC nanowires [71]

One of the fascinating options opened by the OAG method is the possibility of transforming from one type of nanowires to the other. A notable example is the synthesis of oriented SiC nanowires by reacting SiO with aligned carbon nanotubes prepared via the established method of pyrolysis of acetylene over film-like iron/silica substrates [72, 73, 74]. Solid SiO powders (purity 99.9%) were placed in a graphite crucible and covered with a molybdenum grid. The highly aligned carbon nanotubes were placed on the molybdenum grid. The crucible was covered with a graphite lid, placed in the hot zone inside the alumina tube, and held in a flowing argon atmosphere (50 sccm) at 1400 °C for 2 h. After reaction, the aligned carbon nanotube arrays were converted to oriented SiC nanowire arrays. These highly oriented SiC nanowires were similar in appearance to the original aligned carbon nanotubes. The bottom end of the nanowire array is composed of a high density of well-separated and highly oriented nanowire tips (Fig. 22). TEM (Fig. 23a) imaging and diffraction showed that the

Fig. 22. SEM images of oriented SiC nanowire array showing high density of well-separated, oriented nanowire tips [71].

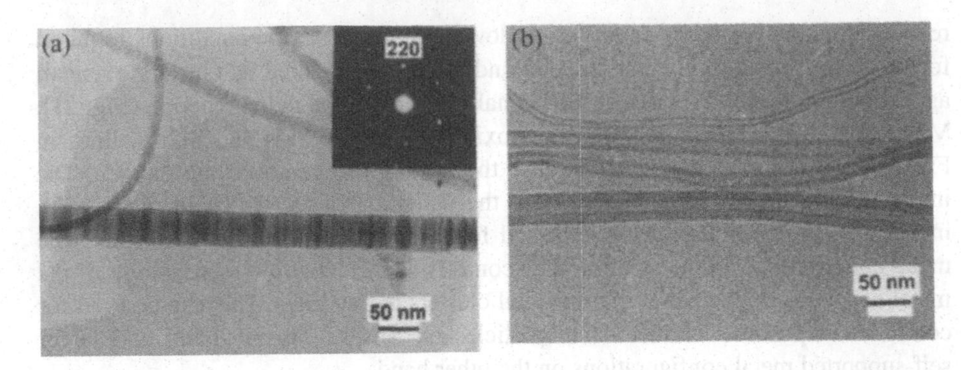

Fig. 23. (a) TEM image of β-SiC nanowires. The SiC nanowires exhibit a high density of stacking faults perpendicular to the wire axes. The inset shows a selected area electron diffraction pattern of the β-SiC nanowires. (b) TEM image of the initial carbon nanotubes. Note the transition from tubes to filled SiC wires [71].

transformed wires are single crystalline β-SiC with the wire axes along the (111) direction and a high density of stacking faults perpendicular to the wire axis. In contrast to the carbon nanotubes (Fig. 23b) the SiC were full wires and not hollow tubes.

5. Characterization of SiNWs

5.1. General

The study of nanomaterials is motivated by the expected small size effects on the nanowire properties. For this reason the characterization of the SiNWs and the establishment of the correlation between their structure, size and properties is a major objective in the science of SiNWs. The nanowires produced may however have a large distribution of shapes and sizes as discussed in the previous sections. The best way to establish the above correlation would therefore be to perform single-wire measurements on well-defined wires. This task is difficult and is only in its very initial stages. Alternatively, one may study the properties of an ensemble of nanowires for which (depending on the specific property investigated) the data will reflect either an average value, or will be dominated by a specific population of the large variety of the shapes in the studied ensemble, which not necessarily represents those we would like to evaluate (it is specially difficult to study small-size wires for which the signal might be much lower than for the larger wires). Most of our studies up till now included ensemble measurements (Raman, PL, field emission, FTIR, I-V for gas sensing) and only a few (STM/STS) were real single-wire measurements.

5.2. Raman and PL of SiNWs

The Raman spectrum of Si nanowires (Fig. 24a) shows an asymmetric peak at 521 cm^{-1} compared to that of a bulk single crystal Si. The peak profile may be associated with the effect of the small size of Si nanocrystals or defects. The presence of

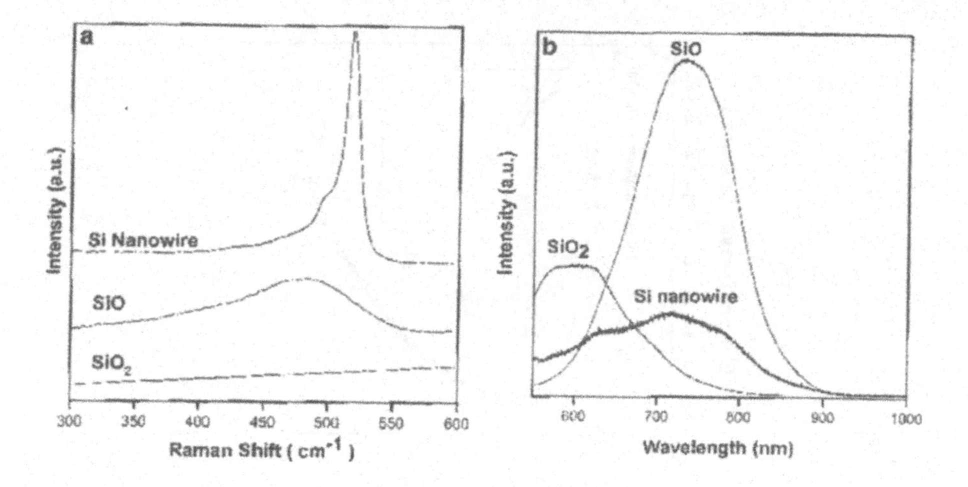

Fig. 24. (a) Raman spectra taken from the as-grown Si nanowires, Si monoxide and fully oxidized Si nanowires; (b) PL spectra taken from the as-grown Si nanowires, Si monoxide and fully oxidized Si nanowires [24].

nonstoichiometric Si sub-oxide may also contribute to the peak asymmetry. For comparison, the spectrum from a SiO film contains a broad peak at 480 cm^{-1}, whereas the fully-oxidized SiNWs (prepared by annealing in air) show no Raman scattering (Fig. 24a).

Si monoxide has a strong photoluminescence (PL) at ~740 nm, while the oxidized nanowire gives a weak PL peak at ~600 nm (Fig. 24b). The PL from the SiNW product is weak and complicated. A typical PL spectrum from SiNWs covers the range of 600–800 nm. Clearly, the SiO and Si sub-oxide components in the nanowires are the main contributors to this spectrum. The SiO generated by thermal evaporation is indeed a mixture of various oxides of Si. Si nanoparticles also co-exist with the SiO generated.

5.3. Field emission from different Si-based nanostructures

It is well known that nanotubes and nanowires with sharp tips are promising materials for applications as cold cathode field emission devices. We have investigated the field emission of different nanowire structures. The first is from SiNWs. SiNWs exhibit well-behaved and robust field emission fitting a Fowler–Nordheim (FN) plot. The turn-on field for SiNWs, which is needed to achieve a current density of 0.01 mA cm^{-2}, was 15 V-μm^{-1} [26]. The field emission characteristics may be improved by further optimization, such as oriented growth or reducing the oxide shell, and may be promising for applications.

The second example of field emission of Si-based nanowires is that of B doped Si nanochains described in Section 3. The SiNCs were attached onto a Mo substrate by a conductive carbon film. The anode-sample separation ranges from 120 to 220 μm. The turn-on field was 6 V-μm^{-1}, and smaller than that (15 V-μm^{-1}) for the SiNWs. The field-emission characteristics of the SiNCs were analyzed according to the

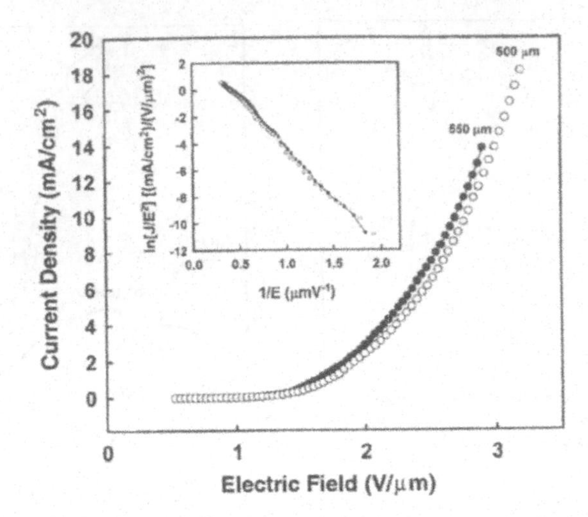

Fig. 25. Emission J-E curves from an oriented SiC nanowire emitter (emitting area 3.65 mm^2). The average turn-on field and threshold field for this sample are about 0.9 V/mm and 2.7 V/mm, respectively. Inset: Fowler-Nordheim plot. The linearity of these curves indicates that the emission of the oriented SiC nanowires agrees with the properties expected for field emission [71].

FN theory [75]. All the FN curves with different anode-sample separations fall nearly in the same region and have similar "Y" intercepts, showing that the SiNCs are uniformly distributed. A stability test showed no obvious degradation of current density and the fluctuation was within ±15%, indicating that the B-doped SiNCs are a promising material for field-emission applications.

The third example of field emission from Si-based nanowires is from the aligned SiC nanowires described in Section 4.5. The field emission measurements [71] were carried out in a vacuum chamber at a pressure of ~5×10^{-7} torr at room temperature. An oriented SiC nanowire array, which was used as the cathode, was stuck to a stainless steel substrate by silver paste with the bottom end of the nanowires facing upward. A copper plate with a diameter of 1 cm, mounted on a precision linear feedthrough, was used as the anode. Field emission current densities of 10 μA/cm^2 were observed at applied fields of 0.7–1.5 V/μm, and current densities of 10 mA/cm^2 were realized at applied fields as low as 2.5–3.5 V/μm, as shown in Fig. 25. These results represent one of the lowest fields ever reported for any field-emitting materials at technologically useful current densities. We attributed this emission behavior to the very high density of emitting tips with a small ratio of curvature at the emitting surface. The fact that when the oriented SiC nanowire array was pressed flat the current density was an order of magnitude lower than for the initial sample under the same electric field strongly supports this point.

5.4. Resistivities of SiNWs in different ambients [76]

I–V measurements performed on an ensemble of SiNWs with a variety of diameters, growth directions, defect densities etc. are expected to yield only averaged behavior, which is dominated by those wires with the lowest resistivity [as in a parallel configuration of wires (resistors)]. While such ensemble measurements cannot be used to study the electrical conduction properties and mechanisms of nanowires, they can however give rough indications to check a variety of possible applications, one being gas sensing.

We have fabricated bundles of SiNWs of two types: (1) as-grown SiO_2 sheathed wires, (2) SiNWs dipped in HF to remove the SiO_2. Silver contacts were glued to the edges of the bundles and their resistivity was measured at different ambient conditions (vacuum (2×10^{-2} torr), air with ~60% humidity, dry N_2, $NH_3:N_2$ 1 : 1000). The resistivity of the oxide-removed bundles was strongly reduced (by more than three orders of magnitude) upon exposure to humid air and to ammonia, but was hardly changed by exposure to dry nitrogen (Fig. 26a). The process was found reversible, i.e. the resistivity increased to the initial value after pumping (Fig. 26b). In contrast, the resistivity of the SiNWs embedded in the SiO_2 sheath did not change when exposed to different ambient environments.

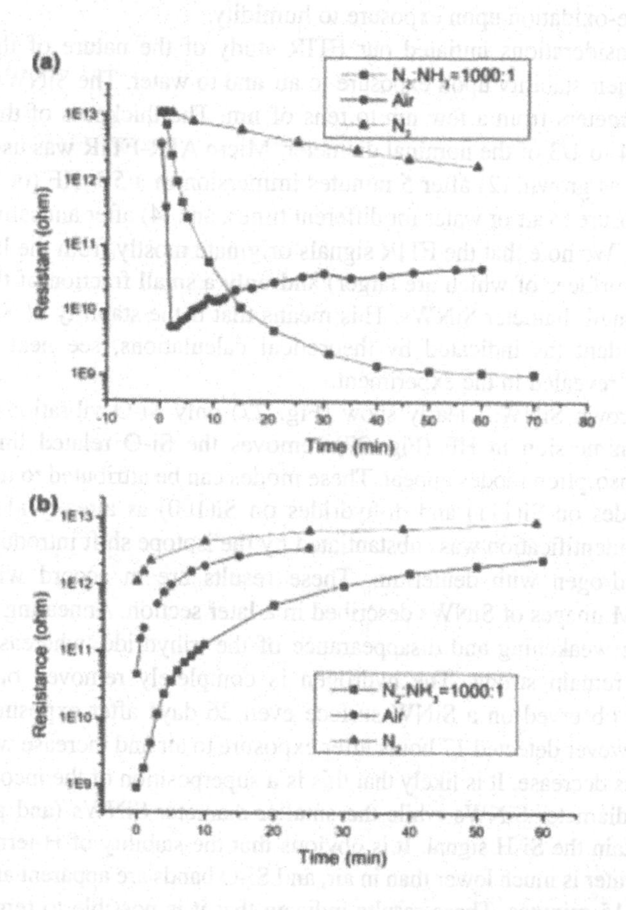

Fig. 26. Electrical responses with time of the Si nanowire bundle to N_2, a mixture of N_2 and NH_3 (NH_3 concentration: 1000 ppm), and air with a relative humidity of 60%; (a) when the gases were introduced into the chamber and (b) when the gases were pumped away [76].

The gas molecules may affect either the contact resistance across two nanowires or the surface resistance along individual wires, e.g. through charge exchange similar to polycrystalline semiconductor SnO_2 sensors. This would not happen for SiO_2 sheathed wires (having a high resistivity) for which gas incorporation has no effect. The chemical sensitivity of HF-etched SiNWs to NH_3 and water vapor exposure indicate their possible use in gas sensing applications.

5.5. FTIR measurements of the stability of H-terminated SiNW surfaces [77]

Silicon-based technology requires the removal of the surface oxide layer and the termination and the stabilization of the Si surfaces. This is conventionally performed by dipping in HF, which not only removes the oxide layer, but provides H-terminated Si surfaces. Examples of the significance of such a treatment are given in the previous and following sections. The stability of the oxide-removed H-terminated Si wafer surfaces has been extensively investigated, and many techniques were developed to suppress its re-oxidation upon exposure to humidity.

These considerations initiated our FTIR study of the nature of the HF-dipped SiNWs and their stability upon exposure to air and to water. The SiNWs had a distribution of diameters from a few nm to tens of nm. The thickness of the oxide layer was about 1/4 to 1/3 of the nominal diameter. Micro ATR-FTIR was used to monitor the wires (1) as grown, (2) after 5 minutes immersion in a 5% HF (or DF) solution, (3) upon exposure to air or water for different times, and (4) after annealing to different temperatures. We note that the FTIR signals originate mostly from the large-diameter SiNWs (the surfaces of which are larger) and only a small fraction of the signal represents the small-diameter SiNWs. This means that if the stability of SiNW surfaces is size-dependent (as indicated by theoretical calculations, see next section), this would not be revealed in the experiment.

The as-grown SiNWs clearly show (Fig. 27) only Si-O vibrations at 1050 and 800 cm^{-1}. Immersion in HF (Fig. 27) removes the Si-O related lines and Si-H_x ($x = 1,2,3$) absorption modes appear. These modes can be attributed to mono-hydrides and tri-hydrides on Si(111) and di-hydrides on Si(100) as already identified for Si wafers. This identification was substantiated by the isotope shift introduced by substitution of hydrogen with deuterium. These results are in accord with atomically resolved STM images of SiNWs described in a later section. Annealing of the SiNWs results in the weakening and disappearance of the trihydride, whereas the monohydride peaks remain strong. The hydrogen is completely removed only at 850 K. Hydrogen is observed on a SiNW surface even 26 days after exposure to air. Si-O bands are however detected 17 hours after exposure to air and increase with time while the SiH bands decrease. It is likely that this is a superposition of the incorporation of O in the large-diameter SiNWs while the smaller-diameter SiNWs (and probably more stable) maintain the Si-H signal. It is obvious that the stability of H-terminated SiNW surfaces in water is much lower than in air, and Si-O bands are apparent after immersion in water for 15 minutes. These results indicate that it is possible to remove the oxide layer from the SiNW surfaces and terminate them by H by immersion in HF, similar to Si wafers. The H-terminated surfaces seem to be stable in air for at least a day,

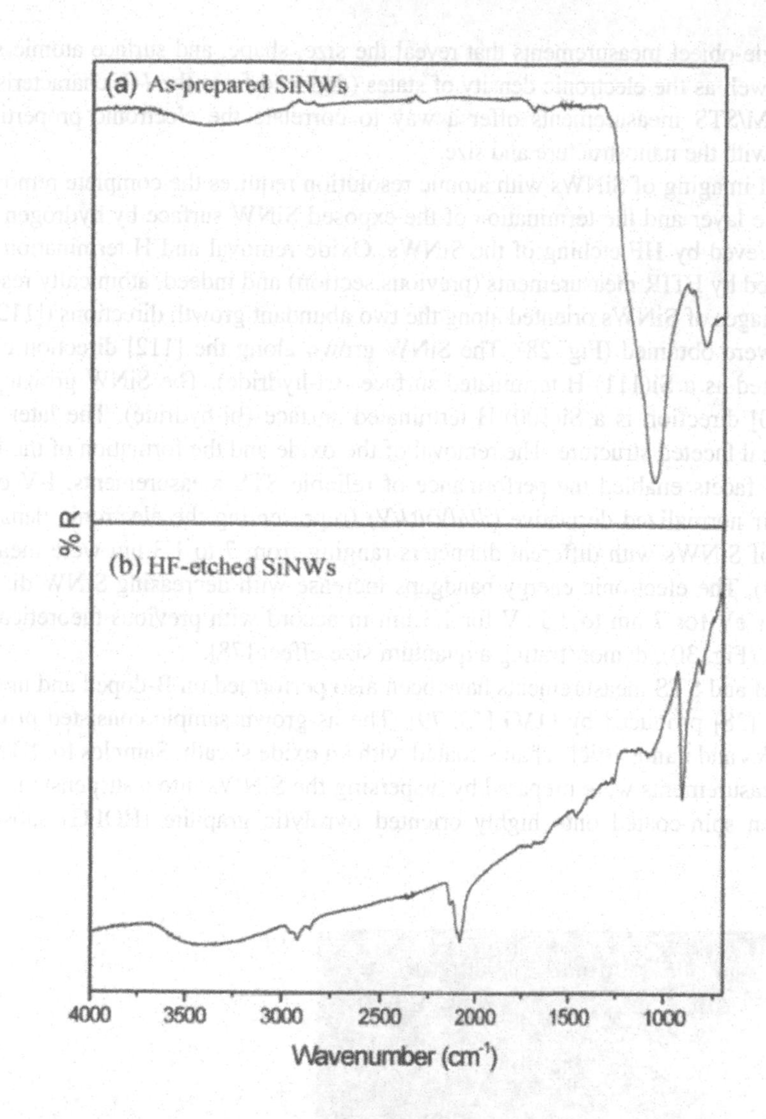

Fig. 27. ATR-FTIR spectra of (a) as-prepared SiNWs and (b) HF-etched SiNWs in the range of 700–4000 cm^{-1}. Note the removal of the oxide absorption line and the formation of Si-H$_x$ absorption lines in the etched SiNW [77].

whereas the stability of small-diameter SiNWs as determined by single-wire STM measurements seems to be considerably better than that of Si wafers.

5.6. STM and STS measurements of SiNWs and B-doped SiNWs

Almost all the characterizations performed by us until now are ensemble characterizations (i.e. probing many nanostructures simultaneously). HRTEM and HRSEM do probe the structure (and elemental composition) of individual nanostructure, but they do not correlate this structure with a specific property. STM and STS measurements are real single-object measurements that reveal the size, shape, and surface atomic structure, as well as the electronic density of states (deduced from the *I–V* characteristics). The STM/STS measurements offer a way to correlate the electronic properties of SiNWs with the nanostructure and size.

STM imaging of SiNWs with atomic resolution requires the complete removal of the oxide layer and the termination of the exposed SiNW surface by hydrogen. This was achieved by HF etching of the SiNWs. Oxide removal and H termination were confirmed by FTIR measurements (previous section) and indeed, atomically resolved STM images of SiNWs oriented along the two abundant growth directions ([112] and [110]) were obtained (Fig. 28). The SiNW grown along the [112] direction can be interpreted as a Si(111)-H terminated surface (tri-hydride). The SiNW grown along the [110] direction is a Si(100)-H terminated surface (bi-hydride). The later has a hexagonal faceted structure. The removal of the oxide and the formation of the H terminated facets enabled the performance of reliable STS measurements. I-V curves and their normalized derivative $(dI/dV)/(I/V)$ (representing the electronic density of states) of SiNWs with different diameters ranging from 7 to 1.3 nm were measured (Fig. 29). The electronic energy bandgaps increase with decreasing SiNW diameter from 1.1 eV for 7 nm to 3.5 eV for 1.3 nm in accord with previous theoretical predictions (Fig. 30), demonstrating a quantum size effect [78].

STM and STS measurements have been also performed on B-doped and undoped SiNWS [38] produced by OAG [23, 79]. The as-grown sample consisted primarily of SiNWs and nanoparticle chains coated with an oxide sheath. Samples for STM and STS measurements were prepared by dispersing the SiNWs into a suspension, which was then spin-coated onto highly oriented pyrolytic graphite (HOPG) substrates.

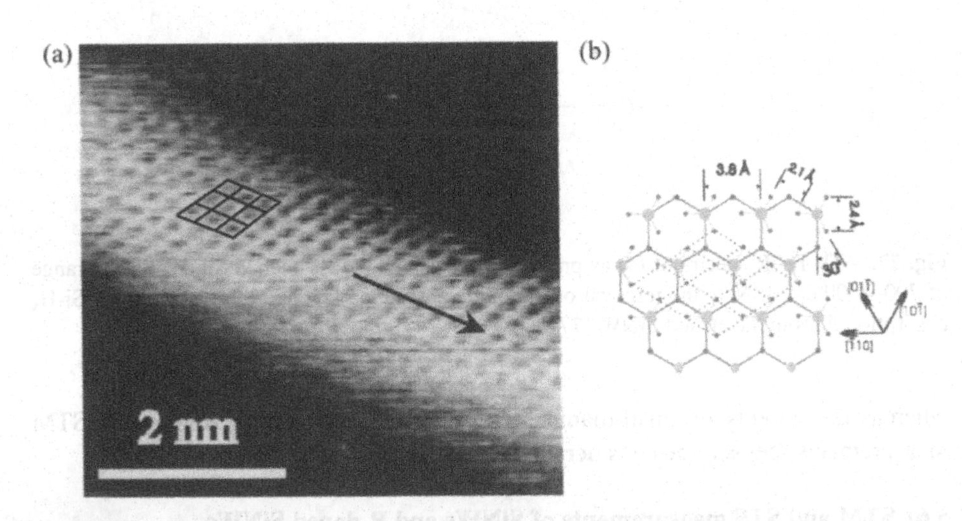

Fig. 28. STM image and schematic view of a SiNW with Si (111) facet. (a) Constant current STM image of a SiNW on a HOPG substrate. The wire's axis is along [112] direction, and (b) schematic view of SiH_3 on Si (111) viewed along the [111] direction. Red and large blue circles represent the H atoms and Si atoms in the SiH_3 radical, respectively. Small blue circles represent Si (111) atoms in the layer below. The crystallographic directions are shown in the inset [78].

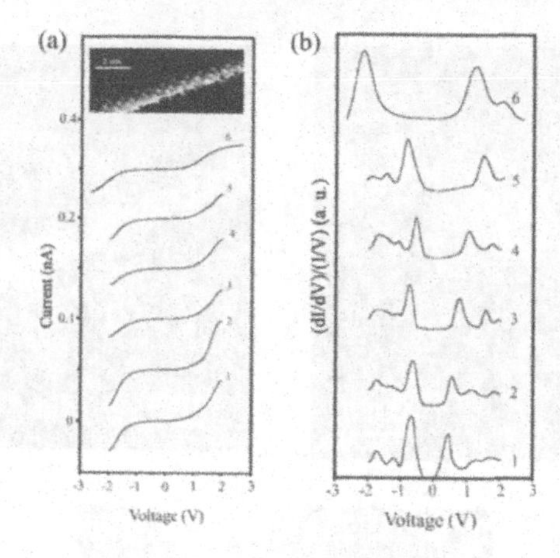

Fig. 29. Electronic properties of the SiNWs surfaces: (a) Current (*I*)-voltage (*V*) curves obtained by STS on six individual SiNWs; the diameter of wires 1 to 6 being 7, 5, 3, 2.5, 2, and 1.3 nm respectively; The inset shows the atomically resolved STM images of the 1.3 nm wire (6) which is the smallest ever reported. (b) The corresponding normalized tunneling conductances, (*dI/dV*)/*I/V*; the curves are offset vertically for clarity [78].

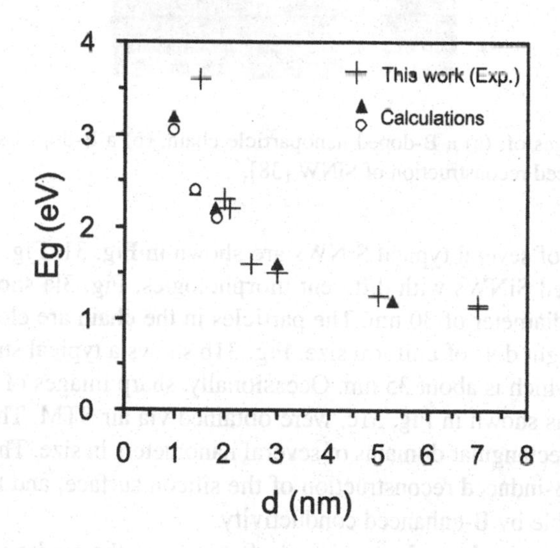

Fig. 30. Experimental bandgap deduced from 29b versus the diameter of wires 1–6 plus additional three wires not shown in Fig. 29a,b. The references of the calculated bandgaps are found in [78].

The presence of nanoparticle chains and nanowires in the B-doped SiNWs sample was observed. Clear and regular nanoscale domains were observed on the SiNW surface, which were attributed to B-induced surface reconstruction. STS measurements have provided current-voltage curves for SiNWs, which showed enhancement in electrical conductivity by boron doping.

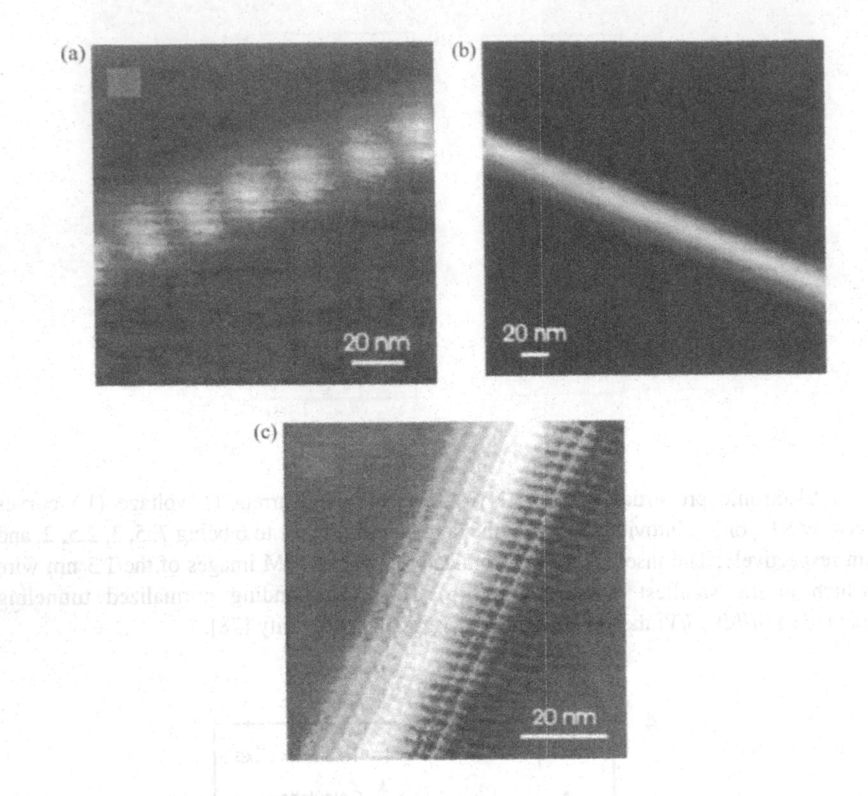

Fig. 31. STM images of: (a) a B-doped nanoparticle chain, (b) a B-doped straight nanowire, and (c) boron-induced reconstruction of SiNW [38].

STM images of several typical SiNWs are shown in Fig. 31. Fig. 31a–c show the images of B-doped SiNWs with different morphologies. Fig. 3la shows a nanoparticle chain with a diameter of 30 nm. The particles in the chain are clearly revealed as evenly spaced bright dots of uniform size. Fig. 31b shows a typical straight nanowire, the diameter of which is about 35 nm. Occasionally, sharp images of SiNW of ca. 40 nm in diameter, as shown in Fig. 31c, were obtained via air STM. The image reveals clearly resolved rectangular domains of several nanometers in size. These domains are associated with B-induced reconstruction of the silicon surface, and the clear images were made possible by B-enhanced conductivity.

STS measurements have been also performed on the undoped and B-doped SiNWs shown in Fig. 31. The *I–V* and the corresponding differentiated *dI/dV* curves of the nanowires reveal several features. First, while the curves for the two B-doped nanowires with different morphologies are quite similar, they are distinctly different from that for the undoped wire. This is because that, as far as tunneling from a STM probe is concerned, the particle in the nanochain is the same as any point on a nanowire with a similar doping concentration and oxide sheath. Second, the steeper rise in the *I–V* curves and the higher values of *dI/dV* for the B-doped SiNWs are consistent with the expected B-induced conductivity enhancement. As *dI/dV* values can be regarded as a measure of local density of states (LDOS), the low value between -1

and +1 V in the *dI/dV* curve of the undoped wire indicates relatively little LDOS within the gap, while the higher *dI/dV* in the same region of the doped wires is in accord with the presence of the B dopants. The minima of the curves are at 0.3 V, indicating that the Fermi level in the B-doped SiNWs lies 0.3 eV closer to the valence band relative to the undoped wire. The position of the Fermi level corresponds to a hole carrier concentration of 1.5×10^{15} cm^{-3} in the boron-doped nanowires.

5.7. Periodic array of SiNW heterojunctions [80]

The OAG of SiNWs using thermal evaporation of SiO at 1200 °C was studied in detail as previously discussed and the growth at ~900 °C leads to straight, uniform (in diameter) nanowires with a crystalline silicon core embedded in a SiO_2 sheath. STM analysis was performed on such SiNWs that were dipped into HF to remove the oxide sheath and then dispersed on a HOPG substrate and loaded to the UHV STM system. Among the less abundant forms of SiNWs, ~1% have a zigzag shape, composed of many bends or junctions.

Fig. 32 shows the structure of such a typical zigzag shaped SiNW with a diameter of 3 nm and a length of several microns. The wire is composed of a periodic array of long (~10 nm) and short (~5 nm) segments denoted A and B respectively. The angle between the segments is ~30°. The junctions repeat themselves regularly so that the length of the different segments is fairly constant along the entire wire. We have discussed in a previous section the growth directions of SiNWs in the OAG, and have reported that the two most abundant directions are [112] and [110]. We thus speculate that the zigzag shape originates from a periodic transition between two growth directions (which is 30° between [112] and [110] and 35° between [110] and [111]).

STS of these wires performed on several of each segment indicates that the *I–V* curves are almost the same along identical segments (i.e. along all As or along all Bs),

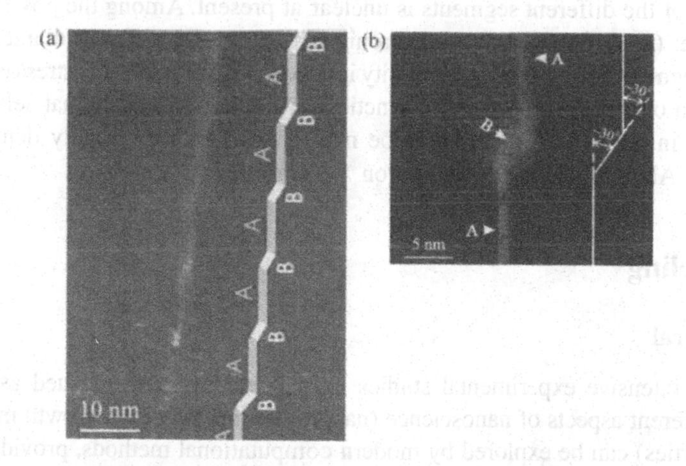

Fig. 32. STM images of a SiNW showing the periodic array of Si segments: (a) a medium magnification showing six sequences as further illustrated by the schematic sketch denoting segments A (~10 nm long) and B (~5 nm long) and (b) a higher magnification showing an image of a pair of junctions (ABA), also denoting the angle between the segments at ~30° [80].

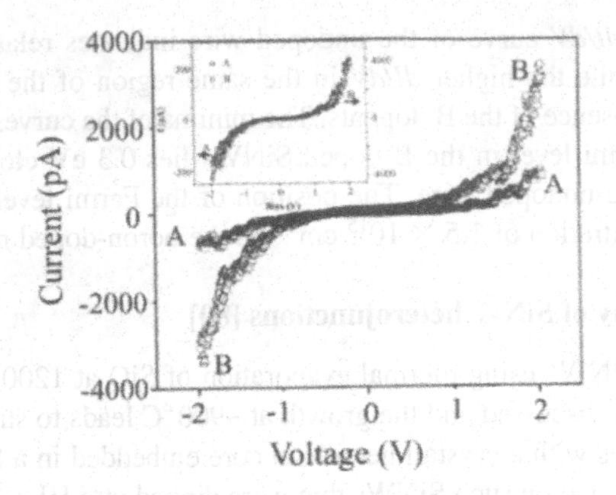

Fig. 33. *I–V* STM curves of segments A and B. Note the significant difference between the two *I–V* curves indicating different electronic properties. The inset shows the same *I–V* curves for A (squares) and B (circles and triangles) in which the right-hand side scale (for segment B) is the same as in Fig. 32 but the left-hand side scale (for segment A) was increased (50 pA per bar instead of 1000 pA per bar) until the *I–V* curves of the two segments overlapped. It is evident that the shape of the *I–V* curves of the two different segments is similar (the electronic energy gap is almost the same) but the I values differ by a factor of ~20. The *I–V* curve is the same for all A segments or B segments and along each segment. The change from one *I – V* curve to the other along the junction (going from A to B or from B to A) is however very sharp [80].

but very different for the two types of segments as indicated in Fig. 33. The transition between one type of *I–V* curve and the other along the segment interface (junction) is very sharp. We can conclude that periodic arrays of Si intramolecular junctions were grown in a single SiNW growth process (segmented growth was previously reported by a periodic change of the growth conditions). The difference in the electronic properties of the different segments is unclear at present. Among the possible origins we include: (1) different surface electronic structures, (2) different diameters of the different segments, (3) defect and impurity induced variations, and (4) stress effects. The observation of arrays of 66 pairs of junctions per micron indicate that self assembly of SiNWs in the OAG process may be manipulated to grow highly dense devices (1.1×10^4 ABA "transistors" per micron $\times$ micron).

6. Modeling

6.1. General

While intensive experimental studies have been/are being pursued as described above, different aspects of nanoscience (nanomatrerials structure, growth mechanisms and properties) can be explored by modern computational methods, providing insight where experimental methods find difficulty and predicting properties which guide experimentalists. Here, we review the computational works performed in CityU that address different aspects of science and technology of silicon based nanostructures, and demonstrate the power of computational tools in nanoscience and nanotechnology.

6.2. Silicon oxide vapor

Silicon oxide is a critical source material in the oxide-assisted growth (OAG) as described above. It also plays important roles as is well known in many fields such as electronics, optical communications, and thin-film technology. Our recent finding of silicon oxide in the synthesis of silicon nanowires as we reviewed in the previous part of this chapter would extend further the important new application of silicon oxide.

We have studied the silicon-oxide-assisted formation of Si nanostructures based on quantum-mechanical calculations of Si_nO_m (n, m = 1–8) clusters [51, 52, 81]. We found that most of the structures contain planar or buckled ring units. Pendent silicon atoms bonded only to a single oxygen atom are found in silicon-rich clusters. Oxygen-rich clusters have perpendicular planar rings, while silicon monoxide-like clusters usually form a large buckled ring. Structures made up of tetrahedrally bonded units are found only in two clusters. Furthermore, the energy gap and net charge distribution for clusters with different Si:O ratios have been calculated. We further found that: (i) energetically the most favorable small silicon-oxide clusters have O to Si atomic ratios at around 0.6 (see Fig. 34); and (ii) remarkably high reactivity at the Si atoms exists in silicon suboxide Si_nO_m clusters with $2n > m$ (see Figs 4 and 35). The results show that the formation of Si-Si bond is preferred and thus facilitates the nucleation of Si nanostructures when silicon suboxide clusters come together or stack to a substrate. Based on these findings, the mechanism of oxide-assisted nucleation of silicon nanowires has been drawn clearly [53] as we reviewed in Sections 2.3 and 2.4 of this chapter.

6.3. Pure silicon nanostructured materials

For application of the SiNWs in advanced areas, their oxide sheath has to be removed. If the silicon core were not saturated, the stability of the structure would be very poor. Demonstrations of their stability would be useful for understanding the

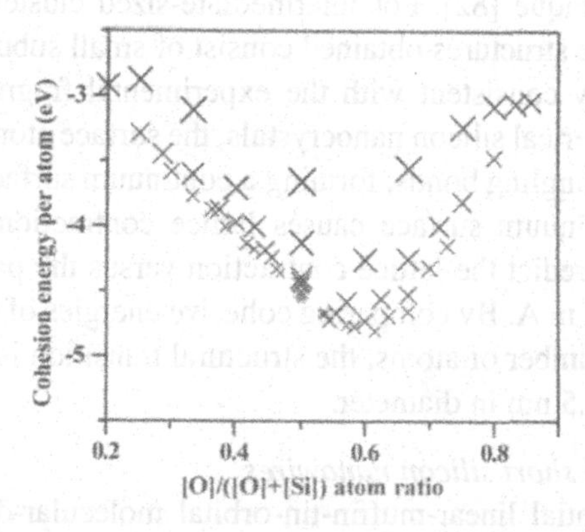

Fig. 34. Cohesion energy per atom of Si_nO_m (n, m = 1–8) clusters as a function of O ratio based on total energy calculations with B3LYP/3-21G : Si; 6-31G* : O. The decreasing size of the symbol × is related to increasing cluster size (n + m) [52].

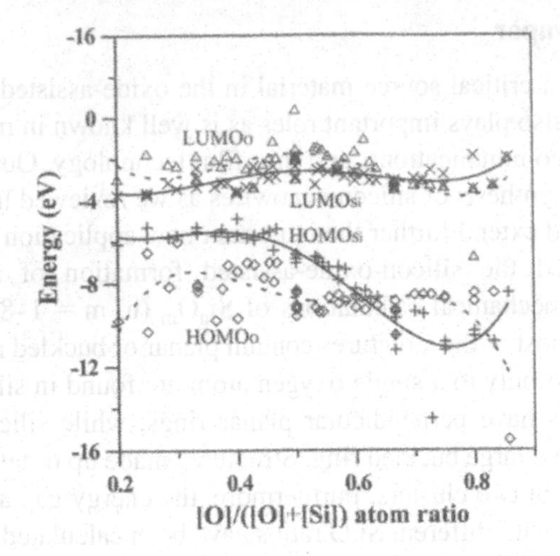

Fig. 35. $LUMO_{Si}$ ("×"), $LUMO_O$ ("Δ"), $HOMO_{Si}$ ("+") and $HOMO_O$ ("◇") of Si_nO_m clusters determined based on the calculations using B3LYP/3-21G: Si; 6-31G*: O. Their fittings using four-order polynomials are shown with solid ciirves for $LUMO_{Si}$ (upper) and $HOMO_{Si}$ (lower) while dashed curves represent $LUMO_O$ (upper) and $HOMO_O$ (lower).

related problems. Other issues of pure silicon nanostructures including the structure and property of thinnest nanowire and the stability of silicon nanotubes are also interesting. Summarized below are our computational efforts regarding these issues.

6.3.1. Structural transition in silicon nanostructures

As is well known, small silicon clusters do not have any structural feature similar to that of bulk silicon (tetrahedral). Between the small silicon clusters and bulk silicon, there may be structural transition from amorphous to ordered tetrahedral structure. The structural transition to bulk diamond structure in nanosized silicon clusters has been studied by tight-binding molecular dynamics method combined with a simulated annealing technique [82]. For intermediate-sized clusters (<200 atoms), the energetically favorable structures obtained consist of small subunits like Si_{10} and Si_{12} (Fig. 36), qualitatively consistent with the experimental fragmentation behavior of these clusters. For spherical silicon nanocrystals, the surface atoms reconstruct to minimize the number of dangling bonds, forming a continuum surface (Fig. 37). The large curvature of the continuum surface causes lattice contraction in the nanocrystals. Present calculations predict the lattice contraction versus the particle radius as $\Delta a = 0.38/R$, with Δa and R in Å. By comparing cohesive energies of the two sorts of structures with the same number of atoms, the structural transition is estimated to occur at about 400 atoms, or 2.5 nm in diameter.

6.3.2. Thinnest stable short silicon nanowires

Using a full-potential linear-muffin-tin-orbital molecular-dynamics method, we have studied the geometric and electronic structures of thin short silicon nanowires consisting of tri-capped trigonal prism Si_9 sub-units and uncapped trigonal prisms, respectively [83]. Comparing to other possible structures, these structures are found to

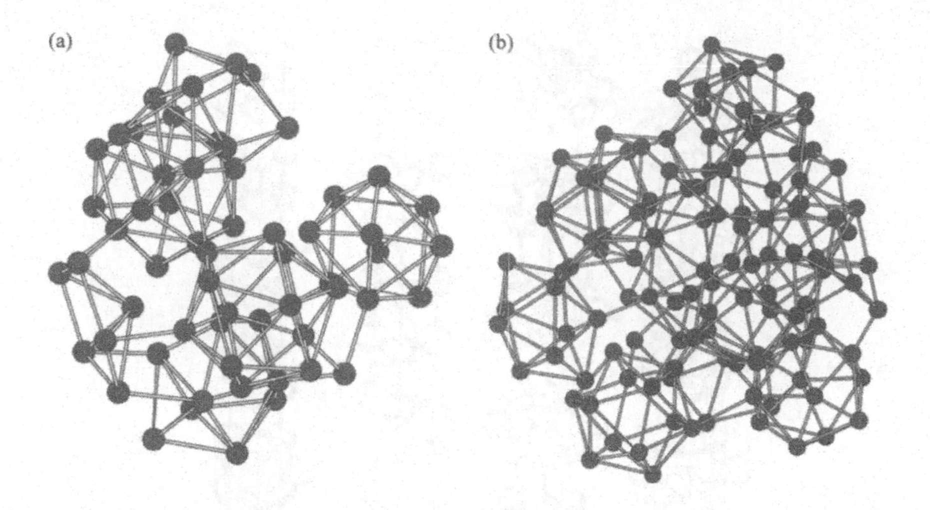

Fig. 36. Structures of (a) Si_{60} and (b) Si_{123} after annealing. All bonds below 2.8 Å are drawn out. The structures are fully relaxed, with the root-mean-square force to be 0.015 eV/Å [82].

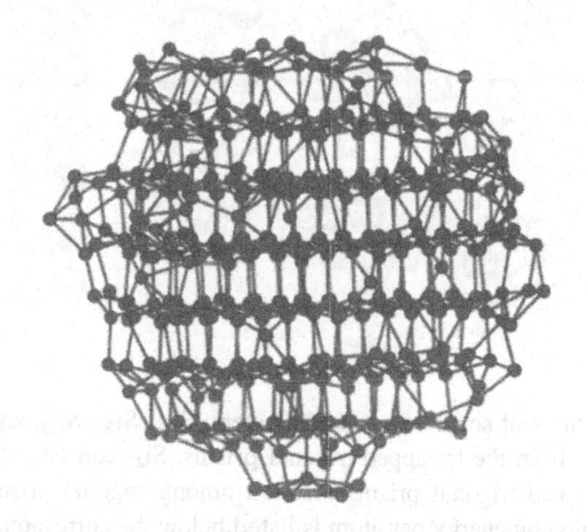

Fig. 37. Structure of the 417-atom Si nanocrystal with reconstructed surface. All bonds below 2.8 Å are drawn out. The structure is fully relaxed, with the root-mean-square force to be 0.015 eV/Å [82].

be the thinnest stable silicon nanowires, being particularly much more stable than the silicon nanotubes built analogously to small carbon nanotubes (Fig. 38). As for their electronic structures, these silicon wires show very small gaps of only a few tenths of an eV between the lowest unoccupied energy level and the highest occupied energy level, and the gaps decrease as the stacked layers increase. The results provide guidance to experimental efforts for assembling and growing silicon nanowires.

6.3.3. *Silicon nanotubes*

In contrast to the synthesis of large quantity of SiNWs, no Si nanotube has ever been observed experimentally, indicating that silicon is an element very different

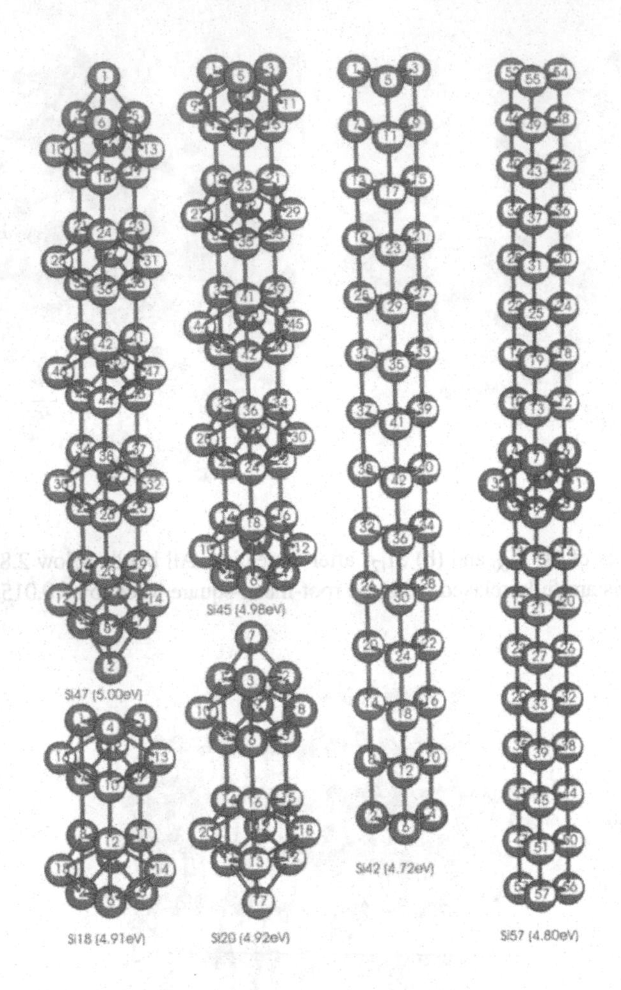

Fig. 38. Stable structures of some selected Si clusters. Si_{18}, Si_{20}, Si_{45}, and Si_{47} correspond to the stacked structures from the tricapped trigonal prisms. Si_{42} consists of the trigonal prisms. Si_{57} refers to the stacked trigonal prisms inserted among trigonal prisms by one tricapped trigonal prism. The binding energy per atom is listed below the corresponding structure [83].

from carbon in forming nanotublar structure although they are in the same group of Period Table. The difficulty in the synthesis of silicon nanotubes is widely attributed to the property of sp^3 hybridization in silicon. How, to what extent such hybridization affects the tubular structural formation still needs further clarification. To understand the reason(s) for the hitherto unsuccessful synthesis of silicon nanotubes we have studied [84] the differences in the structures and bonding between cubic (diamond-like) and tubular nanostructures of carbon and silicon, and their relative stabilities in terms of their characteristic electronic structures. Our calculated results indicated that when the dangling bonds at the open ends of the tubular structure are properly terminated, Si nanotubes with a severely puckered structure can in principle be formed. Such computationally stable, energetically minimized, and geometrically optimized Si nanotube structures may serve as models for the design and synthesis of silicon nanotubes.

6.4. Hydrogenated silicon nanostructures

As reviewed in Section 5, the SiNWs could be etched using HF solution so that the surface oxide sheath is removed and the exposed silicon surface is saturated with hydrogen. The stability of the oxide removed H terminated Si wafer surfaces has been investigated and many techniques were invented to suppress its re-oxidation upon exposure to humidity. For the case of hydrogenated silicon nanostructured, its chemical stability would be different from the case of wafer as has been demonstrated in our recent computational works. One of our predictions that the hydrogenated SiNWs possess better stability than that of silicon wafer could find a number of supportive evidences from our experiments as described above.

6.4.1. Structural properties of hydrogenated silicon nanocrystals and nanoclusters

The structures of hydrogenated Si nanocrystals and nano-clusters were studied using the empirical tight-binding optimizations and molecular dynamics simulations [85].

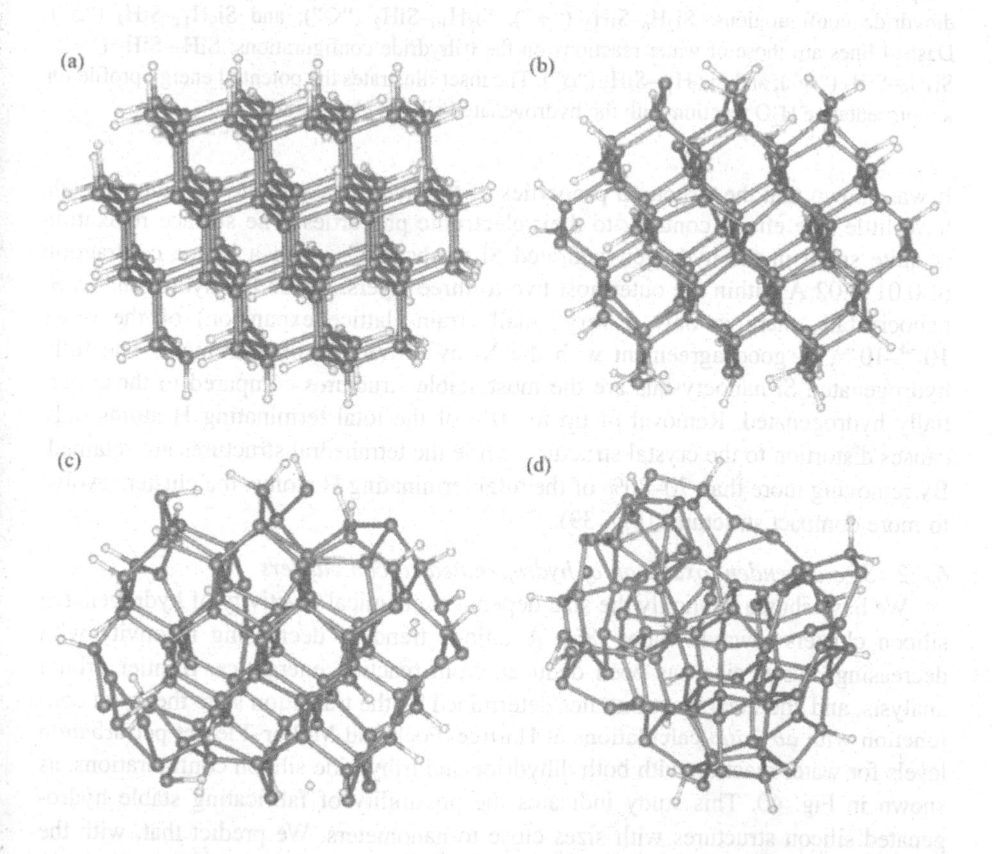

Fig. 39. Ball-and-stick diagrams of the structures of $Si_{100}H_x$ clusters obtained from simulated annealings. (a) Fully H saturated $Si_{100}H_{86}$; (b) $Si_{100}H_{60}$; (c) $Si_{100}H_{40}$; and (d) $Si_{100}H_{20}$ [85].

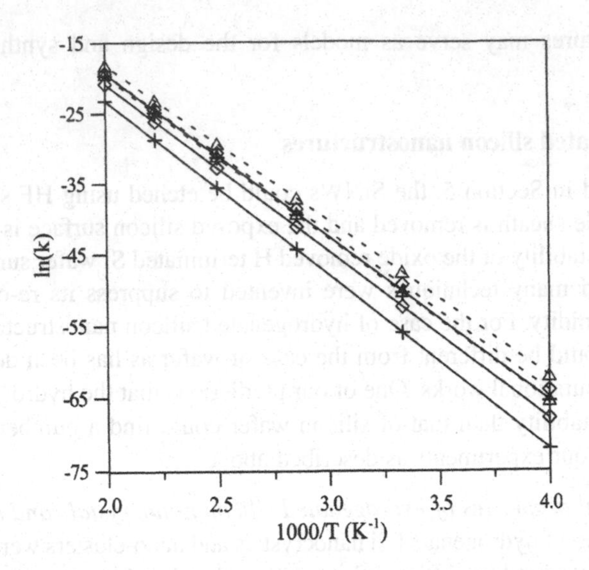

Fig. 40. Calculated total rate constants ($cm^3 mol^{-1} s^{-1}$) at the pressure of 1 atm in a temperature range between 250 and 500 K. Solid lines are those of water reactions on the dihydride configurations: Si_2H_6–SiH_2 ("+"); Si_5H_{10}–SiH_2 ("◇"); and Si_9H_{14}–SiH_2 ("Δ"). Dashed lines are those of water reactions on the trihydride configurations: SiH_3–SiH_3 ("+"); Si_4H_9–SiH_3 ("◇"); and $Si_{10}H_{15}$–SiH_3 ("Δ"). The inset illustrates the potential energy profile for a representative H_2O reaction with the hydrogenated silicon cluster [86].

It was shown that the structural properties of the hydrogen-saturated Si nanocrystals have little size effect, contrary to their electronic properties. The surface relaxation is quite small in the hydrogen-saturated Si nanocrystals, with a lattice contraction of 0.01–0.02 Å within the outermost two to three layers. Inside the hydrogenated Si nanocrystals, there is only a very small strain (lattice expansion) of the order 10^{-4}–10^{-3}, in good agreement with the X-ray diffraction measurement. The fully hydrogenated Si nanocrystals are the most stable structures compared to those partially hydrogenated. Removal of up to 50% of the total terminating H atoms only causes distortion to the crystal structure, while the tetrahedral structures are retained. By removing more than 70–80% of the total terminating H atoms, the clusters evolve to more compact structures (Fig. 39).

6.4.2. Size-dependent oxidation of hydrogenated silicon clusters

We have shown explicitly the size-dependent chemical reactivity of hydrogenated silicon clusters towards water [86]. A unique trend of decreasing reactivity with decreasing cluster size has been deduced from reaction energetics, frontier orbital analysis, and chemical reaction rates determined by the transition state theory in conjunction with *ab initio* calculations at Hartree–Fock and Møller–Plesset perturbation levels for water reaction with both dihydride and trihydride silicon configurations, as shown in Fig. 40. This study indicates the possibility of fabricating stable hydrogenated silicon structures with sizes close to nanometers. We predict that, with the nanosized hydrogenated silicon structures, it is possible to fabricate nonreactive, stable nanodevices.

7. Oxide-Assisted Growth of Other (Not Containing Si) Semiconducting Nanowires

The OAG method has a general nature and can be applied to a variety of materials other than Si. Based on the OAG method, we have synthesized nanowires of a wide range of semiconducting materials including Ge [39], GaN [40, 41], GaAs [42, 43], GaP [45], SiC [44], and ZnO [46]. The actual OAG process was activated by laser ablation, hot-filament chemical-vapor-deposition (HFCVD) or thermal evaporation.

Similar to the production of Si nanowires, we used laser ablation of a mixed GeO_2/Ge target to synthesize Ge nanowires at 830 °C. In comparison, Morales and Lieber [21] used metal-containing targets for the same purpose. We show that Ge nanowires are obtained by the OAG method in an analogous way to SiNWs. TEM studies (Fig. 41) show that the structure of the Ge nanowires is similar to that of SiNWs, with a crystalline Ge core and a thick amorphous oxide shell. The diameters of the Ge nanowires in this particular experimental setup had a larger size distribution than that of SiNWs (ranging from 16 to 370 nm).

Gallium arsenide nanowires with zinc-blende structure were fabricated by laser ablation of GaAs powders mixed with Ga_2O_3 (no metal catalyst used). SEM observation (Fig. 42a) shows that the product consists of wire-like structures with lengths up

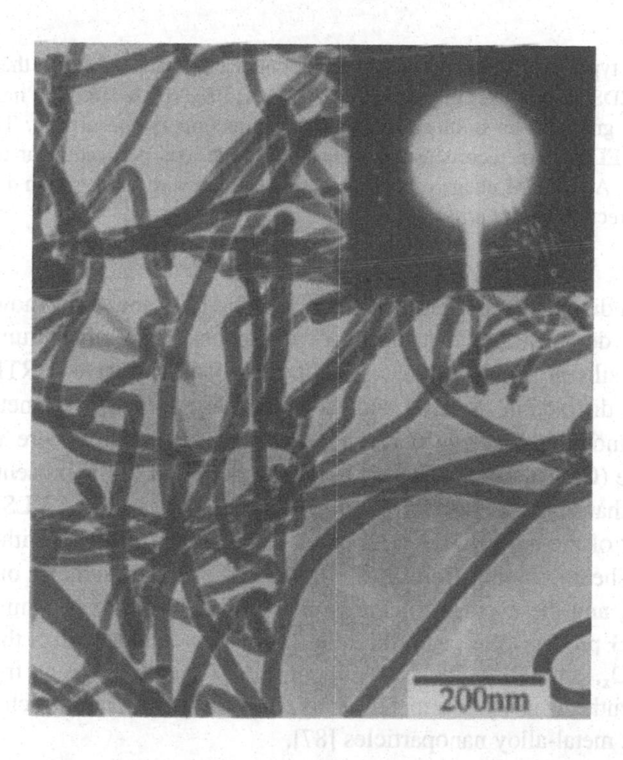

Fig. 41. A TEM image of Ge nanowires and a selected-area electron-diffraction pattern (inset) [39].

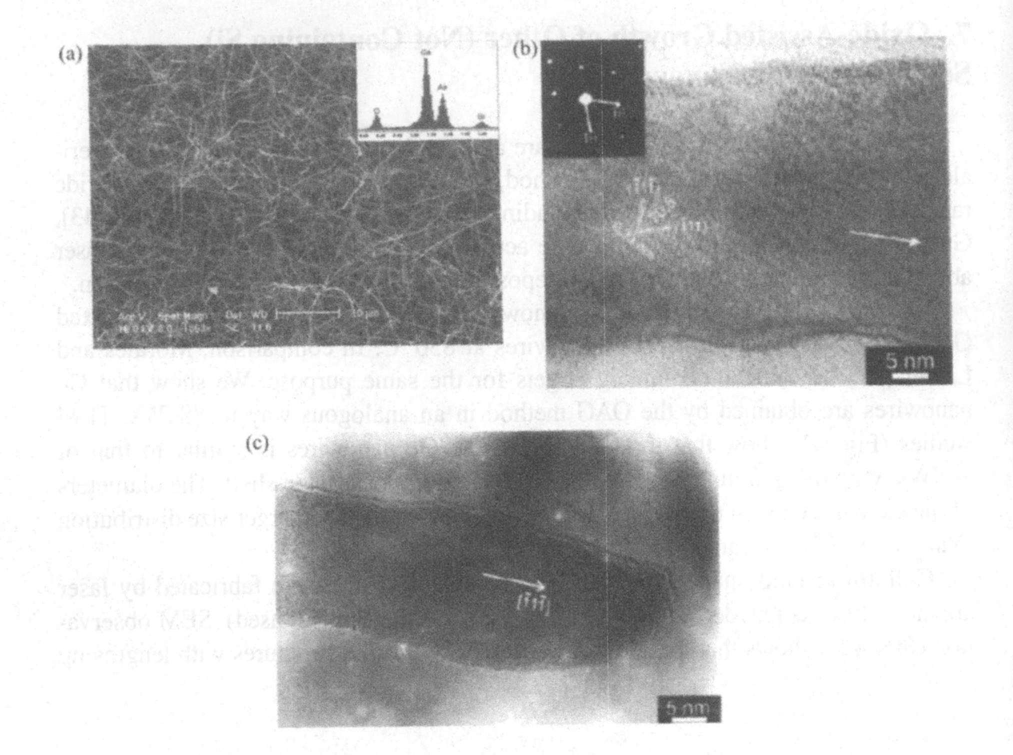

Fig. 42. (a) A typical SEM image of the GaAs nanowires synthesized by the oxide-assisted method. The EDS in the inset indicates Ga, As, O and Si; (b) A HRTEM image of a GaAs nanowire. The growth axis is close to the $[\bar{1}1\bar{1}]$ direction (white arrow). The inset is the corresponding ED pattern recorded along the [110] zone axis perpendicular to the nanowire growth axis; (c) A HRTEM image of the tip of a GaAs nanowire. The growth direction is close to the $[\bar{1}1\bar{1}]$ direction (white arrow) [42].

to 10 μm and diameters on the order of 50 nm. The EDS spectrum shown in the inset of this figure demonstrates that the nanowires consist only of gallium, arsenic and oxygen. The silicon signal originates from the silicon substrate. HRTEM, selected-area electron diffraction (SAED) and electron energy-loss spectrometry (EELS) of individual nanowires (Fig. 42b) revealed a zinc-blende GaAs core enclosed in a gallium oxide (GaO_x) sheath. The [111] growth direction of the present nanowires is the same as that of GaAs nanowires grown by a metal-catalyzed VLS process [87]. The diameter of the crystalline-GaAs cores range from 10 to 120 with the thickness of the outer sheath ranging from 2 to 10 nm. The average diameter of the core was about 60 nm, and the average thickness of the outer sheath was 5 nm. As expected from an OAG process, the crystalline GaAs tip was coated with a thin amorphous layer of GaO_x, similar to the SiO_x tip of SiNWs, and different from the GaAs nanowires synthesized by the metal-catalyzed VLS growth, in which the tips were terminated at metal-alloy nanoparticles [87].

We can thus suggest that the oxide-assisted nucleation and growth of GaAs nanowires advances through the following reactions: (1) laser-induced decomposition of GaAs into Ga and As, (2) reaction of $4Ga + Ga_2O_3 = 3Ga_2O$ at the high-temperature zone, (3) transport of volatile Ga_2O and As to the low-temperature-deposition zone, (4) reaction of $3Ga_2O + 4As = 4GaAs + Ga_2O_3$ at the low-temperature-deposition zone leading to the nucleation and growth of the GaAs nanowires.

The above model similarly applies to the successful synthesis of GaN and GaP and other binary compounds. Figs. 43 and 44 show the typical HRTEM images of GaP and GaN nanowires, respectively. A TEM image of GaP nanowires and a SEM image of GaN nanowires are shown as insets in Figs. 43 and 44, respectively. Again, both kinds of nanowires had a core of crystalline GaP or GaN wrapped in a thin layer of amorphous gallium oxide (GaO_x). The formation of GaP and GaN nanowires is similar to that of GaAs nanowires, replacing As by P and N, respectively. The critical reactions responsible for the formation of GaP and GaN nanowires are, $3Ga_2O + 4P = 4GaP + Ga_2O_3$ and $3Ga_2O + 4N = 4GaN + Ga_2O_3$, respectively.

Fig. 43. A typical HRTEM image of GaP nanowires grown at about 750 °C on a silicon (100) substrate. The growth direction is close to the $[0\bar{1}1]$ direction. The inset is a TEM image of GaP nanowires [44].

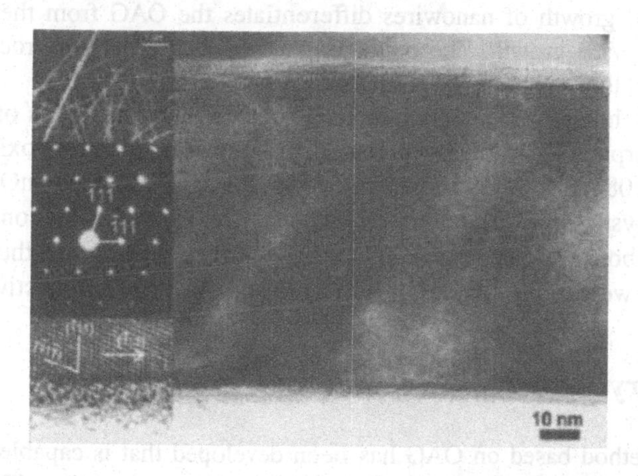

Fig. 44. A HRTEM image of a GaN nanowire grown at about 900 °C on a silicon (100) substrate. The insets are a SEM image (top), the corresponding ED pattern (middle) recorded along the [110] zone axis perpendicular to the nanowire growth axis, and the local enlarged HRTEM image (bottom). The growth direction is along the $[\bar{1}11]$ direction (white arrow in bottom inset) [40].

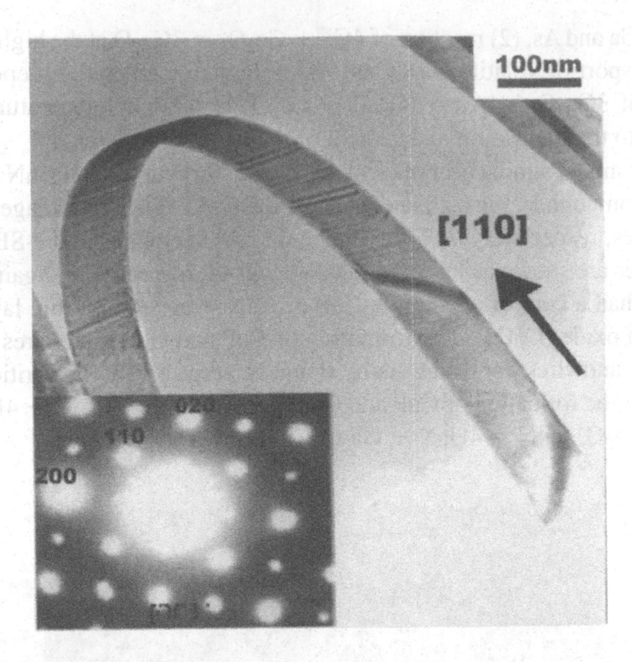

Fig. 45. TEM image of a single SnO_2 nanoribbon with [110] growth direction, inset showing the SAED pattern along the [001] axis [89].

It is thus obvious that for all the cases described above the oxide reacted with the element to form a volatile oxide. A chemical reaction (oxidation-reduction reaction where the oxide serves as the reactant) induced nucleation and growth of a nanowire embedded in an oxide sheath follows [22–27, 88]. The chemical reaction induced nucleation and growth of nanowires differentiates the OAG from the conventional metal-catalyst VLS growth. The oxide-assisted nanowire growth process is free of metal catalysts thus enabling the formation of pure nanowires.

Similar to the growth of silicon nanoribbons, 2D nanostructures of SnO_2 with a ribbonlike morphology were also prepared in large scale via rapid oxidation of elemental tin at 1080 °C [89]. As shown in Fig. 45, the as-synthesized SnO_2 nanoribbons were single crystals and had preferred [110] and [203] growth directions. The lengths of the nanoribbons were up to several hundreds of micrometers, and the typical width and thickness were in the range of 30–150 nm and 10–30 nm, respectively.

8. Summary

A new method based on OAG has been developed that is capable of producing high-quality and bulk-quantity of various semiconductor nanowires. The presence of oxides in the target is a common and essential ingredient for the synthesis using laser ablation or thermal evaporation, so that the targets are capable of generating semiconductor oxides in the vapor phase. Subsequent decomposition of the vapor phase oxides at high temperature and defect structures play crucial roles in the nucleation and growth of high-quality nanowires. The developed OAG approach has been applied to grow a host of semiconducting nanowires such as Si, Ge, GaN, GaAs, GaP, SiC, and ZnO, by either laser-ablation, HFCVD or thermal evaporation. Large-area, aligned, and long SiNWs via flow control, and diameter and morphologies control by temperature have been achieved. High-quality SiNW-based nanocables and Si ribbons from SiO-assisted growth have been additionally grown. An approach was developed for oxide stripping, dispersion and assembly in order to apply the various semiconducting nanowires in device fabrications. The morphologies of SiNWs were characterized by STM and consequently the STS of SiNWs was measured. The STM characterization has been extended to B-doped SiNWs. The nanochemistry of SiNWs has further been researched by performing surface reactions with Ag and Cu ions on the SiNWs. Nanosized ligated metal clusters on SiNWs surface were realized to achieve good metal-nanowire contact. Our synthesized highly-oriented SiC nanowires were shown to be an excellent field emitter with large field emission current densities at very low electric turn-on and threshold fields. Modeling of SiNW structures, nucleation and growth processes and properties done parallel to the experimental work is highly valuable in giving additional insight into the nature of the oxide-assisted growth and in explaining our experimental results.

Acknowledgment

We are thankful to our colleagues (I. Bello, C. S. Lee, D. S. Y. Tong, N. Wang, N. B. Wong), post-doctoral fellows (J. Q. Hu, Y. Jiang, D. D. D. Ma, Z. W. Pan, H. Y. Peng, W. S. Shi), graduate students (F. Au, C. P. Li, X.M. Meng, Y. H. Tang), and visiting professors (T. K. Sham, B. K. Teo, D. P. Yu, Y. F. Zhang) to COSDAF for their contributions to the work described in this chapter. The work described herein was supported by the financial support from the Research Grants Council of Hong Kong SAR [Project No. 8730016 (e.g. CityU 3/01C); 9040459; 9040533 (e.g. CityU 1033/00P); and 9040633 (e.g. CityU1011/01P)] and City University of Hong Kong.

... *Preparation and Properties*, CRC Press, Boca Rat... (1997).
3. M. Dresselhaus, G. Dresselhaus, P. Eklund and R. Saito, *Phys. World* **33** (1998).
4. J. W. G. Wildoer et al., *Nature* **391** (1998) 59.
5. P. M. Ajayan, *Chem. Rev.* **99** (1999) 1787.
6. C. Dekker, *Phys. Today* **22** (1999).
7. M. M. J. Treacy, T. W. Ebbesen and J. M. Gibson, *Nature* **381** (1996) 678.
8. S. J. Tans, R. M. Verschueren and C. Dekker, *Nature* **393** (1998) 40.
9. J. Kong et al., *Science* **287** (2000) 622.
10. P. G. Collins, K. Bradley, M. Ishigami and A. Zettl, *Science* **287** (2000) 1801.
11. Q. H. Wang et al., *Appl. Phys. Lett.* **72** (1998) 2912.
12. S. Fan et al., *Science* **283** (1999) 512.

13. A. C. Dillon et al., *Nature* **386** (1997) 377.
14. J. W. G. Wildoer, L. C. Venema, A. G. Rinzler, R. E. Smalley and C. Dekker, *Nature* **391** (1998) 59.
15. T. Odom, J. Huang, P. Kim and C. Lieber, *Nature* **391** (1998) 62.
16. L. Langer, V. Bayot, E. Grivei, J. P. Issi, J. P. Heremans, C. H. Olk, L. Stockman, L. C. Van Haesendonck and Y. Bruynseraede, *Phys. Rev. Lett.* **76** (1996) 479.
17. T. W. Ebbesen, H. Lezec, H. Hiura, J. W. Bennett, H. F. Ghaemi and T. Thio, *Nature* **382** (1996) 54.
18. A. Kasumov, I. I. Khodos, P. M. Ajayan and C. Colliex, *Europhys. Lett.* **34** (1996) 429.
19. S. Frank, P. Poncharal, Z. L. Wang and W. A. de Heer, *Science* **280** (1998) 1744.
20. A. Batchtold, C. Strunk, J. P. Salvetat, J. M. Bonard, L. Forro, T. Nussbaumer and C. Schonenberger *Nature* **397** (1999) 673.
21. A. M. Morales and C. M. Lieber, *Science* **279** (1998) 208.
22. Y. F. Zhang, Y. H. Tang, N. Wang, D. P. Yu, C. S. Lee, I. Bello and S. T. Lee, *Appl. Phys. Lett.* **72** (1998) 1835.
23. S. T. Lee, Y. F. Zhang, N. Wang, Y. H. Tang, I. Bello, C. S. Lee and Y. W. Chung, *J. Mater. Res.* **14** (1999) 4503.
24. S. T. Lee, N. Wang and C. S. Lee, *Mater. Sci. Eng. A* **286** (2000) 16.
25. S. T. Lee, N. Wang, Y. F. Zhang and Y. H. Tang, *MRS Bull.* **24** (1999) 36.
26. N. Wang, Y. F. Zhang, Y. H. Tang, C. S. Lee and S. T. Lee, *Appl. Phys. Lett.* **73** (1998) 3902.
27. N. Wang, Y. F. Zhang, Y. H. Tang, C. S. Lee and S. T. Lee, *Phys. Rev. B* **58** (1998) R16024.
28. Y. H. Tang, Y. F. Zhang, H. Y. Peng, N. Wang, C. S. Lee and S. T. Lee, *Chem. Phys. Lett.* **314** (1999) 16.
29. Y. F. Zhang, Y. H. Tang, H. Y. Peng, N. Wang, C. S. Lee, I. Bello and S. T. Lee, *Appl. Phys. Lett.* **75** (1999) 1842.
30. Y. Cui, L. J. Lauhon, M. S. Gudiksen, J. Wang and C. M. Lieber. *Appl. Phys. Lett.* **78** (2001) 2214.
31. Y. Y. Wu and P. D. Yang, *J. Am. Chem. Soc.* **123** (2001) 3165.
32. H. Y. Peng, N. Wang, W. S. Shi, Y. F. Zhang, C. S. Lee and S. T. Lee, *J. Appl. Phys.* **89** (2001) 727.
33. H. Y. Peng, Z. W. Pan, L. Xu, X. H. Fan, N. Wang, C. S. Lee and S. T. Lee, *Adv. Mater.* **13** (2001) 317.
34. W. S. Shi, H. Y. Peng, N. Wang, C. P. Li, L. Xu, C. S. Lee, R. Kalish and S. T. Lee, *J. Am. Chem. Soc.* **123** (2001) 11095.
35. Y. H. Tang, Y. F. Zhang, C. S. Lee and S. T. Lee, *Chem. Phys. Lett.* **328** (2000) 346.
36. K. W. Wong, X. F. Zhou, F. C. K. Au, H. L. Lai, C. S. Lee and S. T. Lee, *Appl. Phys. Lett.* **75** (1999) 2918.
37. F. C. K. Au, K. W. Wong, Y. H. Tang, Y. F. Zhang, I. Bello and S. T. Lee, *Appl. Phys. Lett.* **75** (1999) 1700.
38. D. D. D. Ma, C. S. Lee and S. T. Lee, *Appl. Phys. Lett.* **79** (2001) 2468.
39. Y. F. Zhang, Y. H. Tang, N. Wang, C. S. Lee, I. Bello and S. T. Lee, *Phys. Rev. B* **61**(7) (2000) 4518.
40. W. S. Shi, Y. F. Zheng, N. Wang, C. S. Lee, S. T. Lee, *Chem. Phys. Lett.* **345** (2001) 377.
41. H. Y. Peng, X. T. Zhou, N. Wang, Y. F. Zheng, L. S. Liao, W. S. Shi, C. S. Lee and S. T. Lee, *Chem. Phys. Lett.* **327** (2000) 263.
42. W. S. Shi, Y. F. Zheng, N. Wang, C. S. Lee and S. T. Lee, *Adv. Mater.* **13** (2001) 591.
43. W. S. Shi, Y. F. Zheng, N. Wang, C. S. Lee and S. T. Lee, *Appl. Phys. Lett.* **78** (2001) 3304.
44. W. S. Shi, Y. F. Zheng, H. Y. Peng, N. Wang, C. S. Lee and S. T. Lee, *J. Am. Ceram. Soc.* **83** (2001) 3228.
45. W. S. Shi, Y. F. Zheng, N. Wang, C. S. Lee and S. T. Lee, *J. Vac. Sci. & Tech. B* **19** (2001) 1115.

46. J. Q. Hu, X. L. Ma, Z. Y. Xie, N. B. Wong, C. S. Lee and S. T. Lee, *Chem. Phys. Lett.* **344** (2001) 97.
47. D. P. Yu, Z. G. Bai, Y. Ding, Q. L. Hang, H. Z. Zhang, J. J. Wang, Y. H. Zou, W. Qian, G. C. Xiong, H. T. Zhou and S. Q. Feng, *Appl. Phys. Lett.* **72** (1998) 3458.
48. N. Wang, Y. H. Tang, Y. F. Zhang, D. P. Yu, C. S. Lee, I. Bello and S. T. Lee, *Chem. Phys. Lett.* **283** (1998) 368.
49. U. Setiowati and S. Kimura, *J. Am. Ceram. Soc.* **80** (1997) 757.
50. G. Hass and C. D. Salzberg, *J. Opt. Soc. Am.* **44** (1954) 18.
51. T. S. Chu, R. Q. Zhang and H. F. Cheung, *J. Phys. Chem. B* **105** (2001) 1705.
52. R. Q. Zhang, T. S. Chu, H. F. Cheung, N. Wang and S. T. Lee, *Phys. Rev. B* **64** (2001) 113304.
53. R. Q. Zhang, T. S. Chu, H. F. Cheung, N. Wang and S. T. Lee, *Mater. Sci. & Eng. C* **16** (2001) 31.
54. J. P. Borel, *Surf. Sci.* **106** (1981) 1.
55. C. P. Li, C. S. Lee, X. L. Ma, N. Wang, R. Q. Zhang and S. T. Lee, *Adv. Mat.* **15** (2003) 607.
56. T. Y. Tan, S. T. Lee and U. Göasele, *Appl. Phys. A: Mater. Sci. & Proc.* **74** (2002) 423.
57. J. T. Hu, T. W. Odom and C. M. Lieber, *Acc. Chem. Res.* **32** (1999) 435–445.
58. S. T. Lee et al. (unpublished).
59. E. I. Givargizov, *J. Cryst. Growth* **20** (1973) 217.
60. G. Hass, *J. Am. Ceram. Soc.* **33** (1950) 353.
61. Y. Cui, X. F. Duan, J. T. Hu and C. M. Lieber, *J. Phys. Chem. B* **104** (2000) 5213.
62. Y. H. Tang, Y. F. Zhang, N. Wang, I. Bello, C. S. Lee, and S. T. Lee, *J. Appl. Phys.* **85** (1999) 7981.
63. W. S. Shi, H. Y. Peng, Y. F. Zheng, N. Wang, N. G. Shang, Z. W. Pan, C. S. Lee and S. T. Lee, *Adv. Mater.* **12** (2000) 1343.
64. W. Z. Li, S. S. Xie, L. X. Qian, B. H. Chang, B. S. Zhou, W. Y. Zhou, R. A. Zhao and G. Wang, *Science* **274** (1996) 1701.
65. [illegible] and R. G. Treuting, *J. Appl. Phys.* **32** (1961) 2490.
66. [illegible] Y. Peng, L. Xu, N. Wang, Y. H. Tang and S. T. Lee, *Adv. Mater.* **12** [illegible]
[illegible] C. Colli[illegible]
69. X. H. Sun, H. [illegible] T. K. Sham, *J. A*[illegible]
70. Y. F. Zhang, L. S. [illegible] Lee, R. Sammynaiken and T. K. Sham, *Phys. Rev. B.* **61** (2000) 8298.
71. Z. W. Pan, H. L. Lai, Frederick C. K. Au, X. F. Duan, W. Y. Zhou, W. S. Shi, N. Wang, C. S. Lee, N. B. Wong, S. T. Lee and S. S. Xie, *Adv. Mater.* **12**(16) (2000) 1186.
72. W. Z. Li, S. S. Xie, L. X. Qian, B. H. Chang, B. S. Zhou, W. Y. Zhou, R. A. Zhao and G. Wang, *Science* **274** (1996) 1701.
73. Z. W. Pan, S. S. Xie, B. H. Chang, C. Y. Wang, L. Lu, W. Liu, W. Y. Zhou, W. Z. Li and L. X. Qian, *Nature* **394** (1998) 631.
74. Z. W. Pan, S. S. Xie, B. H. Chang, L. F. Sun, W. Y. Zhou and G. Wang, *Chem. Phys. Lett.* **299** (1999) 9.
75. R. H. Fowler and L. W. Nordheim, *Proc. R. Soc. London, Ser. A* **119** (1928) 173.
76. X. T. Zhou, J. Q. Hu, C. P. Li, D. D. D. Ma, C. S. Lee and S. T. Lee, *Chem. Phys. Lett.* **369** (2003) 220.
77. X. H. Sun, S. D. Wang, N. B. Wong, D. D. D. Ma, S. T. Lee and B. K. Teo, *Inorg. Chem.* **42** (2003) 2398.

78. D. D. D. Ma, C. S. Lee, F. C. K. Au, S. Y. Tong and S. T. Lee, *Science* **299** (2003) 1874.
79. Y. H. Tang, X. H. Sun, F. C. K. Au, L. S. Liao, H. Y. Peng, C. S. Lee, S. T. Lee and T. K. Sham, *Appl. Phys. Lett.* **79** (2001) 1673.
80. D. D. D. Ma, C. S. Lee, Y. Lifshitz and S. T. Lee, *Appl. Phys. Lett.* **81** (2002) 3233.
81. R. Q. Zhang, T. S. Chu and S. T. Lee, *J. Chem. Phys.* **114** (2001) 5531.
82. D. K. Yu, R. Q. Zhang and S. T. Lee, *Phys. Rev. B* **65** (2002) 245417.
83. B. X. Li, R. Q. Zhang, P. L. Cao and S. T. Lee, *Phys. Rev. B* **65** (2002) 125305.
84. R. Q. Zhang, S. T. Lee, C. K. Law, W. K. Li and B. K. Teo, *Chem. Phys. Lett.* **364** (2002) 251.
85. D. K. Yu, R. Q. Zhang and S. T. Lee, *J. Appl Phys.* **92** (2002) 7453.
86. R. Q. Zhang, W. C. Lu and S. T. Lee, *Appl. Phys. Lett.* **80** (2002) 4223.
87. X. Duan, J. Wang and C. M. Lieber, *Appl. Phys. Lett.* **76** (2000) 1116.
88. N. Wang, Y. H. Tang, Y. F. Zhang, C. S. Lee, I. Bello and S. T. Lee, *Chem. Phys. Lett.* **299** (1999) 237.
89. J. Q. Hu, X. L. Ma, N. G. Shang, Z. Y. Xie, N. B. Wong, C. S. Lee and S. T. Lee, *J. Phys. Chem. B* **106** (2002) 3823.

Index